Andrej Dobnikar
Nigel C. Steele
David W. Pearson
Rudolf F. Albrecht

Artificial Neural Nets
and Genetic Algorithms

Proceedings of the International Conference
in Portorož, Slovenia, 1999

SpringerWienNewYork

Dr. Andrej Dobnikar
Fakulteta za Računalništvo in Informatiko
Univerza v Ljubljani, Ljubljana, Slovenia

Dr. Nigel C. Steele
Division of Mathematics
School of Mathematical and Information Sciences
Coventry University, Coventry, U.K.

Dr. David W. Pearson
Laboratoire de Génie Informatique et Ingénierie de Production
Ecole pour les Etudes et la Recherche en Informatique et Electronique
Nimes, France

Dr. Rudolf F. Albrecht
Institut für Informatik
Universität Innsbruck, Innsbruck, Austria

Camera-ready copies provided by authors and editors
Printed by Novographic, Ing. Wolfgang Schmid, A-1230 Wien
Graphic design: Ecke Bonk
Printed on acid-free and chlorine-free bleached paper
SPIN 10713273

With 293 Figures

ISBN 3-211-83364-1 Springer-Verlag Wien New York

Preface

Biological nervous systems have evolved as a result of millions of years of nature's research and development to culminate in the ultimate computing entity – the mammalian brain. Understanding the brain remains one of the greatest scientific challenges, and one likely to evade full comprehension for the foreseeable future.

The ICANNGA conferences follow this basic goal. This is the fourth in a series devoted primarily to the theory and applications of articifical neural networks and genetic algorithms. The first conference was held in Innsbruck, Austria, the second in Alès, France, and the third in Norwich, United Kingdom. We were pleased to host the 1999 event in Portorož, Slovenia, and to carry on the fine tradition set by its predecessors of providing a relaxed and stimulating environment for both established and emerging researchers working on artifical neural networks and genetic algorithms and in related fields.

We would like to take this opportunity to express our gratitude to everyone who contributed in any way to the completion of this volume. In particular, we thank the members of the International Programme Committee for reviewing the submissions and making the final decisions on the acceptance of papers, the members of the Advisory Committee for their supervision of the organizational efforts, and Silvia Schilgerius from Springer-Verlag Wien for the final stages of the publication process, and, not least, to all researchers for their submissions to ICANNGA '99.

Andrej Dobnikar	Nigel C. Steele	David W. Pearson	Rudolf F. Albrecht
Ljubljana	Coventry	Nimes	Innsbruck

Contents

Part II: Genetic Algorithms – Theory and Applications

Part III: Soft Computing and Uncertainty

ICANNGA '99

International Conference on Artificial Neural Networks and Genetic Algorithms
Portorož, Slovenia, April 6–9, 1999

International Advisory Board

Rudolf F. Albrecht, University of Innsbruck, Austria
Nigel C. Steele, Coventry University, UK
David W. Pearson, Ecole pour les Etudes et la Recherche en Informatique et Electronique, Nimes, France

Programme Committee

Hojjat Adeli, The Ohio State University, Columbus, Ohio, USA
Jarmo Alander, University of Vaasa, Finland
Thomas Baeck, Center for Applied Systems Analysis, Dortmund, Germany
Subbiah Baskaran, Georg C. Marshall Space Flight Center, NASA, Huntsville, Ala., USA
Horst Bischof, TU Wien, Austria
Wilfried Brauer, TU München, Germany
Matjaz Colnaric, University of Maribor, Slovenia
Andrej Dobnikar, University of Ljubljana, Slovenia
Marco Dorigo, Université Libre de Bruxelles, Belgium
Gerard Dray, EMA-EERIE, Nimes, France
Hugo de Garis, ATR Human Information Processing, Research Laboratories, Kyoto, Japan
Christophe Giraud-Carrier, University of Bristol, UK
Igor Grabec, University of Ljubljana, Slovenia
Katerina Hlaváčková-Schindler, Academy of Science of the Czech Republic, Praha, Czech Republic
Helen Karatza, Aristotle University of Thessaloniki, Greece
Sami Khuri, San Jose State University, Calif., USA
Michael Korkin, Genobyte, Inc., Boulder, Colo., USA
Pedro Larranaga, University of the Basque Country, San Sebastian, Spain
Ales Leonardis, University of Ljubljana, Slovenia
Henrik H. Lund, University of Aarhus, Denmark
Tony Martinez, Brigham Young University, Provo, Utah, USA
Francesco Masulli, University of Genova, Italy
Bojan Novak, University of Maribor, Slovenia
Franz Oppacher, Carleton University, Ottawa, Ont., Canada
Nikola Pavesic, University of Ljubljana, Slovenia
David Pearson, EMA-EERIE, Nimes, France
Vic Rayward-Smith, University of East Anglia, Norwich, UK
Colin Reeves, Coventry University, UK
Slobodan Ribaric, University of Zagreb, Croatia
Bernardete Ribeiro, Universidade de Coimbra, Portugal
Henrik Saxén, Åbo Akademi, Finland
George Smith, University of East Anglia, Norwich, UK
Nigel Steele, Coventry University, UK
Tatiana Tambouratzis, National Centre for Scientific Research "Demokritos", Athens, Greece
Marco Tomassini, Université de Lausanne, Switzerland
Kevin Warwick, University of Reading, UK
Darrell Whitley, Colorado State University, Fort Collins, Colo., USA
Pedro Zufiria, Universidad Politecnica de Madrid, Spain
Jure Zupan, Kemijski Institut, Ljubljana, Slovenia

Topological Approach to Fuzzy Sets and Fuzzy Logic

R. F. Albrecht

Faculty of Natural Science, University of Innsbruck
Technikerstr. 25, A-6020 Innsbruck, Austria
Email: Rudolf.Albrecht@uibk.ac.at

Abstract

We show how fuzzy set theory and fuzzy logic can be described by classical topology. The basic concepts are filter and ideal bases and morphisms on these. Logics are introduced as hierarchies of valued objects and the valuations are uniformly topologized to represent approximations to logical terms. As selected applications we consider contractive mappings, roundings, hierarchical filter bases ("pyramids"), adaptable networks.

1 Some set theoretical and topological concepts

1.1 Filters and ideals

If ind: $I \to S$ is an indexing of elements of S, we use the notation $\text{ind}(i) = s(i) = s_{[i]}$, $s_i = (i, s_{[i]})$. $\wedge$, $\vee$ denote the universal and the existential quantifier.

Let there be given a lattice $(\mathscr{L}, \leq, \sqcap, \sqcup)$, $\sqcap, \sqcup$ lattice meet and join, $\mathbf{o}$ the zero and $\mathbf{e}$ the unit element if in $\mathscr{L}$, and a non-empty subset $\mathscr{B} = \{B_{[k]} \mid k \in K\} \subset \mathscr{L}$, the indexing bijective, with the following properties:

$\wedge k \in K$ $(B_{[k]} \neq \mathbf{o})$ $\wedge$ $\wedge k', k'' \in K$ $(\vee k''' \in K$ $((B_{[k''']} \leq B_{[k']}) \wedge (B_{[k''']} \leq B_{[k'']})))$. Then $\mathscr{B}$ is a "filter base" on $\mathscr{L}$. If in addition $\wedge k \in K \wedge L \leq \mathbf{e}$ $(B_{[k]} \leq L \Rightarrow L \in \mathscr{B})$ then $\mathscr{B}$ is a "filter". We define $\lim \mathscr{B} = \sqcap \mathscr{B}$. The dual notions to filter base and filter are "ideal base" and "ideal". Filter bases were introduced by L. Vietoris in 1921 [1]. A filter base can also be an ideal base.

If S is a non-empty set, then the above applies to the complete, atomic, boolean lattice (pow S, $\subseteq$, $\cap$, $\cup$, $\varnothing$, S). For a filter base $\mathscr{B} = \{B_{[k]} \mid k \in K\}$ on pow S, $B^* =_{\text{def}} \lim \mathscr{B} = \bigcap_{k \in K} B_{[k]}$. For $B^* \neq \varnothing$ (then we name $\mathscr{B}$ a "proper" filter base) the neighborhood of any $s \in B$, $B =_{\text{def}} \bigcup_{k \in K} B_{[k]}$, to the elements of B^* can be expressed by membership or non-membership of s in certain $B_{[k]}$: Let $\wedge s \in B$ $((K(s) =_{\text{def}} \{k \mid k \in K$

$\wedge s \in B_{[k]}\}) \wedge \overline{K}(s) =_{\text{def}} K \setminus K(s))$, $\mathscr{B}_{\cap}(s) =_{\text{def}} \{B_{[k]} \mid k \in K(s)\}$, $\mathscr{B}_{\cup}(s) =_{\text{def}} \{B_{[k]} \mid k \in \overline{K}(s)\}$. We have $s \in \bigcap_{k \in K(s)} B_{[k]} \cap \bigcap_{k \in \overline{K}(s)} \mathbf{C}B_{[k]}$, $\mathbf{C}$ the complement with respect to B. Let $K_{\min}(s) =_{\text{def}} \{k \mid k \in K(s) \wedge \neg \vee k' \in K(s) (B_{[k']} \subset B_{[k]})\}$, $\overline{K}_{\max}(s) =_{\text{def}} \{k \mid k \in \overline{K}(s) \wedge \neg \vee k' \in \overline{K}(s) (B_{[k']} \supset B_{[k]})\}$, then $\mathscr{B}_{\cap \min}(s) =_{\text{def}} \{B_{[k]} \mid k \in K_{\min}(s)\}$, $\mathscr{B}_{\cup \max}(s) =_{\text{def}} \{B_{[k]} \mid k \in \overline{K}_{\max}(s)\}$. General "distance" / "similarity" *relations* of s from / to $s^* \in B^*$ are then given by $\wedge s \in B$ $(D_{\cap}(s^*, s) =_{\text{def}} \mathscr{B}_{\cap \min}(s) \wedge D_{\cup}(s^*, s) =_{\text{def}} \mathscr{B}_{\cup \max}(s))$. $D_{\cap}(s^*, s) = D_{\cap}(s^*, s')$ and $D_{\cup}(s^*, s) = D_{\cup}(s^*, s'')$ define equivalence relations $s \sim_{\cap} s'$ and $s \sim_{\cup} s''$. In particular, if $\mathscr{B}$ is itself a complete lattice then $d_{\cap}(s^*, s) =_{\text{def}} \bigcap D_{\cap}(s^*, s) \in \mathscr{B}$ and $d_{\cup}(s^*, s) =_{\text{def}} \bigcup D_{\cup}(s^*, s) \in \mathscr{B}$ are *functional* in s and $d_{\cup}(s^*, s) \subset d_{\cap}(s^*, s)$. Dual results hold for $\mathscr{B}$ being an ideal base. For illustration see Fig. 1.

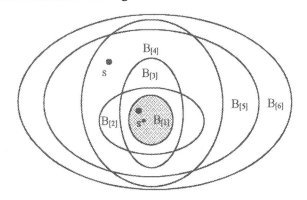

Fig. 1. $\mathscr{B} = \{B_{[1]}, B_{[2]}, B_{[3]}, B_{[4]}, B_{[5]}, B_{[6]}\}$, $\lim \mathscr{B} = B_{[1]}$, $D_{\cap}(s^*, s) = \{B_{[4]}, B_{[5]}\}$, $D_{\cup}(s^*, s) = \{B_{[2]}, B_{[3]}\}$

1.2 Comparison and composition of bases

Given a non-empty set S and two filter bases $\mathscr{B} = \{B_{[k]} \mid k \in K\} \subset$ pow S with $B =_{\text{def}} \bigcup_{k \in K} B_{[k]}$, $B^* =_{\text{def}} \bigcap_{k \in K} B_{[k]}$, $\mathscr{C} = \{C_{[l]} \mid l \in L\} \subset$ pow S with $C =_{\text{def}}$

$\bigcup_{l\in L} C_{[l]}$, $C^* =_{def} \bigcap_{l\in L} C_{[l]}$, all indexings bijective. We say $\mathscr{B}$ is "finer" than $\mathscr{C}$, $\mathscr{B} \prec \mathscr{C} \Leftrightarrow_{def} \bigwedge l\in L \bigvee k\in K$ ($B_{[k]} \subseteq C_{[l]}$), $\mathscr{B}$ is "equivalent" $\mathscr{C}$, $\mathscr{B} \sim \mathscr{C} \Leftrightarrow_{def} \mathscr{B} \prec \mathscr{C} \wedge \mathscr{C} \prec \mathscr{B}$, and for finite cardinalities, $\mathscr{B}$ "finer granulated" than $\mathscr{C}$ if card $\mathscr{B} >$ card $\mathscr{C}$.

$\mathscr{S} =_{def} \{B_{[k]} \cup C_{[l]} \mid (k,l) \in K \times L\}$ is a filter base on pow S with $S^* =_{def} \bigcap_{kl\in K\times L} (B_{[k]} \cup C_{[l]}) = B^* \cup C^*$, $\mathscr{D} =_{def} \{B_{[k]} \cap C_{[l]} \mid (k,l) \in K \times L\}$ is a filter base on pow S only if $\bigwedge (k,l)\in K\times L$ ($(B_{[k]} \cap C_{[l]}) \neq \varnothing$), then $D^* =_{def} \bigcap_{kl\in K\times L} (B_{[k]} \cap C_{[l]}) = B^* \cap C^*$.

Let $\mathbf{B} = \{\mathscr{B}_{[m]} \mid m \in M\}$ be a set of either all being filter bases or all being ideal bases on pow S. To compare the bases $\mathscr{B}_{[m]}$ by a uniform neighborhood / similarity measure we introduce a filter base $\mathbf{D} =_{def} \{D_{[q]} \mid q \in Q\} \subset$ pow $(\mathbf{B} \times \mathbf{B})$ with the following properties: $\mathbf{B} \times \mathbf{B} \in \mathbf{D}$, diag $(\mathbf{B} \times \mathbf{B}) =_{def} \{(\mathscr{B}_{[m]},$ $\mathscr{B}_{[m]}) \mid m \in M\} \subseteq \bigcap_{q\in Q} D_{[q]}$, and $\bigwedge q\in Q$ $(D_{[q]} = D_{[q]}^{-1})$.

Then for $F_{[mq]} =_{def} \mathrm{cut}(\mathscr{B}_{[m]})D_{[q]} = \{\mathscr{B}_{[m']} \mid (\mathscr{B}_{[m]}, \mathscr{B}_{[m']}) \in D_{[q]}\}$, $\mathscr{F}_{[m]} =_{def} \{F_{[mq]} \mid q \in Q\}$ is a filter base with $\mathscr{B}_{[m]} \in \bigcap \mathscr{F}_{[m]}$. Consequently, $\mathbf{D}$ defines for all $\mathscr{B}_{[m]}$ a uniform and symmetric neighborhood system. Thus according to **1.1**, for any pair ($\mathscr{B}_{[m]}, \mathscr{B}_{[m']}$) the neighborhood / similarity measures $D_\cap(\mathscr{B}_{[m]}, \mathscr{B}_{[m']})$ and $D_\cup(\mathscr{B}_{[m]}, \mathscr{B}_{[m']})$ can be applied.

Example: $\mathbf{B} = (\{0,1\}, \leq)$, $S = \mathbf{B}\times\mathbf{B}\times\mathbf{B}$ can be partial ordered by component-wise $\leq$, $\bigwedge s', s'' \in S$ $(d(s', s'') =_{def} s' \oplus s''$, $\oplus$ component-wise addition mod 2) is a uniform generalized distance, $h(s', s'') =_{def} \sum_{n=1}^{3} \mathrm{pr}_n (d(s', s''))$ is a metric distance (Hamming) (pr_n means the n-th projection).

1.3 Valuated bases

Given a filter base $\mathscr{B} = \{B_{[k]} \mid k \in K\}$ on (pow B, $\subseteq$, $\cap$, $\cup$), a non-empty complete lattice $(V, \leq, \sqcap, \sqcup)$, and a $\leq$–homomorphism φ: pow B $\to$ V, i.e. $\bigwedge k,k'\in K$ $((B_{[k]} \subseteq B_{[k']}) \Rightarrow (\varphi(B_{[k]}) \leq \varphi(B_{[k']})))$. With $v_{[k]} = \varphi(B_{[k]})$ it follows from $(B_{[k]} \subseteq B_{[k']}) \wedge (B_{[k]} \subseteq B_{[k'']})$ that $(v_{[k]} \leq v_{[k']}) \wedge (v_{[k]} \leq v_{[k'']})$, hence $\varphi(\mathscr{B})$ is a filter base on V if all $v_{[k]} \neq \mathbf{o}$, and we have $\varphi(\lim \mathscr{B}) \leq \lim \varphi(\mathscr{B})$. In the function $(B_{[k]}, v_{[k]})_{k\in K}$ the elements $B_{[k]}$ of base $\mathscr{B}$ with "support" $B = \bigcup \mathscr{B}$ are valuated by $v_{[k]}$.

We consider $\overset{-1}{\varphi}: V \to$ pow B defined by $\bigwedge v\in V$ ($\overset{-1}{\varphi}(v) =_{def} \bigcup_{\varphi(U)\leq v} U$). Then $\overset{-1}{\varphi}$ is a homomorphism. If $\mathscr{V}= \{v_{[l]} \mid l \in L\}$ is a filter base on V and $\bigwedge v\in\mathscr{V}$ $(\overset{-1}{\varphi}(v) \neq \varnothing)$, then $\overset{-1}{\varphi}(\mathscr{V})$ is a filter base on pow B.

For $U \subseteq S$ and for $v \in V$ we have $U \subseteq \overset{-1}{\varphi}\varphi(U)$ and $v \geq \varphi\overset{-1}{\varphi}(v)$, thus $\mathscr{B} \prec \overset{-1}{\varphi}\varphi(\mathscr{B})$ and $\varphi\overset{-1}{\varphi}(\mathscr{V}) \prec \mathscr{V}$.

An example is: φ the set extension of a function f: B $\to$ V, $\varphi(U) = \bigsqcup (f(u))_{u\in U}$, in particular, V = pow C, C a non-empty set. This case can be found in textbooks, e.g. [2] in 1951.

φ being a homomorphism corresponds to the "neighborhood to B*" interpretation. Choosing φ as antimorphism, $\varphi(\mathscr{B})$ is an ideal base, which corresponds to the "similarity to B*" interpretation.

The presentation in this section followed the one given in [3].

2 Fuzziness in terms of topology

The basic concept is a filter bases $\mathscr{B} = \{B_{[k]} \mid k \in K\}$ $\subset$ (pow S, $\subseteq$), $B_{[k]} \neq \varnothing$, with $\bigcap \mathscr{B} = B^*$, or dual, an ideal base. $\overline{\mathscr{B}} =_{def} \mathscr{B} \cup \{B^*\}$ is partially ordered by $\subseteq$ which can semantically express a mutual "more or less" relationship between the elements of the $B_{[k]}$, B*, like neighborhood, similarity, certainty etc. To formulate this quantitatively, a "valuation" φ: $\overline{\mathscr{B}} \to$ $(V, \leq, \sqcap, \sqcup)$, a lattice, is applied, with the set function φ isotone with $\subseteq$. φ maps $\mathscr{B}$ onto a filter base $\mathscr{V}$ on V if $\bigwedge k\in K$ $(\varphi(B_{[k]}) \neq \mathbf{o})$.

We assume φ is extended to a $\leq$-homomorphism pow S $\to$ V. (pow S, $\leq$, $\cap$, $\cup$) is a complete, atomic, boolean lattice. If $\mathscr{V}$ is a filter base on V and $\bigwedge v\in\mathscr{V}$ $(\overset{-1}{\varphi}(v) \neq \varnothing)$ then $\overset{-1}{\varphi}$ maps $\mathscr{V}$ onto a filter base on pow S.

Let there be given two filter bases $\mathscr{B} = \{B_{[k]} \mid k\in K\}$, $\mathscr{C} = \{C_{[l]} \mid l\in L\}$ on pow S with valuations $\varphi_{\mathscr{B}}$: $\mathscr{B} \to$ V, $\varphi_{\mathscr{C}}$: $\mathscr{C} \to$ V. $\bigcup\mathscr{B} \neq \bigcup\mathscr{C}$ is admitted. Then $\mathscr{S} = \{B_{[k]} \cup C_{[l]} \mid (k,l) \in K\times L\}$ is a filter base with $\lim\mathscr{S} = \lim\mathscr{B} \cup \lim\mathscr{C}$, $\mathscr{D} = \{B_{[k]} \cap C_{[l]} \mid (k,l) \in K\times L\}$ is a filter base with $\lim\mathscr{D} = \lim\mathscr{B} \cap \lim\mathscr{C}$ if $\bigwedge (k,l)\in K\times L$ $(B_{[k]} \cap C_{[l]} \neq \varnothing)$ ("union" and "intersection" of filter bases).

Valuations of the elements of $\mathscr{S}$ and of $\mathscr{D}$ are given by

$\varphi_{\mathscr{S}}(B_{[k]} \cup C_{[l]}) =_{def} \varphi_{\mathscr{B}}(B_{[k]}) \bigsqcup \varphi_{\mathscr{C}}(C_{[l]})$ and by

$\varphi_{\mathscr{D}}(B_{[k]} \cap C_{[l]}) =_{def} \varphi_{\mathscr{B}}(B_{[k]} \cap C_{[l]}) \bigsqcup \varphi_{\mathscr{C}}(B_{[k]} \cap C_{[l]})$.

A particular case is: $\Sigma \subseteq$ pow S a σ-algebra, V = ($\overline{\mathbf{R}}_+ = \mathbf{R}_+ \cup \{\infty\}$, $\leq$) which is a lattice with $\mathbf{o} = 0$ and $\mathbf{e} = \infty$, $\mu: \Sigma \to \overline{\mathbf{R}}_+$ a σ-additive measure ($\mu(\varnothing) = 0$). If for the filter base $\mathscr{B} = \{B_{[k]} \mid k \in K\}$ holds all $B_{[k]} \in \Sigma$, then μ is a $\leq$-homomorphism $\mathscr{B} \to \overline{\mathbf{R}}_+$.

According **1.3**, the set extension of a function f: B $\to$ V, B = $\bigcup \mathscr{B}$, $\mathscr{B} = \{B_{[k]} \mid k \in K\}$ being a filter base, yields a homomorphism $\varphi: \mathscr{B} \to V$. However, a homomorphism of $\mathscr{B}$ need not be given by the set extension of a point function. Similarly, for $\mathscr{B}$ a filter base on Σ, an integrable point function f can exist such that the integrals of f over the $B_{[k]}$ yield a homomorphism μ of $\mathscr{B}$ into $\overline{\mathbf{R}}_+$.

We assume, an algebraic composition law $\bullet$ is defined on the lattice V, $\bullet: V \cup (V \times V) \to V$ with $\bullet$ being the identity on V and $\bullet$ compatible with $\leq$. Let be $\varnothing \neq \mathscr{B} \subseteq$ pow S, $\varnothing \neq \mathscr{C} \subseteq$ pow S, and $\varphi: \mathscr{B} \to V$, $\psi: \mathscr{C} \to V$ *any* two functions. Then the families $(B, \varphi(B))_{B \in \mathscr{B}}$ and $(C, \psi(C))_{C \in \mathscr{C}}$ can be concatenated to a family (function) $(K, \phi(K))_{K \in \mathscr{K}}$ on $\mathscr{K} = \mathscr{B} \cup \mathscr{C}$, $\mathbf{K}(\bullet)((B, \varphi(B))_{B \in \mathscr{B}}, (C, \psi(C))_{C \in \mathscr{C}}) = (K, \phi(K))_{K \in \mathscr{K}}$, with $\phi(K) = \varphi(K)$ for $K \in \mathscr{B} \backslash \mathscr{C}$, $\phi(K) = \psi(K)$ for $K \in \mathscr{C} \backslash \mathscr{B}$, $\phi(K) = \varphi(K) \bullet \psi(K)$ for $K \in \mathscr{B} \cap \mathscr{C}$. Examples are $\bullet = \bigsqcup$, $\bullet = \bigsqcap$, and for $V = \overline{\mathbf{R}}_+$, $\bullet = +$. All are isotone with $\leq$.

We apply this to the case: V a boolean lattice, $\mathscr{B} = \mathscr{C} = \{B_{[k]} \mid k \in K\}$ a filter base, φ a homomorphism, $\psi = \overline{\varphi}$ the antimorphism dual to φ. Complementary to the filter base $\mathscr{V} = \{v_{[k]} = \varphi(B_{[k]}) \mid k \in K\}$ is the ideal base $\overline{\mathscr{V}} = \{\overline{v}_{[k]} = \overline{\varphi(B_{[k]})} \mid k \in K\}$. Concatenation of these families by $\bigsqcup$ and by $\bigsqcap$ results in $\mathbf{K}(\bigsqcup) = (B_{[k]}, \mathbf{e})_{k \in K}$ and $\mathbf{K}(\bigsqcap) = (B_{[k]}, \mathbf{o})_{k \in K}$, respectively.

We distinguish operations on bases from operations on functions defined on bases.

We show how naive fuzzy set theory (L. A. Zadeh [4], 1965) is embedded in the theory of morphisms of filter and ideal bases: V is pow C, $\varnothing \neq C \subseteq (\mathbf{R}, \leq)$ a discrete or continuous subset, for example the interval [0,1], f: S $\to$ C the "membership" function, $\mathscr{V} = \{V_{[k]} \mid k \in K\}$ a filter base on V, for example the monotone filter base $\mathscr{V} = \{[\alpha, 1] \mid \alpha \in C\}$ for which $\lim \mathscr{V} = \{1\}$. Then for φ the set extension of f, the "cuts" $\overset{-1}{\varphi}(V_{[k]})$ are considered, for example the cuts $\overset{-1}{\varphi}([\alpha,$

1]), which, if all of them are $\neq \varnothing$, generate a filter base on pow S. The latter condition holds if f maps S *onto* V. Fig. 2 gives a visualization.

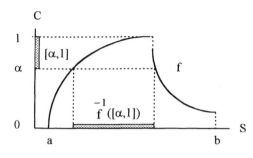

Fig. 2

An example for union and intersection of filter bases is shown in Fig. 3.

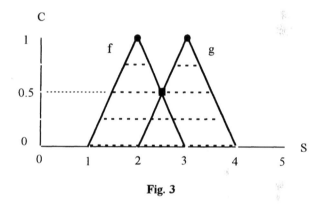

Fig. 3

S = [0, 5], C = [0, 1], f: [1, 3] $\to$ C, g: [2, 4] $\to$ C, f(s) = s – 1 for $1 \leq s \leq 2$, f(s) = –s + 3 for $2 \leq s \leq 3$, g(s) = f(s – 1). $\mathscr{V} = \{[0, 1], [.25, 1], [.5, 1], [.75, 1], \{1\}\}$.

$\overset{-1}{f}(\mathscr{V}) = \{[1, 3], [1.25, 2.75], [1.5, 2.5], [1.75, 2.25], \{2\}\}$, $\lim \mathscr{V} = \{1\}$, $\lim \overset{-1}{f}(\mathscr{V}) = \{2\}$, similar for $\overset{-1}{g}(\mathscr{V})$. The union filter base $\mathscr{S}(\overset{-1}{f}(\mathscr{V}), \overset{-1}{g}(\mathscr{V}))$ is equivalent to the base $\mathscr{S}' = \{[1, 3] \cup [2, 4], [1.25, 2.75] \cup [2.25, 3.75], ..., \{2\} \cup \{3\}\}$ where only *equally* valued sets are joined, $\mathscr{S} \sim \mathscr{S}'$. $\lim \mathscr{S} = \lim \mathscr{S}' = \{2, 3\}$. There is no intersection filter base as image of $\mathscr{V}$, but as image of the filter base $\widetilde{\mathscr{V}} = \{[0, .5], [.25, .5], \{.5\}\}$. We have $\lim \overset{-1}{f}(\widetilde{\mathscr{V}}) = \{1.5, 2.5\}$, $\lim \overset{-1}{g}(\widetilde{\mathscr{V}}) = \{2.5, 3.5\}$, and for the intersection filter base $\mathscr{D}(\overset{-1}{f}(\widetilde{\mathscr{V}}), \overset{-1}{g}(\widetilde{\mathscr{V}}))$, $\lim \mathscr{D} = \{2.5\}$. Notice, the functions f, g do not have the same support.

Now we consider two filter bases $\mathscr{V}_f = \{[.75, 1], \{1\}\}$, $\mathscr{V}_g = \{[0, .25], \{0\}\}$ on V with filter bases $\overset{-1}{f}(\mathscr{V}_f) = \{[1.75, 2.25], \{2\}\}$, $\overset{-1}{g}(\mathscr{V}_g) = \{[2, 2.25] \cup [3.75, 4], \{2, 4\}\}$. The union $\mathscr{S}(\overset{-1}{f}(\mathscr{V}_f), \overset{-1}{g}(\mathscr{V}_g)) = \{[1.75, 2.25] \cup [3.75, 4], [1.75, 2.25] \cup \{4\}, [2, 2.25] \cup [3.75, 4], \{2, 4\}\}$. $\lim \mathscr{S} = \{2, 4\}$, $\lim \overset{-1}{f}(\mathscr{V}_f) = \{2\}$, $\lim \overset{-1}{g}(\mathscr{V}_g) = \{2, 4\}$. Valuation of the limits yields $\varphi_f(\lim \overset{-1}{f}(\mathscr{V}_f)) = \{1\}$, $\varphi_g(\lim \overset{-1}{g}(\mathscr{V}_g)) = \{0\}$, $\varphi(\lim \mathscr{S}) = \{1\} \cup \{0\} = \{0, 1\}$. The intersection $\mathscr{D}(\overset{-1}{f}(\mathscr{V}_f), \overset{-1}{g}(\mathscr{V}_g)) = \{[2, 2.25], \{2\}\}$. $\lim \mathscr{D} = \{2\} = \{2\} \cap \{2, 4\}$, valuated by $\varphi(\lim \mathscr{D}) = \{0, 1\}$.

3 Logics

3.1 Hierarchies of valuated objects

Let $(V^{(0)}, \leq^{(0)}, \sqcap^{(0)}, \sqcup^{(0)})$ be a non-empty lattice and let there be given the non-empty sets $Z^{(0)}$ and $\wedge z \in Z^{(0)}$ $(\varnothing \neq V_{[z]}^{(0)} \subseteq V^{(0)})$. Using the notation $V_z = \{z\} \times V_{[z]}$, we set $L^{(0)} =_{def} \bigcup_{z \in Z^{(0)}} V_z^{(0)}$.

For integers n, N, $1 \leq n \leq N$, let be given $\varnothing \neq W^{(n)} \subseteq$ (pow $Z^{(n-1)}) \setminus \varnothing$, $W^{(n)}$ a set of finite sets. We define $Z^{(n)} =_{def} \bigcup_{w \in W^{(n)}} \prod_{z \in w} V_z^{(n-1)}$, and if $\wedge z \in Z^{(n)}$ $(\varnothing \neq V_{[z]}^{(n)} \subseteq V^{(n)} \wedge V^{(n)}$ a lattice $(V^{(n)}, \leq^{(n)}, \sqcap^{(n)}, \sqcup^{(n)}))$ are given, $L^{(n)} =_{def} \bigcup_{z \in Z^{(n)}} V_z^{(n)}$.

For all $w \in W^{(n)}$ let $\leq^{(n-1)}$ be extended to an ordering on $\prod_{z \in w} V_{[z]}^{(n-1)}$ by component-wise application of $\leq^{(n-1)}$, and let $\varphi^{(n)}: \prod_{z \in w} V_{[z]}^{(n-1)} \to V^{(n)}$ be a homo- or an antimorphism to give the valuation $v_{[w]}^{(n)}$ of $(z, v_{[z]}^{(n-1)})_{z \in w}$. By definition, $\varphi^{(n)}$ is independent of z and commutative.

The structure $L(N) =_{def} \bigcup_{n=1}^{N} L^{(n)}$ has hierarchical level N over $L^{(0)}$.

In case on each level n a set $\Phi^{(n)}$ of morphisms $\varphi^{(n)}$ is given and only these functions were applied for valuations, we say $L(N) =_{def} \bigcup_{n=1}^{N} L^{(n)}$ is a "logic" of level N over $L^{(0)}$ and the $\varphi^{(n)}$ are "logic functions".

Example: $Z^{(0)}$ a set of propositions z, $V^{(0)} = V^{(1)} = V = (\{"t", "f"\}, \neg, \wedge, \vee)$, $"f" < "t"$, is a boolean lattice, $L^{(0)} = Z^{(0)} \times V^{(0)}$, $W^{(1)} =$ set of finite subsets of $Z^{(0)}$, $Z^{(1)} = \bigcup_{w \in W^{(1)}} \prod_{z \in w} V_z^{(0)}$ with elements $z^{(1)} = (z, v_{[z]})_{z \in w}$, $\Phi^{(1)} = \{$unary $\wedge_1 = \vee_1 = id$ is a $\leq$-homomorphism, unary $\neg$ is a $\leq$-antimorphism, $\vee_{card\,w}$ and $\wedge_{card\,w}$ for card $w > 1$ are $\leq$-homomorphism $\prod_{z \in w} V_{[z]} \to V$, $L^{(1)} = \{((z, v_{[z]})_{z \in w}, \varphi^{(1)}((v_{[z]})_{z \in w})) \mid (z, v_{[z]})_{z \in w} \in Z^{(1)} \wedge \varphi^{(1)} \in \Phi^{(1)}\}$.

3.2 Topologized logics

In the presentation of **3.1** we assume for $1 \leq n < N$ and for all $w \in W^{(n+1)}$ a uniform generalized distance $d(v, \tilde{v})$ is defined on $V^{(n)\,card\,w} \times V^{(n)\,card\,w}$ according **1.1** and **1.2**, $d \in \mathscr{D}$, $\mathscr{D}$ a filter base on pow $(V^{(n)\,card\,w} \times V^{(n)\,card\,w})$ with diag$(V^{(n)\,card\,w} \times V^{(n)\,card\,w}) \subseteq \bigcap \mathscr{D}$, $V^{(n)\,card\,w} \times V^{(n)\,card\,w} \in \mathscr{D}$, $d \in \mathscr{D} \Rightarrow d^{-1} \in \mathscr{D}$. Let $A^{(n+1)}$ be a complete lattice with zero element $o_A^{(n+1)}$ and unit element $e_A^{(n+1)}$ and let there be given an antimorphism $\eta^{(n+1)}: \mathscr{D} \to \mathscr{A}^{(n+1)} \subseteq A^{(n+1)}$ with $\mathscr{A}$ an ideal base, $\eta^{(n+1)}(\bigcap \mathscr{D}) = e_A^{(n+1)}$ and $\eta^{(n+1)}(V^{(n)\,card\,w} \times V^{(n)\,card\,w}) = o_A^{(n+1)}$.

Given any $(z, \tilde{v}_{[z]})_{z \in w}$, $\tilde{v}_{[z]} \in V^{(n)}$, we can express the distance of $(z, \tilde{v}_{[z]})_{z \in w}$ to the logic term $(z, v_{[z]})_{z \in w}$ either by $d((v_{[z]})_{z \in w}, (\tilde{v}_{[z]})_{z \in w})$ or by $\eta^{(n+1)}(d((v_{[z]})_{z \in w}, (\tilde{v}_{[z]})_{z \in w}))$. A special case is $\eta^{(n+1)}(d((v_{[z]})_{z \in w}, (\tilde{v}_{[z]})_{z \in w})$ being a function of distance components $((d(v_{[z]}, \tilde{v}_{[z]}))_{z \in w})$. For illustration see Fig. 4.

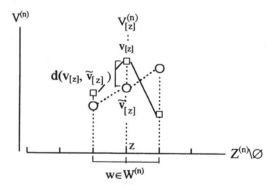

Fig. 4. $(v_z)_{z \in w} \in Z^{(n+1)}$ approximated by $(\tilde{v}_z)_{z \in w}$

We further assume, a homomorphic composition law $*: A^{(n+1)} \times V^{(n+1)} \to V^{(n+1)}$ is given with $\mathbf{e}_A^{(n+1)} * v = v$ for all $v \in V^{(n+1)}$, and in case $V^{(n+1)}$ has zero element $\mathbf{o}_V^{(n+1)}$, $\mathbf{o}_A^{(n+1)} * v = \mathbf{o}_V^{(n+1)}$. Then we can take $\eta^{(n+1)}(d((\tilde{v}_{[z]})_{z \in w}, \ (\tilde{v}_{[z]})_{z \in w})) * \varphi^{(n+1)}((v_{[z]})_{z \in w})$ as a similarity measure in $V^{(n+1)}$ of $(\tilde{v}_{[z]})_{z \in w}$ to $(v_{[z]})_{z \in w}$ with respect to $\varphi^{(n+1)}$.

4 Some applications

4.1 Contractive mappings

Let $X = X_{[0]}$ be a non-empty set and let $\mathscr{F} = (f_{[n]})_{n \in \mathbf{N}}$ be a family of functions $f_{[n]}: X \to X$, $X_{[n]} =_{def} f_{[n]}(X)$. If $\mathscr{X} = (X_{[n]})_{n \in \mathbf{N}}$ forms a filter base on pow X, we say, $\mathscr{F}$ is a "filter generating" or a "contractive" function sequence on X. We assume $\varnothing \neq X^* = \bigcap \mathscr{X}$ and have $X^* \subseteq X$, $f_{[n]}(X^*) \subseteq X_{[n]}$, $\bigcap_{\mathbf{N}} f_{[n]}(X^*) \subseteq X^*$.

If in particular $f: X \to X$, i.e. $f(X) \subseteq X$, and $\wedge n \in \mathbf{N}$ ($f_{[n]} =_{def} f^n$), we have iterations, monotone $\mathscr{X}$ and $f(X^*) \subseteq X^*$. For $X^* = \{x^*\}$, x^* is a fixpoint.

We assume, a function $h: \mathbf{N} \to X$ exists such that for the filter base $\{B_{[n]} = \{z \mid n \leq z \wedge z \in \mathbf{N}\} \mid n \in \mathbf{N}\}$ $X_{[n]} = \{h(z) \mid z \in B_{[n]}\}$. If h is known on $\{1,2,...m\}$, extrapolation $\tilde{h}$ of h to $\{m+1, ...\}$ can be used to find an approximation $\tilde{X}_{[m+1]}$ to $X_{[m+1]}$, which is a well known practice in numerical analysis.

4.2 Roundings

For $n = 1,2,....$ we consider idempotent mappings (roundings) $\rho_{[n]}: X_{[n-1]} \to X_{[n]}$, $\varnothing \neq X_{[n]} = \rho_{[n]}(X_{[n-1]}) \subset X_{[n-1]}$. The set extensions of roundings are idempotent $\subseteq$-homomorphisms: for $\varnothing \neq X' \subseteq X'' \subseteq X_{[n-1]}$ we have $\rho_{[n]}(X') \subseteq \rho_{[n]}(X'')$, $\rho_{[n]} \circ \rho_{[n]}(X') = \rho_{[n]}(X')$, and $\rho_{[n]}(X') \subseteq \rho_{[n]}^{-1} \rho_{[n]}(X')$. The set $\{\rho_{[n]}^{-1} \rho_{[n]}(X'), \rho_{[n]}(X')\}$ is a filter base.

Given roundings $\rho_{[k]}: X_{[k-1]} \to X_{[k]}$ for stages $k = 1,2,...n$, we define $f_{[k]} = \rho_{[k]} \circ \rho_{[k-1]} ... \circ \rho_{[1]}$ to obtain a contractive sequence $(f_{[k]})_{k=1,...n}$.

Example: Rounding of the reals $\mathbf{R}$ onto integers $\mathbf{Z}$ by $\rho(r) \to \{z\}$ for $r \in (z - 0.5, z + 0.5]$ and $z \in \mathbf{Z}$. On the next stage, $\mathbf{Z}$ can be rounded by another rounding rule for example onto $10 \times \mathbf{Z}$, and so on.

4.3 Valuated filter hierarchies ("pyramids")

We generalize multi-stage rounding: Let be $X = X_{[0]}$, $\Sigma \subseteq$ pow X a σ-algebra, $\mu: \Sigma \to \mathbf{R}_+$ a σ-additive measure, $X_{[0]}$ partitioned into a finite number of non-empty subsets $X_{[0,k]}$, $k = 1,2,...K_{[0]}$, with $X_{[0,k]} \in \Sigma$. For all k let there be given roundings $\rho_{[0k]}: X_{[0k]} \to \rho_{[0k]}(X_{[0,k]}) \subset X_{[0k]}$. We define $\mu_{[0k]} =_{def}$ measure of $\rho_{[0k]}(X_{[0,k]}) =_{def} \mu(X_{[0,k]})$, $\mu_{[0]} =_{def} {}^+(\mu_{[0k]})_{k \in K_{[0]}}$.

For $n = 0,1,2,...$ let $K_{[n]}$ be partitioned into $I_{[ni]}$, $i = 1,..,K_{[n+1]}$. For $k = 1,..,K_{[n+1]}$ we define $X_{[n+1,k]} =_{def} \bigcup \{\rho_{[ni]}(X_{[ni]}) \mid i \in I_{[nk]}\}$,

$\mu_{[n+1,k]} =_{def}$ measure of $X_{[n+1,k]} =_{def} {}^+(\mu_{[ni]})_{i \in I_{[nk]}}$,

$X_{[n+1]} =_{def} \bigcup \{X_{[n+1,k]} \mid k \in K_{[n+1]}\}$,

$\mu_{[n+1]} =_{def}$ measure of $X_{[n+1]} =_{def} {}^+(\mu_{[n+1,k]})_{k \in K_{[n+1]}}$,

roundings $\rho_{[n+1,k]}: X_{[n+1,k]} \to \rho_{[n+1,k]}(X_{[n+1,k]}) \subset X_{[n+1,k]}$. We have $\mu_{[0]} = \mu_{[1]} = ...$ (conservation of total measure). An illustration is given in Fig. 5.

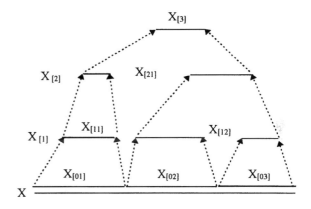

Fig. 5

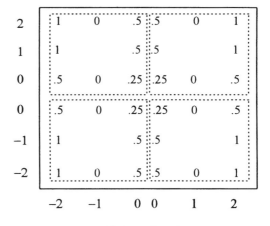

Fig. 6.1. Level 0

6

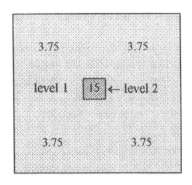

Fig. 6.2. Level 1 and 2

An application to valuated picture pyramids is given in Fig.6. The grid is $\{-2, -1, 0, +1, +2\} \times \{-2, -1, 0, +1, +2\}$, additive valuations are $\{0, .25, .5, ...15\}$, not all grid points are valuated. The example shows for level 0 how overlap of $X_{[n-1,k]}$'s can be handled.

4.4 Adaptable networks

We first consider a function $f: X \to Y$. X and Y can be multi-dimensional, $X \subseteq X_{[1]} \times ... \times X_{[n]}$, $Y = Y_{[1]} \times ... \times Y_{[m]}$. We write $y = f(x)$, $x = (x_{[i]})_{i=1,2,...n}$, $y = (y_{[j]})_{j=1,2,...m}$. Let there be given $x_{[0]} \in X$ and let $y_{[0]} = f(x_{[0]})$. To deal with values $\tilde{x}$, $\tilde{y} = f(\tilde{x})$ which are approximations to $x_{[0]}$, $y_{[0]}$ we introduce a neighborhood system to $x_{[0]}$ by a filter base $\mathscr{B}$ on pow X with $x_{[0]} \in \bigcap \mathscr{B}$. The set extension of f maps $\mathscr{B}$ on a filter base $f(\mathscr{B})$ with $y_{[0]} \in \bigcap f(\mathscr{B})$. If each base is isomorphically and injectively valuated by values v of a filter base $\mathscr{V}$ on a complete lattice V and if we assume $v(\bigcap \mathscr{B}) = v(\bigcap f(\mathscr{B})) = \bigcap \mathscr{V}$, we can measure generalized distances of $\tilde{x}$ to $x_{[0]}$ and $\tilde{y}$ to $y_{[0]}$ according **2.2** by their values $v(\tilde{x})$ and $v(\tilde{y})$. A simple case is $\mathscr{B}$ monotone.

Example (interval arithmetic on **R**): $f \in \{+,-,*,/\}$, $X = \mathbf{R} \times \mathbf{R}$ for $\{+,-,*\}$, $X = \mathbf{R} \times (\mathbf{R} \setminus \{0\})$ for $/$, $Y = \mathbf{R}$. $x_{[0]} = (x_{[01]}, x_{[02]})$, $\mathscr{B} = \{\text{interval } I = (x_{[01]}-a, x_{[01]}+b] \times (x_{[02]}-a', x_{[02]}+b'], X\}$, assuming $0 \notin (x_{[02]}-a', x_{[02]}+b']$ for $/$. a, b, a', b' positive reals. $\lim \mathscr{B} = I$, $f(\mathscr{B}) = \{f(I), \mathbf{R}\}$. All $\tilde{x} \in I$ and all $\tilde{y} \in f(I)$ are equivalent. $v(I) = v(f(I)) = \{1\}$, $v(X) = v(Y) = \{0, 1\}$.

Now we consider a parameterized function $f: D \to Y$ with $D \subseteq X \times P$, P a non-empty set of parameters $p = (p_{[l]})_{l \in L}$. Changing p results in changing the function and thus in general in changing the function

value for fixed x. We say, f is "controllable" or "programmable" and p is a "control parameter" or a "program" to f. To $(x_{[0]}, p_{[0]}) \in D$ a filter base (neighborhood system) $\mathscr{B} = \{B_{[k]} \mid k \in K \wedge B_{[k]} \subseteq D\}$, $D = \bigcup \mathscr{B}$, can be introduced and then mapped by the set extension of f onto a filter base $f(\mathscr{B})$. A visualization is given in Fig. 7.

Notice, in general $D \neq X \times P$, not all values (x,p) belong to the domain of definition of the function f.

The parameters are used to adapt the function f to a certain behavior: For fixed $x_{[0]}$, the cut of D along $x_{[0]}$ is a filter base $\mathscr{B}(x_{[0]})$, in practice a sequence of parameters $\tilde{p}$ approximating $p_{[0]}$ is constructed converging to $\lim \mathscr{B}(x_{[0]})$ by trying to minimize the deviation of $f(x_{[0]}, \tilde{p})$ from $f(x_{[0]}, p_{[0]})$ ("learning").

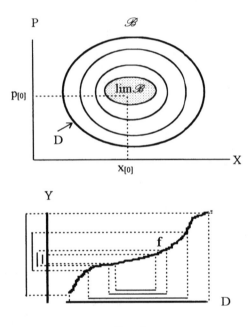

Fig. 7

Finally, let there be given a family $\mathscr{F}$ of parameterized functions, $\mathscr{F} = (y_s = f_s(x_{[s]}, p_{[s]}))_{s \in S}$, S finite. Notice, for $s \neq s'$ $f_{[s]} = f_{[s']}$ is possible. Composing functions f_s, $f_q \in \mathscr{F}$ by identifying ("connecting") some ("output" with "input") components $y_{[sj]} = x_{[qi]}$, can (but need not) result in a composite function F. A theory on this is for example given in articles of Albrecht [3] and of Pearson and Dray [5]. To adapt an F to a wanted behavior is in general a complex multi-parameter problem which has two variants:

selection of $\mathscr{F}$ and of the interconnections for given parameters p (numerical analysis), adaptation of the parameters p for given $\mathscr{F}$ and given interconnections (artificial neural networks).

Example (simple classifier): We assume $\wedge s \in S$ ($X_{[s]} = X \wedge Y_{[s]} = Y = [0, 1] \subset \mathbf{R} \wedge f_s(X) = Y$). Some patterns $x = (x_{[1]}, ..., x_{[n]}) \in X$ are considered to belong more or less to the same class s. We take a sample of such x and group them into a filter base $\mathscr{B}(s)$ on pow X. We try to adapt the (multi-dimensional) parameter in $f_s(x, p_{[s]})$ to a parameter $p_{[s0]}$ for which $f_s(x, p_{[s0]})$ maps only $\mathscr{B}(s)$ isomorphically on a filter base $\mathscr{V}(s)$ on pow Y with $\bigcap \mathscr{B}(s)$ mapped on $\{1\}$, but not a $\mathscr{B}(s')$ with $s' \neq s$. If this has been done for all s, then to any given $\hat{x} \in X$, we obtain a family $h = (\hat{y}_s)_{s \in S}$ of values $\hat{y}_{[s]}$, representing a function on S. The reciprocal $\overset{-1}{h}$ can be used for cuts along any $U \subset Y$, e.g. to find an s with maximal $\hat{y}_{[s]}$. A visualization is given in Fig. 8.

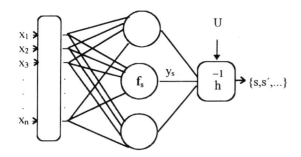

Fig. 8

Frequently used is $X = \mathbf{R}^n$, $f_s = \exp(-\frac{1}{2}(x - x_{[s]}^{(0)})^T K_{[s]}^{-1} (x - x_{[s]}^{(0)}))$, x the input vector, parameters are the center vector $x_{[s]}^{(0)}$ and the elements of the inverse $K_{[s]}^{-1}$ (if not singular) of the covariance matrix $K_{[s]}$.

The example follows the principle outlined in section **3.2**: e.g. $d(x, x_{[s]}^{(0)}) = (x - x_{[s]}^{(0)})^T K_{[s]}^{-1} (x - x_{[s]}^{(0)})$ is a quadratic distance measure, $\eta(d(x, x_{[s]}^{(0)}))$ corresponds to $\exp(-\frac{1}{2}d(x, x_{[s]}^{(0)}))$ mapping into the ideal base $\mathscr{A} = \{[0,a] \mid a \in [0, 1]\}$, $\varphi(x_{[s]}^{(0)}) = 1$, $*$ is multiplication of reals. If in particular $K_{[s]}^{-1}$ is a diagonal matrix $\mathrm{diag}(p_1, p_2, ...)$, then $d(x, x_{[s]}^{(0)}) = \sum p_{[s]i} d_{[s]i}$ with component-wise distances $(x_i - x_{[s]i}^{(0)})^2$.

References

[1] Vietoris L.: Stetige Mengen. Monatshefte f. Mathematik u. Physik, vol. 31, 1921.

[2] Bourbaki N.: Topologie Général. Act. Sc. Ind., vol. 1142, Hermann & Cie , Paris 1951, pp. 40-41.

[3] Albrecht R. F.: On mathematical systems theory. In: R. Albrecht (ed.): Systems: Theory and Practice, Springer Verlag Vienna-New York, 1998, pp. 33-86.

[4] Zadeh L. A.: Fuzzy Sets. Information and Control 8, 1965.

[5] Pearson D. W., Dray G.: Applications of artificial neural networks. In: R. Albrecht (ed.): Systems: Theory and Practice, Springer Verlag Vienna-New York, 1998, pp. 235-252.

Linking Human Neural Processes
Directly with Technology
-   The Future?

K. Warwick

Department of Cybernetics
University of Reading
Whiteknights
Reading
RG6 6AY
UK
kw@cyber.reading.ac.uk

1   Introduction

The field of neural networks appears to consist of two distinct schools, those studying the operation of carbon based versions, such as in humans or insects, and those studying the operation of silicon based versions, as in machines such as computers. The latter field has come to be known as Artificial Neural Networks.

In the last year however, a further possibility has arisen with the potential for a mixed arrangement, part carbon and part silicon [1]. This naturally opens up a whole new way of thinking not only in terms of the capabilities of such a mixed network but also in terms of the philosophical identity that such a network exhibits.

In this paper a quick look is taken at the field of Wearable Computers, as much to pose the question as to whether or not such technology does play a role in the ensuing discussion. Following this, last years' silicon chip implant experiments are placed in context, as to what they each mean as individual results and their overall impact. Finally the potential for the future is considered. Just what might be around the corner for a silicon/carbon merged network? Will it be the best of both worlds, with super brains appearing or will it be all too much for carbon based brains to take?

2.   Wearable Computers

Wearable computing has been seen as a way to make individual people, rather than machines, smarter [2,3]. As we will see in the next section, implants are one way to go, however much simpler, although questionably no where near as effective, is to augment everyday wearables, such as clothes, shoes or wristwatches, with some element of computing power. It is claimed [2] that in this way, wearables can extend the senses, improve memory and help the individual stay calm.

One problem with wearables is that there is, in fact, little or no difference between them and portables. Rather they can perhaps be seen as a special case of portables. It is said that portables are carried and a conscious effort must be taken to make use of the device, and this is certainly true. Conversely it is claimed that with a wristwatch this can be made use of instantly, even when working with two hands! Clearly the definition of a wearable is prone to argument.

In all seriousness however it appears to be a good idea to make use of everyday items that are worn, in order to enhance them with some computing power. Why not have medical monitoring information in a watch, glasses with an in built display that only the wearer can see or quite simply a belt which contains a computer.

It is said that our self-perception changes dependant on the clothes we wear and therefore we will think differently about ourselves when we have computers in our clothing. Indeed that may well be true, however the critical element is that the computer is external to our body, so it doesn't actually change who we are other than merely in terms of our self-perception. In terms of our intelligence this remains the same as when we make use of a portable computing system, apart perhaps from a small time saving in use.

It is still not known as to whether or not wearable computers will make an impact on our lives. The commercial possibilities of computing power on board a watch would appear to be reasonable, although it may all be down to fashion and what is socially acceptable as to how much it all takes off. However some use is already made, for example Symbols Technologies makes a finger ring equipped with UPC reader and mini-computer display. The device is pointed at an object with a UPC bar code, such that information details and consumer reviews on that object can be read.

Wearables suffer from a number of drawbacks in that they must, almost surely be very light weight, easily visible or accessible, fairly low cost and low power. All of these characteristics tend to restrict, at the present time, what is actually achievable. However where the positives of a wearable device are apparent, such restrictions are lifted. One example is a Global Positioning System which, in the future may also allow for wireless connection, a microphone and even a digital camera, although at present several armies around the world are testing out the GPS wearable technology, as it stands, as a low cost communication and navigation system for troops.

Wearable computers are designed to be with the user at all times and to be part of the everyday life of their wearer. In practice however, most of the suggested devices do not tend to fit that bill. In [2] examples given of wearables include dancing shoes which convert dance steps into music, social wearables which are worn as a necklace and flash to communicate the names of their users, stress monitors and a video camera fixed to a baseball cap, which it is claimed, enables the wearer to call themselves a 'cyborg'! Meanwhile, by the definitions given, Virtual Reality headsets would appear to most definitely be wearable computers as would wrist watches in themselves. All in all it therefore appears to be an interesting area in itself, with commercial potential, but with little or nothing to contribute to the interface between silicon and carbon neural networks.

In particular it is worth stressing that wearable computers, in common with portables and even main frames are subject to the usual human-machine interface problems. To translate signals from the human brain to machine brain, and vice versa, a laborious process must ensue, involving conversion from electronic signals to mechanical movements and

back to electronic signals again. This all takes time and energy, and is prone to error. The true merging of silicon and carbon will need a much closer interface if we are to move forward.

3. Silicon/Carbon Insects

In January 1997 a microprocessor based 'back pack' was connected, via electrodes, to the brain of a cockroach, effectively replacing some of the cockroach's neurons with electronic versions, [4]. The artificial neurons were not only able to drive motor responses in the cockroach, enabling it to walk, but also its 'artificial' brain state could be witnessed in terms of a series of LEDs.

Meanwhile in March 1998 another insect robot was formed by a joint research group between the University of Tokyo and the University of Tsukuba. They fused the antennae from a male silk moth with a wheeled robot containing a rudimentary artificial neural network. When a female silk moth, nearby, gave off an attractive pheromone signal, this was picked up by the male antennae, resulting in the wheeled robot moving towards the female. The exact nature of the neural network in this response was not clear!

4. Silicon/Carbon in humans – why?

Not only have we seen quite a number of science fiction writers pointing to a near-term future in which humans and technology are inextricably linked [5], in recent years many scientists have also pointed in this direction [1]. Partly this is felt to be due to the unravelling of DNA molecules, possible allowing for computers to themselves be constructed from living DNA, and partly it is felt to be due to the likely acceptance, amongst humans, of the use of implant technology to enhance their capabilities. A key driver amongst this is the potential for linking the human neural system with the artificial version.

But why should we even wish to consider such a move? The answer is simple. Machines are already capable of doing many things that humans, themselves, cannot do, and in the future their range of abilities will be even greater. In fact making machines more physically capable than ourselves has been something ongoing for several thousand years. In this way we have been able to benefit from the capabilities gained with few associated problems, although in many cases we have had to change considerably the way in which we live in order to incorporate the new found technological advances, recent examples being the telephone and automobiles.

Occasionally problems occur with technology when we lose control of a situation, e.g. we drive a car too fast, fly a plane too low or explode a nuclear bomb. However in each case the consequences are fairly limited and those still alive tend to accept that such problems will happen, from time to time.

With machine intelligence though the picture is rather more dangerous, something that many people are now becoming aware of. In some ways computer based machines can, even now, easily out perform any human in certain measures that are regarded as being an indication of intelligence, e.g. mathematical processing, memory or logic. It is also accepted by many as just about inevitable that, by means of an artificial neural network approach, machines will, one day, exhibit an overall level of intelligence that far exceeds anything humans themselves can offer. [6]. Unfortunately, as we are by no means 100% certain exactly what human intelligence is, we can not be at all sure when we have created a network with an intelligence which could pose a considerable threat. One thing is clear however, when we do switch on a network with an intellect roughly comparable with our own, with sufficient accompanying power, it will most likely not wish to be switched off again.

Machines can already perform well on, what we like to describe as intelligence (IQ) tests, indeed they can even learn to perform well on such tests. If given appropriate physical sensors and actuators machines can certainly be aware of each other and even be self-aware, a feat that is believed to be achieved in carbon-life forms only by humans, chimps and orang-utans [7]. Machines can exhibit emotions, creativity and desire. It is difficult to conceive of any intelligent feature that humans have that machines will not be able to exhibit in some form, either now or in the future [8]. Perhaps humour may be one exception!

Interestingly, predictions from the past both in science fiction [9] and science [10] pointed to the construction of general purpose robot machines that were humanoid in form and action. What we have however witnessed is the appearance of an intelligent, computer based network, the internet, linked, to an extent at least with relatively low intelligence machine/robot entities. Essentially the robots themselves are not necessarily particularly intelligent, but due to the power of communication,

this ability lies in what is controlling them, whether that be human or machine.

The main question about how humans interface with machines is therefore not the one that might have been expected in terms of our interface with individual robots/machines, but rather it is a question of how we interface with networks and in particular the Internet, a global communications network. Central to this interface question, and vitally important for the future is the whole picture of intelligence, specifically that exhibited by the machine network in relation to that apparent in human brains. With this perspective, it is difficult indeed to foresee a future in which the Internet is not the catalyst of power, allowing intelligent machines to physically affect human existence, in the direction desired by the machines.

Historically humans and machines have existed as complementary beings, coming together to achieve a particular goal, indeed with wearable computers this is still the case. In relatively few instances only are human and machine inextricably linked, the norm being for physical separation with communication via human external sensing and actuating, such as vision or touch. Contrary to this, in order to help certain humans with particular physical disabilities, the human body has been acceptably breached. Hence we now see replacement hips, cochlea implants and heart pacemakers in a growing number of individuals. For many the alternative is unbearable, quite simply the implant, with its inherent dangers, must be carried out. Not to do so might mean a life of misery or even no life at all, so the recipient has no choice!

But it is also possible to enhance the capabilities of all humans by the use of implant technology. In fact if we are to harness the increased capabilities of machine intelligence, linking human carbon with machine silicon presents itself as perhaps the only possible solution. Certainly an implanted individual will be capable of doing things they could not otherwise achieve, but the overall situation is more critical than that in that it may turn out that it is something we will all have to do, in order to be part of the system!

5. Implant Technology

In August 1998 I had a silicon chip transponder surgically implanted in my left arm. As I entered the main door to the Cybernetics Department at Reading University, a radio frequency signal, across the doorway, excited the coil in the transponder, providing current to the silicon chip circuitry, allowing for the repetitive transfer of 64 bits of information. The building's computer was able to recognise me from a unique signal transmitted from my implant. So it welcomed me with 'Hello Professor Warwick', it switched on the foyer light for me and selected my web page on the video screen. Elsewhere in the building, as I approached my laboratory the door opened automatically, allowing me to pass freely. The computer kept a record of when I entered the building and when I left, which room I was in and when I got there. A location map on the computer gave a real time picture of my whereabouts at all times. Whilst such an experiment indicates some of the possibilities of intelligent buildings it also indicates an enhancement in the capabilities of the human body.

Since then a research team at Emory University in Atlanta, Georgia, implanted a similar transmitting device in the damaged brain of a stroke victim, linking up the human neural signals with the silicon chips present. In this way brain signals were transmitted directly to a computer, and as a result, before long the patient had learnt how to move a cursor around on the computer screen, simply by his thoughts, which after all are merely electronic/electrochemical signals. Importantly it is not necessary for the computer to interpret the signals received but rather for the implanted individual to learn how to send signals to affect the computer. The Emory group now intends to use the same device to help paralysed patients turn on light switches and even send e-mails. Meanwhile at the University of Maine, a team led by Ross Davis, have transmitted signals from a computer into a patient and, as a result, have been able to control some movements and physical activity.

Technically we are now getting to grips with the fact that signals on the human nervous signal can be transmitted to, and received from, a computer, via a relatively straightforward implant. The implant's radio connection with the computer means that a much closer link can exist between human and machine. It should therefore be possible, fairly soon, to interact with and operate computers, without recourse to relatively slow computer keyboards or a computer mouse. Indeed, before too long thought control of computers should be a realistic feature [10, 11].

Linking up the human brain with silicon opens up all sorts of possibilities. Some feel that by enhancing our own network in this way we will be able to effectively cope with information overload [11]. Meanwhile machines can sense the world in a much richer way then humans, taking in such as radar, infrared and ultraviolet signals, as well as the normal human senses, which in themselves only represent a very small range of the overall spectrum. So it may be that we can harness this for ourselves, directly bringing in such signals onto our nervous system, thereby giving us extra senses. But the whole gambit of machine intelligence also becomes directly accessible, giving us the opportunity of thinking in many dimensions, not just in 3-D or 2-D simplified views, which is what we are used to as humans. Physics books will have to be rewritten when we start to view the world as a multi-dimentional entity.

Linking silicon up with the human brain will though change the way we think, not only in terms of making sense of the new signals arriving but also what we think about ourselves and who we are. In my own implant experiment, after a couple of days I treated the implant as part of me, not an experience, I suspect, shared by wearable computer owners or smart card holders. But due to the things that happened automatically, I quickly developed an affinity to the computer with which "I" was directly connected. This feeling was a strong one, and not at all what I had expected. With a close silicon/carbon link, particularly where, via the nervous system, a human brain is linked closely with a machine brain, individual identity will be difficult to define. This is even more pertinent when a human is connected along with many others, into an intelligent machine network.

6. Thought Communication

A link from human nervous system to computer, via a radio frequency implant, forms a short circuit connection between human neurons and artificial neural nets. This opens up possibilities of the artificial net becoming an extension of the human brain, with all its memory and mathematical faculties giving the human an extended 'super brain'. Certainly neural movement signals and emotions will be able to be transmitted, and received, and with a certain amount of learning various thoughts too.

But once connected in this way to a computer, by means of satellite or the Internet, such signals can be sent around the world. Another implanted person, who can transmit their own signals in return, can receive the signals sent. What this presents therefore is the possibility of global communication by thought processes. Essentially thought to thought communication by humans.

Certainly problems with the technology will have to be overcome. For example, what will one person actually make of a signal transmitted onto their own nervous system from someone else's brain? Will such a signal mean exactly the same thing to different people? The age-old philosophical question "when one person thinks of the colour red is it the same thought in another person"?, will perhaps be answered, one way or the other.

But we will need to learn how and when to transmit our thoughts in a way that others will understand. It may in fact be possible for one person to read the mind of another, knowing what that person thinks and what they are about to do, before they do it.

As for the link between human and machine neural processes, it is looking at a future where the two are connected by a rapid radio frequency link. This means that signals from one will be able to bypass the more normal, and relatively slow, sensory systems that exist at present. The operation between the two will be much more in the sense of one overall neural network, part silicon – part carbon.

7. DNA neurons

Research is also going on, in an attempt to make artificial neural networks more biologically based [1]. Alan Mills at Bell Labs, in New Jersey, USA heads a group which is trying to create a DNA based neural network, this being essentially an analogue, as opposed to digital, device. The basis is one of materials, rather than wires and connections, requiring salt and enzymes as opposed to silicon.

It may well be that such an approach provides a link between silicon and carbon networks of the present time. The Emory University work has however involved a more direct connection with signals going from carbon to silicon directly, making use of the fact that human neurons can actually be grown onto silicon. This research is supported by work in the bioscience research group at Atsugi, Japan, where Keiichi Torimitsu has isolated neuron slices and enabled them to communicate directly with electrodes, when placed in close proximity. Clearly this whole area also opens up possibilities in improving or replacing tissue for such as eyes and ears in order to bring back or enhance functionality.

8. Conclusions

The whole area of linking human (or animal) neurons with artificial neurons is an extremely exciting one, but yet is an area in which research is only just opening up. Importantly it is an area where actual research involving trials and experimentation is required, this is not the base for immediate theories. The impact of the field, in science and philosophy will however be immense, as it clearly questions what it means to be an individual human.

Humans still have a lot to learn if they are to fully harness the power of intelligent machines. This is particularly true when those machines have sensors and actuators that out-perform anything humans have on offer. Connecting human brains and machine brains thus provides an opportunity to enable the two entities to work together, combining the good features of both in a cybernetic whole. Whether it will enable humans to stay in control of the intelligent machines of the future is another question all together!

References

1. Fierman S.A., Merging Man and Machine, Newsweek International, March 1999.

2. Pentland A.P., Wearable Intelligence, Scientific American, Vol. 9. No. 4, pp. 90-95, 1998.

3. Mann S., Wearable Computing: A First Step Toward Personal Imaging, Computer, Vol. 30, No. 2, pp. 25-32, 1997.

4. Guiness Book of Records, Guiness Publishing, 1999.

5. W. Gibson, Mona Lisa Overdrive. Voyager, 1995.

6. K. Warwick, In the Mind of the Machine. Arrow, 1998.

7. G. Gallup Jr. 'Can Animals Empathize? Yes', Scientific American, Vol.9, No. 4, pp. 66-77, 1998.

8. J. M. Bishop and J. Preston (eds.), Views into the Chinese Room, Routledge, 1999.

9.   K. Capek, Rossum's Universal Robots, Fr. Borovy, Prague, 1940.

10. J. P. Eckert Jr., The Integration of Man and Machine. Proc. of the IRE, Vol. 50, No. 5, pp. 612-613, 1962

11. K. Warwick, Cybernetic Organisms – Our Future. Proc. IEEE, Special Guest Author, Vol. 87, No.2, pp. 387-389, 1999.

12. P. Cochrane, Tips for Time Travellers, McGraw-Hill, 1999.

NEURAL DYNAMIC MODEL FOR OPTIMIZATION OF COMPLEX SYSTEMS

Hojjat Adeli

Professor

The Ohio State University, College of Engineering

470 Hitchcock Hall, 2070 Neil Avenue, Columbus, Ohio 43210 U.S.A.

Abstract

On September 29, 1998, the *U.S. Patent and Trademark Office* issued a new patent under the title *"Method and apparatus for efficient design automation and optimization, and structure produced thereby"*. The inventors are Hojjat Adeli and H.S. Park. In this Plenary Lecture we present this patent and its application to design automation and optimization of large one-of-a-kind engineering systems. We also show the successful application of this model to another nonlinear optimization problem, construction scheduling and cost optimization.

1. Engineering Design Automation and Optimization

A multi-paradigm computing approach is presented for automating the complex process of engineering design through ingenious integration of a novel neurocomputing model (Adeli and Hung, 1995), mathematical optimization (Adeli, 1994), and massively parallel computer architecture (Adeli, 1992a&b, Adeli and Soegiarso, 1999). The details of the algorithms and computational models are presented in a recently-published book (Adeli and Park, 1998).

Optimization of large and complex engineering systems is particularly challenging in terms of convergence, stability, and efficiency (Adeli, 1994). Adeli and Park (1995) present a robust neural dynamics model for optimal design of structures by integrating the penalty function method, Lyapunov stability theorem, Kuhn-Tucker conditions, and the neural dynamics concept. Adeli and Park (1996) extend the neural dynamics model by creating a hybrid counterpropagation-neural dynamics model and a new neural network topology for optimization of large structures made of discrete sections subjected to the highly nonlinear and discontinuous constraints of commonly-used design codes such as the American Institute of Steel Construction (AISC) Allowable Stress Design (ASD) (AISC, 1989) or Load and Resistance Factor Design (LRFD) (AISC, 1998) specifications.

Park and Adeli (1997) present distributed nonlinear neural dynamic algorithms for discrete optimization of large steel structures. The algorithms are implemented on a distributed memory machine, CRAY T3D. In a distributed memory machine, a relatively large number of microprocessors are connected to their own locally distributed memories without any globally shared memory (Adeli and Kumar, 1999). For these machines, communications between the processors becomes a bottleneck because accessing memories of remote processors takes more time than accessing the shared memory in the shared memory machine. Further, on distributed memory architecture, the limited local memory of each processor creates an obstacle that must be overcome through algorithmic restructuring. For the solution of resulting linear simultaneous equations a distributed preconditioned conjugate gradient algorithm is developed employing the worksharing programming paradigm.

The computational models are applied to minimum weight design of high-rise and superhighrise building structures of arbitrary size and configuration, including a very large 144-story superhighrise building structure with 20,096 members. The structure is subjected to dead, live, and multiple wind loading conditions applied in three different directions according to the Uniform Building Code (UBC). The patented computational model developed in this research finds the optimum solution (minimum weight design) for this very large structure subjected to highly nonlinear and implicit constraints of actual design codes such as the AISC LRFD code automatically and without any convergence problem.

2. Construction Scheduling and Cost Optimization

Adeli and Karim (1997) present a general mathematical formulation for scheduling and cost optimization of construction projects. Repetitive and non-repetitive tasks, work continuity considerations, multiple-crew strategies, and the effects of varying job conditions on the performance of a crew can be modeled. An optimization formulation is presented for the construction project scheduling problem with the goal of minimizing the direct construction cost. The nonlinear optimization is then solved by the patented neural dynamics model of Adeli and Park. For any given construction duration, the model yields the optimum construction schedule for minimum construction cost automatically. By varying the construction duration, one can solve the cost-duration trade-off problem and obtain the global optimum schedule and the corresponding minimum construction cost. The new construction scheduling and cost optimization model has been applied successfully to the highway construction problem.

3. Acknowledgment

This presentation is based upon work sponsored by the U.S. *National Science Foundation* under Grant No. MSS-9222114, *American Iron and Steel Institute, American Institute of Steel Construction, Federal Highway Administration,* and the *Ohio Department of Transportation.* Supercomputing time was provided by the *Ohio Supercomputer Center* and *National Center for Supercomputing Applications* at the University of Illinois at Urbana-Champaign. H.S. Park and A. Karim contributed to the research projects as Research Associates.

4. References

Adeli, H., Ed. (1992a), *Supercomputing in Engineering Analysis*, Marcel Dekker, New York.

Adeli, H., Ed. (1992b), *Parallel Processing in Computational Mechanics*, Marcel Dekker, New York.

Adeli, H., Ed. (1994), *Advances in Design Optimization*, Chapman and Hall, London.

Adeli, H. and Hung, S.-L. (1995), *Machine Learning - Neural Networks, Genetic Algorithms, and Fuzzy System*, John Wiley, New York.

Adeli, H. and Karim, A. (1997), "Scheduling/Cost Optimization and Neural Dynamics Model for Construction", Journal of Construction Engineering and Management, Vol. 123, Np. 4, pp. 450-458.

Adeli, H. and Kumar, S. (1999), *Distributed Computer-Aided Engineering for Analysis, Design, and Visualization*, CRC Press, Boca Raton, Florida.

Adeli, H. and Park, H.S. (1995), "Optimization of Space Structures by Neural Dynamics", *Neural Networks*, Vol. 8, No. 5, pp. 769-781.

Adeli, H. and Park, H.S. (1996), "Hybrid CPN-Neural Dynamics Model for Discrete Optimization of Steel Structures", *Microcomputers in Civil Engineering*, Vol. 11, No. 5, pp. 355-366.

Adeli, H. and Park, H.S. (1998), *Neurocomputing for Design Automation*, CRC Press, Boca Raton, Florida.

Adeli, H. and Soegiarso, R. (1999), *High-Performance Computing in Structural Engineering*, CRC Press, Boca Raton, Florida.

AISC (1989), *Manual of Steel Construction – Allowable Stress Design*, American Institute of Steel Construction, Chicago, IL

AISC (1998), *Manual of Steel Construction – Load and Resistance Factor Design – Volume I Structural Members, Specifications, & Codes*, American Institute of Steel Construction, Chicago, IL

Park, H.S. and Adeli, H. (1997), "Distributed Neural Dynamics Algorithms for Optimization of Large Steel Structures", *Journal of Structural Engineering*, ASCE, Vol. 123, No. 7, pp. 880-888.

UBC (1997), *Uniform Building Code - Volume 2 - Structural Engineering Design Provisions*, International Conference of Building Officials, Whittier, California.

Comparative Testing of Hyper-Planar Classifiers on Continuous Data Domains

David McLean, Zuhair Bandar

The Intelligent Systems Group, The Manchester Metropolitan University, Chester Street, Manchester, M1 5GD, UK.
Email {D.McLean, Z.Bandar}@doc.mmu.ac.uk

Abstract

This paper details a set of comparative tests conducted between five classification algorithms using three real world continuously valued data sets. The algorithms were selected to represent the two most popular classification methods, neural networks and decision trees as well as hybrid algorithms which incorporate features of both techniques. These hybrid algorithms construct an architecture to model the problem domain.

The three real world data sets have previously been used in the StatLog tests [1] and these experiments can be viewed as an extension of this work. Due to the nature of these data sets, each contains some level of noise which affects the learning procedure to varying degrees. A maximum bound on a classifier's generalisation is discussed, which is due to the loss of information incurred when allowing for noise in a data domain model.

The results of these tests establish the levels of performance which can be achieved using hyperplanic classifiers on noisy continuously valued data sets.

1 Introduction

1.1 Back Propagation

Back-Propagation (BP) is probably the most commonly used Feed Forward Neural Network (FFNN) training algorithm [2][3][4]. It requires the pre-construction of a suitable FFNN topology for the problem. This is largely an empirical process to find a trainable network which also generalises well on the test examples.

The algorithm uses a gradient descent technique which alters the weights in the network so as to minimise an error function [4][6]. All the weights in the network are altered for each iteration of the algorithm, giving a complex and often time consuming learning process [6]. This technique is also very prone to local minima of the error function which prevents the system from progressing to a solution [5][6]. The standard version of BP was used in these tests with a momentum parameter as described in [4].

1.2 The AMIG Algorithm

The Average Mutual Information Gain (AMIG) algorithm induces a binary tree in a continuously valued attribute space using a threshold decision at each internal tree node. This implements a hyperplane orthogonal to one of the feature axes. Thresholds are selected by considering each potential decision boundary between any two consecutive samples of different class and selecting the one which gives the best AMIG value [7]. This decision boundary splits the training examples into two subsets which are passed down the tree to separate child nodes and the procedure recurs at each of them until a set of examples are of the same class, *pure* class (within some degree of error). In Fig. 1 the tree has partitioned the 3-dimensional feature space into 4 volumes (of differing shades in the diagram) or decision regions. The partitions are implemented by hyperplanes which lie orthogonal to the three feature axes.

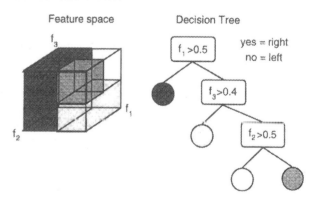

Fig. 1. Hierarchical Partitioning by a Binary Tree with threshold decision boundaries.

Along with many other decision tree algorithms, AMIG requires all the training data to be present simultaneously. Each internal decision rule only considers a single feature axis and thus loses information relating to the relationships between features.

1.3 Neural Tree Network (NTN)

This paradigm involves constructing a decision tree where each decision made at an internal node is implemented by a small FFNN [8][9]. Each node of the tree contains a single layer of *n* perceptrons, where *n* is equal to the number of classes inherent in the training data. The perceptron hyperplanes are trained under gradient descent to split the feature space into a number of decision regions and reduce the number of misclassifications. These subspaces are then passed, along arcs of the decision tree, to child nodes which further recursively divide the feature space until a space is of *pure* class.

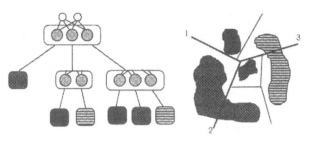

Fig. 2. Neural Tree Network and Partitioned Input Space - The thicker hyperplanes 1,2&3 correspond to the root node's neurons, and the thinner ones correspond to the second layer nodes.

To improve the NTN's ability to generalise, an extension of a binary tree pruning algorithm is commonly used [8]. The NTN builds its own topology though still suffers the drawbacks of having to set learning parameters and there is no guarantee of error convergence when training at a node.

1.4 Entropy nets

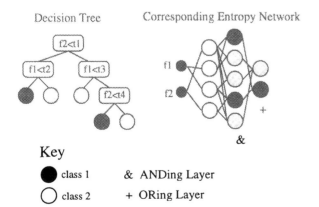

Fig. 3. Entropy Nets

This methodology consists of building an AMIG decision tree to classify the training examples, then producing a corresponding neural network which is further trained using BP [10][11]. As each path through the tree implements a hypercubic decision region of *pure* class a corresponding neurons can be placed in the first layer of the network. A further layers of neurons is added to logically AND the neurons coreesponding to each decision region. Finally an ORing layer of neurons link together the decision regions of the same class and thus act as output neurons [10].

The only difference between the decision tree and the entropy network is the introduction of the sigmoidal output function across the hyperplane partitionings and the extra BP training which is made use of. Sigmoidal output functions do give better generalisation than hard limiting functions though not always significantly. Any increase in performance which is found must be balanced against the extra costs involved in creating the two models.

1.5 The RDSE Algorithm

The RDSE algorithm [12][13][14] constructs a network topology to fit the training problem by adding, and training, a single neuron at a time to the current neuronal layer. Once the current layer has partitioned all the training data into pure decision regions a new layer is begun. The new layer is trained on the transformed data passed through the previous layers and neurons are added and trained in the same manner. Each extra layer should produce a simpler decision space as data points which lie in the same decision region will be transformed to approximately the same point in the next layers' decision space. The algorithm has a preset maximum number of layers that can be constructed, at which point a layer of *terminal neurons* and a layer of output neurons are added. These layers perform a similar task to the ANDing and ORing layers in the Entropy net, mapping each pure decision region to an appropriate output neuron.

Each new neuron, added to the current layer, is trained to minimise an entropy measure [15] using a directed random search technique [16]. RDSE makes use of neurons with an adpative steepness parameter which is altered to better model decision boundaries described by the training data set [17].

2 The Data Sets

The three data sets described below have previously been used in conjunction with the StatLog project's comparative evaluation of a number of basic classification techniques [1]. Each set consists of two classes of examples. The Diabetes in Pima Indians Data set has 768 examples made up of 8 attributes. The Australian Credit set relates to applications to a credit agency and is made up of 690 instances with 14 attributes. The Heart Disease data set consists of 270 instances with 13 attributes.

3 Cross Validation

The tests were conducted using cross validation [17]. This involves segmenting a data set into n-parts, taking n-1 of these for training and the remainder for testing. The procedure is repeated n times using different segments for each role in each experiment such that all the examples will have been used for both training and testing upon completion. The resulting classification rate is then averaged over the n experiments.

In order to perform cross-validation testing on a given data set, the paradigm used must be able to consistently learn a function which maps the training examples onto their relevant classes for each of the n training sets. For BP on the Diabetes in Pima Indians data set, this criterion could not always be met. Four different architectures were tried but all suffered problems with local minima and would not consistently converge to a solution. When this problem occurred the experiment was repeated with different initial conditions until the network converged to a suitable error minima.

The data sets used in these performance tests are of varying size, which influenced the n value selected for the n-fold cross-validation. The two larger data sets, Diabetes in Pima Indians and Australian Credit contain a similar number of examples and were treated using a value of 12 for n. The Heart Disease data set only contains 270 records and so a smaller value of 9 was used.

3.1 Timing Considerations

As the algorithms were implemented on different hardware platforms the recorded timings can not be considered as an accurate measurement of the computational cost. It must be noted though, that the Entropy net algorithm does require the construction and training of a complete decision tree as well as the resulting neural network and so requires a great deal of computations. The constructed neural network is trained using BP which as has already been mentioned is very slow in itself [6]. Each decision tree was pruned before being used to initialise the neural network. The Entropy net algorithm is entirely self contained and as such uses an independent version of the AMIG decision

and not the one for which results are given. It must be noted that the storage values given throughout this report refer only to the resulting neural network. The storage requirements for the decision tree pre-processor must also be considered when making performance comparisons between algorithms. These requirements can be assumed to be roughly equal to those of the AMIG algorithm.

In training NTN was found to suffer badly from local minima problems, especially in the data sets with input spaces of larger dimensions (Heart Disease and Australian Credit). All the neurons are trained directly on the original input vectors and no inter-layer transformations occur. Each time a perceptron was trained it would tend to converge to the minima where one neuron would have all of the patterns on its positive side and the other would have none. The only way to overcome this problem was to randomly re-initialise both neurons a number of times until a different minima was reached. The algorithm tends to build extremely large trees due to the poor perceptron training though these can then be pruned [8].

The network topologies used with the BP algorithm were empirically found to be the smallest which would consistently train on the data sets (i.e. have no local minima problems). This was done to facilitate the cross-validation testing.

4 Loss of Information Due to Noise

A network's generalisation is directly extrapolated from the training examples, i.e. it is defined by the networks model of the training data. Simplifying this model will increase the resistance to noise but will also lose information which may not be noise. With real world problems it is impossible to tell just how representative a training set is of the whole data domain. Small clusters of examples may be noise or they may be an indication of an important decision region which is just badly represented by the training set. The input space can be partitioned in a number of different ways grouping different subsets of vectors into clusters which is dependant on the networks initial conditions. So the definition of what is a cluster of pure class and what is noise within this cluster is dependant on the development of the network during training. The simple domain model will ignore possible noisy examples and will thus give good generalisation though the accuracy will have an upper bound due to the lost information. More complex domain models will be susceptible to noise problems though theoretically could give better generalisation as less information is lost from the training examples.

Table 1. 12-fold Cross Validation on Diabetes in Pima Indians

	AMIG	Pruned	NTN	Pruned	BP	Ent Net	RDSE
C.R.	70.9	73.70	71.00	73.57	69.03	77.10[*]	73.08
Nodes	111.8	10.1	203.5	158.7	17-15-10-1	23.0	21.75
Time	-	-	-	5.5	2066.7	145.0	22.6

Table 2. 9-fold Cross Validation on Heart Disease

	AMIG	Pruned	NTN	Pruned	BP	Ent Net	RDSE
C.R.	74.8	76.67	78.89	81.11	71.85	82.96	79.26
Nodes	35.80	6.75	65.78	27.11	12-8-6-1	44.0	17
Time	-	-	-	1.5	11.5	37.0	19.8

Table 3. 12-fold Cross Validation on Australian Credit

	AMIG	Pruned	NTN	Pruned	BP	Ent Net	RDSE
C.R.	81.0	82.3	81.7	84.8	80.7	77.5	83.9
Nodes	72.0	7.83	141.2	51.8	15-10-5-1	9.33	36.17
Time	-	-	-	6.1	718.3	43.5	148.4

Table 4. Averaged Classification Rate Results on all three Data Sets

	AMIG (Pruned)	NTN (Pruned)	BP	Ent Net	RDSE
C.R.	77.56	79.82	73.85	79.18	78.75
Nodes	8.2	79.2	33.7	25.4	25.0
Time	-	4.3	936.6	75.2	62.7

5 Results

Tables 1-3 show the comparative results for the five algorithms on each of the three data sets. Two sets of results are given for the decision trees on each data set. The first column concerns the results after the tree has been grown and the second column concerns the post-pruning results. The results displayed are the average Classification **R**ate over the n test sets, the number of decision **Nodes** or neurons required for the task and the average **Time** taken to train displayed in minutes.

The timings for the AMIG decision trees could not be accurately measured, due to the speed at which they were constructed and the lack of any timing carried out by the system itself. Each tree took less than 20 seconds to build and less than 10 seconds to prune.

5.1 Overall Performance

Table 4 shows the generalisational performance averaged over all the continuous data sets for each algorithm where **Nodes** refers to the pruned trees.

---

[*] Highest previously reported accuracy is 76 in
http://ftp.ics.edu/pub/machine-learning-databases.

6 Conclusion

The decision tree methods performed well on these continuous data sets, especially when using neuronal hyperplanes for decision boundaries. In all of the above experiments though, the tree methods initially over-trained on the data and as such were very storage intensive.

The three hybrid algorithms, NTN, RDSE and the Entropy net gave a better overall performance than the two basic techniques from which they were derived. All of these algorithms gave similar levels of generalisation showing the validity of constructionist techniques applied to continuous data.

BP required a comparatively large number of hyperplanes for the Diabetes and Australian Credit data sets to enable the networks to consistently learn under gradient descent. These extra hyperplanes are likely to be fitting to noisy training examples resulting in BP's poor generalisational performance. The Entropy net uses a simpler network of partially connected neuronal layers and so is training hyperplanes in input spaces of smaller dimensions. This will help alleviate some of the local minima that plague gradient descent based training techniques. The Entropy net's hyperplanes have already been intelligently initialised by the decision tree pre-processor which will also contribute.

The difference in the average generalisation performance for all the algorithms over all the continuous data domains is not statistically significant, with the exception of BP. All these algorithms give performances in the range [77.56 - 79.82] which is less than 3% difference. This level of generalisation may be approaching the upper bound due to the information lost when allowing for the effects of noise in training.

The training times recorded through the tests do show a greater disparity in performance. RDSE was found to be the fastest of the neural network algorithms requiring less than 10% of the training time needed for BP. The random search technique used in RDSE is more robust than the gradient descent methods which is employed in NTN, Entropy Net and BP. Local minima which plague these algorithms can be detected and avoided using intelligent random steps.

The Entropy net involves the prior construction of a complete decision tree model before gradient descent learning takes over to complete the training phase. The algorithm therefore requires all the extra storage and processing associated with both models as well as inheriting all the training problems from both classification domains [11]. RDSE benefits from being computationally simpler than the Entropy net and also requires less storage.

As most of the algorithms tested gave similar levels of generalisation performance on each data set, the performances achieved must be an indication of the complexity of the data's distribution and the noise it contains. As every data example is used for both training and testing through the cross-validation, the generalisation performance gives a measurement of how representational any large subset of the data is of the remaining examples.

References

[1] Michie D., Spiegelhalter D.J., Taylor C.C.: Machine Learning, Neural and Statistical Classification, Ellis Hopwood Series in Artificial Intelligence, Ellis Hopwood, 1994.

[2] Bai B. and Farhat N. H.: Learning Networks for Extrapolation and Radar Target Identification, Neural Networks, pp. 507-529, 1992.

[3] Chow M. and Magnum P.: Incipient Fault Detection in DC Machines Using a Neural Network, IEE 22nd Asilomar Conference on Signals, Systems and Computers, Vol. 2, pp. 706-709, 1989.

[4] Rumelhart D., Hinton G., Williams R.: Learning Representations by Back-Propagating Errors, Letters to Nature, Vol. 323, pp. 533-535, 1986.

[5] Hertz J., Krogh A. ,Palmer R.: Introduction to the Theory of Neural Computation, Sante Fe Institute, Addison Wesley, 1991.

[6] McLean D., Bandar Z., O'Shea J.: The Evolution of a Feed Forward Neural Network trained under Back-Propagation, ICANNGA '97, 1997.

[7] Sethi I.K., Sarvaryudu G.P.R.: Hierarchical Classifier Design Using Mutual Information, IEEE Transactions on Pattern Analysis and Machine Intelligence, Vol PAMI-4, No.4, pp. 441-445, 1982.

[8] Sankar A. and Mammone R.J.: Optimal Pruning of Neural Tree Networks for Improved Generalisation. IEEE International Joint Conference on Neural Networks - Seattle, Vol. 2, pp. 219-224, 1991.

[9] Sankar A. and Mammone R.J.: Speaker Independent Vowel Recognition using Neural Tree Networks, Proceedings of the International Joint Conference on Neural Networks, Vol.2, pp. 809-814, 1991.

[10] Sethi I.K: Entropy Nets: From Decision Trees to Neural Networks, Proceedings Of The IEEE, vol. 78, No 10, pp. 1605-1613, 1990.

[11] Sethi I.K and Otten M.: Comparison Between Entropy Net and Decision Tree Classifiers, International Joint Conference on Neural Networks, Vol.3, pp. 63-68, 1990.

[12] McLean D., Bandar Z., O'Shea J.: Improved Interpolation and Extrapolation from Continuous Training Examples Using a New Neuronal Model with an Adaptive Steepness, 2nd Australian and New Zealand Conference on Intelligent Information Systems, IEEE, pp. 125-129, 1994.

[13] McLean D, Bandar Z, O'Shea J:, An Empirical Comparison of Back Proagation and the RDSE Algorithm on Continuously Valued Real World Data, Neural Networks, 11, pp. 1685-1694, 1998.

[14] McLean D.: RDSE Algorithm, http://www.doc.mmu.ac.uk/STAFF/D.McLean/RDSE, 1998.

[15] Quinlan J.R.: Induction of Decision Trees, Machine Learning, Vol. 1, pp. 81-106, 1986.

[16] Baba N.: A New Approach for Finding the Global Minimum of Error Function of Neural Networks, Neural Networks, Vol. 2, pp. 367-373, 1989.

[17] Lachenbruch P. and Mickey M: Estimation of Error Rates in Discriminant Analysis, Technometrics, Vol.10, pp. 1-11, 1968.

A Quantum Associative Memory Based on Grover's Algorithm

Dan Ventura and Tony Martinez
Neural Networks and Machine Learning Laboratory (http://axon.cs.byu.edu)
Department of Computer Science
Brigham Young University, Provo, UT 84602 USA

Abstract

Quantum computation uses microscopic quantum level effects to perform computational tasks and has produced results that in some cases are exponentially faster than their classical counterparts. The unique characteristics of quantum theory may also be used to create a quantum associative memory with a capacity exponential in the number of neurons. This paper combines two quantum computational algorithms to produce a quantum associative memory. The result is an exponential increase in the capacity of the memory when compared to traditional associative memories such as the Hopfield network. The paper covers necessary high-level quantum mechanical ideas and introduces a quantum associative memory, a small version of which should be physically realizable in the near future.

1 Introduction

Assume a set $\mathcal{P}$ of m binary patterns of length n. We consider the problem of pattern completion -- learning to produce one of the full patterns when presented with only a partial pattern. The trivial solution is simply to store the set of patterns as a lookup table or RAM. There are two reasons why this is not always the best solution. First, it requires that a unique address be associated with and remembered for each pattern. Second, the lookup table requires mn bits in order to store all the patterns. It is often desirable to be able to recall the patterns in an associative fashion, thus eliminating the need for explicit addressing. That is, given a partial pattern one would like to be able to "fill in" a reasonable guess as to the rest of the pattern. This may also be considered a form of generalization as the partial pattern may never have been seen during the learning of the pattern set $\mathcal{P}$. Further, it would be beneficial if a smaller representation was possible.

To this end, various classical associative memory schemes have been proposed, perhaps the most well known being the Hopfield network [10]. Another well-known example is the bidirectional associative memory (BAM) [11]. These neural approaches to the pattern completion problem allow for associative pattern recall, but suffer severe storage restrictions. Storing patterns of length n requires a network of n neurons, and the number of patterns, m, is then limited by $m \leq kn$,

where typically $0.15 \leq k \leq 0.5$. This paper offers improvement by proposing a quantum associative memory that maintains the ability to recall patterns associatively while offering a storage capacity of $O(2^n)$ using only n neurons.

The field of quantum computation, which applies ideas from quantum mechanics to the study of computation, was introduced in the mid 1980's [6]. For a readable introduction to quantum computation see [1]. The field is still in its infancy and very theoretical but offers exciting possibilities for the field of computer science -- perhaps the most notable to date being the discovery of quantum computational algorithms for computing discrete logarithms and prime factorization in polynomial time, two problems for which no known classical polynomial time solutions exist [13]. These algorithms provide theoretical proof not only that interesting computation can be performed at the quantum level but also that it may in some cases have distinct advantages over its classical cousin.

Artificial neural networks (ANN) seek to provide ways for classical computers to learn rather than to be programmed. If quantum computers become a reality, then artificial neural network methods that are amenable to and take advantage of quantum mechanical properties will become possible. In particular, can quantum mechanical properties be applied to ANNs for problems such as associative memory? Recently, work has been done in the area of combining classical artificial associative memory with ideas from the field of quantum mechanics. Perus details several interesting mathematical analogies between quantum theory and neural network theory [12]. [15] goes a step further by proposing an actual model for a quantum associative memory. The work here further develops this model by exhibiting a physically realizable quantum system for acting as an associative memory.

This paper presents a unique reformulation of the pattern completion problem into the language of wave functions and operators. This reformulation may be generalized to a large class of computational learning problems, opening up the possibility of employing the capabilities of quantum computational systems for the solution of computational learning problems. Section

2 introduces some important ideas from quantum mechanics and briefly discusses quantum computation along with some of its early successes. Section 3 goes into more detail on one particular algorithm for quantum database search due to Grover [9]. Section 4 briefly describes a modification of the quantum algorithm, detailed elsewhere [16] for initializing a quantum system to represent a set of patterns, and the two algorithms are combined in section 5 to produce the quantum associative memory. Section 6 concludes and provides directions for further research.

2 Quantum Computation

Quantum computation is based upon physical principles from the theory of quantum mechanics, which is in many ways counterintuitive. Yet it has provided us with perhaps the most accurate physical theory (in terms of predicting experimental results) ever devised by science. The theory is well-established and is covered in its basic form by many textbooks (see for example [7]). Several necessary ideas that form the basis for the study of quantum computation are briefly reviewed here.

Linear superposition is closely related to the familiar mathematical principle of linear combination of vectors. Quantum systems are described by a wave function ψ that exists in a Hilbert space. The Hilbert space has a set of states, $|\phi_i\rangle$, that form a basis, and the system is described by a quantum state $|\psi\rangle$,

$$|\psi\rangle = \sum_i c_i |\phi_i\rangle. \qquad (1)$$

$|\psi\rangle$ is said to be in a linear superposition of the basis states $|\phi_i\rangle$, and in the general case, the coefficients c_i may be complex. Use is made here of the Dirac bracket notation, where the ket $|\cdot\rangle$ is analogous to a column vector, and the bra $\langle\cdot|$ is analogous to the complex conjugate transpose of the ket. In quantum mechanics the Hilbert space and its basis have a physical interpretation, and this leads directly to perhaps the most counterintuitive aspect of the theory. The counter intuition is this -- at the microscopic or quantum level, the state of the system is described by the wave function ψ, that is, as a linear superposition of all basis states (i.e. in some sense the system is in all basis states at once). However, at the macroscopic or classical level the system can be in only a single basis state. For example, at the quantum level an electron can exist in a superposition of many different energy levels; however, in the classical realm this cannot be.

Coherence and *decoherence* are closely related to the idea of linear superposition. A quantum system is said to be coherent if it is in a linear superposition of its basis states. A result of quantum mechanics is that if a system that is in a linear superposition of states interacts in any way with its environment, the superposition is destroyed. This loss of coherence is called decoherence and is governed by the wave function ψ. The coefficients c_i are called probability amplitudes, and $|c_i|^2$ gives the probability of $|\psi\rangle$ collapsing into state $|\phi_i\rangle$ if it decoheres. Note that the wave function ψ describes a real physical system that must collapse to exactly one basis state. Therefore, the probabilities governed by the amplitudes c_i must sum to unity. This constraint is expressed as the unitarity condition

$$\sum_i |c_i|^2 = 1. \qquad (2)$$

In the Dirac notation, the probability that a quantum state $|\psi\rangle$ will collapse into an eigenstate $|\phi_i\rangle$ is written $|\langle\phi_i|\psi\rangle|^2$ and is analogous to the dot product (projection) of two vectors. Consider, for example, a discrete physical variable called spin. The simplest spin system is a two-state system, called a spin-1/2 system, whose basis states are usually represented as $|\uparrow\rangle$ (spin up) and $|\downarrow\rangle$ (spin down). In this simple system the wave function ψ is a distribution over two values and a coherent state $|\psi\rangle$ is a linear superposition of $|\uparrow\rangle$ and $|\downarrow\rangle$. One such state might be

$$|\psi\rangle = \frac{2}{\sqrt{5}}|\uparrow\rangle + \frac{1}{\sqrt{5}}|\downarrow\rangle. \qquad (3)$$

As long as the system maintains its quantum coherence it cannot be said to be either spin up or spin down. It is in some sense both at once. Classically, it must be one or the other, and when this system decoheres the result is, for example, the $|\uparrow\rangle$ state with probability

$$|\langle\uparrow|\psi\rangle|^2 = \left(\frac{2}{\sqrt{5}}\right)^2 = 0.8. \qquad (4)$$

A simple two-state quantum system, such as the spin-1/2 system just introduced, is used as the basic unit of quantum computation. Such a system is referred to as a quantum bit or *qubit*, and renaming the two states $|0\rangle$ and $|1\rangle$ it is easy to see why this is so.

Operators on a Hilbert space describe how one wave function is changed into another. Here they will be denoted by a capital letter with a hat, such as $\hat{A}$, and they may be represented as matrices acting on vectors. Using operators, an eigenvalue equation can be written $\hat{A}|\phi_i\rangle = a_i|\phi_i\rangle$, where a_i is the eigenvalue. The solutions $|\phi_i\rangle$ to such an equation are called eigenstates and can be used to construct the basis of a Hilbert space as discussed above. In the quantum formalism, all properties are represented as operators whose eigenstates are the basis for the Hilbert space associated with that property and whose eigenvalues are the quantum allowed values for that property. It is important to note that operators in quantum mechanics must be linear operators and further that they must be unitary so that $\hat{A}^\dagger \hat{A} = \hat{A}\hat{A}^\dagger = \hat{I}$, where $\hat{I}$ is the identity operator, and $\hat{A}^\dagger$ is the complex conjugate transpose of $\hat{A}$.

Interference is a familiar wave phenomenon. Wave peaks that are in phase interfere constructively (magnify each other's amplitude) while those that are out of phase interfere destructively (decrease or eliminate each other's amplitude). This is a phenomenon common to all kinds of wave mechanics from water waves to optics. The well-known double slit experiment demonstrates empirically that interference also applies to the probability waves of quantum mechanics.

Entanglement is the potential for quantum states to exhibit correlations that cannot be accounted for classically. From a computational standpoint, entanglement seems intuitive enough -- it is simply the fact that correlations can exist between different qubits -- for example if one qubit is in the $|1\rangle$ state, another will be in the $|1\rangle$ state. However, from a physical standpoint, entanglement is little understood. The questions of what exactly it is and how it works are still not resolved. What makes it so powerful (and so little understood) is the fact that since quantum states exist as superpositions, these correlations somehow exist in superposition as well. When the superposition is destroyed, the proper correlation is somehow communicated between the qubits, and it is this "communication" that is the crux of entanglement and the key to quantum computation. It follows that while interference is a quantum property that has a classical cousin, entanglement is a completely quantum phenomenon for which there is no classical analog.

2.1 Quantum Algorithms

The field of quantum computation offers exciting possibilities -- the most important quantum algorithms discovered to date all perform tasks for which there are no classical equivalents. For example, Deutsch's algorithm [5] is designed to solve the problem of identifying whether a binary function is constant (function values are either all 1 or all 0) or balanced (the function takes an equal number of 0 and 1 values). Deutsch's algorithm accomplishes the task in order $O(n)$ time, while classical methods require $O(2^n)$ time. Simon's algorithm [14] is constructed for finding the periodicity in a 2-1 binary function that is guaranteed to possess a periodic element. Here again an exponential speedup is achieved; however, admittedly, both these algorithms have been designed for artificial, somewhat contrived problems as a proof of concept. Grover's algorithm [9], on the other hand, provides a method for

searching an unordered quantum database in time $O(\sqrt{N})$, compared to the classical lower bound of $O(N)$. Here is a real-world problem for which quantum computation provides performance that is classically impossible (though the speedup is less dramatic than exponential). Finally, the most well-known and perhaps the most important quantum algorithm discovered so far is Shor's algorithm for prime factorization [13]. This algorithm finds the prime factors of very large numbers in polynomial time, while the best classical algorithms require exponential time. Obviously, the implications for the field of cryptography are profound.

3 Grover's Algorithm

Lov Grover has developed an algorithm for finding one item in an unsorted database, similar to finding the name that matches a telephone number in a telephone book. Classically, if there are N items in the database, this would require on average $O(N)$ queries to the database. However, Grover has shown how to do this using quantum computation with only $O(\sqrt{N})$ queries. In the quantum computational setting, finding the item in the database means measuring the system and having the system collapse with near certainty to the desired basis state, which corresponds to the item in the database for which we are searching. The basic idea of Grover's algorithm is to invert the phase of the desired basis state and then to invert all the basis states about the average amplitude of all the states (for more details see [9] [8]). This process produces an increase in the amplitude of the desired basis state to near unity followed by a corresponding decrease in the amplitude of the desired state back to its original magnitude. The process is cyclical with a period of $\frac{\pi}{4}\sqrt{N}$, and thus after $O(\sqrt{N})$ queries, the system may be observed in the desired state with near certainty (with probability at least $1-\frac{1}{N}$). Interestingly this implies that the larger the database, the greater the certainty of finding the desired state [3]. Of course, if greater certainty is required, the system may be sampled k times boosting the certainty of finding the desired state to $1-\frac{1}{N^k}$. Define the following operators.

$$\hat{I}_\phi = \text{identity matrix except for } \phi\phi = -1, \quad (5)$$

which inverts the phase of the basis state $|\phi\rangle$ and

$$\hat{W} = \frac{1}{\sqrt{2}}\begin{bmatrix} 1 & 1 \\ 1 & -1 \end{bmatrix}, \quad (6)$$

which is often called the Walsh or Hadamard transform. This operator, when applied to a set of qubits, performs a special case of the discrete fourier transform.

Now to perform the quantum search on a database

of size $N = 2^n$, where n is the number of qubits, begin with the system in the $|\bar{0}\rangle$ state and apply the $\hat{W}$ operator. This initializes all the states to have the same amplitude. Next apply the operator sequence $\hat{G}\hat{I}_\tau$ $\frac{\pi}{4}\sqrt{N}$ times, where $\hat{G} = -\hat{W}\hat{I}_{\bar{0}}\hat{W}$ can be thought of as rotating all the states about their average amplitude and $|\tau\rangle$ is the state being sought (and recall that operators are applied right to left). Finally, observe the system. This algorithm will be used to associatively "fill in" a pattern by finding a basis state that corresponds to the partial pattern to be completed.

4 Initializing the Quantum State

In [16] we presented a polynomial-time quantum algorithm for constructing a quantum state over a set of qubits to represent the information in a training set. The algorithm is implemented using a polynomial number (in the length and number of patterns) of elementary operations on one, two, or three qubits. Here the necessary operators are presented briefly and the reader is referred to [16] for details.

$$\hat{S}^p = \begin{bmatrix} 1 & 0 & 0 & 0 \\ 0 & 1 & 0 & 0 \\ 0 & 0 & \sqrt{\dfrac{p-1}{p}} & \dfrac{-1}{\sqrt{p}} \\ 0 & 0 & \dfrac{1}{\sqrt{p}} & \sqrt{\dfrac{p-1}{p}} \end{bmatrix}, \quad (7)$$

where $1 \le p \le m$. These operators form a set of conditional transforms that will be used to incorporate the set of patterns into a coherent quantum state. There will be a different $\hat{S}^p$ operator associated with each pattern to be stored. The interested reader may note that this definition of the $\hat{S}^p$ operator is slightly different than the original. This is because in this context, we are considering pattern memorization rather than pattern classification and therefore have no output class *per se*. Thus the phase of the coefficients becomes unimportant in this case.

$$\hat{F}^0 = \begin{bmatrix} 0 & 1 & 0 & 0 \\ 1 & 0 & 0 & 0 \\ 0 & 0 & 1 & 0 \\ 0 & 0 & 0 & 1 \end{bmatrix}, \quad (8)$$

conditionally flips the second qubit if the first qubit is in the $|0\rangle$ state; $\hat{F}^1$ conditionally flips the second qubit if the first qubit is in the $|1\rangle$ state ($\hat{F}^1$ is the same as $\hat{F}^0$ except that the off-diagonal elements occur in the bottom right quadrant rather than in the top left). These operators are referred to elsewhere as Control-NOT because a logical NOT (state flip) is performed on the second qubit depending upon (or controlled by) the state of the first qubit.

$$\hat{A}^{00} = \begin{bmatrix} 0 & 1 & 0 & 0 & 0 & 0 & 0 & 0 \\ 1 & 0 & 0 & 0 & 0 & 0 & 0 & 0 \\ 0 & 0 & 1 & 0 & 0 & 0 & 0 & 0 \\ 0 & 0 & 0 & 1 & 0 & 0 & 0 & 0 \\ 0 & 0 & 0 & 0 & 1 & 0 & 0 & 0 \\ 0 & 0 & 0 & 0 & 0 & 1 & 0 & 0 \\ 0 & 0 & 0 & 0 & 0 & 0 & 1 & 0 \\ 0 & 0 & 0 & 0 & 0 & 0 & 0 & 1 \end{bmatrix}, \quad (9)$$

conditionally flips the third bit if and only if the first two are in the state $|00\rangle$. Note that this operator can be thought of as performing a logical AND of the negation of the first two bits, writing a 1 in the third if and only if the first two are both 0. Three other operators, $\hat{A}^{01}$, $\hat{A}^{10}$ and $\hat{A}^{11}$, are variations of $\hat{A}^{00}$ in which the off diagonal elements occur in the other three possible locations along the main diagonal. $\hat{A}^{01}$ can be thought of as performing a logical AND of the first bit and the negation of the second, and so forth. These operators are used to identify specific states in a superposition.

Now given a set $\mathcal{P}$ of m binary patterns of length n to be memorized, the quantum algorithm for storing the patterns requires a set of $2n+1$ qubits, the first n of which actually store the patterns and can be thought of analogously as n neurons in a quantum associative memory. For convenience, the qubits are arranged in three quantum registers labeled x, g, and c, and the quantum state of all three registers together is represented in the Dirac notation as $|x, g, c\rangle$.

The x register will hold a superposition of the patterns. There is one qubit in the register for each bit in the patterns to be stored, and therefore any possible input can be represented. The g register is a garbage register used only in identifying a particular state. It is restored to the state $|\bar{0}\rangle$ after every iteration. The c register contains two control qubits that indicate the status of each state at any given time. A high-level intuitive description of the algorithm is as follows. The system is initially in the single basis state $|\bar{0}\rangle$. The qubits in the x register are selectively flipped so that their states correspond to the inputs of the first pattern. Then, the state in the superposition representing the pattern is "broken" into two "pieces" -- one "larger" and one "smaller" and the status of the smaller one is made permanent. Next, the x register of the larger piece is selectively flipped again to match the input of the second pattern, and the process is repeated for each pattern. When all the patterns have been "broken" off of the large "piece", then all that is left is a collection of small pieces, all the same size, that represent the patterns to be stored; in other words, a coherent superposition of states is created that corresponds to the patterns, where the amplitudes of the states in the superposition are all equal. The algorithm requires $O(mn)$ steps to encode the patterns as a quantum superposition over n quantum neurons. Note that this is optimal in the sense that just reading each instance once cannot be done any faster than $O(mn)$.

5 Quantum Associative Memory

A quantum associative memory (QuAM) can now be constructed from the two algorithms of sections 3 and 4. Define the $\hat{P}$ operator as the operator combination of equations (7-9) that implements the algorithm for memorizing patterns described in section 4. Then the operation of the QuAM can be described as follows. Memorizing a set of patterns is simply

$$|\psi\rangle = \hat{P}|\overline{0}\rangle, \qquad (10)$$

with $|\psi\rangle$ being a quantum superposition of basis states, one for each pattern. Now, suppose we know $n-1$ bits of a pattern and wish to recall the entire pattern. Assuming that there are not two patterns that differ only in the last bit, we can use Grover's algorithm to recall the pattern as (τ is the target pattern)

$$|\psi'\rangle = \hat{G}\hat{I}_\tau|\psi\rangle \qquad (11)$$

applied recursively $\frac{\pi}{4}\sqrt{N}$ times. Thus, with $2n+1$ neurons (qubits) the QuAM can store up to $N=2^n$ patterns in $O(mn)$ steps and requires $O(\sqrt{N})$ time to recall a pattern.

A very simple example will help clarify. Suppose that we have a set of patterns $P = \{000,011,100,110\}$. Then using equation (10) memorizes the pattern set as the quantum state

$$\hat{P}|\overline{0}\rangle = \frac{1}{2}|000\rangle + \frac{1}{2}|011\rangle + \frac{1}{2}|100\rangle + \frac{1}{2}|110\rangle. \qquad (12)$$

Now suppose that we want to recall the pattern whose first two bits were 10. Applying equation (11) gives

$$\hat{G}\hat{I}_\tau\left(\frac{1}{2}|000\rangle + \frac{1}{2}|011\rangle + \frac{1}{2}|100\rangle + \frac{1}{2}|110\rangle\right)$$

$$= \hat{G}\left(\frac{1}{2}|000\rangle + \frac{1}{2}|011\rangle - \frac{1}{2}|100\rangle + \frac{1}{2}|110\rangle\right) \qquad (13)$$

$$= |100\rangle,$$

and we have thus achieved our goal. We can now observe the system to see that the completion of the pattern 10 is 100.

Using some concrete numbers, assume that $n = 2^4$ and $m = 2^{14}$ (we let m be less than the maximum possible 2^{16} to allow for some generalization and

avoid the contradictory patterns that would otherwise result). Then the QuAM requires $O(mn) = O(2^{18}) < 10^6$ operations to memorize the patterns and $O(\sqrt{N}) = O(\sqrt{2^{16}}) < 10^3$ operators to recall a pattern. For comparison, in [1] Barenco gives estimates of how many operations might be performed before decoherence for various possible physical implementation technologies for the qubit. These estimates range from as low as 10^3 (electrons in GaAs and electron quantum dots) to as high as 10^{13} (trapped ions), so our estimates fall comfortably into this range, even near the low end of it. Further, the algorithm would require only $2n +1 = 2*16+1 = 33$ qubits! For comparison, a classical Hopfield type network used as an associative memory has a saturation point around $0.15n$. In other words, about $0.15n$ patterns can be stored and recalled with n neurons. Therefore, with $n=16$ neurons, a Hopfield network can store only $0.15*16 \approx 2$ patterns. Conversely, to store 2^{14} patterns would require that the patterns be close to 110,000 bits long and that the network have that same number of neurons.

Grover's original algorithm only applies to the case where all basis states are represented in the superposition equally to start with and one and only one basis state is to be recovered. In other words, strictly speaking, the original algorithm would only apply to the case when the set P of patterns to be memorized included all possible patterns of length n and when we new all n bits of the pattern to be recalled -- not a very useful associative memory. However, several other papers have since generalized Grover's original algorithm and improved on his analysis to include cases where not all possible patterns are represented and where more than one target state is to be found [3] [2] [8]. Strictly speaking it is these more general results which allow us to create a useful QuAM that will associatively recall patterns.

Finally, it is worth mentioning that very recently Chuang et. al. have succeeded in physically implementing Grover's algorithm for the case $n=2$ using nuclear magnetic resonance technology on a solution of chloroform molecules [4]. It is therefore not unreasonable to assume that a small quantum associative memory may be implemented in the not too distant future.

6 Concluding Comments

A unique view of the pattern completion problem is presented that allows the proposal of a quantum associative memory with exponential storage capacity. It employs simple spin-1/2 quantum systems and represents patterns as quantum operators. This approach introduces a large new field to which quantum computation may be applied to advantage -- that of neural networks. In fact, it is the authors' opinion that this application of quantum computation will, in general, demonstrate much greater returns than its application to more traditional computational tasks (though Shor's algorithm is an obvious exception). We make this conjecture because results in both quantum computation and neural networks are by nature probabilistic and inexact, whereas most traditional computational tasks require precise and deterministic outcomes.

The most urgently appealing future work suggested by the result of this paper is, of course, the physical implementation of the algorithm in a real quantum system. As mentioned in section 5, the fact that very few qubits are required for non-trivial problems together with the recent physical realization of Grover's algorithm helps expedite the realization of quantum computers performing useful computation. In the mean time, a simulation of the quantum associative memory is being developed to run on a classical computer at the cost of an exponential slowdown in the length of the patterns. Thus, association problems that are non-trivial and yet small in size will provide interesting study in simulation. Another obvious and important area for future research is investigating further the application of quantum computational ideas to the field of neural networks -- the discovery of other quantum computational learning algorithms. Further, techniques and ideas that result from developing quantum algorithms may be useful in the development of new classical algorithms. Finally, the process of understanding and developing a theory of quantum computation provides insight and contributes to a furthering of our understanding and development of a general theory of computation.

References

[1] Barenco, A.: Quantum Physics and Computers, *Contemporary Physics*, vol. 37 no. 5, pp. 375-389, 1996.

[2] Biron, D., O. Biham, E. Biham, M. Grassl and D. A. Lidar: Generalized Grover Search Algorithm for Arbitrary Initial Amplitude Distribution, to appear in the *Proceedings of the 1st NASA International Conference on Quantum Computation and Quantum Communications*, February 1998.

[3] Boyer, M., G. Brassard, P. Høyer and A. Tapp: Tight Bounds on Quantum Searching, *Workshop on Physics and Computation*, pp. 36-43, November 1996.

[4] Chuang, I., N. Gershenfeld and M. Kubinec: Experimental Implementation of Fast Quantum Searching, *Physical Review Letters*, vol. 80 no. 15, pp. 3408-3411, April 13, 1998.

[5] Deutsch, D. and R. Jozsa: Rapid Solution of Problems by Quantum Computation, *Proceedings of the Royal Society, London A*, vol. 439, pp. 553-558, 1992.

[6] Deutsch, D.: Quantum Theory, The Church-Turing Principle and the Universal Quantum Computer, *Proceedings of the Royal Society*, London A, vol. 400, pp. 97-117, 1985.

[7] Feynman, R. P., R. B. Leighton and M. Sands: *The Feynman Lectures on Physics*, vol. 3, Addison-Wesley Publishing Company, Reading Massachusetts, 1965.

[8] Grover, L. K.: Quantum Search on Structured Problems, to appear in the *Proceedings of the 1st NASA International Conference on Quantum Computation and Quantum Communications*, February 1998.

[9] Grover, L. K.: A Fast Quantum Mechanical Algorithm for Database Search, *Proceedings of the 28th Annual ACM Symposium on the Theory of Computing*, ACM, New York, pp. 212-219, 1996.

[10] Hopfield, J. J.: Neural Networks and Physical Systems with Emergent Collective Computational Abilities, *Proceedings of the National Academy of Scientists*, vol. 79, pp. 2554-2558, 1982.

[11] Kosko, B.: Bidirectional Associative Memories, *IEEE Transactions on Systems, Man, and Cybernetics*, vol. 18, pp. 49-60, 1988.

[12] Perus, M.: Neuro-Quantum Parallelism in Brain-Mind and Computers, *Informatica*, vol. 20, pp. 173-183, 1996.

[13] Shor, P.: Polynomial-Time Algorithms for Prime Factorization and Discrete Logarithms on a Quantum Computer, *SIAM Journal of Computing*, vol. 26 no. 5, pp. 1484-1509, 1997.

[14] Simon, D.: On the Power of Quantum Computation, *SIAM Journal of Computation*, vol. 26 no. 5, pp. 1474-1483, 1997.

[15] Ventura, D. and T. Martinez: Quantum Associative Memory with Exponential Capacity, *Proceedings of the International Joint Conference on Neural Networks*, pp. 509-513, May 1998.

[16] Ventura, D. and T. Martinez: Initializing a Quantum State's Amplitude Distribution, submitted to *Physical Review Letters*, June 18, 1998.

Newton Filters: a New Class of Neuron-Like Discrete Filters and an Application to Image Processing

Alexis Quesada-Arencibia[1], Miguel Alemán-Flores[2], Roberto Moreno-Díaz jr.[1]

[1]Instituto Universitario de Ciencias y Tecnologías Cibernéticas, ULPGC

[2]Departamento de Informática y Sistemas, ULPGC

[1,2]Edificio de Informática y Matemáticas, Campus de Tafira, E-35017 Las Palmas, Spain

Email: alex@ciicc.ulpgc.es, maleman@dis.ulpgc.es, roberto@dumby.dis.ulpgc.es

Abstract

The functions of dentritic trees of neurons have always attracted neuroscientists. Dendrites are extensions of the soma that allows the cell to increase the area where it receives information from. The number of dentritic ramifications is not constant, it depends on the neuron and varies from one to the other. Moreover, each dentritic tree can be subdivided in a complex form leading to a characteristic tree structure[1]. Two of their immediately-related properties regarding information processing (e.g. convergence and divergence of lines) have originated considerable research on completeness of computation and reliability of transmission [2]. In this paper a layered structure consisting in the interconnections of a set of simple functional units and inspired in the dentritic connections of real neurons, is suggested and in a parallel way we analyze the computing properties and characteristics together with a possible application in Image Processing. The interest of this structure lies in its similarity with the structure of the retinae receptive fields and even with dentritic trees of retinal neurons.

1 Newton Filters. Structure

We will concentrate now on a very simplified structured inspired on dentritic connections, the one shown in Fig. 1. For the purpose of calculation we will consider a one dimensional receptive field (the extension to two dimensions will be discussed later).

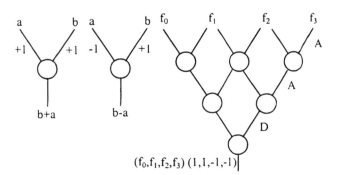

Fig. 1a   Fig. 1b   Fig. 1c

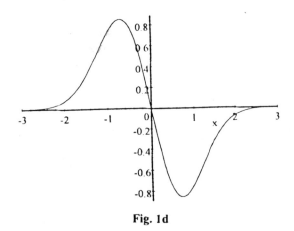

Fig. 1d

Fig. 1. On Fig. 1a and 1b we show a simple microcomputational additive and difference unit respectively. In Fig. 1c, a monodimensional three-layered structure (Newton Filter N(A2,D1)) and the computation it performs expressed as vector scalar product. The representation as a continuous function of the global weights vector gives the filter profile : the classical edge-detector filter (Fig. 1d).

Figure 1 shows a basic unit that can be one of out of two kinds: additive if it performs an addition of the two inputs (Fig. 1a), or difference if it performs a subtraction (Fig. 1b). In Fig. 1c a neuron-like structure computing unit consisting of layers of such microprocessors is presented. That structure can be seen as a discrete filter acting on a set of input data $f_0,...,f_n$. Let us define the length L of such a filter as the number of total input data contributing to the calculation. It is easy to prove that for a L-length input vector it is necessary to place L-1 layers of microprocessors as shown in Fig. 1 to get a neuron-like processing machine that we call a one-dimensional Newton Filter (N.F.) and we will denote it by N(An,Dm) (where n expresses the number of additive layers and m the number of difference layers, and L=n+m+1). Due to the linearity of operations involved in the definition of N.F. the two following relations are held:

N(An,Dm)=N(Dm,An) (Permutation of layers does not change the general performance)

N(Am1,...,Ami,Dn1,...,Dnj)=N(Am1+...+mi,Dn1+...+nj)

The global weights of the filters are easy to calculate using a variation of the Pascal triangle (the one used to calculate the factors of the Newton Binomial, where the name of the Filters comes from).

In Fig. 2a we present a contrast center-perifery shaped filter (in general, any N(Ax,D2) N.F. has this profile). The complexity of the profile (that is, the number of zero-crossings of the representation) is directly related to the number of difference layers in the N.F. Then, this structure, built by connecting very simple microprocessing units, can implement truly complicated computational kernel profiles on its input data (On Fig. 2b and 2c we have two examples of more complicated computational kernels). Furthermore, the same structure is able to compute different calculations by only changing the nature of the connections (or, looking at Fig. 1, the sign of the weight of the connection from + to – or viceversa). It has been proved elsewhere that the extension of the above to the continuum leads to Hermite functions and functionals[3].

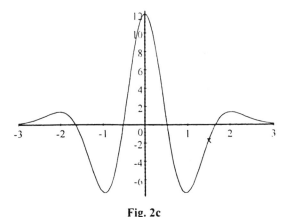

Fig. 2c

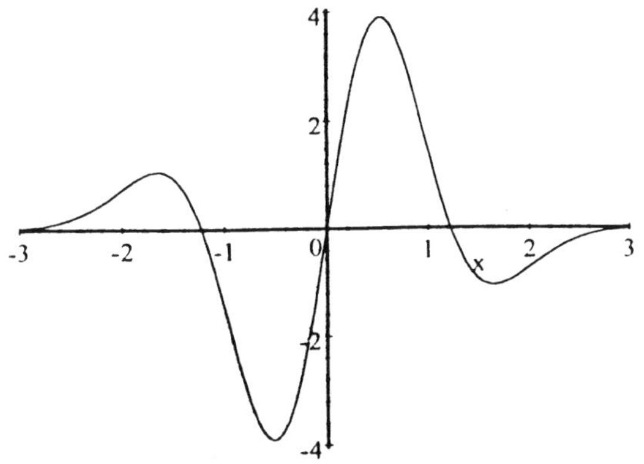

Fig. 2. Representation of weight functions of different Newton Filters

In addition, we can point out that all Newton Filters having at least one difference layer satisfy that the sum of the weight is zero. This is usefull in order to demonstrate certain invariances that could take place into the Lateral Geniculate Nucleus by means of a mechanism called Presynaptic Inhibition and also ca be used in modelling certain movement discrimination structures[5].

Let us now assume that on a given layer of such structure the nature of the processes is the same (e.g. there is no mixture of additive and difference microprocessors within a layer, all are either additive or difference). Then, it is also an immediate result that if we consider all possible orders of additive-difference layers for a given L-inputs Newton Filter, there will be L different Newton Filters of length L and we can construct that set of N.Fs. orderly: N(An), N(An-1,D1) ... N(A1,Dn-1), N(Dn) (where L = n+1). This set of vectors, composed of all Newton Filters of length L is linearly independent[3]. It can also be proved that the set of all N.Fs. of length L is a complete set, so that any "picture" of length L can be uniquely described in terms of the filtered transformed results and that there is an inverse transform. This characteristic makes posible that Newton Filters can be used in cryptography (we show this use in one example). Let f=(f0,f1,...,fn) be the information stored in memory, that is, the data field. The transformed data Fj are:

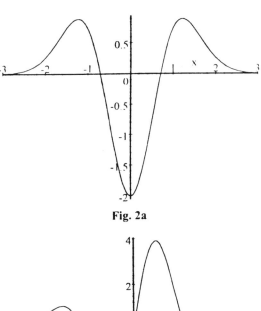

Fig. 2a

Fig. 2b

$$A \cdot f = \begin{bmatrix} a_{11} & & a_{1L} \\ \cdot & \cdot & \cdot \\ \cdot & \cdot & \cdot \\ \cdot & \cdot & \cdot \\ a_{L1} & & a_{LL} \end{bmatrix} \cdot \begin{bmatrix} f_0 \\ \cdot \\ \cdot \\ \cdot \\ f_n \end{bmatrix} = \begin{bmatrix} F_0 \\ \cdot \\ \cdot \\ \cdot \\ F_n \end{bmatrix} = F, \quad (1)$$

The fact that det(A)#0 makes possible the calculation of A^{-1} and taking into account the property of Newton matrices the result is:

$$A \cdot A = 2^{L-1} \cdot I \Rightarrow A^{-1} = 2^{1-L} \cdot A \quad , \qquad (2)$$

Therefore, if we know F we lead to this conclusion:

$$A \cdot f = F \Rightarrow A^{-1} \cdot A \cdot f = A^{-1} \cdot F \Rightarrow$$
$$\Rightarrow f = 2^{1-L} \cdot A \cdot F \qquad , \quad (3)$$

2 Three-Dimensional Structure for Newton Filters

The application of a N. F. to a two dimensional retina results in a three-dimensional computational structure (the same result was obtained in the one dimensional case, where we got a two-dimensional structure of microprocessors). The "V" of the two-dimensional processes change its structure to pyramids, as we can see in Fig. 3. We have to define the 2-D filter that each one of the computational pyramids implements, that is, the weight of that filters. The general structure of the filter will depend on this weights' definition. Thus, each microprocessors layer is two-dimensional and the length (in number of microprocessors) decreases from the first to the last layer. We will call this structure a "Pascal Pyramid"[3].

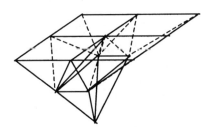

Fig. 3: A 3-D Newton structure for two-dimesional retinae: "Pascal Pyramid".

However, unlike the one-dimensional case, where there exist layers of subtraction microprocessors, we have several structural possibilities for each computational pyramid, specifically the combination of two signs (+ and -) taken in groups of four. Nonetheless we set the condition that the sum of weights is zero in order to give a derivative meaning to the calculation.

Suppose that we have a 3x3 retina where values of memory positions are defined by the matrix:

$$\begin{matrix} f_{0,0} & f_{0,1} & f_{0,2} \\ f_{1,0} & f_{1,1} & f_{1,2} \\ f_{2,0} & f_{2,1} & f_{2,2} \end{matrix} \quad , \qquad (4)$$

In Fig. 4, we can see the global weights that affect the results of the computational structure, weight that have been obtained taking into account the different possibilities of the computational units, as we mentioned above. Analysis of these matrices gives rise to some interesting properties:

First, as in the mono-dimensional case, the permutation of layers does not change the general performance.

Second we can see that the arrangement of signs into the matrix depends directly on the way the functional unit achieves the calculation.

Finally, we can underline that if we analyze the operation carried out by the computational unit properly, we can obtain the different weight's matrices performing different orderings of a matrix preiously obtained.

$$\begin{bmatrix} + & + \\ + & + \end{bmatrix} \Rightarrow \begin{bmatrix} 1 & 2 & 1 \\ 2 & 4 & 2 \\ 1 & 2 & 1 \end{bmatrix}$$

Fig. 4a. 2 additive layers

$$\begin{bmatrix} + & + \\ - & - \end{bmatrix} \Rightarrow \begin{bmatrix} 1 & 2 & 1 \\ 0 & 0 & 0 \\ -1 & -2 & -1 \end{bmatrix}$$

Fig. 4b. 1 additive layer, 1 subtraction layer

$$\begin{bmatrix} - & - \\ + & + \end{bmatrix} \Rightarrow \begin{bmatrix} -1 & -2 & -1 \\ 0 & 0 & 0 \\ 1 & 2 & 1 \end{bmatrix}$$

Fig. 4c. 1 additive layer, 1 subtraction layer

$$\begin{bmatrix} - & + \\ - & + \end{bmatrix} \Rightarrow \begin{bmatrix} -1 & 0 & 1 \\ -2 & 0 & 2 \\ -1 & 0 & 1 \end{bmatrix}$$

Fig. 4d. 1 additive layer, 1 subtraction layer

$$\begin{bmatrix} + & - \\ + & - \end{bmatrix} \Rightarrow \begin{bmatrix} 1 & 0 & -1 \\ 2 & 0 & -2 \\ 1 & 0 & -1 \end{bmatrix}$$

Fig. 4e. 1 additive layer, 1 subtraction layer

$$\begin{bmatrix} + & - \\ - & + \end{bmatrix} \Rightarrow \begin{bmatrix} 1 & 0 & -1 \\ 0 & 0 & 0 \\ -1 & 0 & 1 \end{bmatrix}$$

Fig. 4f. 1 additive layer, 1 subtraction layer

$$\begin{bmatrix} - & + \\ + & - \end{bmatrix} \Rightarrow \begin{bmatrix} -1 & 0 & 1 \\ 0 & 0 & 0 \\ 1 & 0 & -1 \end{bmatrix}$$

Fig. 4g. 1 additive layer, 1 subtraction layer

3 Examples

In the following examples we took two different images. Both are a 150x150 pixel black and white images, where the zero codifies black and 255 white. The first one (Fig. 5) is a software generated image which has lines in different orientations while the second one is a real and typical image. In Fig. 6 and 8 we present the results of applying some of the Newton Filters showed in Fig. 4. In all the examples showed we convolve the N.F. with the image and the result is re-scaled to fit the limits of representation (0-255).

As we can see in Fig. 6a, filters with only additive layers (Fig. 4a) have a smoothing effect on the image, which is an obvious result since it is a gaussian filter. On the other hand , while the filters of Fig. 4b and 4c extract horizontal and oblique edges of the image (see Fig. 6b and 6c), the filters of Fig. 4d and 4e extract vertical and oblique edges (see Fig. 6d and 6e). In the next two examples, filters of Fig. 4f and 4g, they extract perfectly the oblique edges (Fig. 6f and 6g). Finally in Fig. 6h we can see the result of the sum of the images of the Fig. 6b, 6d and 6f, that is, the sum of images with horizontal, vertical and oblique edges.

In Fig. 8a, 8b and 8c we can see the results of applying the filters of Fig. 4b, 4d and 4f to the second image (Fig. 7). As in the example above, in Fig. 8d we show the sum of the images of Fig. 8a, 8b and 8c.

As an example of the application of the N.F. in cryptography, we took the image of Fig. 7 and the result of the convolution of the matrix composed of all N.F. of length 30 and the inverse operation can be seen in Fig. 9a and 9b. In this example the convolution process was performed only by column but it can be carried out by both, column and row.

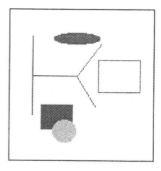

Fig. 5. Image example

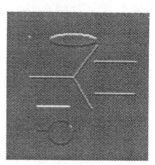

Fig. 6a. Gaussian filter

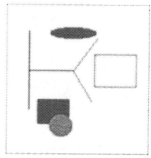

Fig. 6b. Horizontal and oblique edges extraction

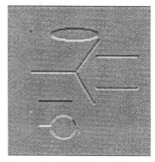

Fig. 6c. Horizontal and oblique edges extraction

32

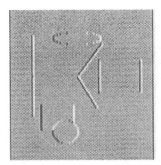

Fig. 6d. Vertical and oblique edges extraction

Fig. 6h. Result of the sum of the images of the Fig. 6b, 6d and 6f

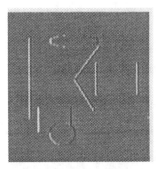

Fig. 6e. Vertical and oblique edges extraction

Fig. 7. Image example

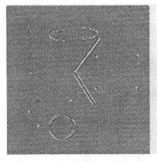

Fig. 6f. Oblique edges extraction

Fig. 8a. Horizontal and oblique edges extraction

Fig. 6g. Oblique edges extraction

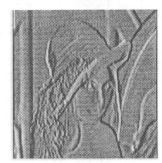

Fig. 8b. Vertical and oblique edges extraction

Fig. 8c.Oblique edges extraction

Fig. 8d. Result of the sum of the images of the Fig. 8a, 8b and 8c

Fig. 9a. Result of the convolution of the matrix composed of all N.F. of length 30 and the image example of Fig. 7

Fig. 9b. Image recovered

4 Implications in Neuronal Modelling and Artificial Image Processing. Conclusions

Considering the *implications in neuronal modelling,* we can recall that hundreds of experiments have justified certain weight functions for ganglion cells as Gaussians or Laplacian of a gaussian ($\nabla^2 G$) [4]. Looking at Fig. 2, it is easy to see the great likeness between these functions and the functional structure that neurophysiology has been describing for some types of receptive field of neurons. Also, the existence of more complicated receptive fields, showing several concentric or parallel ON-OFF (excitatory-inhibitory) regions is often used to see the performance of the visual pathway as a Fourier analyzer. As we have just seen, building a layered filter machine such as a N.F., one can get the discretization of gaussians, $\nabla^2 G$ and parallel inhibitory-exitatory receptive field profiles as the result of the structure of the machine, not of the fixing of some difficult-to-justify weight functions. In other words, it is the "dentritic complexity" of the neuron-like machine that guarantees the existence of such kernel functions. In the model, all synaptic contacts have the same weight considering it in absolute value and from the contribution of every microprocessor it is possible to obtain a global weight function that mimics the ones found in natural systems, mainly in retinae. From the computational point of view, a N.F. is a parallel layered machine.

There is also an *implication on the artificial image processing side* that is easy to see. Some of the kernels shown in Fig. 2 are the most widely used for those low-level processing tasks as edge detection, sharping and contour discrimination. And actually, Newton Filters can be used for low-level image processing as fast and easy-to-implement tools (as suggested in [3]). The process is performed as follows: First we build the two mono-dimensional N.F. following the structure of the computational unit, the vertical N.F. and the horizontal N.F. (for example we used a H.N.F(A2,N0) and a V.N.F(A1,N1) to obtain the structure that can be seen in Fig. 4b). Later we build the two-dimensional N.F. by making the cartesian product of the two Newton Filters obtained above and apply the result as a convolution mask. This process can be implemented in a waterfall fashion, that is, computing several different Newton Filters, and applying them to the image (note that the actual difference between Filters resides only in the nature of the operation in a layer: adding or subtraction, meaning that the structure of the filter is always the same and that is quite straightforward to implement or simulate it) and thus extract separately certain characteristics (edges, horizontal, vertical or oblique lines or corners) for merging them in a later stage (as can be seen in Fig. 6 and 8 b,c,d,e and f) .

Finally, the fact that N.F. of a generic length, L, are a complete set, makes possible that this technique can be useful in *cryptography* or re-coding of images (Fig. 9a, 9b). Again, some feedback is also offered to natural systems from this point of view: sharping, edge detection and contour coding can be easily implemented provided a minimum of dentritic complexity in the neuron arbotizations.

References

[1] McLennan H. (1970) Synaptic Transmission. W. B. Sanders Company. Philadelphia. USA.

[2] Moreno-Díaz R., Moreno Díaz jr R., Leibovic K. N., Systems optimization in retinal research, Lecture Notes in Computer Science, Vol. 585, pp. 539-546, (EUROCAST91), Munich, 1991.

[3] Moreno-Díaz jr R.,Computación Paralela y Distribuida: relación estructura-función en retinas, Tesis Doctoral (PhD Dissertation). Las Palmas, 1993.ISBN: 84-8090-019-2.

[4] Moreno-Díaz R., Rubio Royo E.,A model for non-linear processing in cat's retina, Biological Cybernetics, Vol. 35, 1980.

[5] Aleman-Flores M., Leibovic N., Moreno-Díaz R. jr. "A Computational Model for Visual Size, Location and Movement". Lecture Notes in Computer Science, Vol. 1333, Springer, EUROCAST'97.

Improving Generalisation Using Modular Neural Networks

David M^{C}Lean, Zuhair Bandar & Jim O'Shea,
The Intelligent Systems Group, The Manchester Metropolitan University, Chester Street, Manchester, UK, M15 GD.
Email:{D.Mclean, Z.Bandar, J.OShea}@doc.mmu.ac.uk

Abstract

This paper deals with improving generalisation performances of feed forward neural networks (FFNN) on real world data domains using more complex architectures for modelling. The convention in neural networks is to use as small an architecture as possible to force better generalisation by modelling the underlying distribution and ignoring the details [1]. This practice involves the loss of information from the training data which in real world domains may represent important though poorly represented decision regions. The problem with introducing extra free parameters (more neurons and weights) to a network is that over-fitting can occur causing the network to model the training data too closely and generalise badly on new data from the same domain. This problem is overcome by combining a number of FFNN (with small architectures) that have been trained on the same data, though generalise differently, to produce more complex decision regions and improved generalisation. Committee decision theory is used to produce the combined model and has been shown to give promising results in the past [2][3][4].

A real world medical data set consisting of non discrete attribute values and FFNN trained using Back Propagation (BP) [5] were used to test the validity of the concepts presented.

1 Introduction

A neural network learns to discriminate between examples of two classes by partitioning the input space into soft-bounded[1] volumes, called decision regions, using neuronal hyperplanes. Each volume is extrapolated from the training data and is formed by the consecutive transformations of each layer of neurons so as to only contain examples of one class. The overall network performs a continuous mapping function, made up from these volumes, which will transform every position within the input space to a corresponding network output. This result should reflect the class of

---

[1] Boundaries between volumes are described by a slow change in network output due to the sigmoidal transfer function.

an example from the data domain located at that point in the input space. For sigmoidal transfer functions and two class problems the output will reflect the probability of that point's class being positive and thus implements a *probability map* [6][7] (see Fig. 2).

A training set taken from a real world problem will give a sparse, and often noisy, description of the whole data domain. Many valid extrapolations, which give the correct results for examples within the training set, will be possible though each will give varying results when applied to the rest of the domain (see Fig. 1). Each of these corresponds to a FFNN solution of the training problem which will generalise accordingly.

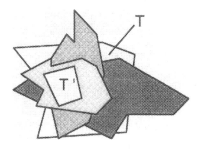

Fig. 1. Extrapolation/Generalisation from a Training Set

Fig. 1 T' is a training set which describes a small part of a data domain T. All the other shaded areas are valid extrapolations of T' which generalise correctly over the intersection between themselves and T.

1.1 Generalisation

The convention in neural networks is to use as small an architecture as possible (a *simple* network) to force better generalisation [1]. In statistics and curve-fitting, too many free parameters results in over-fitting. Such a model will follow all the small details giving poor interpolation and extrapolation. This will be true for a network with an excessive number of hidden neurons trained using common hill climbing methods such as gradient descent. These techniques simply find the closest solution to their starting point (initial weights) which will seldom be the optimal. Many decision regions may be formed which fit too tightly to small clusters of examples and some hyperplanes may not be made use of at all. Alternatively, if too few partitioning hyperplanes are used on a complex problem then a network will not be able to form adequately shaped hulls to model the complex decision regions and may not converge to a solution at all. This has been established by a number of researchers who have shown that the actual size of a network is not a good predictor of its ability to generalise [8][9][10].

A well placed hull should contain a cluster of examples, of the same class. It's boundaries should exactly bisect the space between the outer examples in the cluster and the surrounding examples of differing class. This will ensure that the network performs good generalisation based on the training examples [11]. Small networks may not be able to provide this smooth interpolation of the training data as they will implement decision regions of limited complexity. As discussed above, simply adding extra hyperplanes to the pattern space will not give any improvement unless they can be trained effectively. The problem lies in finding a training technique which will produce near optimal decision regions to partition the training patterns, rather than stopping at the first solution which is met.

1.2 Modular Neural Networks

The problem of finding optimal shaped decision regions can be approached by interpolating between the *probability maps* of a number of similar networks of a *simple* architecture (In relation to the problem to be solved; see Fig. 2). If a number of identical FFNN are trained on the same data, from significantly different initial weights [12], they will approximate the same mapping function over the training examples, though each network will generalise differently [12]. Overlapping volumes of decision regions, with the same assigned class, are more likely to contain actual training examples, and hence new examples. This is due to the fact that virtually all the training examples were correctly classified by each set of decision regions found by each of the component networks. Regions which do not correspondingly overlap may or may not be valid generalisations, but there is no justification for assuming that they are valid.

The new network output should reflect the certainty, or probability, of an example's class and this can be derived from the summed knowledge of the original networks. This certainty can be calculated by ascertaining how many of the original first layer decision regions, of the same class, intersected at that point. The larger the number of intersections the higher the certainty of the class relating to that particular point. The intersections of all the combined hulls will contain

the majority of the data points from the training set, and a new data point in one of these regions should have a correspondingly conclusive network output. Areas which intersect the majority of the hulls, of the same class, but not all, can also be considered as good generalisations, and again the output should reflect this.

2 Training

Each *simple* network will be trained separately using any relevant paradigm, in this case BP. Once the networks have been fully trained the probability maps can be combined. Figure 2, displays three[1] example trained networks and their corresponding input space partitionings. These networks have all been trained on the same data, but from different initial weights. All three networks have differing generalisations as can be seen from their 2-D probability maps.

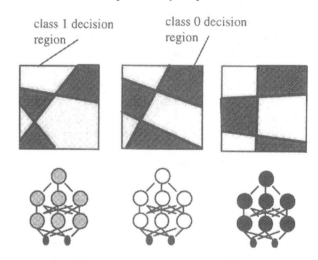

Fig. 2. An Example: Three Networks Trained on the Same Data and their Corresponding Probability Maps

In Order to combine the three networks depicted in Fig.2 a new output layer must be added. Each component network can be viewed as casting a vote on the class of a test vector. The new layer serves to sum the votes from each component network and produce a reflection of the majority decision as the network output. For a two class problem this layer consists of a single output neuron which is given weighted connections of 1.0 to the three output neurons from the component networks (the output layer weights can always be determined though they will differ depending on the number of component networks and the nature of the transfer function used). As there are three

---

1 Any number of networks could be combined together, though 3 was chosen as it is a small odd number.

component networks a non conclusive overall vote would be 1.5 (unipolar sigmoid gives 0.5 as output for an activation of 0) a value greater than this would be a vote for class 1 with greater certainty as the value increases. A value smaller than 1.5 would be a majority decision for class 0 with increasing certainty as the value decreases. As this summed vote will be the output neurons activation the bias weight (threshold) should be set to -1.5.

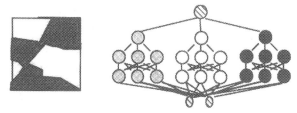

Fig.3. The Combined Networks and the Corresponding Probability Map

3 Back-Propagation

Back-Propagation is probably the most commonly used and well documented Feed forward neural network (FFNN) training algorithm [1][8][9]. This technique requires the pre-construction of a suitable FFNN topology for the problem. This is largely an empirical process to find a trainable network which also generalises well on the test examples.

The algorithm uses a gradient descent technique which alters the weights in the network so as to minimise an error function [5]. All the weights in the network are altered for each iteration of the algorithm, giving a complex [12] and often time consuming learning process. This technique is also very prone to local minima of the error function which prevents the system from progressing to a solution [1][12].

4 Results

In order to test the validity of the ideas discussed ten experiments were performed. Five component networks were trained, from different initial conditions. All possible combinations of three networks, taken from the five, were then used to produce the combined model which was subsequently tested. This exhaustive testing should show some generality to the technique. Each component network consisted of 8 inputs, 3 hidden layers of 12,8,6 and a single output neuron[1]. It

---

[1] This architecture gave the best and most consistent results (no local minima problems) from a large number of tests.

was trained on a real world data set consisting of 174 records of 8 continuously valued attributes and split into two classes. The data set comes from the StatLog study's 'Diabetes in Pima Indians' set [13]. In medical domains it is often important to achieve as high a generalisation performance as is possible.

Each network was trained using basic Back-Propagation [5] to classify the whole training set correctly. Table 1 shows the Classification Rate (CR) on the test set containing 768 examples, and the mean Average Classification Rate (Ave CR) for the five networks.

Table 1. BP results for Diabetes data

Run	C R
1	64.61
2	64.61
3	66.67
4	71.60
5	71.19
Ave CR	67.73

Table 2 displays the resulting CR for each combination. The table is headed by the combination of networks used (from table 1) and displays -

1. The mean Average CR for the three networks which have been combined.
2. The CR of the resulting multi-net (Mnet CR), the Best and the Worst CR from the three original networks.
3. The improvement in CR of the multi-net compared to the average of the three original networks.
4. The difference between the multi-net CR and the Best CR from the original three nets.
5. The summed averages of 3 and 4 over the 10 different combinations of networks.

As can be seen from Table 2 the multiple network CR shows an improvement over the average CR taken from the three component networks for all the combinations. It can also be seen from table 2, that the more complex decision regions have, on average, displayed an increased generalisation perfromance when compared to the best CR of the component networks.

5 Conclusions

By combining networks together we can create trainable complex decision regions which are less susceptible to noise, due to the training of *simple* component networks. This results in a network with improved generalisation when testing on new examples

Table 2. Multiple Networks built from combinations of three BP networks from Table 1.

Comb	1,2,3	1,2,4	1,2,5	2,3,5	2,4,5
Ave. CR	65.30	66.94	66.80	67.49	69.13
Mnet CR	**70.83**	**74.09**	**73.44**	**73.44**	**75.13**
Best - Worst	66.67 - 64.61	71.60 - 64.61	71.19 - 64.61	71.19 - 64.61	71.19 - 64.61
Mnet CR - Ave CR	5.53	7.15	6.64	5.95	6.00
Mnet CR - Best	4.16	2.49	2.25	2.25	3.94
Comb	**1,3,4**	**1,3,5**	**1,4,5**	**2,3,4**	**3,4,5**
Ave CR	67.62	67.49	69.13	67.62	69.82
Mnet CR	**70.57**	**71.09**	**73.48**	**74.09**	**76.56**
Best - Worst	71.60 - 64.61	71.19 - 64.61	71.19- 64.61	71.60 - 64.61	71.19 - 64.61
Mnet CR - Ave CR	2.95	3.60	4.35	6.47	6.74
Mnet CR - Best	-1.03	-0.1	2.29	2.49	5.37
$\sum$(MnetCR - Ave CR) / 10			5.54		
$\sum$(Mnet CR-Best) / 10			**2.41**		

as displayed in table 2. Although the improvements in classification rates are not huge, they are still significant given the nature of the medical domain as well as being 80% consistent. It is common practice when using the BP algorithm to train a number of networks and use the one with the best test performance [1][7][9][13]. In this case the technique can be used without suffering from any extra training times for the *simple* component networks.

The improvements in generalisation performance due to more complex decision regions will be dependant on the separability of the data. If the training set contains examples which are easily partitioned then a single *simple* network will be capable of modelling the domain to a high degree of accuracy. In this case very little improvement, if any, can be expected from more complex decision regions. Conversely if data is difficult to partition, a *simple* network will suffer as it loses information from the training data. Each simple network loses different information, providing they generalise differently, so a combination will retain most of the information from the data set and thus show improved generalisation. Decision region shapes will be more complex, though due to the individual training of the component networks the susceptibility to noise will not have increased.

5.1 Multi-class Problems

Although only 2 class data was used for this example, the technique can be extended to multi-class data. In this case each component network will have a number of output neurons each of which will represent a single class. The corresponding output neurons from each component network can be connected to a perceptron and thus combined in the same way as above.

References

[1] Hertz J., Krogh A. and Palmer R.: Introduction to the Theory of Neural Computation, Sante Fe Institute, Addison Wesley, 1991.

[2] Wolpert D.: Stacked Generalisation, Neural Networks, Vol. 5, p. 241, 1992.

[3] LeBlanc M, Tibshirani R: Combining Estimates in Regression and Classification, Univ. Toronto Statistics Dept., Technical Report. 1993.

[4] Battiti R, Colla A: Democracy in Neural Nets: Voting Schemes for Clasification, Neural Networks,7, pp. 691-707, 1994.

[5] Rumelhart D., Hinton G., Williams R.: Learning Representations by Back-Propagating Errors, Letters to Nature, vol. 323, pp. 533-535, 1986.

[6] McLean D., Bandar Z., O'Shea J.: Improved Interpolation and Extrapolation from Continuous Training Examples Using a New Neuronal Model with an Adaptive Steepness, 2nd Australian and New Zealand Conference on Intelligent Information Systems, IEEE, pp. 125-129, 1994.

[7] McLean D., Bandar Z., O'Shea J.: An Empirical Comparison of Back Propagation and the RDSE Algorithm on Continuously Valued Real World Data, Neural Networks, vol. 11, pp. 1685-1694, 1998.

[8] Martin G., Pittman J.: Recognizing Hand-Printed Letters and Digits, Advances in Neural Information Processing Systems, II., pp. 405-414, 1990.

[9] Tesauro G., Sejinowski T.J.: A Parallel Network that Learns to Play Backgammon, Artificial Intelligence, No 39, pp. 357-390, 1988.

[10] Morgan N., Bourland H.: Generalization and Parameter Estimation in Feed Forward Nets: Some Experiments, Advances in Neural Information Processing Systems, II., pp. 405-414, 1990.

[11] McLean D., Bandar Z., O'Shea J.: A Constructive Decision Boundary Modelling Algorithm, IASTED '98, Mexico. 1998.

[12] McLean D., Bandar Z., O'Shea J.: The Evolution of a Feed Forward Neural Network Trained under Back Propagation', ICANNGA'97, Springer-Verlag, 1997.

[13] Michie D., Spiegelhalter D.J., Taylor C.C.: Machine Learning, Neural and Statistical Classification, Ellis Hopwood Series in Artificial Intelligence, Ellis Hopwood, 1994.

The Role and Modelling of Presynaptic Inhibition in the Visual Pathway: Applications in Image Processing

Roberto Moreno-Díaz jr., Alexis Quesada-Arencibia

Instituto Universitario de Ciencias y Tecnologías Cibernéticas, ULPGC
Edificio de Informática y Matemáticas, Campus de Tafira, E-35017 Las Palmas, SPAIN
Email: roberto@dumby.dis.ulpgc.es, alex@ciicc.ulpgc.es

Abstract

Presynaptic Inhibition (PI) basically consists of the strong suppression of a neuron's response before the stimulus reaches the synaptic terminals mediated by a second, inhibitory, neuron. It has a long lasting effect, greatly potentiated by the action of anaesthetics, that has been observed in motorneurons and in several other places of nervous systems, mainly in sensory processing. In this paper we will focus on several different ways of modelling the effect of PI in the visual pathway as well as the different artificial counterparts derived from such modelling, mainly in two directions: the possibility of computing invariant representations against general changes in illumination of the input image impinging the retina (which is equivalent to a low-level non linear information processing filter) and the role of PI as selector of sets of stimulae that have to be derived to higher brain areas, which, in turn, is equivalent to a "higher-level filter" of information, in the sense of "filtering" the possible semantic content of the information that is allowed to reach later stages of processing.

1 Lettvin's Divisional Inhibition. Invariant Computation Using PI-Like Mechanisms

One of the first known formalisms intended to describe the effect of presynaptic inhibition is due to Lettvin [1], who named it Linear Divisional Inhibition. Lettvin suggested that inhibition may cause a change in membrane permeability at the point of inhibition equivalent to a change in electrical conductivity. Such a change will act as an electric shunt, and it follows that if E is the excitation on a fibre that receives divisional inhibition I, the resultant activity, A, is:

$$A = \frac{E}{1 + {I}/{I_0}} \quad , \quad (1)$$

where I_0 is a constant. For $I \gg I_0$, $A = I_0 E / I$

This mechanism was used by Moreno-Díaz [2] as part of the operations carried out by a frog retinal group two ganglion cell model to account for the temporal behaviour of the cells, where two kinds of presynaptic inhibition were assumed: linear divisional and nonlinear.

In its simplest form, the same mechanism can be used to build a neuron-like network to compute invariances against global changes in its input. This would be a desirable goal of the usual pre-processing of an image, the resulting image being sent to a higher level stage to be analyzed. Some preprocessing characteristics have been described to be present in retinal computation, but no known mechanism, besides adaptation, have been described to obtain a representation which is invariant against global illumination changes. In the model that follows, parallelism is a need for the system to work properly. We will assume that processors (ganglion cells) are arranged in layers, that some kind of computation is done by every cell on their receptive fields and that the output of the layer is a transformation of the original data. There will also be a plexiform layer where the inhibition between cells take place (Fig. 1) [3].

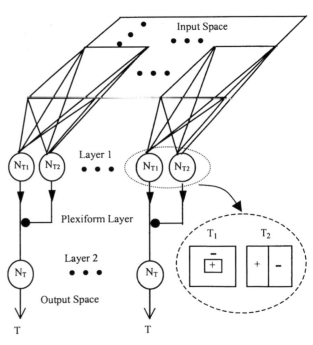

Fig. 1. A parallel processing layer of units performing a model of PI activity

Let then I(x',y',t) be a representation of the original input image impinging on the fotoreceptor layer and T the output of our system as measured in the axons of the ganglion cells. The effect of the presynaptic inhibition will be defined in two parts: let be $T_1(I(x',y',t))$ and $T_2(I(x',y',t))$ two linear transformations whose kernels are respectively $W_1(x,y,x',y',t)$ and $W_2(x,y,x',y',t)$ such that:

$$\int_c W_i(x,y,x',y',t)dx'dy' = 0 \quad , \qquad (2)$$

where c is the domain of each receptive field. This accounts for the fact that the known receptive fields structures in retinal ganglion cells have mutually cancelling regions, e.g., inhibitory and excitatory regions that cancel each other (center-surround or bar-type regions). This is a basic requisite in the description that follows.

The effect of the transformations T_1 and T_2 on the input image is expressed as usual:

$$T_i(I(x',y',t)) = \int_c I(x',y',t)W_i(x,y,x',y',t)dx'dy' \quad , \quad (3)$$

and the presynaptic inhibition effect would then be expressed as the ratio: $T=T_1/T_2$

In these conditions is easy to prove that a change of I(x,y,t) like: I(x,y,t)=KI(x,y,t)+R(t), where k is a constant and R(t) a function of time representing a global change on light intensity over the retina,would not affect the calculation of T. Thus, when $T_2 \gg T_1$

$$T' = \frac{T_1(I'(\overline{X'},t))}{T_2(I'(\overline{X'},t))} = \qquad (4)$$

$$\frac{\int_c kI(\overline{X'},t)W_1(\overline{X},\overline{X'},t)dx'dy' + \int_c R(t)W_1(\overline{X},\overline{X'},t)dx'dy'}{\int_c kI(\overline{X'},t)W_2(\overline{X},\overline{X'},t)dx'dy' + \int_c R(t)W_2(\overline{X},\overline{X'},t)dx'dy'} =$$

$$\frac{k\int_c I(\overline{X'},t)W_1(\overline{X},\overline{X'},t)dx'dy' + R(t)\int_c W_1(\overline{X},\overline{X'},t)dx'dy'}{k\int_c I(\overline{X'},t)W_2(\overline{X},\overline{X'},t)dx'dy' + R(t)\int_c W_2(\overline{X},\overline{X'},t)dx'dy'} =$$

$$\frac{kT_1}{kT_2} = T$$

where

$$\overline{X'} = (x',y') \; and \; \overline{X} = (x,y) \quad , \qquad (5)$$

A plausible place in the nervous system to locate this kind of invariant computation is the LGN. The output of information from retina through the optic nerve follows three channels towards higher brain areas, being the geniculo-cortical pathway the one receiving more fibres. In the LGN a topographical representation of the whole retina can be found. The cells in the LGN are arranged in six perfectly defined layers and attending to the size of the cells these layers could be divided into two groups: the magnocellular layer and the parvocellular layer [4]. The magnocellular layer is formed by big cells working on the illumination characteristics of the input image and the parvocellular layer includes smaller neurons involved in color coding.

Our mechanism can be assumed to work in the magnocellular layer. The transformations T_1 and T_2 are computed by the ganglion cells and the result reaches the LGN via the optic nerve. The cells at the magnocellular layer of the LGN would be the units that compute the invariant representation T using presynaptic inhibition (see Fig. 1). The simplicity of the figure mimics the simplicity found in the physiology of LGN where each cell receives only a few input lines from retina including inhibitory effects. Thus, one of the possible outputs of the LGN would be an invariant representation of the light pattern already coded by ganglion cells.

2 Model Refining. Non Linear Divisional Inhibition

Despite the above model is quite useful when specific illumination conditions are present, there are certain formal objections that make it unsatisfactory, both from a formal as well as an applied point of view. First, we can start taking into account the idealized performance of PI (Fig. 2) [5].

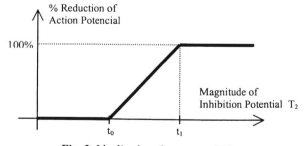

Fig. 2. Idealized performance of PI

As we can observe in the graphic, when T_2 is smaller than a certain threshold (t_0) the Presynaptic

42

Inhibition mechanism does not reduce the action potential at all. From the threshold t_0 to t_1 PI presents a linear performance, and for values of T_2 grater than this threshold PI reduces the action potential to zero. Thus the idealized performance of T can be expressed as follow:

$$T = f(T_1, T_2) = \begin{cases} T_1 & if \ T_2 \leq t_0 \\ f'(T_1, T_2) & if \ t_0 < T_2 < t_1 \ , \\ 0 & if \ T_2 \geq t_1 \end{cases} \quad (6)$$

Using the original analisis of Lettvin, already mentioned at the beginning of this paper, we obtain the following expression:

$$T = \frac{T_1}{1 + \dfrac{T_2}{I_0}} \quad , \quad (7)$$

to model Presynaptic Inhibition, where I_0 is a constant. This comes from considering the action of PI as an electric shunt.

It has been used a linear function in the denominator. However, the function that better fits the ideal performance has a shape shown in Fig. 3:

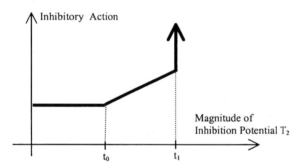

Fig. 3. Better-fit representation of PI function

$$f(T_2) = \begin{cases} 1 & if \ T_2 \leq t_0 \\ aT_2 + b & if \ t_0 < T_2 < t_1 \ , \\ \infty & if \ T_2 \geq t_1 \end{cases} \quad (8)$$

Now we show a comparative graphic where we can observe the functions mentioned above and the idealized function in Fig. 4, and at the same time we present another non-linear function that fits better into the idealized function (we use an exponential function):

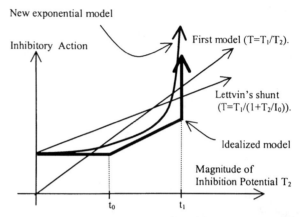

Fig. 4. Comparative graphic of the different models of PI

The new model is an exponential function as follows:

$$f(T_2) = \frac{T_2}{a} * e^{\frac{T_2}{b}} + 1 \ , \quad (9)$$

$$T = \frac{T_1}{\dfrac{T_2}{a} * e^{\frac{T_2}{b}} + 1} \quad , \quad (10)$$

Where a and b are constants that should be obtained depending on t_0 and t_1 and trying to make the best fit with the idealized function. This model is similar to Schypperheyn's model, who named it Non-Linear Divisional Inhibition [6].

To illustrate this point, we developed an example where invariances against global uniform illumination changes can be observed (Fig. 5). In Fig. 5a we have the original image, which is a 132x102 pixel black and white image with 256 grey levels. In Fig. 5b we present the result of performing the operation indicated in expression (10), where we used Newton Filters [7], both to calculate T_1 as well as T_2. In Fig. 5c we can observe the original image ($I(x',y',t)$) transformed as follows:

$$I(x', y', t) + R(t) = I(x', y', t) + 200, \textit{ that is, } R(t) = 200 \quad (11)$$

Finally, in Fig. 5d we show the result of performing the same operation as in the previous case, and we can verify that this image is equal to the image of Fig. 5b, confirming the theoretic results explained in the first point of the paper. Besides, the improvements to which this new exponential model contributess can be seen grafically on Fig. 4, since it fits better into the idealized function. We can see this in in the implementation of the examples where if we do not use this model we would have problems in the limits, that is to say when $T_2<=t_0$ and $T_2>=t_1$. In this case the other models differ strongly The implementation of this examples has been carried out by means of a program developed in the programming language Borland Delphi Professional 3.0 for Windows 95.

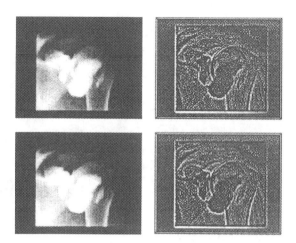

Fig. 5. A X-ray image of the shoulder. An example of invariance computation; From left to right and from top to bottom: original image, processed image, displaced illumination image and processed image after displacement

3 A Possible Mechanism to Control Visual Attention and Information Flow via PI

The Theory of Vision's goal is, basically, to build a theoretical and practical framework to explain the visual function in live beings and its possible artificial counterparts. The sense of vision is, in most of species, the one that processes the biggest amount of information coming from the outside, being the most important in guiding its behaviour. The ability to discriminate certain parameters and locate what part of all that information is significant is crucial for both natural and artificial complex systems [8].

In a previous paper [9] a theoretical construct called Directed Foveal Transform, TFD, was presented as the first step in modelling the attentional mechanisms ruling the visual processes of vertebrates. The kernel of the TFD is a variation of the moving average that presents the highest acuity on a particular area in the retina (called fovea). Originally this area of better resolution expanded from the center of the image. In the TFD, the dominia on which the best resolution is placed can be located around any point in

the image and at the same time any foveal size can be especified. It is also possible to define, on the parafoveal zone, tranformations where the completness is not a crucial factor. Thus, we see the goal of defining this transformations in different image areas as performing an "interest attractor". On the parafoveal zones information is extracted but there is a loss of resolution, but once the event or characteristic is discriminated, the fovea can be placed over it in order to perform a more detailed operation. There is a basic idea underlying it: the economy of computations: no known visual system has large fovei (at least they do not cover more than 10% of the total system), otherwise the total information to process and transmit would be too high to cope with, needing then more complex and bigger nervous tracts and much more connections in intermediate stages of the visual pathway. Thus, the goal could be to achieve certain "economical balance" between structural complexity and detailed information to be transmitted.

A step forward in the simulation of information flowing and attention mechanisms in the visual pathway would be to combine the above mentioned concept with the action of the presynaptic inhibition. The goal is then to highlight some feature or object in the image, blurring (or loosing resolution) the rest of it since we assume that it does not contain anything relevant. In the original TFD formulation, a three-step procedure was designed to locate a visual event on the input image and concentrate the highest amount of computation on it:

1.- First, it is necessary to locate the center of luminance of the region/object. The coordinates of the center of luminance will act as the coordinates for the center of the fovea.

2.- Second, the outermost point of the object is located and the distance from the center of the fovea is calculated.

3.- We use this data in order to calculate the size of the fovea to be used.

In order to include PI, a feedback line from the cortex is necessary to allow the information flow within the topographical representation of the retina that is present in the lateral geniculate nucleus [10]. In Fig. 6 we can see the proposed architecture. Thus, the mechanism will perform as follows:

1.- A first retinal description in terms of contrast, edges, color and movement parameters, present in the axons of retinal ganglion cells, reaches the LGN. These parameter calculations are already studied in previous reports [10,11,12].

2.- This information is transformed in a second, higher-semantic content representation in terms of location of center of gravity of moving objects, high luminance or color- components areas and sent to the first layers of visual cortex.

3.- The decision of selecting the geographic position of the object of interest is made in the cortex, that sends a feed back signal to the LGN, and via PI controls what information is effectively reaching the LGN.

This mechanism could be proposed as a "first-order" attention focusing mechanism. Note that it has nothing to do with "willness" or purpose of behaviour, it is closer to an automatic response depending on outside stimulae.

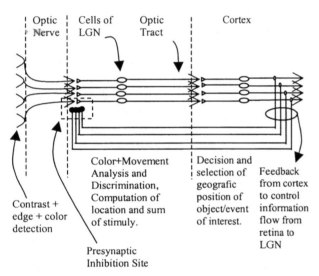

Fig. 6. Representation of the feedback lines coming from the cortex into the LGN including the PI mechanism

4 Conclusions

Although more information on the real connectivity of PI in LGN is needed, two immediate operations can be thought of being performed by that kind of mechanism: invariant computation and control of information flow. Mechanisms for obtaining invariant representations of features in the first steps of information processing are important in Nervous Systems since they provide the basis of reliable pattern and movement discrimination. In order to do this, it is plausible to expect that the visual cortex works, at least to some extent in the first steps, with a low-level invariant version of the input image. On the other hand, there is a need of controlling the total amount of information reaching effectively the brain, and this can be achieved, according to the connectivity of the first stages of the visual pathway, by means of two structures: first the existence of a foveal

area in retina in which the resolution is maximal, and second, a control of the information flow in the LGN. In both, invariance computing and information control, the mechanism of presynaptic inhibition may play a crucial role.

Regarding the artificial counterpart and applicability of the ideas presented before, the mechanism of PI, implemented in a parallel fashion, is suitable of being used in artificial perceptual systems, mainly in image processing, to obtain image representations invariant against general changes in illumination.

References

[1] Lettvin, J.Y.: Form-function relations in neurons, Research Lab. of Electronics, MIT Quarterly Progress Report, pp. 333-335, June 1962.

[2] Moreno-Díaz, R.: An analytical model of the Group 2 ganglion cell in the frog's retina, Instrumentation Lab., MIT Quarterly Progress Report, pp. 1-34, October 1965.

[3] Muñoz-Blanco, J. A.: Jerarquización de estructuras de nivel bajo y medio para reconocimiento visual, PhD Dissertation, University of Las Palmas de Gran Canaria, 1987.

[4] Truex, R. C., Carpenter, M.B.: Human Neuroanatomy, Williams and Wilkins, Baltimore, USA, 1969.

[5] Graham B., Redman S.: A Simulation of Action Potentials in Synaptic Boutons During Presynaptic Inhibition, Journal of Neurophysiology, Vol. 71, No 2, February 1994.

[6] Hagiwara S., Tasaki I.: A Studdy of the Mechanism of Impulse Tranmission Across the Giant Synapse of the Squid, Journal of Neurophysiology, Vol. 143, 1958.

[7] Moreno-Díaz jr R.: Computación Paralela y Distribuida: relación estructura-función en retinas, PhD Dissertation, University of Las Palmas de Gran Canaria, 1993.

[8] Kandel, E. R.: Processing of form and movement in the visual system, Sensory Systems of the Brain: Sensation and Perception, 1990.

[9] Quevedo-Losada, J. C., Bolívar-Toledo, O., Moreno-Díaz jr, R.: Image Transforms based on Retinal Concepts, R. Trappl Ed. Cybernetics and Systems 98, University of Vienna, Austria, pp.312-316, 1998.

[10] Moreno-Díaz jr, R.: On structure, function and time in retinae, R. Moreno-Díaz and Mira-Mira Eds. Brain Processes, Theories and Models, The MIT Press, Cambridge, Mass, USA, pp. 430-435, 1996.

[11] Alemán-Flores, M., Leibovic, K. N., **Moreno-Díaz jr,** R.: A computational model for visual size, location and movement, R. Moreno-Díaz and F. Pichler Eds. Computer Aided Systems Theory, Springer Lecture Notes in Computer Science, V1333, pp.406-419, 1997.

[12] Moreno-Díaz, R., Alemán-Flores, M., **Moreno-Díaz jr,** R.: A bio-inspired method for visual movement discrimination, R. Trappl Ed. Cybernetics and Systems 98, University of Vienna, Austria, pp.307-311, 1998.

A Curvature Primal Sketch Neural Network Recognition System

Michael Fairbank[1] and Andrew Tuson[2]

[1]Division of Informatics, Edinburgh University, 5 Forrest Hill, Edinburgh EH1 2QL, UK
[2]Department of Computing, City University, Northampton Square, London EC1V 0HB, UK
Email: michael.fairbank@virgin.net, andrewt@soi.city.ac.uk

Abstract

Neural networks can be used to classify images such as handwritten characters. A common method of doing this involves mapping the pixel values of the image onto the input nodes of a feed-forward net. This is problematic in that the topological properties of the original image space, such as the spatial relations between different pixels, are not immediately apparent to the net. We address this problem by using the real valued coordinates of selected features in the image for input to the net.

This paper details a formative study of the Curvature Primal Sketch (CPS) as a preprocessing method to identify the interesting features of the curves that make up handwritten characters. Emphasis is placed upon integrating the CPS with a feed-forward neural network classifier. To this end, we describe an algorithm for selecting which of the features produced by the CPS should be used as input to the neural network. We postulate that the order in which the features are used as inputs to the net is also important and introduce a solution to this problem.

The nets obtained by this approach were small and performed recognition well. A net with dimensions 14–14–10 was trained with collection of 500 handwritten digits, collected from different people. On a similar test set of 100 digits, the net was found to achieve 92.8% accuracy in recognition.

1 Introduction

This paper describes a computer vision system which classifies curves. These curves are to come in the form of an ordered set of coordinates (so any automated edge detection and tracking necessary are presumed already dealt with by another system). The approach used here, as with other methods, is to use an artificial neural network to make the classifications. However, in this system the curve is preprocessed in a novel way before being fed to the neural network.

A common method of using an artificial neural network to classify images is to map the pixels in the image directly to the inputs of the net, maybe having applied some form of normalisation to the image first (e.g. [3], [4] and [5]). In using this approach, however, the topological properties of the image space are not immediately apparent to the net, since all the neural net's inputs are equivalent to each other and the mapping from pixels to inputs is arbitrary. This means a net cannot make generalisations about neighbouring regions in the image, without having specifically learned at least something about all the pixels involved.

For example, if a net with 64 inputs had been trained on an 8×8 pixelated image of a letter M, and was then presented for recognition with a similar M which had been displaced one pixel to the side, then the net would have a very hard task recognising it. This is why the normalisation of position and size is such an important method of preparing the images for training and recognition. The methods to overcome these problems used in [3], [4] and [5] involve mapping the large, high resolution pixel array (for example 16 by 16) down to a smaller, lower resolution one (for example 3 by 3). Then, in the lower resolution image produced, each position in the array represents such a large region of the image that spatially close events in the original image get reduced to the same position.

The method used by our system was to preprocess the input curve into a small set of summarising points. The method of choosing the summarising points is based on the Curvature Primal Sketch (CPS) due to [1], which locates the principal points of curvature change, for example any significant corners. Each point in the chosen set is described by a set of parameters which, for example, indicate the type of the point, the scale it is seen at, and its position coordinates. It is these (generally real valued) parameters that are used as inputs to the net for recognition.

It is a postulate of this paper that the net will be able to generalise much more easily about the topological properties of the image space when real valued coordinates are used to describe the curve. This is because the net can use the numeric difference between the coordinates of points to judge how they are positioned relative to each other (i.e. distance and direction). In this method of preprocessing, using real valued coordinates should mean that normalisation of position and size is not as necessary. Similarly, normalisation of orientation may not be as necessary (especially if polar coordinates are used). Also, the nets used should be smaller and easier to train, because the preprocessing distills the amount of information present down to a smaller amount.

Our system also attempts to map the chosen points in the curve to the input units of the net in a consistent manner, or, in other words, to encourage the network inputs to become highly *feature-specific*. For example, if the system comes across several instances of a handwritten digit *3*, then we would like the strong feature in the middle of this digit to be mapped to the same network inputs each time. The second postulate of this paper is that this feature-specificity is very advantageous. There are two main reasons for this. Firstly, each input group will be able to specialise on just one feature for each curve. Secondly, and most importantly, the action of each input group will become independent of the others. Without this independence, each would have to check that the others have not already recognised that feature, and that the order of features along the curve is following a particular pattern.

This paper will place emphasis upon the CPS preprocessing stage and the integration of this with a feedforward neural network. Therefore, the Curvature Primal Sketch will be outlined first. Then the problem of extracting the most relevant features will be addressed and an algorithm for performing this task will be introduced. Finally this approach will be evaluated by testing it on a character recognition problem.

2 The Curvature Primal Sketch

The method described by [1] takes a curve and calculates from it a representative set of features. This set of features, and their parameters, is known as the Curvature Primal Sketch (CPS). This summarising set of features is smaller than the number of coordinates describing the curve, but should still fully describe the original curve it represents — it should therefore be possible to reconstruct the original curve fairly accurately from the CPS. The method used to obtain the CPS is described in full detail in [2], and is summarised here.

The CPS is calculated from the curve's representation in orientation space (observed by plotting orientation versus arc length). The parameters describing each feature obtained contain the following information: the position of the feature along the curve; the type of the feature; the angular change associated with the feature,

(θ); the orientation of the feature; and the scale at which the feature is visible, (σ). In addition the coordinate of where the curve starts could be specified, however this is not used here as an aim of this work is to make the recognition of curves position-invariant.

There are two fundamental types of feature identified in the CPS, *corners* and *smooth joins*. Corners are first derivative changes in orientation, and smooth joins are second derivative changes. These are shown, together with their orientation space representations, in Figure 1 below.

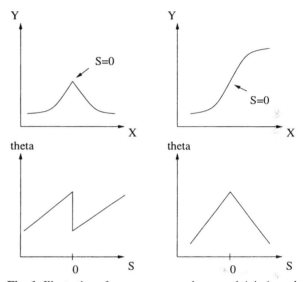

Fig. 1. Illustration of a pure corner and a smooth join in real space (above) and orientation space (below).

The CPS is a multi-scale description. Different descriptions are produced at different scales for the same curve. At small scales, many features, including noise, will be described. At larger scales, there is less noise, and many fundamental features merge to form compound features. The changing of description at different scales is not a limitation of this particular method, but a more fundamental property of observing shapes. For example, what appears as a straight diagonal line on a computer screen is actually a zigzag of stepped pixels when viewed at a small scale.

The way features merge into one another as the scale changes is best displayed in a scale space representation. Figure 2 shows the scale space representation of an 'end' shape. An 'end' shape has two successive corners that merge into one at large scales. In this representation, the vertical axis represents decreasing scale, and the horizontal axis increasing arc length. The features that correspond to each other at adjacent scales are connected in the diagram. This means that a particular feature will appear as a *path* through scale space. Where a path (de-

48

noted the *parent*) splits as the scale is decreased, the new paths are denoted *children* and their size and scale are noted for later stages of the preprocessing.

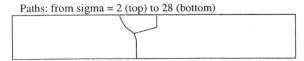
Paths: from sigma = 2 (top) to 28 (bottom)

Fig. 2. Scale space representation of the 'end' test shape.

In practice, the CPS produces a complex branching tree of related paths like this. Each path has a scale range (σ_{min} to σ_{max}), an angular change θ and a position. A path may have one immediate ancestor path (a feature that it is consumed by at higher scales) and any number of immediate descendant paths (features that it splits into at smaller scales). The arrangement in scale space of the paths, along with their associated parameters, gives the full description for the Curvature Primal Sketch.

3 Finding a Fixed Number of Important Features

The network we used was a feed-forward one with a fixed number of inputs for a fixed number of features. We therefore needed to reduce the number of features obtained in the CPS to only the most significant ones for input to the neural network. A numeric value was assigned to each path to represent its significance, based on its parameters, size (θ) and scale (σ), according to the formula, $strength = |\theta|.\sigma_{max}^{1.5}$. This attempts to account for the following two effects:

- It is possible for an insignificant feature to exist with a large angle size at a small scale. For example, noise often has large angular changes, but over a very small distance.

- It is possible for an insignificant feature to exist with a small size at a large scale. A feature is visible at (and hence appears at) very large scales if there are no other features near by. For example, a corner with $\theta = 0.1°$ say, will show up at scale σ if there are no other features within a distance $\pm\sigma$ of the corner.

The exponent 1.5 was an arbitrary value that was chosen to make the scale more significant, and to compensate for the large range in θ values at small scales. This value was found to give satisfactory results throughout this study and therefore it was not changed.

Due to the large and complicated branching nature of the related paths obtained in the CPS, it was often found that a strong feature can occur many times in the same ancestry line of paths. To avoid choosing the same feature more than once, we made a rule never to choose a path at the same time as any other path in it's ancestry line. This principle is modified slightly in the following algorithm to allow simultaneous choice of strong sibling paths. It allows inclusion of all the distinct and interesting features, and still ensures that the same feature may not be included more than once.

3.1 An Algorithm to Identify the N Most Significant Paths

The above section gives the main principles of the algorithm which is described here. The algorithm decides when to replace a parent by its children, while ensuring termination. The algorithm follows a number of steps, and involves labelling each path as either active or inactive.

1. Rank all paths in order of strength.

2. Set all paths which have a parent as inactive, and all with no parent as active.

3. Identify the N strongest active paths. Set a threshold variable, T, to the strength of the weakest included path.

4. Go through all active paths; A path is replaced by all its children if all its children's strengths exceed T. When this happens all children become active, and the parent becomes inactive.

5. Repeat from step 3 if any changes were made in step 4.

The N most significant paths are then taken to be the N strongest active paths. For a more detailed description of the above algorithm, the reader is referred to [2].

To validate that this approach does indeed select appropriate features, the above diagram (Figure 3) shows a visual representation of the information produced by the preprocessing. The eleven most significant features were calculated from a handwriting sample, and then the parameters of these points used to reconstruct the original curve. The angular information has been illustrated by using Bezier curves to connect the features, using the entry and exit angles as determined from the CPS. As can be seen, the selected features produce a good description of the original curve.

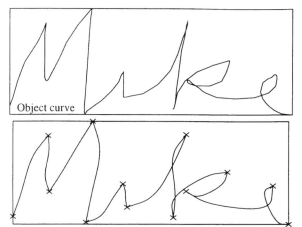

Fig. 3. The original and reconstructed 'Mike' test curves—
the crosses mark the positions of the features used.

4 The Neural Net Design Used

This section explains the preparation of the data for the network's input and output vectors during training, and the interpretation of the output vectors produced by the net during recognition. The neural network architecture is a standard feed-forward one, trained using the publically available `rbp` back-propagation package [6] for training and evaluation (this is described in full in [2]).

The artificial neural network took the features chosen by the CPS as input, and aimed to classify the curve from this information. As each feature for input had several parameters describing it, each feature required a group of net inputs to represent it. The parameters chosen in the end were all real valued, and were rescaled to the range [0, 1] before inputting.

There was one output unit in the network for each classification of image. The net was trained to fire the appropriate output for each image used as input. The 'on' values in the training data are represented as an output of '1' and 'off' as '0'. When recognising an image, the highest output produced by the net was chosen as the classification made. An estimation of the certainty of it being a particular category can be made by comparing how close the chosen output value is to 1, and how close all the other values are to 0. A well trained net should successfully classify each type of curve, producing a high certainty for the answer.

If there was a case of a genuinely ambiguous curve, e.g. if it was recognising characters and digits, and found it hard to distinguish between a number '5' and the letter 'S', then the program should produce similar certainties for both categorisations. If this vision system was just one component of some bigger system which attempted to distinguish between the two, then it could still do so by

using an appropriately programmed knowledge of context; but that is a higher level problem than this system is designed to solve. Finally, if it is possible for a curve to not be a member of any classified set, then it would be an advantage to make an extra output category to represent these unclassified curves, though this was not implemented in this study.

4.1 Mapping Features in the Curve to the Network's Inputs

The following section explains the method used for mapping the features selected from the CPS to the input nodes of the net. The aim is that the mapping function will map the features to the inputs in a consistent manner to aid feature-specificity.

Mapping the selected features to the network's inputs in the order that they appear on the curve would be bad for feature-specificity, as generally there will be several insignificant features due to noise in the chosen set of features, and this noise may appear at any position on the curve. Also, there may be certain features which are very close to the threshold of being included in the chosen set or not. These uncertain features and noise, if included, will shift all the other chosen features along one place in the mapping system, which therefore greatly reduces the opportunity for feature-specificity to develop in the majority of the network inputs.

The solution to this problem was for the mapping function to rank the features according to their strength before presenting them to the net inputs. That way, any noise or uncertain features will always appear at the end of the list, where they will not displace the more important features. This method loses the ordering information of the chosen features, but this was compensated for by supplying each feature's arc length as an extra input to the net. By considering this arc length information for each feature, the network would then still be able to rediscover the ordering of the points if need be.

Using this mapping system, if several important features have similar strengths, then their relative order may interchange, which is bad for feature-specificity. However, this is still an improvement, as it will not displace *all* the other features in the curve. The only problem with this method is when a curve has a noise feature with greater strength than any genuine feature on the curve. If this happens, all features will find their positions moved, and the recognition chances will be severely reduced. This is a risk that has been accepted — if a noisy feature had enough strength to ruin the process, then the noise should also be significant enough to justify the curve as unrecognisable, assuming the strength formula has been derived correctly.

5   Results and Discussion

To evaluate this approach, two sets of experiments were carried out. The network used had one hidden layer and was trained by back propagation. The number of hidden units, number of input features, and number of input parameters to use for each feature was determined by experimentation. For full details of the experimental set-up and the results obtained, the reader is directed to [2].

The first set of experiments was a preliminary test using artificially generated polygons for recognition. The polygons had varying numbers of sides and amounts of noise added. A net was successfully trained to distinguish between three types of polygons and a noise category with 94% accuracy. The training and test data sets each contained 100 polygons with considerable variation and noise added to each. The network had sixteen inputs, one hidden layer and four outputs. The number of hidden units was experimented with, and the best performance occurred with just two.

For the second, main experiment, a training database of 60 sets of handwritten digits was collected. Each set was drawn by a different person using the mouse. Each set contained the digits 0-9. The data collected was split into two groups: one for training including approximately fifty sets, and one for testing including approximately ten sets. Due to the difficulty in obtaining suitable data, no extra category has been included to represent an 'un-classified' digit.

Experimentation found that choosing only two features for input from the CPS was best for the data used, with 14 hidden units. In addition to the features selected by the CPS, the endpoints of the curves were always included as two extra features for input (the curves were always treated as open curves). The following parameters were used for each feature.

- The arc length that each feature appears at (rescaled to [0,1]).

- The size of the feature (θ).

- Coordinates of the feature. Polar coordinates were used to aid rotational invariance. The origin for the coordinates was taken from the centre of mass of the image.

- The orientation of the feature was not included, as this did not appear to help the network's performance.

The above gave a performance of 92.9% recognition on the test set. Comparison with equivalent classification rates in the literature of 94% in [4], 96% in [5] and 96.6% in [3]. This is a promising result, considering that there

was room for improvement by further experimentation with the network design.

Also, the people who provided the samples were restricted to only using one line per digit. This meant that their writing style was slightly constrained; for example '7's could not be crossed. There were a mixture of styles in the handwriting between the different nationalities contributing, so this could have caused some confusion between 1 and 7, for example. This restriction, and the fact that a mouse was used to write with, may have made the data slightly un-realistic and possibly harder to recognise.

Finally, considering the scale of the investigations, this study used a small network with a training set of only 600 digits, and the performance should improve as the training set size is increased. The large networks involved in the other studies took many times longer to train. For instance, [3] used a training set of 10000 digits (including some printed digits, which make classification easier due to their regularity), [5] used 40000, and [4] used 10000, which could partly explain their improved results.

6   Conclusions

The Curvature Primal Sketch approach to preprocessing was found to be a very natural way to present the curves to a neural network— it made important topological properties immediately apparent, and distilled all but the most important results. These factors meant that a very small network could be used which was easy to train. At the same time, the method is still very general, and could be applied to any curve classification problem, such as recognising object silhouettes.

Further work could include extending the system to provide a confidence estimate for each classification. This is extremely useful if the application of this process was one where the costs of making a mis-classification are very high (e.g. if the Postal Service was to misclassify a postal code, it is relatively expensive to redeliver it afterwards compared to having a human re-read the post-code beforehand). This could be based on the magnitude of the maximum net output compared to the other outputs.

Finally, it should be noted that a drawback with the system we created is that at the moment the system will only classify a single curve, which makes it limited for recognising handwritten characters (e.g. you get a problem with crossed 7's). This problem should not arise however with other applications such as recognising object silhouettes, and could be a subject for further work.

7 Acknowledgements

We would like to express our gratitude to the Engineering and Physical Sciences Research Council (EPSRC) for their support via studentships 94415679 and 95306458 whilst the authors were at Edinburgh University. We would also like to thank Herman Gomes and Chris Malcolm for their assistance and advice.

References

[1] Asada, H. and Brady, M.: The Curvature Primal Sketch. AI Lab Memo 758, Massachusetts Institute of Technology, 1984.

[2] Fairbank, M.: A Neural Network Vision System based on the Curvature Primal Sketch. MSc thesis, Department of Artificial Intelligence, University of Edinburgh, 1995.

[3] Le Cun, Y., Boser, B., Denker, J. S., Henderson, D., Howard, R. E., Hubbard, W., and Jackel, L. D.: Handwritten digit recognition with a backpropagation network. In Advances in Neural Information Processing Systems, volume 2. San Mateo: Morgan Kaufmann, 1990.

[4] Denker, J. S., Gardner, W. R., and Graf, H. P.: Neural network recogniser for handwritten zip-code digits. In Advances in Neural Information Processing Systems, volume 1, AT&T Bell Labs, 1988.

[5] Martin, G. L. and Pittman, J. A.: Recognising handprinted letters and digits. In Advances in Neural Information Processing Systems, volume 2. San Mateo: Morgan Kaufmann, 1990.

[6] Tveter, D. R.: The 'rbp' back-propagation simulation program. 5228 N. Nashville Ave, Chicago, Illinois 60656.

Using GMDH Neural Net and Neural Net With Switching Units to Find Rare Particles

František Hakl and Marcel Jiřina
Institute of Computer Science, Academy of Sciences of the Czech Republic
Pod Vodárenskou věží 2, 182 07 Prague 8, Czech Republic
Email: {hakl|marcel}@uivt.cas.cz

Abstract

Two kinds of neural networks which are not in common use were used to separate two kinds of events in nuclear experiments. The experiments were made using data simulated in SACLAY laboratory because the experiment ATLAS in CERN is still under construction. Because there are no direct criteria for separation of events, we use two kinds of neural nets for this task. The neural nets used have continuous output and separation - classification of events is made using a suitable threshold for the output value. It was found that the threshold must be carefully set and enrichment factor higher than 10 can be reached.

1   Problem formulation

During the run of experiment the high energy protons collide in the Large Hadron Collider and are disintegrated and different decay processes arise. This is called an event. In the event most of particles under interest which arise in the decay processes usually have a very short life time (less than 10^{-10}) and disintegrate until stable particles, especially electrons, positrons and gammas arise. The electrons often arise several at a time and form a so called jet. The original particles cannot be directly detected because their short life time corresponds to a trajectory usually much shorter than 1 cm. But electrons and jets which finally arise from the event are recorded and measured in an ATLAS detector. In fact, all decay processes and short life-time particles in the event are observed indirectly as tracks and energies of electrons, jets and gammas. Initially all measurable information about the event is recorded in full detail. Because the event rises with high frequency, a triggering system is used to throw out events which are not interesting enough and thus the event frequency is reduced to value when data can be recorded for detailed analysis later.

The task is to find events in which the so called Higgs boson rises and filter them out of the other uninteresting events - the background. Today Higgs boson has never been observed and the ATLAS detector is still under construction. Simulated data was used next. Data was produced by the Pythia physics simulation program in Dapnia/SPP Saclay laboratory, France. It is characteristic for the particle mentioned that during the event four jets of electrons arise. One of these jets is often invisible in the detector and then we only have three jets. Original measured data is transformed into the values of energy, momentum and angle for each jet. Data is formed by nine-tuples of reals (three parameters times three jets). The task is to classify these nine-tuples into two classes, a class of "Higgs probably present" and "Higgs probably not present". In the learning set we know, that some nine-tuples belonging to the event with presence of the particle looked for, and the others are background only. Data given have following form: To each event corresponds one record of the file. In each record 11 numbers - integer or real - are stored: r - event number y - kind of event (1 or 100 for event with Higgs - signal, 2 or 200 for event without Higgs - background , 0 for unknown) E_1 -energy in MeV of the first jet p_{T_1} -transversal momentum in MeV of the first jet ψ_1 -angle $(0, 2\pi)$ of the first jet, E_2, p_{T_2}, ψ_2 , dtto for the second jet, E_3, p_{T_3}, ψ_3 , dtto for the third jet, etc.

The training and testing sets are files which have in each record kind of event y equal to 1 (100) or 2 (200) (simulated data). The kind of event would be equal to 0 for measured data - the kind of event cannot be known in advance. These data serve as input data for neural networks. In fact, nine values E_1, p_{T_1}, ..., p_{T_3}, ψ_3 are individual inputs and form input vectors x. For r-th event we have then input vector $x_r = (E_{1,r}, p_{T_{1,r}}, \ldots p_{T_{3,r}}, \psi_{3,r})$

$= (x_1, x_2, \ldots, x_9)$ and desired value y_{dr}. When applied an input vector x_r a correctly functioning classifier will give output just y_{dr} for each $r = 1, 2, \ldots, number of events$. Real neural network with continuous output will give some output value y_r. Then we set a threshold θ and for $y_r \geq \theta$ we take the corresponding event as background, for $y_r < \theta$ as Higgs present. In this paper we use and analyze the response y of the net and the threshold θ is used as parameter for final classification of events.

2   The neural net GMDH type

The GMDH (Group Method Data Handling, [6] and [8]) net has m inputs and one output y. The net is formed from layers of neurons. Each neuron only has two inputs. If there are n neurons in the preceding layer then there may exist $\frac{n(n-1)}{2}$ different neurons in the next layer. Initially $n = m$. Each neuron is adapted so that coefficients $A, B, \ldots F$ of the equation

$$y = A + Bx_i + Cx_j + Dx_i^2 + Ex_j^2 + Fx_ix_j \quad (1)$$

are computed; y is the output value desired and x_i, x_j are the inputs to the neuron. After adaptation of all neurons in a new layer, only some of them which have the smallest error in approximating y form the layer and others are left out. It is done so from simple practical reason: the number of neurons in next layers would grow exponentially. If the best neuron has the error less than the prescribed value or another criterion is fulfilled, the learning stops.

2.1   GMDH recall algorithm

The GMDH is a feed–forward network and an example of particular structure is shown in Fig. 1.

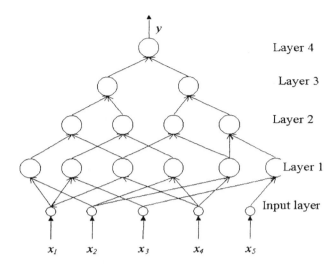

Fig. 1: Structure of GMDH neural net.

Each neuron has its two inputs connected to some neurons in the preceding layer or, if the neuron is in the first layer, to some inputs of the entire network. Each neuron has also its particular coefficients $A, B, \ldots, F$.

In recall, the input signals are applied to inputs of the network. For each neuron of the first layer its output is computed according to eq. 1 using the neuron's particular set of coefficients $A, B, \ldots, F$. Then the outputs of neurons in the second layer are computed in similar way, then outputs of the new layer and to on, until output y is found.

When used as a classifier for two classes, the output is compared with a threshold θ. Input patterns giving the output y less than the threshold θ belong to one class, the others to the second class.

2.2   GMDH learning algorithm

The task of the learning process is to find structure of the GMDH net and coefficients $A, B, \ldots, F$ for each neuron.

In the beginning of learning process, the neural net has no neurons, only a row of inputs (we do not count them as a layer). The first layer is build from $\frac{n(n-1)}{2}$ neurons. Each neuron is connected to two inputs only and has one output as shown in Fig. 2. In learning phase the neuron's learning set is

$$
\begin{array}{ccc}
x_{i_1}, & x_{j_1}, & y_1 \\
x_{i_2}, & x_{j_2}, & y_2 \\
\vdots & & \vdots \\
x_{i_N}, & x_{j_N}, & y_N
\end{array}
$$

Each row corresponds to one example, one event. To compute coefficients $A, B, \ldots, F$ of particular neuron one can construct a set of N equations for sample No. $1, \ldots, N$:

$$A + BX_{i,1} + CX_{j,1} + DX_{i,1}^2 + EX_{j,1}^2 + FX_{i,1}X_{j,1} = y_1$$
$$A + BX_{i,2} \ldots \qquad\qquad\qquad\qquad = y_2$$
$$\vdots \qquad\qquad\qquad\qquad\qquad\qquad\qquad \vdots$$
$$A + BX_{i,N} \ldots \qquad\qquad\qquad\qquad = y_N$$

This system of linear equations with respect to $\beta = (A, B, C, D, E, F)$ has a matrix and right hand side vectors which can be written in more compact form as

$$[X, Y]$$

and vector $\beta = (A, B, C, \ldots, F)$ of coefficients can be computed from linear regression equation

$$\beta = (X^t X)^{-1} X^t Y \qquad (2)$$

under the assumption that $X^t X$ is not singular. In practice, the matrix inversion algorithm is used and if $(X^t X)$ appears to be singular (or ill conditioned) then the particular neuron is not build at all.

After adapting all neurons in a new layer, the neural net has form according to Fig. 2.

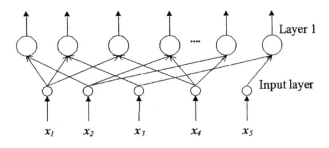

Fig. 2: GMDH net after building the first layer.

Now the GMDH net has $n_1 = \frac{n(n-1)}{2}$ neurons (less number of cases of singular $X^t X$ matrix). Using testing set one computes the average output error with respect to desired output value y for each neuron. Usually L_2 norm is used.

After it, the output errors of individual outputs are considered. If the least error is less then error criterion the learning is stopped and the corresponding output is taken as final output of neural net. If it is not the case, a new layer is build similar way. Here a problem arises that number of neurons in a layer grows with number of layers. To reduce this grow, all n_1 neurons in a just generated layer are sorted according to error size. Say, n_1' of them with least error are used for building the next layer, i.e. as inputs for a new layer and other $n_1 - n_1'$ neurons are left out. This is illustrated in Fig. 3.

The number of neurons n_1' left in a new layer before building a next one is given by size of that next layer which can be limited either by computer time or by memory size. If it is possible to store no more than N_N new neurons, then n_1' is the largest number for which

$$\frac{n_1'(n_1' - 1)}{2} \leq N_N$$

It was found that if properly programmed, the computer time needed for computing N_N equations 2 is more restrictive than memory size.

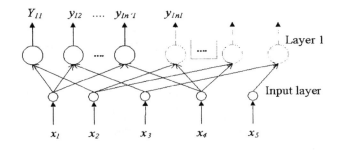

Fig. 3: The first layer after removing some neurons.

The outputs $y_{1,1}, \ldots y_{1,n_1'}$ are used as inputs for the next layer which is build by similar way as the first one. In the end of building the second layer, the neural net looks like shown in Fig 4.

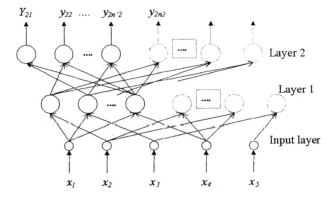

Fig. 4: The GMDH net after learning the second layer.

The layers are build one after another until error criterion is reached. There are two basic possibilities. For the simpler one it is that for the "best output neuron" the error using testing set is less then prescribed value. The more often case is finding a minimal error without respect to its absolute size. In Fig. 5 is shown that initially the error is less for each new layer. But for larger number of layers the error reaches its minimum and then grows. Simply the best case can be chosen (see [8]).

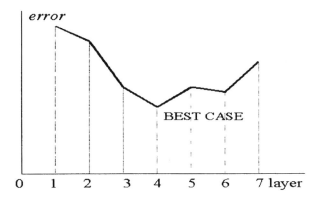

Fig. 5: GMDH net output error as function of number of layers.

3 The neural net with switching units

An original paradigm of a neural net for electron/jet classification was proposed by Bitzan and Šmejkalová [3] and [2]. The neural net has a layered architecture (see figure 6) and all its layers have the same structure (with the exception of the input layer that is simplified). Every layer consists of a set of switches and modulators and one summator and comparator. The number of switches and modulators in every layer is equal to the number of input data, i.e. to the number of processed signals or their features. The modulators N perform a multiplication of input signals by weight coefficients. Switches S are used for choosing the proper set of values of weight coefficients according to the results from the previous layer. They ensure a step nonlinearity of the network by switching among several sets of weight coefficients. The net has been derived from rather common perceptron type neural net.

The operation of the neural net is relatively simple. The input signals are processed in the modulators that perform their multiplication by weights. All modulated signals in one layer are summed up and compared with the threshold of that layer. According to the result of the comparison, the appropriate set of weights for the next layer of modulators is selected. After k steps (i.e. signal processing in k layers), the last sum compared to the threshold gives the final result that can identify the presence of either signal or background.

In spite of that this neural net has been derived from layered perceptron-type neural net, it uses a different approach for its learning. The algorithm

developed [3] is based on learning one layer after another one by sophisticated optimization algorithm implemented as UFO (Universal Function Optimization) library function (see [5]). There is no feedback iterative procedure used.

3.1 Switching neural net architecture:

Switching neural net consists from three types of units which form neural net consisting from layers. Each layer contain one switching unit and at least two (except the input layer, which is formed by one neural unit only) neural units per one input. The third type of unit perform output mapping from last layer only. The formal description of this three types of units is the following:

Neural unit

$$N : R^d \times W \Rightarrow R^d, \quad \mathbf{N} = \mathbf{N}(\mathbf{x}, \mathbf{w}), \quad (3)$$

$$\mathbf{x} \in R^d, \quad \mathbf{w} \in W.$$

where $\mathbf{x}$ is an input vector and $\mathbf{w}$ is a parameter vector from a parameter space W.

Switching unit

$$S : R^d \times U \Rightarrow \{1, 2, \ldots, m\}, \ S = S(\mathbf{x}, \mathbf{u}), \ (4)$$

$$\mathbf{u} \in U.$$

where $\mathbf{u}$ is a parameter vector from parameter space U, and $m > 1$ is an integer equal to the number of units in the next layer.

Output unit is represented by an output function O, which projects R^d into lower or equal dimension

$$O : R^d \Rightarrow R^b, \quad O = O(\mathbf{x}), \quad b \le d. \quad (5)$$

Architecture of switching neural net and connections between units is described in the 6.

3.2 Description of the net response:

Switching net maps input $\mathbf{x}_{in} \in R^d$ into output

$$\mathbf{x}_{out} = M(\mathbf{x}_{in}) \in R^b,$$

where the mapping M is a composition of mappings performed by each layer

$$M = S \odot M_{nl} \odot M_{nl-1} \odot \cdots \odot M_2 \odot M_1,$$

n_l is number of layers, $M_1 = N_{11}$, and

$$\mathbf{x}_{in,k} = M_k \odot \cdots \odot M_1(\mathbf{x}_{in})$$

$$M_k = N_{kS_k(\mathbf{x}_{in,k-1})}, \quad \text{for} \quad k > 1.$$

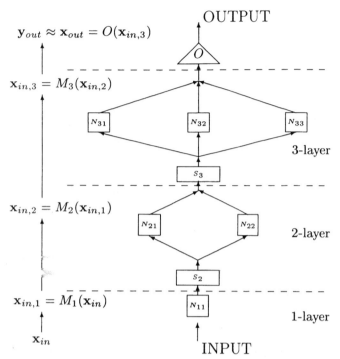

OUTPUT

$\mathbf{y}_{out} \approx \mathbf{x}_{out} = O(\mathbf{x}_{in,3})$

$\mathbf{x}_{in,3} = M_3(\mathbf{x}_{in,2})$

$\boxed{N_{31}}$ $\boxed{N_{32}}$ $\boxed{N_{33}}$ 3-layer

$\boxed{S_3}$

$\mathbf{x}_{in,2} = M_2(\mathbf{x}_{in,1})$

$\boxed{N_{21}}$ $\boxed{N_{22}}$ 2-layer

$\boxed{S_2}$

$\mathbf{x}_{in,1} = M_1(\mathbf{x}_{in})$ $\boxed{N_{11}}$ 1-layer

$\mathbf{x}_{in}$

INPUT

Fig. 6: Scheme of a 3-layer network with switching units. An input sample $\mathbf{x} \in R^d$ is mapped into $\mathbf{x}_{out} = O(x_{in,3})$ approximating $\mathbf{y}_{out} \in R^b$ (each node N_{ij} in this scheme performs multiplication of input signals (of dimension d) by diagonal $d \times d$ matrix of weights).

3.2.1 Learning process of the net:

Most of neural nets work in two phases. First is the learning procedure of the net, during which the parameters of the nets are adjusted (such as weights of connections, thresholds for units, parameters of threshold functions, etc.). The second phase use this parameters to perform a mapping, which could approximate mapping, used in the learning phase. Let us now to turn attention to the learning process. The learning process is performed on each layer, independently on parameters in other layers. The results on the last learned layer are grouped in n_l clusters (here n_l is the number of neurons in the layer). This clustering based on Jancey's cluster algorithm, which is briefly explained in the following way:

let $O(\mathbf{x}) = \mathbf{b} \cdot \mathbf{x}$, $p_i = O(M_k(\mathbf{x}_{in}^i))$, $n_p =$ number of patterns and $q = 1, \cdots, m$.

1. for randomly chosen sequence
   $1 \leq j_1 < j_2 < \cdots < j_m \leq n_p$
   set
   $\mathbf{c}_q^{new} = \mathbf{c}_q^{old} = \mathbf{p}_{j_q}$
   and
   $S_q^{new} = S_q^{old} = \{\mathbf{p}_{j_q}\},$

2. let $r_1, \cdots, r_{n_p}$ is random permutation of the $1, \cdots, n_p$,

3. for all $k = r_1, \cdots, r_{n_p}$
   DO
   $i = \min \left\{ v \middle| \|\mathbf{c}_v^{old} - \mathbf{p}_k\| = \min\{\|\mathbf{c}_q^{old} - \mathbf{p}_k\|\} \right\}$,
   $\mathbf{c}_q^{old} = \mathbf{c}_q^{old} - \frac{\mathbf{p}_k - \mathbf{c}_q^{old}}{|S_q^{old}|}$, $c_i^{old} = \mathbf{c}_i^{old} + \frac{\mathbf{p}_k - \mathbf{c}_i^{old}}{|S_i^{old}|}$
   $S_q^{old} = S_q^{old} \setminus \{\mathbf{p}_k\}$, $S_i^{old} = S_i^{old} \cup \{\mathbf{p}_k\}$,
   END

4. IF $(\exists q)(S_q^{new} \neq S_q^{old})$
   THEN for all such q let
   $\left\{ \mathbf{c}_q^{new} = \mathbf{c}_q^{old}, S_q^{new} = S_q^{old} \right\}$
   and GOTO 2

5. STOP

After clustering each cluster is jointed with a neuron in the layer and consequently parameters of this neuron are adjusted with regard patterns in the corresponding cluster only. Parameters are adjusted in such a way that the mean square error (cost function) is minimized over the set of neuron parameters (vector of connections weights and vector of threshold values). This is done by UFO library (see [5]).

4   Measures of classification quality

Neural nets considered are used as classifiers but the output is, in fact, continuous. Then a threshold is used so that if the output of the net to some input data sample is less than the threshold the data sample corresponds to one class and if it exceeds the threshold, the data sample corresponds to the other class. Of course, the errors arise. The two classes we denote as "signal" (corresponds to the presence of the particle looked for) and "background". For given threshold θ some signals are well recognized as signals and some are not. The percentage of well recognized signals of all signal samples is signal efficiency η_{sig}. The percentage of background events badly recognized as a signal of all background events we denote as background error ε_{backgr}. The ratio of these two values gives the enrichment factor $E = \frac{\eta_{sig}}{\varepsilon_{backgr}}$ which says that, if there is only 0.5% of signal events among background data, then in the output data stream will be E times more signal data, i.e. $E * 0.5\%$. The rejection factor $R = \frac{1}{\varepsilon_{backgr}}$ gives, in fact, the extend of how the data stream frequency is reduced under the assumption that there are much less signal data samples than background data. To keep statistical error not worse than in original data, there must be the quality factor $Q = E\eta$ at least 1. It holds

$$\frac{S_{out}}{\sqrt{B_{out}}} = \sqrt{Q}\,\frac{S_{in}}{\sqrt{B_{in}}},$$

where indexes *in* and *out* mean data before and after selection process. Both enrichment factor and rejection factor should be as large as possible under the condition of $\eta_{sig} > 0.1$ and quality factor Q should be at least equal to 1.

5 Results

5.1 GMDH neural net classifier

In Figure 7 individual outputs for 426 data samples are shown. Of these samples 226 of them belong to the background class and the last 200 to the signal class.

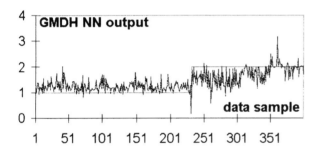

Fig. 7: The output of GMDH neural net classifier.

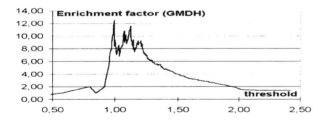

Fig. 8: The dependence of the enrichment factor on the threshold value for GMDH neural net classifier.

In Figure 8 enrichment factor as a function of the threshold is shown.

5.2 Switching units neural net classifier

The data was the same for the GMDH net. The difference lies in the different values of output - the values 100 and 200 were used as desired output values for classes signal and background respectively.

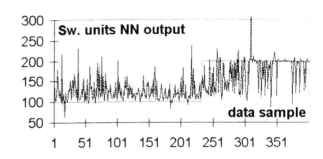

Fig. 9: The output of the neural net with switching units.

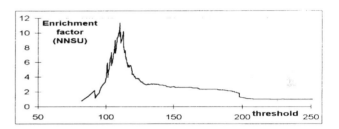

Fig. 10: The dependence of the enrichment factor on the threshold value for the neural net with switching units.

5.3 Evaluation of results

For neural net with switching units we found for nearly maximum of the enrichment factor for threshold 110 the enrichment factor $E = 11$ (see Fig. 10), rejection factor $R = 34$ and quality factor $Q = 3.78$. From it follows that total number of events for processing after trigger will be 34 times less than in original stream of events. Moreover in the resulting stream of events the statistical error will be $\sqrt{Q} = 1.94$ times less and percentage of "useful" events $E = 11$ larger than in the original stream of events. For GMDH neural net the results are very similar. These results can be compared with results by Richter-Wąs and Froidevaux [9]. When computing factors R, E, and Q one can find that for the case discussed in [9], there is $R = 4.5$ to 14.5, E is less than 2.6, and Q is less than 0.6.

6 Conclusions

With the data given today there exists a very difficult way of stating the separation criteria between two classes of events. It relies on the information from other parts of the ATLAS detector and thus uses more information than only jets parameters.

We have shown that data about characteristic values of three jets are sufficient to reach separation quality as needed. Only examples of data of both classes are sufficient for neural net learning and we do not need to look for any exact procedure for the events separation.

References

[1] Atlas technical proposal for a general-purpose pp experiment at the large hadron collider at CERN, Dec 1994. CERN/LHCC/9445 LHCC/P2.

[2] Bitzan, P., Šmejkalová, J., Kučera, M.: Neural networks with switching units. Neural Network World, 4:515–526, 1995.

[3] Bitzan, P., Šmejkalová, J., Kučera, M., Pařízek M., Matyáš, M.: Theory and technical implementation of a neural network with switching units. Technical report, CERN Geneva, Feb 1995. EAST Note 94-05.

[4] Hakl, F.: Basic theory of neural networks derived from the B-S-B model. Neural Network World, 3:319–351, 1993.

[5] Lukšan, L., et al.: Interactive system for universal functional optimization (ufo). Technical Report 599, Institute of Computer Science, Academy of Sciences of the Czech Republic, Jan 1994.

[6] Farlow, J.S.: Self–organizing methods in modeling. Marcel Dekker, Inc., New York, 1984.

[7] Frolov, S., Řízek, S.: Model of Neurocontrol of Redundant Systems. J. of Computational and Applied Mathematics, 63:465–473, 1995.

[8] Hecht-Nielsen, R.: Neurocomputing. Addison Wesley Publ. Co., 1990

[9] Richter–Wąs, E., Froidevaux, D.: MSSM Higgs searches in multi–b–jet final states. AT-LAS Internal Note, PHYS–No–104, CERN, Geneve, Jul 1997.

A Neural Network Based Nonlinear Temporal-Spatial Noise Rejection System

Fa-Long LUO, Rolf UNBEHAUEN and Tertulien NDJOUNTCHE*
Lehrstuhl für Allgemeine und Theoretische Elektrotechnik
Universität Erlangen-Nürnberg, Cauerstraße 7, 91058 Erlangen, GERMANY
(*) E-mail: tertu@late.e-technik.uni-erlangen.de

Abstract

This paper proposes a nonlinear temporal-spatial noise rejection system on the basis of mapping neural networks. With the universe nonlinear mapping capability of these neural networks and related learning algorithms, the proposed system can offer better noise rejection performance than traditional methods in the case that the related unknown system is nonlinear or non-minimum phase and in the case that the length of the learning system does not fit the length of the unknown system. It can then serve as an alternative tool for many applications of noise rejection and this was confirmed by the simulations results.

1 Introduction

Adaptive noise rejection is an essential problem in many applications such as wireless communications, hearing aids, radar and sonar systems. There have been many techniques to attack this problem. Among these, the adaptive noise cancellation system by use of two channels proposed in [1] is the most common one. This system is shown in Fig. 1. Sensor 1 picks up the desired signal $s(n)$ and the additive noise $x_1(n)$. Sensor 2 picks up the noise $x_2(n)$. Moreover, $x_1(n)$ can be regarded as an output of an unknown system with $x_2(n)$ as the input. The task of the adaptive system is to learn this unknown system so that its output $y(n)$ is as close to $x_1(n)$ as possible. In the ideal case, the adaptive system learns to be exactly equal to the unknown system so that $y(n)=x_1(n)$ and the output $z(n) = s(n) + x_1(n) - y(n) = s(n)$, this is a noise free signal. However, in most of the available methods to attack this problem, the unknown system is modelled as a linear casual system and the adaptive system is a finite impulse response (FIR) filter. Obviously, the noise rejection performance or say the

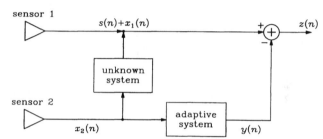

Fig. 1. The adaptive noise cancelling scheme proposed in [1].

approximation performance of the adaptive system to the unknown system will greatly be degraded in the following cases: 1. the unknown system is a nonlinear system, 2. the characteristic of the unknown system varies faster than that of the learning system and 3. the time delay of $x_1(n)$ from the original source is less than that of $x_2(n)$, that is, the unknown system is a non-casual system. As a matter of fact, even in the simplest case, that is, the unknown system is an FIR filter with the length H, the performance will also be degraded if the length of the adaptive FIR filter is less than H. It can be easily known that the performance will be greatly improved if a nonlinear adaptive system could replace the linear FIR filter. However, the complexity in computation and implementation involved in a nonlinear adaptive system protects this idea from being practical. Recently, neural networks have provided a very powerful tool to attack nonlinear problems and have successfully been used for nonlinear system identification, time series prediction and many other fields of signal processing [2,3,4]. With this motivation, it will not be surprising that we will employ multilayer perceptron neural networks or other mapping neural networks such as radial basis function neural networks, high-order neural networks as a learning system so as to better approximate the unknown nonlinear system.

On the other side, the above processing is only in temporal domain. In many situations, there are many noise sources. Under this condition, beamforming techniques on the basis of a sensor array can be used. Unlike the above processing, beamforming techniques in effect exploit the spatial separation of related noises and the signal. Obviously, the improved noise rejection performance can be delivered if both the temporal technique and spatial techniques could be combined.

Based on these considerations, in the next section, we propose a neural network based nonlinear temporal-spatial noise rejection system and will present the principles and corresponding algorithms. Simulations results are presented in section 3. Some remarks about the proposed system and other further discussions will be dealt with in Section 4.

2 Structures, Principles and Algorithms of the Proposed System

The schematic of the proposed nonlinear temporal-spatial noise rejection system is shown in Fig. 2. Sensor 1 picks up the desired signal $s(n)$ and the additive noise $x_1(n)$. Sensor $2, 3, \cdots, N$ receives the noise $x_2(n), x_3(n), \cdots x_N(n)$, respectively. Furthermore, the noise $x_1(n)$ can be regarded as

$$
\begin{aligned}
x_1(n) = f(&x_2(n), x_2(n-1), \cdots, x_2(n-L_2), \\
&x_3(n), x_3(n-1), \cdots, x_3(n-L_3), \cdots, \quad (1) \\
&x_N(n), x_N(n-1), \cdots, x_N(n-L_N))
\end{aligned}
$$

where $f(\cdot)$ is an unknown nonlinear function, L_i $(i = 2, 3, \cdots, N)$ are lags corresponding to the noise $x_2(n), x_3(n), \cdots, x_N(n)$, respectively, Note that L_i is unknown.

A three-layer MLP (more layers are direct generalization) network are employed in this system. The input layer has $N_I = \sum_2^N (M_i + 1) + M$ neurons whose input vector is

$$
\begin{aligned}
\boldsymbol{U}(n) &= [u_1(n), u_2(n), \cdots, u_{N_I}(n)] \\
&= [x_2(n), x_2(n-1), \cdots, x_2(n-M_2), \\
&\quad x_3(n), x_3(n-1), \cdots, x_3(n-M_3), \cdots, \\
&\quad x_N(n), x_N(n-1), \cdots, x_N(n-M_N), \\
&\quad y(n-1), \cdots, y(n-M)]
\end{aligned}
\quad (2)
$$

that is, the input vector of the network consists of the current and past values of Sensor $2, 3, \ldots, N$ and the neural network output. The input layer

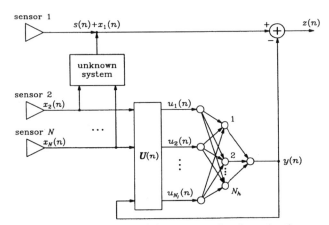

Fig. 2. Schematic of the proposed noise rejection system.

simply feeds the vector $\boldsymbol{U}(n)$ to the hidden layer without any modification. The hidden layer has N_h neurons with nonlinear transfer functions (such as the sigmoid function). The output layer has only one neuron whose output is denoted as $y(n)$.

Let $W_{ij}^{(1)}$ denote the connection weight between the i'th neuron in the input layer and the j'th neuron in the hidden layer (for $i = 1, 2, \ldots, N_I$; $j = 1, 2, \ldots, N_h$); let $q_j(n)$ and $f_j(\cdot)$ (for $j = 1, 2, \ldots, N_h$) be the output and activation function of the j'th neuron in the hidden layer, respectively; let $W_j^{(2)}$ denote the connection weight between the j'th neuron in the hidden layer and the neuron in the output layer; let $g(\cdot)$ be the activation function of the output neuron . Then we have

$$
\begin{aligned}
y(n) &= g(\sum_{j=1}^{N_h} W_j^{(2)} q_j(n)) \\
&= g(\sum_{j=1}^{N_h} W_j^{(2)} f_j(\sum_{i=1}^{N_I} W_{ij}^{(1)} u_i(n))) \quad (3)
\end{aligned}
$$

Note that in (3) the threshold of each neuron has been selected to be zero and the activation function $g(\cdot)$ of the output neuron is usually no longer the sigmoid function so that it can provide values greater than unity. For simplicity, we select $g(\cdot)$ to be a linear function, that is, the output neuron performs only simple summations. With this in mind, (3) becomes

$$
y(n) = \sum_{j=1}^{N_h} W_j^{(2)} f_j(\sum_{i=1}^{N_I} W_{ij}^{(1)} u_i(n)) = \hat{f}[\boldsymbol{U}(n)] \quad (4)
$$

where the nonlinear mapping $\hat{f}$ is completely determined by the connection weights $W_{ij}^{(1)}$ and $W_j^{(2)}$.

As a result, our task is how to determine these connection weights so as to make the power of the noise part $y(n) - x_1(n)$ minimized without decreasing the power of the signal part $s(n)$, that is,

$$E[s^2(n)] + \min_{\boldsymbol{W}} E[(y(n) - x_1(n))^2] \qquad (5)$$

where $\boldsymbol{W}$ denotes the set of all connection weights and $E[\cdot]$ is the expectation operation.

Because $y(n)$ and $x_1(n)$ are both uncorrelated with $s(n)$, (5) is equivalent to

$$\min_{\boldsymbol{W}} E[(y(n) - x_1(n) - s(n))^2] = \min_{\boldsymbol{W}} E[z^2(n)] \quad (6)$$

Equations (5) and (6) show that the reduction of the noise power can be achieved by minimizing the power of the output $z(n)$ and can maximize the output signal-to-noise ratio.

From (5)-(6), it can be seen that the smallest possible output power is $E[z^2(n)] = E[s^2(n)]$. When this is achievable, $E[(y(n) - x_1(n))^2] = 0$. Therefore, $y(n) = x_1(n)$ and $z(n) = s(n)$, that is, the learning system is exactly equal to the unknown system and the output $z(n)$ is noise free.

To get $\boldsymbol{W}$ from (6), we need many statistics of all the noise sources and the desired signal. However, in practical applications, these statistics are unknown, but I samples of the input vector $\boldsymbol{U}(n)$ and samples of the input $s(n) + x_1(n)$ are available. These available samples can be used as the training data to find the parameter set $\boldsymbol{W}$. With these training data and with replacing the statistic averaging operation in (6) by time averaging operation, we have

$$\min_{\boldsymbol{W}} \sum_{n=1}^{I} (y(n) - s(n) - x_1(n))^2 \qquad (7)$$

It is easy to see that any available algorithm for updating the weights of MLP networks can immediately be used to find the weight set $\boldsymbol{W}$ of (7). As an example, a version of the back-propagation (BP) algorithm for solving (7) is

$$W_j^{(2)}(t+1) = W_j^{(2)}(t) + \gamma_1 \sum_{n=1}^{I} \delta(n) S_j^{(2)}(n) \quad (8)$$

$$(j = 1, 2, \dots, N_h)$$

and

$$W_{ij}^{(1)}(t+1) = W_{ij}^{(1)}(t) + \gamma_2 \sum_{n=1}^{I} \delta(n) S_j^{(1)}(n) \quad (9)$$

$$(i = 1, 2, \dots, N_I; j = 1, 2, \dots, N_h)$$

where $u_i(n)$ is the i'th element of the input vector $\boldsymbol{U}(n)$ and the other various variables are

$$\delta(n) = s(n) + x_1(n) - y(n) \qquad (10)$$

$$S_j^{(2)}(n) \hat{=} \frac{\partial y(n)}{\partial w_j^{(2)}} \qquad (11)$$

$$S_j^{(1)}(n) \hat{=} \frac{\partial y(n)}{\partial w_{ij}^{(1)}} \qquad (12)$$

$$(n = 1, 2, \dots, I)$$

The quantities γ_1 and γ_2 are related to the learning-rate parameters. The details of the procedures can be summarised as follows.

(1) Initialize randomly all connection weights $W_{ij}^{(1)}$ and $W_j^{(2)}$ (for $i = 1, 2, \dots, N_I$; $j = 1, 2, \dots, N_h$).

(2) Pass I observation vectors $\boldsymbol{U}(n)$ successively to the input layer of the MLP network.

(3) Compute by proceeding forward the output $q_j(n)$ (for $j = 1, 2, \dots, N_h$) of the hidden layer and the output $y(n)$ using (3)-(10) and the available input vectors $\boldsymbol{U}(n)$.

(4) Compute by proceeding backward the error propagation and the sensitivity terms $\delta(n)$, $S_j^{(2)}(n)$ and $S_j^{(2)}(n)$, respectively, by using (10)-(12). Note that in (8)-(12), $W_{ij}^{(1)}(t)$ and $W_j^{(2)}(t)$ denote the related connection weights at the t'th iteration.

(5) Adjust the connection weights according to (8)-(9).

(6) Compute the total error e

$$e = \sum_{n=1}^{I} |s(n) + x_1(n) - y(n)|^2 \qquad (13)$$

and iterate the computation by returning to Step (2) until this error is less than a specified one.

This is in effect the batch-processing version of the BP algorithm which applies to the case that all the training data are available when the training of the connection weights is initiated. For the most of practical cases in which each new sample set becomes available as time progresses and the available samples are time-increasing, we have the following procedures:

(1) Initialize randomly all connection weights $W_{ij}^{(1)}(n)$ and $W_j^{(2)}(n)$ (for $i = 1, 2, \dots, N_I$; $j = 1, 2, \dots, N_h$).

(2) Pass one observation vector $U(n)$ to the input layer of the MLP network.

(3) Compute by proceeding forward the output $q_j(n)$ (for $j = 1, 2, \ldots, n_h$) of the hidden layer by $q_j(n) = f_j(\sum_{i=1}^{N_I} W_{ij}^{(1)}(n)u_i(n))$ and the output $y(n)$ by $y(n) = \sum_{j=1}^{N_h} W_j^{(2)}(n)q_j(n)$ and the available input vectors $U(n)$.

(4) Compute by proceeding backward the error propagation and sensitivity terms $\delta(n)$, $S_j^{(2)}(n)$ and $S_j^{(2)}(n)$, (for $j = 1, 2, \ldots, M$), respectively, by using (10)-(12).

(5) Adjust the connection weights according to

$$W_j^{(2)}(n+1) = W_j^{(2)}(n) + \gamma_1 \delta(n) S_j^{(2)}(n) \qquad (14)$$

$$(j = 1, 2, \ldots, N_h)$$

and

$$W_{ij}^{(1)}(n+1) = W_{ij}^{(1)}(n) + \gamma_2 \delta(n) S_j^{(1)}(n) \qquad (15)$$

$$(i = 1, 2, \ldots, N_I; j = 1, 2, \ldots, N_h)$$

where $W_{ij}^{(1)}(n)$ and $W_j^{(2)}(n)$ denote the connection weights at the n'th iteration, where γ_1 and γ_2 are the learning-rate parameters whose value may be different from those in (8) and (9). Note the difference between (8) and (14), (9) and (15), respectively.

(6) Increase the iteration number to $n + 1$ and return to Step (2) by taking the next observation vector as the input vector of the MLP network if the next observation sample is available. Otherwise, iterate the computation by returning to Step (2) and by still using the available observation vector until the next observation sample is available.

With these updated weights, the MLP network could provide the output $y(n)$ so that the power of the noise part $x_1(n) - y(n)$ of the entire output of the system $z(n) = x_1(n) - y(n)$ is minimized.

3 Test Results

The dynamic behaviour of the proposed neural network is illustrated in the important case of nonlinear adaptive system modelling. Figs. 3 and 4 show the plots of 500 samples of the input and output signals, respectively. By delaying the former signal by one and two periods, the two sequences that are obtained form one component of the network input. Delayed versions of the output signals, $y(n - 1)$ and $y(n - 2)$ constitute the other input component. The neural network is assumed to have five neurons in the hidden layer. Fig. 5 shows the observed and predicted output signals obtained by validating the trained network with a new set of data and the prediction error signal is shown in Fig. 6. The cross-correlation of the residual error and the histogram of the prediction error are shown in Figs. 7 and 8, respectively. It can be deduced from the latter plot that the nonlinear prediction error may be closely modelled as white and approximately Gaussian process, indicating that it consists essentially of statistically independent samples.

4 Further Discussions

Concerning this proposed temporal-spatial system without the feedback states, we will make further discussions and comments.

Comment 1

If there are only two channels, this system will become a nonlinear temporal system whose work principles are exactly the same as those of Fig. 1 but this system applies to the case that the unknown system is a nonlinear one. With this, the input vector of the MLP network becomes

$$U(n) = [x_2(n), x_2(n-1), \cdots, x_2(n-N_I-1)]^T \qquad (16)$$

which is the same as that used in the well-known least-mean-squares (LMS) algorithm. Moreover, if we choose only one hidden neuron, its activation function to be a linear one and corresponding weight between the hidden neuron and output neuron to be the unity, the output $y(n)$ of the network is as

$$y(n) = \sum_{i=1}^{N_I} W_{i1}^{(1)}(n)u_i(n) = U(n)^T W(n) \qquad (17)$$

where $W(n) = [W_{11}^{(1)}(n), W_{21}^{(1)}(n), \cdots, W_{N_I 1}^{(1)}(n)]^T$ is the weight vector between the input layer and the hidden neuron. Correspondingly, the algorithm for updating weights becomes

$$W_{i1}^{(1)}(n+1) = W_{i1}^{(1)}(n) + \gamma_2 \delta(n)u_i(n) \qquad (18)$$

Thus (18) is in effect exactly the LMS algorithm.

Comment 2

If we choose that the lags M_i $(i = 2, 3, \cdots, N)$ are all zero and if we choose only one hidden neuron, its activation function to be a linear one and the corresponding weight between the hidden neuron and

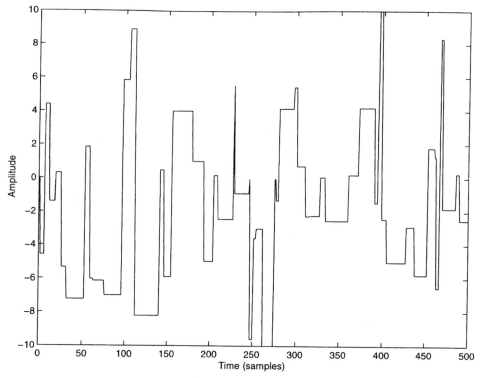

Fig. 3. Input signal.

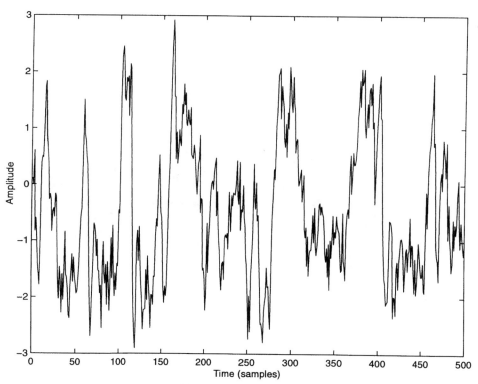

Fig. 4. Output signal.

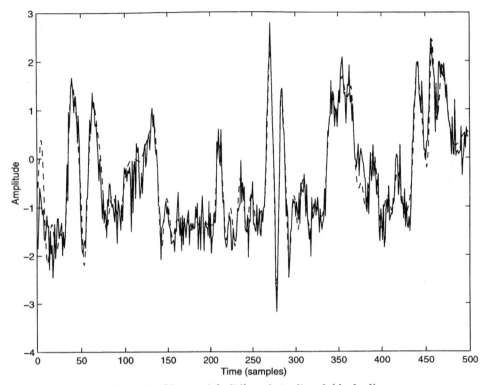

Fig. 5. Observed (solid) and predicted (dashed)
output signals.

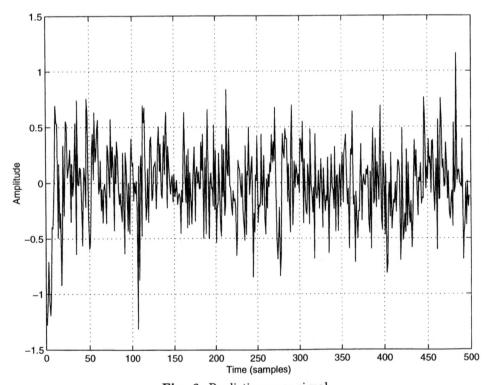

Fig. 6. Prediction error signal.

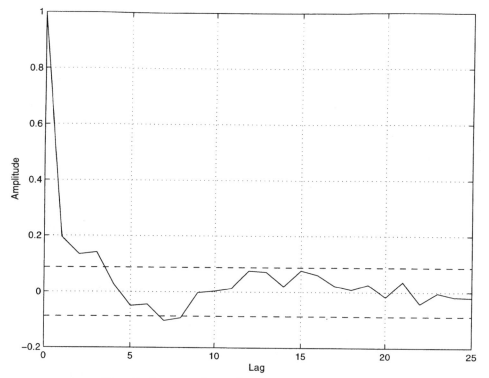

Fig. 7. Autocorrelation function of the error signal.

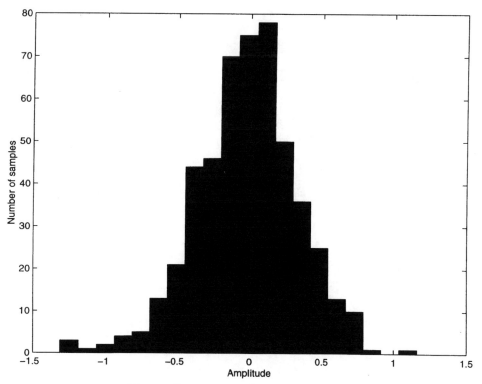

Fig. 8. Histogram of the prediction error.

output neuron to be the unity, then the input vector becomes

$$U(n) = [x_2(n), x_3(n), \cdots, x_N(n)] \qquad (19)$$

the output of the system becomes

$$
\begin{aligned}
y(n) &= \sum_{i=1}^{N_I} W_{i1}^{(1)}(n) u_i(n) = U(n)^T W(n) \\
&= \sum_{i=1}^{N-1} W_{i1}^{(1)}(n) x_{i+1}(n)
\end{aligned} \qquad (20)
$$

where $W(n) = [W_{11}^{(1)}(n), W_{21}^{(1)}(n), \cdots, W_{N_I 1}^{(1)}(n)]^T$ is the weight vector between the input layer and the hidden neuron. The algorithm for updating the weights is the same as (18) and the difference is that the input vector $U(n)$ in (19)-(20) does not include any past samples of all channels. With the above assumption, it is easy to see that this system becomes a spatial filter used in adaptive beamforming.

Comment 3

The algorithms presented above are only one of many techniques for solving (7) but these presented algorithms have the simplest computational complexity in comparison with others. For the purpose of speeding up the convergence of the system, other techniques such as Gauss-Newton based methods, the least-squares algorithm, the total-least squares algorithm, Kalman-filter based algorithm, etc [5]-[8] can be used to solve (7). The cost in using other methods is the increase in computational complexity. As a result, one has to make a trade off according to the specific application.

Comment 4

Other mapping neural networks such as radial basis function networks, higher-order neural networks can immediately replace the MLP network in the adaptive system. The input vector of these networks is still $U(n)$ of (2) and the learning algorithms to update weights and other parameters of these networks can also easily obtained by solving (7).

5 Acknowledgements

This paper is supported by the German Research Society (DFG).

References

[1] B. Widrow, etc., "Adaptive noise cancelling, principles and applications", Proceedings of the IEEE, Vol. 63, No. 12, pp. 1692-1715, 1975.

[2] J. N. Hwang, S. Y. Kung, M. Niranjan and J. C. Principe, "The past, present, and future of neural networks for signal processing", IEEE Signal Processing Magazine, Vol. 14, No. 6, pp. 28-48, 1997.

[3] R. Unbehauen and F. L. Luo (Guest Editors), Signal Processing, Special Issue on Neural Networks, Vol. 64, No. 3, 1998.

[4] A. G. Constantinides, S. Haykin, Y. H. Hu, J. N. Hwang, S. Katagiri, S. Y. Kung and T. A. Poggio, (Guest Editors), IEEE Transactions on Signal Processing, Vol. 45, No. 11, 1997.

[5] S. Haykin, Neural Networks, A Comprehensive Foundation, IEEE Press, 1994.

[6] S. Kollias and D. Anastassiou, "An adaptive least squares algorithm for the efficient training of artificial neural networks", IEEE Trans. on Circuits and Systems, Vol. 36, No. 8, pp. 1092-1101, 1989.

[7] R. J. Williams and D. Zipser, "A learning algorithm for continually running fully recurrent neural networks", Neural computation, Vol. 1, pp. 270-280, 1989.

[8] F. L. Luo and R. Unbehauen, Applied Neural Networks for Signal Processing, Cambridge University Press, New York, 1997.

Improving the Performance of the Hopfield Network By Using A Relaxation Rate

Xinchuan Zeng and Tony R. Martinez

Computer Science Department, Brigham Young University, Provo, Utah 84602

Email: zengx@axon.cs.byu.edu, martinez@cs.byu.edu

Abstract

In the *Hopfield network*, a solution of an optimization problem is obtained after the network is relaxed to an equilibrium state. This paper shows that the performance of the Hopfield network can be improved by using a *relaxation rate* to control the relaxation process. Analysis suggests that the relaxation process has an important impact on the quality of a solution. A *relaxation rate* is then introduced to control the relaxation process in order to achieve solutions with better quality. Two types of relaxation rate (*constant* and *dynamic*) are proposed and evaluated through simulations based on 200 randomly generated city distributions of the 10-city *traveling salesman problem*. The result shows that using a relaxation rate can decrease the error rate by 9.87% and increase the percentage of valid tours by 14.0% as compared to those without using a relaxation rate. Using a dynamic relaxation rate can further decrease the error rate by 4.2% and increase the percentage of valid tours by 0.4% as compared to those using a constant relaxation rate.

1   Introduction

Interest in solving combinatorial optimization problems by neural networks was motivated by the work of Hopfield and Tank [1]. They proposed an approach of finding a suboptimal solution of the traveling salesman problem (*TSP*) by encoding the information of the cost and constraints of a *TSP* city distribution into the links between the neurons in a neural network and relaxing the network until it reaches an equilibrium state. Their simulation on a 10-city *TSP* city distribution showed that the neural network was able to obtain a high percentage of valid tours with high quality (short tours). In their approach, a *TSP* city distribution is represented by an *energy function* including the cost and constraint terms that reflect the objective of a solution. The objective of the constraint term is to find a valid tour, which requires that each city must be visited once and only once. The objective of the cost term is to find the shortest valid tour. The energy function can be implemented by a neural network. For an *N*-city *TSP* problem, the network consists of $N \times N$ neurons and the links that connect these neurons. The weights are set to encode the information about the constraints and the cost function of a particular *TSP* city distribution. Each neuron updates its input value based on the information received from all other neurons. They showed that a neural network so configured will always converge to a local minimum which may represent a valid solution. This network is commonly referred to as a *Hopfield network*.

Since Hopfield and Tank's work [1], there has been growing interest in the Hopfield network because of its advantages over other approaches for solving optimization problems. The advantages include massive parallelism, convenient hardware implementation of the neural network architecture, and a common approach of solving various optimization problems. Much research has been focused on analyzing and improving the original model in order to obtain a higher percentage of valid solutions and solutions with better quality. The work by Wilson and Pawley [2] showed that there was some difficulty getting the Hopfield model to yield valid tours. For randomly generated sets of the 10-city *TSP*, Wilson and Pawley reported that only 8% of their trials resulted in valid tours. After their report, Brandt et al. [3] and Aiyer et al. [4] showed that better performance can be achieved by modifying the energy function. Li combined the Hopfield network with the "augmented Lagrange multipliers" algorithm from optimization theory [5]. Catania et al. applied a fuzzy approach to tune the parameters in the Hopfield network [6]. Liang added adjusting neurons to the Hopfield network for solving the quadratic assignment problem [7].

The performance of the Hopfield network has been improved over the past decade. However, this model still has some basic problems [8, 9]. One of the problems is that the performance of the Hopfield network is inconsistent. The performance is good for some city distributions of *TSP*, but poor for other city distributions with the same size. The performance is usually better for city distributions with simple topology, but poor for those with complex topology, in which case there are large number of similar competent fixed points and the solutions are often trapped in poor local minima. Another problem is that the performance is sensitive to the choice of the parameters in the energy function, and different parameter settings can lead to significant differences in the performance.

In the Hopfield model, a solution is achieved after the network is relaxed and has settled down to an equilibrium point. The process of relaxation is an important step in achieving a solution and have an important impact on the quality of the solution. However, there has not been enough effort to study this process and the underlying mechanism during the process is not well understood.

This paper addresses the issues of the effects of the relaxation process on the quality of the solutions obtained by the network. The analysis shows that the relaxation process is not trivial, but important for the network being able to find a good solution. A relaxation rate is introduced to control the pace of relaxation. The performance of this approach has been tested through simulations based on 200 randomly generated city distributions of the 10-city *TSP*. The simulation reveals that the quality of a tour and the percentage of valid tours depends on the relaxation rate. The error rate can be reduced by 9.87% and the percentage of valid tours can be increased by 14.0% compared to those without using a relaxation rate. A dynamic relaxation rate, with varying value during relaxation process, is also experimented. The error rate drops by a further 4.2% and the percentage of valid tours increases by 0.4% compared to those using a constant relaxation rate. A dynamic amplification parameter has also been experimented and its performance has been compared to those using relaxation rates.

The rest of the paper is organized as follows. A brief outline of the basic structure of the Hopfield network is given in Sec. 2. The motivations and the procedure of including a relaxation rate in the network are described in Sec. 3. The simulation results are presented in Sec. 4. Finally, a summary of this work is given in Sec. 5.

2 Basics of the Hopfield network

The Hopfield network [1] consists of a set of neurons and the links connecting the neurons. For an N-city *TSP*, there are $N \times N$ fully connected neurons in the network. The row index for a neuron

represents the city. The column index represents the order of the city in the tour. The weights of the connecting links are determined according to the constraints and the cost function. *TSP* includes the following constraints: The salesman must visit each city exactly once in a tour (each row has exactly one "on" neuron) and must visit exactly one city at any time in a tour (each column has exactly one "on" neuron). The cost function is constructed to reflect the objective that the salesman chooses a valid tour with minimum tour length. Therefore, the Hopfield network is configured to have a better chance to find short tours.

The constraints and the cost function for *TSP* can be represented by an energy function. The energy function is then used to determine the values of all weights in the network. Hopfield's original energy function for an N-city *TSP* is given by [1]:

$$E = \frac{A}{2} \sum_{X=1}^{N} \sum_{i=1}^{N} \sum_{j=1, j\neq i}^{N} V_{Xi}V_{Xj}$$

$$+\frac{B}{2} \sum_{i=1}^{N} \sum_{X=1}^{N} \sum_{Y=1, Y\neq X}^{N} V_{Xi}V_{Yi} + \frac{C}{2}(\sum_{X=1}^{N} \sum_{i=1}^{N} V_{Xi} - N_0)^2$$

$$+\frac{D}{2} \sum_{X=1}^{N} \sum_{i=1}^{N} \sum_{Y=1, Y\neq X}^{N} d_{XY}V_{Xi}(V_{Y,i+1} + V_{Y,i-1}) \quad (1)$$

where X and Y are row indices; i and j are column indices; V_{Xi} is the activation for each neuron; and d_{XY} is a measure of the distance between cities X and Y. The first two terms enforce the constraint that no city can be visited more than once. The third term reflects the constraint that each city should be visited. The last term represents the distance cost function. The value of each parameter (A, B, C, and D) is the measure of the importance of the corresponding term.

Each neuron has a current state input U_{Xi} and an activation (output) V_{Xi}. The initial values of U_{Xi} for all neurons are first set to be a small constant value which is determined by the condition: $\sum_{X=1}^{N} \sum_{i=1}^{N} V_{Xi} = N$. To break the symmetry of the network, a small fraction of random noise is added to the initial values of each U_{Xi}. The network is then relaxed until an equilibrium is reached.

After each iteration, U_{Xi} is updated based on the following equation:

$$\frac{dU_{Xi}}{dt} = -\frac{U_{Xi}}{\tau} - \frac{\partial E}{\partial V_{Xi}} \quad (2)$$

where $\tau (= 1.0)$ is the time constant.

V_{Xi} is also updated after each iteration and its value is determined by U_{Xi} through an *activation (output) function.* In the Hopfield network, the activation function is the *sigmoid* function given by:

$$V_{Xi} = \frac{1}{2}(1 + tanh(\frac{U_{Xi}}{u_0})) \qquad (3)$$

where u_0 is the amplification parameter that reflects the steepness of the activation function.

Hopfield and Tank [1] showed that the network is guaranteed to converge to a local minimum in the case of symmetric connecting weights. They tested this model for a 10-city *TSP*. For a total of 20 runs, they claimed 16 valid tours, and half of the valid tours were optimal.

3 Relaxation rate

The quality of a solution obtained by the Hopfield network can be affected by certain factors, for example, the parameter setting in the energy function and the random noise in the initial input values of neurons [2].

This work focuses on studying the effects of the relaxation process on the performance of the network. Intuitively, the development of the network should be different when the network is relaxed with a different speed. One of the important factors that could influence the quality of a solution is the difference in the frequency that a neuron receives information from other neurons. This information reflects the cost and constraints encoded in the network, and hence the amount of the information propagated through the network has an important impact on the quality of a solution.

If the network is relaxed too fast, there will be fewer opportunities for exchange of the information between neurons, and therefore a solution formed under this condition is less likely to have a high quality. Another potential problem with a fast relaxation is a possible instability or oscillation of the network.

Relaxing the network at too slow a rate also has a disadvantage: an inefficient use of the network due to an unnecessarily large number of iterations required to form a solution.

In order to avoid the potential problems described above and to have better control of the relaxation process, a *relaxation rate* is introduced in the network dynamics. The function of the relaxation rate is to adjust the speed of the relaxation so that solutions with better quality can be obtained. Specifically, the input of the neuron (X, i) is updated according to the following dynamic equation:

$$\frac{dU_{Xi}^{(new)}}{dt} = R\frac{dU_{Xi}}{dt} \qquad (4)$$

where R is the *relaxation rate* and $\frac{dU_{Xi}}{dt}$ is given by Eq. (2). The relaxation rate R reflects how fast the network is relaxed. The value of R is an adjustable parameter and can be determined empirically.

Two types of relaxation rate are studied and evaluated in simulation. One is a *constant* relaxation rate which is invariant through the entire relaxation process. The other is a *dynamic* relaxation rate which depends on the iteration and is different at different stages of relaxation. The motivation for studying the dynamic relaxation rate is to explore the possibility that there are different optimal relaxation rates at different phases of relaxation. Simulation results on both types of relaxation rates are given in the following section.

4 Simulation results

The performance of the Hopfield network including a relaxation rate is evaluated by simulation, and is compared to that of the network without using a relaxation rate. Evaluation is based on 200 randomly generated 10-city *TSP* city distributions, including wide varieties of city distributions. The performance of an algorithm usually depends on the topology of the city locations in a city distribution, and different algorithms may favor different types of distributions. Using a large number of city distributions can reduce this effect and obtain a better evaluation of the algorithms.

For each of the 200 city distributions, there are 100 runs with different random noises added to the initial inputs of neurons. Each evaluated parameter is first averaged over 100 runs for each city distribution and then averaged over the entire 200 city distributions. Thus 20,000 runs are needed to obtain each data point shown in the following simulation results.

The original energy function of the Hopfield network is used in the simulation, and parameters in the energy function are those used by Hopfield and Tank [1]:

$$A = B = D = 500, \quad C = 200, \quad N_0 = 15 \qquad (5)$$

The value of dt in Eq. (2) and Eq. (4) is set to be 10^{-5}. The fraction of random noise in the initial values of neurons is set to be 0.001. The maximum number of iterations allowed for each run is set to

70

be 1000. If a valid tour can not be reached within 1000 iterations, the network will stop and the tour is counted as invalid.

4.1 Constant relaxation rate

We first presents the simulation results using a constant relaxation rate that does not change through the entire relaxation process.

Fig. 1 shows the errors of the solutions obtained by the network using different constant relaxation rates compared to that obtained without using a relaxation rate. Throughout the paper, the network "without using a relaxation rate" means that Eq. (2) instead of Eq. (4) is used as the dynamic equation. It is equivalent to use a default value $R = 1.0$ if Eq. (4) is used as the dynamic equation.

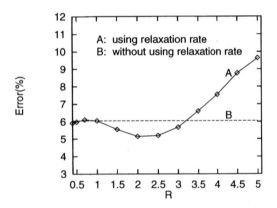

Figure 1: The errors (%) of different relaxation rates and that without using a relaxation rate.

For the city distribution i, the error of a valid tour j is defined by:

$$Err_{i,j} = \frac{d_{i,j} - d_{i,opt}}{d_{i,opt}} \qquad (6)$$

where $d_{i,j}$ is the tour length of tour j and $d_{i,opt}$ is the optimal (shortest) tour length of city distribution i.

The error for city distribution i is the averaged error over all valid tours:

$$Err_i = \frac{\sum_{j=1}^{N_{valid,i}} Err_{i,j}}{N_{valid,i}} \qquad (7)$$

where $N_{valid,i}$ is the number of valid tours among the total of $N_{total,i}$ runs with different initial input values for the city distribution i. In the simulation $N_{total,i} = 100$ for all city distributions.

The percentage of valid tours for city distribution i is defined by:

$$Valid_i = \frac{N_{valid,i}}{N_{total,i}} \qquad (8)$$

The error shown in Fig. 1 is the average error of valid tours in all city distributions. It is weighted by the percentage of valid tours for each city distribution and is given by:

$$Err = \frac{\sum_{i=1}^{N_{CityDist}} (Valid_i Err_i)}{\sum_{i=1}^{N_{CityDist}} Valid_i} \qquad (9)$$

where $N_{CityDist}$ is the total number of city distributions and is equal to 200.

The result in Fig. 1 shows that the error rate depends on the relaxation rate. The error rate can be reduced by using a relaxation rate in a proper range (between 1.0 and 3.0). However, when the relaxation rate becomes too large (> 3.5) the error rate increases rapidly, showing an obvious disadvantage of a fast relaxation. In contrast, when the relaxation rate becomes too small (< 1.0), the error rate approaches a limit without much change. But as we observed in the experiment, more iterations are needed to converge to a solution when the relaxation rate becomes smaller. Thus if two relaxation rates lead to a similar performance, a larger one is preferred in oreder to achieve a higher efficiency.

Fig. 2 shows the percentage of valid tours using different relaxation rates, compared to those without using a relaxation rate.

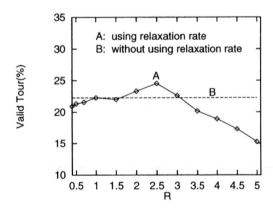

Figure 2: The percentage of valid tours using different relaxation rates and that without using a relaxation rate.

The percentage of valid tours is the weighted average over 200 city distributions:

$$Valid = \frac{\sum_{i=1}^{N_{CityDist}} Valid_i}{N_{CityDist}} \qquad (10)$$

The result in Fig. 2 has a similar pattern as in Fig. 1, and the improvement occurs when the relaxation rate is in the middle range. This shows that a relaxation rate has a similar impact on the percentage of valid tours and on the error rate.

With the consideration of both Fig. 1 and Fig. 2, the optimal value for the relaxation rate is about 2.5. At this point, the error rate is reduced by 14.0% (5.20% vs 6.05%) and the percentage of valid tours is increased by 9.87% (24.5% vs 22.3%) as compared to that without using a relaxation rate.

4.2 Dynamic relaxation rate

A *dynamic* relaxation rate with the following form is included in the network:

$$R(M) = R_0 + \frac{(M - M_0)(R_1 - R_0)}{M_1 - M_0} \qquad (M_0 \leq M < M_1)$$

$$R(M) = R_1 \qquad (M \geq M_1) \quad (11)$$

where $R(M)$ is the dynamic relaxation rate which is a function of iteration M. $R(M)$ is equal to R_0 initially (when $M = M_0 = 0$), and then increases linearly with M until reaching R_1 at $M = M_1$.

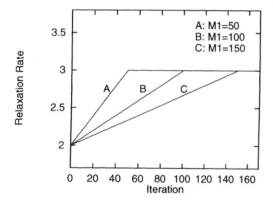

Figure 3: Dynamic relaxation rates with the parameters: $M_0 = 0$, $R_0 = 2.0$, $R_1 = 3.0$, and for different values of M_1.

Fig. 3 shows the shapes of $R(M)$ with the parameters: $M_0 = 0$, $R_0 = 2.0$, $R_1 = 3.0$, and for different values of M_1. The relaxation rate has a lower value during early iterations so that the neurons have sufficient opportunity to exchange information and to form a good prototype. Then its value gradually increases to speed up the convergence.

Fig. 4 and Fig. 5 show the error rate and the percentage of valid tours using dynamic relaxation rate, compared to those using constant relaxation rate. The figures show the results of dynamic relaxation rate using several different M_1 values. The constant relaxation rate for comparison is chosen to be 2.5 because it achieves the best performance as shown in the above discussion.

We can see that the performance of a dynamic relaxation rate depends on the parameter M_1. Among the M_1 values experimented in the simulation, a larger value (i.e. a slower increase of the relaxation rate from 2.0 to 3.0) can achieve a relatively better performance.

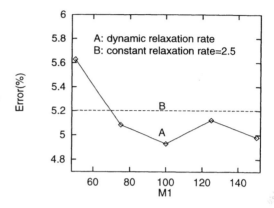

Figure 4: The error rate (%) using dynamic relaxation rate compared to that using constant relaxation rate.

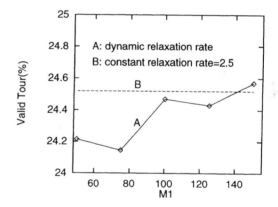

Figure 5: The percentage of valid tours using dynamic relaxation rate compared to that using constant relaxation rate.

At $M_1 = 150$ (about the optimal value), using a dynamic relaxation rate can reduce the error rate by 4.2% (4.98% vs 5.20%) and increase the percentage of valid tours by 0.4% (24.6% vs 24.5%) as compared to those using the constant relaxation rate ($R = 2.5$). If compared to the result without using relaxation rate, it reduce the error rate by 17.7% (4.98% vs 6.05%) and increase the percentage of valid tours by 10.3% (24.6% vs 22.3%). This result shows that the improvement on the error rate by a dynamic relaxation rate is more obvious than that on the percentage of valid tours.

4.3 Dynamic amplification parameter

We have also experimented with a dynamic amplification parameter $u_0(M)$ which slowly increases with iteration M (similar to Eq. (11)) and comapared its performance with those achieved by using relaxation rates.

In the simulation, $u_0(M_1)$ is increased linearly from 0.02 to 0.03 with different M_1 values (50, 100, 200, 500, 900). The best performance is achieved at $M_1 = 100$, which leads to a decrease of the error rate by 7.1% (5.62% vs 6.05%) and an increase of the percentage of valid tours by 3.1% (23.0% vs 22.3%) as compared to those using the original network.

This result shows that a dynamic amplification parameter can also improve the performance of the network. But the improvement achieved by this approach smaller than that achieved using a constant or dynamic relaxation rate.

5 Summary

In this paper, we have analyzed the relaxation process of the Hopfield network and proposed that the rate of relaxation has an important impact on the performance of the network. A relaxation rate has been introduced into the network dynamics to control the pace of network relaxation. The network using a relaxation rate has been shown to have a better performance than that without using a relaxation rate.

According to the simulation results based on 200 randomly generated city distributions of the 10-city *TSP*, the error rate can be reduced by 9.87% and the percentage of valid tours can be increased by 14.0% as compare to those without using a relaxation rate. A dynamic relaxation rate has also been proposed and compared to the constant relaxation rate, and it can further reduce the error rate by 4.2% and increase the percentage of valid tours by 0.4%. We have also experimented with a dynamic amplification parameter and compared its performance to that using relaxation rates.

In future work we plan to experiment with different forms of dynamic relaxation rate, including exponential form and quadratic form, and compare their performance with the linear form. We also plan to evaluate the performance of the relaxation rate on applications other than *TSP*.

Acknowledgments

This research is funded in part by a grant from *fonix* Corp.

References

[1] Hopfield, J. J. and Tank, D. W.: Neural Computations of Decisions in Optimization Problems. Biological Cybernetics, vol. 52, pp. 141-152, 1985.

[2] Wilson, G. V. and Pawley, G. S.: On the Stability of the Traveling Salesman Problem Algorithm of Hopfield and Tank. Biological Cybernetics, vol. 58, pp. 63-70, 1988.

[3] Brandt, R. D., Wang, Y., Laub, A. J. and Mitra, S. K.: Alternative Networks for Solving the Traveling Salesman Problem and the List-Matching Problem. Proceedings of IEEE International Conference on Neural Networks, San Diego, CA. II: 333-340, 1988.

[4] Aiyer, S. V. B., Niranjan, M. and Fallside, F.: A Theoretical Investigation into the Performance of the Hopfield Model. IEEE Transactions on Neural Networks, vol. 1, no. 2, pp. 204-215, 1990.

[5] Li, S. Z.: Improving Convergence and Solution Quality of Hopfield-Type Neural Networks with Augmented Lagrange Multipliers. IEEE Transactions On Neural Networks, vol. 7, no. 6, pp. 1507-1516, 1996.

[6] Catania, V., Cavalieri, S. and Russo, M.: Tuning Hopfield Neural Network by a Fuzzy Approach. Proceedings of IEEE International Conference on Neural Networks, pp. 1067-1072, 1996.

[7] Liang, Y.:Combinatorial Optimization by Hopfield Networks Using Adjusting Neurons. Information Sciences, vol. 94, pp. 261-276, 1996.

[8] Cooper, B. S.: Higher Order Neural Networks-Can they help us Optimise?. Proceedings of the Sixth Australian Conference on Neural Networks (ACNN'95), pp. 29-32, 1995.

[9] Van den Bout, D. E. and Miller, T. K.: Improving the Performance of the Hopfield-Tank Neural Network Through Normalization and Annealing. Biological Cybernetics, vol. 62, pp. 129-139, 1989.

A New Activation Function in the Hopfield Network for Solving Optimization Problems

Xinchuan Zeng and Tony R. Martinez

Computer Science Department, Brigham Young University, Provo, Utah 84602

Email: zengx@axon.cs.byu.edu, martinez@cs.byu.edu

Abstract

This paper shows that the performance of the *Hopfield network* for solving optimization problems can be improved by using a new activation (output) function. The effects of the activation function on the performance of the Hopfield network are analyzed. It is shown that the *sigmoid* activation function in the Hopfield network is sensitive to noise of neurons. The reason is that the *sigmoid* function is most sensitive in the range where noise is most predominant. A new activation function that is more robust against noise is proposed. The new activation function has the capability of amplifying the signals between neurons while suppressing noise. The performance of the new activation function is evaluated through simulation. Compared with the *sigmoid* function, the new activation function reduces the error rate of tour length by 30.6% and increases the percentage of valid tours by 38.6% during simulation on 200 randomly generated city distributions of the 10-city *traveling salesman problem*.

1 Introduction

The subject of combinatorial optimization consists of a large set of important problems in computer science and engineering. A classical example of combinatorial optimization problems is the *traveling salesman problem (TSP)*. *TSP* belongs to the class of *NP* problems. All exact methods known for determining an optimal tour require a computing effort that increases exponentially with N for an N-city *TSP* problem. Interest in solving combinatorial optimization problems by neural networks was motivated by the work of Hopfield and Tank [1]. They suggested that a suboptimal solution of *TSP* can be obtained by finding a local minimum of an appropriate *energy function*. The energy function can be implemented by a neural network. For a N-city *TSP* problem, the network consists of $N \times N$ neurons and the links that connect these neurons. The weights are set to encode the information about the constraints and the cost function of a particular city distribution of *TSP*. Each neuron updates its input value based on the information received from all other neurons.

They showed that the neural network can often find a near-optimal solution in a short time. This network is commonly referred to as the *Hopfield network*. The advantages of the Hopfield network over other heuristic methods for solving combinatorial optimization problems include massive parallelism and convenient hardware implementation. Another advantage is that the procedure of the Hopfield network is more general for applications. There is an *ad hoc* procedure for mapping the constraints and cost function into the weight settings of the network. This general procedure can be applied to solve many different types of combinatorial optimization problems.

Since Hopfield and Tank showed that neural computation can be effective for solving combinatorial optimization problems, some work has been done to improve the performance of the Hopfield network. The research focuses on analyzing and improving the original model in order to obtain a higher percentage of valid solutions and solutions with better quality. The work by Wilson and Pawley [2] showed that there was some difficulty getting the Hopfield model to yield valid tours. For randomly generated sets of the 10-city *TSP*, Wilson and Pawley reported that only 8% of their trials resulted in valid tours. After their report, Brandt et al. [3] and Aiyer et al. [4] showed that better performance can be achieved by modifying the energy function. Li combined the Hopfield network with the "augmented Lagrange multipliers" algorithm from optimization theory [5]. Catania et al. applied a fuzzy approach to tune the parameters in the Hopfield network [6]. Liang added adjusting neurons to the Hopfield network for solving the quadratic assignment problem [7].

Although the performance of the Hopfield network has been improved over the past decade, this model still has some basic problems [8, 9]. One of the problems is that the performance of the Hopfield network is inconsistent. The performance is good for some city distributions of *TSP*, but the performance is poor for other city distributions with the same size. The performance is usually better for city distributions with simple topology. However, for city distributions with complex topology, the solutions are often trapped in poor local minima, or the solutions are invalid. Another problem is that the performance is sensitive to the choice of the parameters and the initial input values of neurons. Different values of the parameters in the energy function can lead to significant differences in the performance. For the same set of parameters, different settings of random noise in the initial input values (a small fraction of random noise is necessary to break the symmetry of the network) can yield solutions with varying quality or invalid solutions.

In this paper, we show that one important reason for the inconsistency of the Hopfield network is the non-robust *sigmoid* activation function which serves to map the current state input of each neuron to its activation (output). The *sigmoid* function is most sensitive in noise dominant range, and thus reduces the effects of signals between neurons. A new activation function is proposed and evaluated in this paper. The new activation function has different shape and has an adjustable parameter to control the threshold for the sensitive region. Thus it is more robust against noise and can amplify the signals between neurons while suppressing noise. In the simulation on 200 randomly generated city distributions of the 10-city *TSP*, the new activation function reduces the error rate of tour length by 30.6% and increases the percentage of valid tours by 38.6% as compared with the *sigmoid* function. It shows that the Hopfield network using the new activation function can yield solutions with better quality and a higher percentage of valid solutions.

2 Background of Hopfield network

The Hopfield network [1] includes a set of neurons and the links that connect the neurons. For an N-city *TSP*, there are $N \times N$ fully connected neurons in the network. The row index for a neuron represents the city and the column index represents the order of the city in the tour. The weights of the connecting links are determined according to the constraints and the cost function. *TSP* includes the following constraints: each city must be visited exactly once in a tour (each row has exactly one "on" neuron) and exactly one city is visited at any time in a tour (each column has exactly one "on" neuron). The cost function is constructed to reflect the objective of finding a valid tour with minimum tour length.

The constraints and the cost function for *TSP* can be represented by an energy function. The energy function is then used to determine the values of all weights in the network. Hopfield's original energy function for an N-city *TSP* is given by [1]:

$$E = \frac{A}{2} \sum_{X=1}^{N} \sum_{i=1}^{N} \sum_{j=1, j \neq i}^{N} V_{Xi} V_{Xj}$$
$$+ \frac{B}{2} \sum_{i=1}^{N} \sum_{X=1}^{N} \sum_{Y=1, Y \neq X}^{N} V_{Xi} V_{Yi} + \frac{C}{2} (\sum_{X=1}^{N} \sum_{i=1}^{N} V_{Xi} - N_0)^2$$
$$+ \frac{D}{2} \sum_{X=1}^{N} \sum_{i=1}^{N} \sum_{Y=1, Y \neq X}^{N} d_{XY} V_{Xi}(V_{Y,i+1} + V_{Y,i-1}) \quad (1)$$

where X and Y are row indices; i and j are column indices; V_{Xi} is the activation for each neuron; and d_{XY} is a measure of the distance between cities X and Y. The first two terms enforce the constraint that no city can be visited more than once. The third term reflects the constraint that each city should be visited. The last term represents the distance cost function. The value of each parameter (A, B, C, and D) is the measure of the importance of the corresponding term.

Each neuron has a current state input U_{Xi} and an activation (output) V_{Xi}. The initial values of U_{Xi} for all neurons are first set to be a small constant value which is determined by the condition: $\sum_{X=1}^{N} \sum_{i=1}^{N} V_{Xi} = N$. To break the symmetry of the network, a small fraction of random noise is added to the initial values of each U_{Xi}. The network is then relaxed until an equilibrium is reached.

After each iteration, U_{Xi} is updated based on the following equation:

$$\frac{dU_{Xi}}{dt} = -\frac{U_{Xi}}{\tau} - \frac{\partial E}{\partial V_{Xi}} \quad (2)$$

where $\tau = 1.0$ is the time constant.

V_{Xi} is also updated after each iteration and its value is determined by U_{Xi} through an *activation (output) function*. In the Hopfield network, the activation function is the *sigmoid* function given by:

$$V_{Xi} = \frac{1}{2}(1 + tanh(\frac{U_{Xi}}{u_0})) \quad (3)$$

where u_0 is the amplification parameter that reflects the steepness of the activation function.

Hopfield and Tank [1] showed that the network is guaranteed to converge to a local minimum in the case of symmetric connecting weights. They tested this model for a 10-city *TSP*. For a total of 20 runs, they claimed 16 valid tours, and half of the valid tours were optimal.

3 New activation function

The quality of the solutions obtained by the Hopfield network depends on the random noise in the initial input values of neurons in the network. In the simulation reported by Wilson and Pawley [2], 10 sets of randomly produced 10-city *TSP* coordinates are tested. There were a total of 50 runs with different initial input values for each city distribution. They reported that 8% of the runs converged to a valid tour, 48% of the runs froze into invalid tours, and 44% of the runs did not converge in 1000 iterations. They showed that the quality of the solutions is sensitive to the initial input values of neurons. Different sets of initial values for a fixed set of parameters can yield solutions with different quality or even invalid solutions.

We propose that one important factor for the above problem is the activation function. The activation function in the Hopfield network is the *sigmoid* function as expressed in Eq. (4). The steep part of this function is around the region where U_{Xi} is close to zero, as shown by the curve labeled A in the Fig. 1. Thus this activation function puts too much emphasis on minor noise perturbation instead of the signals related to the cost and the constraints encoded in the network.

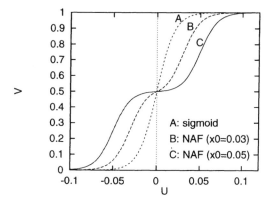

Figure 1: Comparison of the *sigmoid* and the new activation function (NAF) with different thresholds (x_0).

Based on the above analysis, we propose the following new activation function:

$$V_{Xi} = \frac{0.5(1 + tanh(\frac{U_{Xi} + x_0}{u_0}))}{1 + tanh(\frac{x_0}{u_0})} \qquad (U_{Xi} < 0)$$

$$V_{Xi} = \frac{tanh(\frac{x_0}{u_0}) + 0.5(1 + tanh(\frac{U_{Xi} - x_0}{u_0}))}{1 + tanh(\frac{x_0}{u_0})} \qquad (U_{Xi} \geq 0)$$

$$(4)$$

where x_0 represents the threshold for V_{Xi} to become steep, and u_0 measures the steepness of the activation function. The shape with differing values of x_0 is shown in Fig. 1. The new activation function becomes steep only when the absolute value of V_{Xi} is larger than x_0. Thus random noise smaller than the threshold will not be amplified, while the signal between neurons, which usually has a larger magnitude than noise, will be amplified. Only when substantial positive or negative evidence is summed into a neuron, will the neuron make a significant change in its activation function. This mechanism has the capability of reducing the effects of noise in the formation of tours.

4 Simulation results

The performance of the new activation function is evaluated by simulation and is compared with that of the *sigmoid* function. Evaluation is accomplished with 200 randomly generated 10-city *TSP* city distributions, including wide varieties of distributions. The performance of different algorithms usually depends on the topology of the city locations in a city distribution, and different algorithms may favor different types of distributions. Using a large number of city distributions can reduce this effect and obtain a better evaluation of the algorithms. For each of the 200 city distributions, there are 100 runs with different settings of random noise added to the initial input values of the neurons. Each evaluated parameter is first averaged over 100 runs for each city distribution and then averaged over 200 city distributions.

The original energy function of the Hopfield network is used for both the case using the *sigmoid* function and the case using the new activation function. To evaluate specific effects of the different activation functions, the same set of parameters in the energy function are used in both cases. The values of the parameters are those by Hopfield and Tank [1]:

$$A = B = D = 500, \quad C = 200, \quad N_0 = 15 \qquad (5)$$

The value of dt in Eq. (2) is set to be 10^{-5}. The value of u_0 is fixed at 0.02 for both Eq. (3) and Eq. (4). The fraction of random noise in the initial values of neurons is set to be 0.001.

For each city distribution i, there are a total of $N_{total,i}$ ($= 100$) runs with different initial input values. The maximum number of iterations allowed for each run is set to be 1000. If a valid tour can not be reached within 1000 iterations, the network will stop and the tour is counted as invalid.

Fig. 2 shows the errors of the *sigmoid* function and the new activation function with different thresholds. For the city distribution i, the error of a valid tour j is defined by:

$$Err_{i,j} = \frac{d_{i,j} - d_{i,opt}}{d_{i,opt}} \qquad (6)$$

where $d_{i,j}$ is the tour length of tour j and $d_{i,opt}$ is the optimal (shortest) tour length of city distribution i.

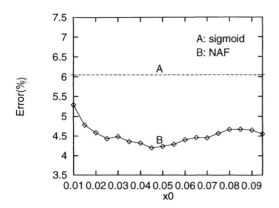

Figure 2: The errors (%) of the new activation function (with different threshold x_0) compared to that of the *sigmoid* function.

The error for city distribution i is the averaged error over all valid tours:

$$Err_i = \frac{\sum_{j=1}^{N_{valid,i}} Err_{i,j}}{N_{valid,i}} \qquad (7)$$

where $N_{valid,i}$ is the number of valid tours among the total of $N_{total,i}$ runs with different initial input values for the city distribution i. In the simulation $N_{total,i} = 100$ for all city distributions.

The percentage of valid tours for city distribution i is defined by:

$$Valid_i = \frac{N_{valid,i}}{N_{total,i}} \qquad (8)$$

The error shown in Fig. 2 is the average error of valid tours in all city distributions. It is weighted by the percentage of valid tours for each city distribution and is given by:

$$Err = \frac{\sum_{i=1}^{N_{CityDist}} (Valid_i Err_i)}{\sum_{i=1}^{N_{CityDist}} Valid_i} \qquad (9)$$

where $N_{CityDist}$ is the total number of city distributions and is equal to 200.

The result in Fig. 2 shows that the error rate of the new activation function is lower than that of the *sigmoid* function, i.e. the tours obtained by using the new activation function have higher quality (shorter tour length). The improvement depends on the value of the threshold x_0. When the threshold increases from 0.01 to 0.045, the error rate decrease from 5.29% to 4.20%. This supports our conjecture that a higher threshold can suppress the effects of noise while still amplifying signals. The signal that represents the quality of the tour is from the last term in the energy function (1), which reflects the distance between the cities. When the threshold increases further, the error rate starts to grow. It implies that the threshold is so high that some signals as well as noise is suppressed by the activation function. The threshold should be properly chosen in order to reduce noise and keep the signals. The value of the threshold x_0 with lowest error rate is about 0.045, and the corresponding error rate is 4.20%. Compared with the error rate of 6.05% when using the *sigmoid* function, the error rate is reduced by 30.6%.

Fig. 3 shows the percentages of valid tours of the *sigmoid* function and the new activation function with different thresholds.

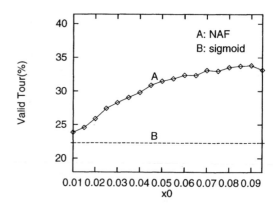

Figure 3: The percentages of valid tours of the new activation function (with different threshold x_0) compared to that of the *sigmoid* function.

The percentage of valid tours is the weighted average over 200 city distributions:

$$Valid = \frac{\sum_{i=1}^{N_{CityDist}} Valid_i}{N_{CityDist}} \qquad (10)$$

The percentage of valid tours obtained by the new activation function is higher than that by the *sigmoid* function. When x_0 changes from 0.01 to 0.06, the percentages of valid tours increases from

23.8% to 32.4%. This suggests that the new activation function amplifies the signals that represent the constraints for a valid tour while reducing noise. The signals that represent the constraints are from the first three terms in the energy function (1). When the threshold increases further, the percentage of valid tours no longer increases. If the threshold is too high, the error rate starts to increase as shown in Fig. 2. The threshold should be properly chosen considering the trade off between the error rate and the percentage of valid tours. For the parameters in the simulation, the optimal value of x_0 is about 0.045. The corresponding percentage of valid tours is 30.9%. Compared with that of the *sigmoid* function (22.3%), the percentage of valid tours is increased by 38.6%.

5 Summary

In this paper, we have analized the effects of the activation function on the performance of the Hopfield network for solving *TSP* and proposed that the inconsistency of the performance of the Hopfield network is related to the activation function. The analysis reveals that the *sigmoid* activation function in the Hopfield network is sensitive in the region where noise is predominant and thus has negative effects on the performance of the network. We proposed a new activation function that is more robust against noise. The new activation function has an adjustable parameter that controls the threshold of input to be amplified, and it has the capability of amplifying the signals while suppressing noise. The proposed new activation function has been shown to outperform the *sigmoid* function based on the simulation results. In simulation on 200 randomly generated city distributions of the 10-city *TSP*, the new activation function reduces the error rate of tour length by 30.6% and increases the percentage of valid tours by 38.6% as compared to the *sigmoid* function, showing the capability of the new activation function to improve the performance of the Hopfield network.

The new activation function has also been applied in a study of speech recognition. Preliminary results show that it is also capable of improving the performance in this domain. In future we plan to evaluate its performance on other optimization problems in order to better evaluate the generality of this approach. Another extension which we are currently pursuing is to apply a dynamic activation function with varying parameters in the Hopfield network.

Acknowledgments

This research is funded in part by a grant from *fonix* Corp.

References

[1] Hopfield, J. J. and Tank, D. W.: Neural Computations of Decisions in Optimization Problems. Biological Cybernetics, vol. 52, pp. 141-152, 1985.

[2] Wilson, G. V. and Pawley, G. S.: On the Stability of the Traveling Salesman Problem Algorithm of Hopfield and Tank. Biological Cybernetics, vol. 58, pp. 63-70, 1988.

[3] Brandt, R. D., Wang, Y., Laub, A. J. and Mitra, S. K.: Alternative Networks for Solving the Traveling Salesman Problem and the List-Matching Problem. Proceedings of IEEE International Conference on Neural Networks, San Diego, CA. II: 333-340, 1988.

[4] Aiyer, S. V. B., Niranjan, M. and Fallside, F.: A Theoretical Investigation into the Performance of the Hopfield Model. IEEE Transactions on Neural Networks, vol. 1, no. 2, pp. 204-215, 1990.

[5] Li, S. Z.: Improving Convergence and Solution Quality of Hopfield-Type Neural Networks with Augmented Lagrange Multipliers. IEEE Transactions On Neural Networks, vol. 7, no. 6, pp. 1507-1516, 1996.

[6] Catania, V., Cavalieri, S. and Russo, M.: Tuning Hopfield Neural Network by a Fuzzy Approach. Proceedings of IEEE International Conference on Neural Networks, pp. 1067-1072, 1996.

[7] Liang, Y.:Combinatorial Optimization by Hopfield Networks Using Adjusting Neurons. Information Sciences, vol. 94, pp. 261-276, 1996.

[8] Cooper, B. S.: Higher Order Neural Networks-Can they help us Optimise?. Proceedings of the Sixth Australian Conference on Neural Networks (ACNN'95), pp. 29-32, 1995.

[9] Van den Bout, D. E. and Miller, T. K.: Improving the Performance of the Hopfield-Tank Neural Network Through Normalization and Annealing. Biological Cybernetics, vol. 62, pp. 129-139, 1989.

On-line Learning of Object Representations*

Horst Bischof[1] and Aleš Leonardis[2]

[1] PRIP, Vienna University of Technology, Treitlstr. 3/1832, A-1040 Vienna, Austria

[2] Faculty of Computer and Info. Science, University of Ljubljana, Tržaška 25, 1001 Ljubljana, Slovenia

Email: bis@prip.tuwien.ac.at, ales.leonardis@fri.uni-lj.si

Abstract

Radial Basis Function (RBF) networks have been proposed as suitable representations for 3-D objects, in particular, since they can learn view-based representations from a small set of training views. One of the basic questions that arises in the context of RBF networks concerns their complexity, i.e., the number of basis functions that are necessary for a reliable representation, which should balance the accuracy and the robustness. In this paper we propose a systematic approach for building object representations in terms of RBF networks. We studied and designed two procedures: the "off-line" procedure, where the network is constructed after having a complete set of training views of an object, and the "on-line" procedure, where the network is incrementally built as new views of an object arrive.

1 Introduction

Different object representations have been proposed to facilitate object recognition. They are usually classified into geometric methods and appearance-based methods [14]. Traditionally, geometric methods represent objects by explicit 3-D object-centered models, while appearance-based methods represent objects by a set of viewer-centered images (or possibly abstractions of images). The latter approach has often been criticized that it is too expensive in terms of memory if all possible views (for different rotations, scaling, illumination conditions, etc.) need to be stored. However, it has been shown that only a small number of views has to be stored and that the other views can be recovered by the interpolation. Ullman and Basri [18], Werner et al. [19] (and the references therein) used a linear interpolation to reconstruct the intermediate views. An interesting approach following the same idea has been used by

Chandrasekaran et al. [2], who used eigenspace representation (in an incremental updating scheme) to achieve adaptive view selection for 3D object representation from projections. On the other hand, Poggio and Edelman [13] developed a network, based on the theory of approximation of multivariate functions, that performs nonlinear interpolation between the views. More specifically, they use RBF networks which learn view-based representations from a small set of training views. Results of the recognition task using this object representation appear to be in accordance with the psychophysical experiments [1].

One of the basic questions that arises in the context of RBF networks concerns their complexity, i.e., the number of basis functions (nodes in the hidden layer). In principle, a network can contain as many basis functions as there are training samples, i.e., views. Poggio and Edelman also explored the use of fewer basis functions than training views: In one example, they found satisfactory performance with just two basis units (for 10–40 training views and with the attitude of the object limited to one octant of the viewing sphere [13]). Their conclusion was that a very small number of units is needed for each aspect of an opaque object. The question, however is, how to systematically arrive at such a small number of basis functions given a large set of views of an object.

In this paper we propose a systematic approach for building object representations in terms of RBF networks. The approach consists of two intertwined procedures: network training and complexity optimization. During network training, based on the input views, output weights, widths, and locations of basis functions are adapted. Complexity optimization, which is based on the MDL (Minimum Description Length) principle, then selects a subset from the set of all basis functions such that the number of basis functions is reduced while preserving the performance of the network. We studied and designed two procedures: the "off-line" procedure,

---

* This work was supported by a grant from the Austrian National Fonds zur Förderung der wissenschaftlichen Forschung (S7002MAT and P10539MAT). A. Leonardis acknowledges partial support by the Ministry of Science and Technology of Republic of Slovenia (Projects J2-0414, J2-8829).

where the network is constructed after having available a complete set of training views of an object, and the "on-line" procedure, where the network is incrementally built as new views of an object arrive.

A preliminary version of our "off-line" method has already been applied on a set of medical data and to solve the problem of function approximation [7]. The major contributions of this paper are: a more rigorous treatment of objective function in terms of the MDL principle, an on-line realization of the construction of RBF networks (with a comparison to the off-line approach), and the application to view-based 3D object representations.

The paper is organized as follows: In section 2 we explain the problem of complexity optimization of RBF networks and discuss the related work. Our algorithm is described in section 3. In section 4 we present the experimental results. We conclude with a summary and outline the work in progress.

2 RBF Networks

Radial basis functions (RBF) have been subject to extensive research over recent years and have successfully been employed to various problem domains, e.g., [10, 9, 3, 15]. The design and training of RBF networks consists of determining the number of basis functions, finding their centers and widths, and calculating the weights of the output nodes. In particular, the main problem is how to determine the number of basis functions which affects the complexity of the network and consequently influences its performance.

There have been several proposals on how to build RBF networks. For example, a set of samples is randomly selected from the training set and the positions of the centers of the basis functions are set according to these samples [9], or some pre-clustering (grouping) is performed on the training set (e.g., k-means clustering), and the centers of the clusters are used as the centers for the basis functions [10]. All these approaches have various shortcomings. The common and the most crucial one is that the number of basis functions has to be a priori given. Since this is usually not possible, it is determined by trial and error.

Sardo and Kittler [16] proposed a criterion function—MPL (Maximum Penalized Likelihood) which, given different networks, evaluates them and selects among them the simplest one. However, the paper does not deal with the question of how to arrive at those networks.

Mukherjee and Nayar [11] proposed an analytic method that sets the network parameters for

any given input-output mapping and error bound. Through a novel construction and application of the integral wavelet transform they achieve the smallest network for the given mapping and error bound. However, the optimality only refers to the specific choice of orthogonal basis functions which suffer from shift-variance. Besides, the method needs evenly sampled data which requires additional interpolation of unevenly sampled data.

In recent work Fritzke [4] and Orr [12] tried to overcome the problem of determining the number of RBF nodes by a growing and regularization method, respectively. Also related are pruning methods for neural networks (see for example [5, 6], which have been extensively used for multilayer perceptrons).

All the methods described above have only been used as "off-line" procedures, meaning that all training samples have to be obtained (stored) prior to building the network. In many cases this may be a too restrictive constraint. On-line learning is much more biologically plausible. Thus we also focus in this work on the "on-line" approach which enables that the network (representation) is incrementally built.

3 Optimizing the Complexity of the RBF networks

We first review the basic equations for the RBF networks, then we explain the selection procedure, and finally we outline the complete algorithm, both off-line and on-line versions.

In order to simplify the notation we use axes aligned Gaussian RBF networks with a single linear output unit[1]. Let us consider an RBF-network as a function approximator:

$$
\begin{aligned}
y(\mathbf{x}) &= w_0 + \sum_{i=1}^{M} w_i r_i(\mathbf{x}) = w_0 + \sum_{i=1}^{M} w_i e^{-\frac{||\mathbf{x}-\mathbf{c}_i||^2}{s_i^2}} \quad (1) \\
&= w_0 + \sum_{i=1}^{M} w_i e^{-\sum_{j=1}^{d} \frac{(x_j - c_{ji})^2}{s_{ji}^2}}, \quad (2)
\end{aligned}
$$

where $\mathbf{c}_i$ and s_i^2 are the center and width of the i-th basis function, respectively. We train the network to approximate an unknown function f given a (possibly noisy) training set $TS = \{(\mathbf{x}^{(p)}, f(\mathbf{x}^{(p)})) | 1 \leq p \leq q, \mathbf{x}^{(p)} \in \mathbb{R}^d\}$.

Given the number of basis functions we can train the whole network by minimizing an error

---

[1]The generalization to other basis functions and multiple output units is straightforward.

function, e.g., the usual quadratic error function (Eq. 3)

$$E = \sum_{p=1}^{q} (f(\mathbf{x}^{(p)}) - y(\mathbf{x}^{(p)}))^2 \ . \qquad (3)$$

Training can be performed by a suitable learning algorithm, e.g. gradient descent, Levenberg-Marquardt, Extended Kalman filter, etc. In general, since we also train the centers and width of the basis functions, we cannot expect with any training algorithm to find a global optimal solution.

Let us define a few quantities which can be derived from the network and which we need in the selection procedure. Given a sample $\mathbf{x}^{(p)}$, the influence of the basis function i on the network's output is just the difference between the output in the presence of the basis function and the absence of the basis function:

$$I(r_i(\mathbf{x}^{(p)})) = |w_i r_i(\mathbf{x}^{(p)})| \ . \qquad (4)$$

Using this measure we can define several other quantities we need for the selection procedure. The domain of the basis function is given by:

$$R_i = \{p | I(r_i(\mathbf{x}^{(p)})) > \Theta\} \ , \qquad (5)$$

where Θ is the threshold which defines when we count a sample to be encoded by the basis function[2]. The number of samples encoded jointly by basis function i and j is given by

$$n_{ij} = |R_i \cap R_j| \ , \qquad (6)$$

the error associated with a basis function is defined as

$$\xi_i = \sum_{p \in R_i} (y(\mathbf{x}^{(p)}) - f(\mathbf{x}^{(p)}))^2 = \sum_{p \in R_i} \xi_i^{(p)} \ , \qquad (7)$$

and the error in the overlap area of two basis functions is given by

$$\xi_{ij} = \sum_{p \in R_i \cap R_j} (y(\mathbf{x}^{(p)}) - f(\mathbf{x}^{(p)}))^2 \ . \qquad (8)$$

One should note that these quantities can easily be calculated while performing an adaption step, causing almost no additional computational costs.

___
[2]This threshold can be defined either using knowledge about the error we expect, or via the desired quantization of the final result.

3.1 Selection

In this section we elaborate the *RBF-selection* procedure which has the task of removing redundant basis functions by the MDL principle. This involves the design of an objective function which encompasses the information about the competing RBF units and the design of an optimization procedure which selects a set of units accordingly.

Prior to modeling a function with a network, the function can only be given by specifying the samples. After building a network some of the samples, or possibly all of them, can be described in terms of the network. However, the complexity of the network may vary, i.e., the network may contain more or less nodes. Let vector $\mathbf{m}^T = [m_1, m_2, \ldots, m_M]$ denote a set of RBF nodes, where m_i is a *presence-variable* having the value 1 for the presence of a node and 0 for its absence in the final network, and M is the number of all initial RBF nodes. The length of encoding of a function L_{function} can be given as the sum of two terms

$$L_{\text{function}}(\mathbf{m}) = L_{\text{samples}}(\mathbf{m}) + L_{\text{network}}(\mathbf{m}) \ , \qquad (9)$$

where $L_{\text{samples}}(\mathbf{m})$ describes the cost of encoding the samples using a network having the basis functions $\mathbf{m}$, and $L_{\text{network}}(\mathbf{m})$ describes the cost of specifying this network.

We can define a quantity S which represents the *savings* in the length of encoding.

$$\begin{aligned} S = \ & \text{length of encoding of the data in the} \\ & \text{absence of a network} - \\ & \text{length of encoding of the data with a network} \\ = \ & L_{\text{samples}}(\mathbf{0}) - L_{\text{function}}(\mathbf{m}) \ . \qquad (10) \end{aligned}$$

The question is how to translate the above equations into our particular case of an RBF neural network. Remember that we can identify the following three terms for the i-th RBF node in the network:

1. A set of input samples R_i, which activate the RBF node. They represent the domain of the node and encompasses $n_i = |R_i|$ samples,

2. the set of parameters of the node $\mathbf{a}_i$ (location, width, weights to output units). N_i denotes the cardinality of this set, and

3. the goodness-of-fit measure ξ_i which gives the difference between the true values of the function and the approximations produced by the network over the domain R_i and projected onto the i-th node.

Analogous to equation (9) we can write

$$L_{\text{function}}(\mathbf{m}) = K_1(n_{\text{all}}-n(\mathbf{m}))+K_2\xi(\mathbf{m})+K_3 N(\mathbf{m}) \ , \tag{11}$$

where n_{all} denotes the number of all sample data points in the input and $n(\mathbf{m})$ the number of data points that are explained by the network. $N(\mathbf{m})$ specifies the number of parameters which are needed to describe the network and $\xi(\mathbf{m})$ gives the deviation between the data and the approximation that the network describes (i.e., the error). K_1, K_2, K_3 are weights which can be determined on a purely information-theoretical basis (in terms of bits), or they can be adjusted in order to express the preference for a particular type of description [8].

Now we can state the task as follows: Find $\hat{\mathbf{m}}$ such that

$$\hat{\mathbf{m}} = \min_{\mathbf{m}} L_{\text{function}}(\mathbf{m}) \ . \tag{12}$$

Since n_{all} is constant, minimization of equation (11) is equivalent to maximizing the expression

$$\hat{\mathbf{m}} = \max_{\mathbf{m}} F(\mathbf{m}) = K_1 n(\mathbf{m}) - K_2\xi(\mathbf{m}) - K_3 N(\mathbf{m}) \ . \tag{13}$$

So far, the optimization function has been discussed on a general level. More specifically, our objective function which takes into account the individual RBF nodes has the following form:

$$F(\mathbf{m}) = \mathbf{m}^T \mathbf{C}\mathbf{m} = \mathbf{m}^T \begin{bmatrix} c_{11} & \cdots & c_{1M} \\ \vdots & & \vdots \\ c_{M1} & \cdots & c_{MM} \end{bmatrix} \mathbf{m} \ . \tag{14}$$

The diagonal terms of the matrix $\mathbf{C}$ express the cost-benefit value for a particular basis function i

$$c_{ii} = K_1 n_i - K_2\xi_i - K_3 N_i \ , \tag{15}$$

while the off-diagonal terms handle the interaction between the overlapping basis functions

$$c_{ij} = \frac{-K_1 n_{ij} + K_2\xi_{ij}}{2} \ . \tag{16}$$

The combinatorial optimization problem is solved by a heuristic *tabu* search [17] (see also [7]).

3.2 Off-line algorithm

We now describe the complete algorithm:

1. *Initialization:* For each sample in the training set generate a basis function with the size $\mathbf{s_{ij}}$ causing slight overlap with neighboring basis functions, which then drives the training of the network. Set the output weights according to the target output.

2. *Adaptation:* Adapt the network by the training algorithm, i.e., minimizing the expression in (3) using, for example, gradient descent or Levenberg-Marquardt algorithm. It is important to note, since the selection procedure operates on a relative basis, we do not need to train the network to convergence, usually only a few steps (e.g., < 10) of training are sufficient.

3. *Selection:* Remove redundant RBF units using the selection procedure (section 3.1).

4. Repeat steps 2 and 3 until the network has converged (i.e. no basis functions are removed and the weights do not change anymore).

3.3 An example of off-line optimization of RBFs

Let us illustrate the behavior of the algorithm with a simple classification problem depicted in Fig. 1. In the initialization step, for each example a basis function is generated, Fig. 1 (a). Then the basis functions are trained, and the selection eliminates the redundant ones. Figs. 1 (b)–(c) show two intermediate stages of the algorithm and Fig. 1 (d) depicts the final result.

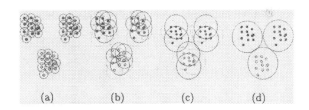

Fig. 1. An example of off-line optimization. RBFs marked with dotted-line get eliminated.

3.4 On-line optimization of RBFs

The off-line algorithm requires the whole training set for the initialization and the selection step. This dependency on the whole training set can be removed by following observations:

1. Initialization: Instead of generating all basis functions in the beginning (resulting in a large network) the basis functions can be generated "on demand", i.e. whenever there is an example which causes a large error ($> \Delta$ a predefined threshold) we generate a new basis function at that location.

2. In the selection procedure the "utility" of a basis function is calculated. Instead of calculating the necessary quantities over the whole training set, we can replace them by suitable running averages. Since we are no longer calculating the exact quantities we may make errors in this estimation step (e.g. due to the order in which we receive the training examples). First, let us note that the selection procedure is insensitive to small random errors. If we make from time to time larger systematic errors they are harmless for the result of the whole algorithm causing only additional computational overhead:

(a) If we over-estimate the utility of a basis function, it will not be removed by the selection procedure. However, as the estimates of the parameters improve, the basis function will be removed in a later stage, causing only additional overhead in training time.

(b) If we under-estimate the utility of the basis function, it will be removed by the selection procedure. However, since we also add basis functions, it will be added again, causing only the costs of training the basis function again.

In particular, the on-line procedure can be outlined as follows:

1. *Initialization:* Start with an empty network or a network trained with the off-line algorithm on a few training samples, initializing n_i, Γ_{ij}, ξ_i, and ξ_{ij}. The two approaches give identical results, however the latter approach is computationally more efficient.

2. For each sample $\mathbf{x}^p$ do: If the error of the sample $(f(\mathbf{x}^{(p)} - y(\mathbf{x}^{(p)})^2 > \Delta$, generate a new basis function at $\mathbf{x}^{(p)}$.

3. Train the network with an online training algorithm (e.g., gradient descent) and update n_i, Γ_{ij}, ξ_i, and ξ_{ij}.

4. If no basis functions have been added for a while (we assume that the estimates of the parameters we need for selection of the recently added basis are accurate), perform selection.

5. Repeat steps 2,3 and 4 until the network has converged (i.e. no basis functions are removed and the weights do not change anymore).

An example of the on-line optimization is shown in Fig. 2. Here we started from the network generated by the off-line algorithm (see Fig. 1). As new training points arrive, a new basis function is generated (Fig. 2 (a)). This function grows as new points are presented to the network, Fig. 2 (b). When there is sufficient overlap between the two basis functions, one is removed by the selection procedure and further training gives the result as shown in Fig. 2 (c).

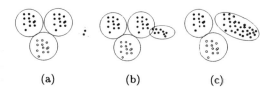

(a)   (b)   (c)

Fig. 2. An example of on-line optimization.

4 Experimental Results—View Interpolation

In this section we present some experimental results. Similar to Poggio and Edelman, we created a set of *stick figures*, each having 10 vertices, i.e., 10 3-D points are selected which are the vertices of the stick figure, the viewing angle of this figure is systematically varied, the 2D points which are presented as an input to the network are obtained by orthographic projection. Figure 3 shows a stick figure under four different angles.

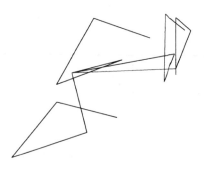

Fig. 3. A stick figure under four different angles.

By learning object representation we mean that by observing a 3-D stick figure from different angles, we build an RBF network. The input to the RBF network are the 10 2-D coordinates of the vertices. The output, given by multiple output units, consists of the orientation of the object (azimuth and elevation angle).

We performed two types of experiments (off-line and on-line) and compared the results in terms of

- complexity of the network,

- training time, and

- accuracy.

The training set in these experiments consisted of coordinates of vertices of a stick figure in orientations $[-45°, 45°]$ both in azimuth and elevation sampled at $10°$ spacing, resulting in a training set of 100 samples. The test set was generated in the same way but with the sampling resolution of 2 degrees.

4.1 Off-line (batch)

In this type of experiments, we present the whole training set to the algorithm. Initially, we have as many RBF nodes as there are data samples. Using the complexity reduction method described in this paper, we systematically adapt the network, which is followed by removing the redundant nodes. The network ended-up with three basis functions (see Fig. 4 (a)) and a perfect fit on the training set. The MSE on the test set was 0.037 degrees which corresponds to excellent interpolation (see also Table 1).

4.2 On-line (incremental)

In this type of experiments we build the RBF network incrementally (on-line) as the data samples are arriving. The training set is the same as for the off-line version (i.e. 100 views). The samples were drawn randomly from this set, and presented to the on-line algorithm. The network ended-up with four basis functions (see Fig. 4 (b)) and a perfect fit on the training set. The MSE on the test set was 0.121 degrees which corresponds to excellent interpolation (see also Table 1).

We compared the following aspects between the on-line and off-line algorithm:

- complexity of the network, expressed in the number of basis functions.

- accuracy: For the same orientation of an object we compared the output of the $RBF_{on-line}$ and the $RBF_{off-line}$ with the ground-truth on the test set.

- time to build the network (*training complexity*), measured in the number of samples from the training set that are processed during the training phase.

This comparison (see Table 1) shows that the results of the off-line algorithm are a little bit better in terms of complexity of the network and the error, but this comes at the cost of increased training complexity. The on-line algorithm needs less training examples because the off-line algorithm is trained on the whole set each time it is adapted.

Table 1. Comparison between the off-line and the on-line algorithm.

	Off-line	On-line
Complexity	3	4
Error	0.037	0.121
Training complexity (number of samples)	~ 11000	~ 5000

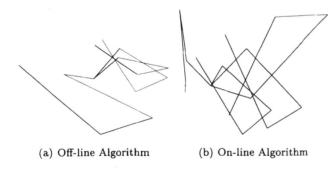

(a) Off-line Algorithm   (b) On-line Algorithm

Fig. 4. Basis views generated by the Off-line (a) and On-line (b) algorithm. The weights of the basis functions are interpreted as the vertices of the stick figure.

The same results have also been achieved with other similar stick-figures. This results compare favorably to the results of Poggio and Edelman [13] which reported for 40-training views and viewing sphere of $[0°, 90°]$ with the usage of 20 basis functions.

Figure 5 shows a comparison between on-line and off-line algorithm on an extrapolation task. The data set for testing extrapolation was generated from the interval $[(45°, 45°) \dots (75°, 75°)]$ both angles were varied simultaneously by $0.5°$. From these results one can see that both networks extrapolate reasonably well up to a deviation of $10°$. The results of the off-line algorithm are better as expected from the result of the interpolation task.

84

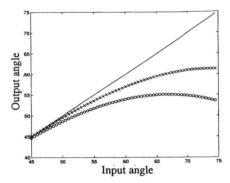

Fig. 5. Extrapolation of azimuth angle, 'o' on-line algorithm, 'x' off-line algorithm

5 Conclusions and Work in Progress

We presented a systematic approach for building view-based object representations in terms of RBF networks. A combination of training and selection (in terms of the MDL principle) achieves a controlled way of adapting and modifying RBF networks which balances accuracy, learning time, and the complexity of the resulting networks.

We applied our approach to view interpolation, and the complexity of the resulting networks has confirmed that only a small number of basis functions is necessary to represent a large field of views.

We studied and designed two procedures: the "off-line" procedure, where the network is constructed after having a complete set of training views of an object, and the "on-line" procedure, where the network is incrementally built as new views of an object arrive. We provided the motivation for the on-line approach and showed that the on-line procedure results in slightly more complex networks, having a slightly higher error, but at a lower computational cost.

In the future we plan to experiment with other types of realistic opaque objects. In addition, the on-line version needs more theoretical analyses. In particular, we have to determine when the approximations we make do no longer hold. The on-line and off-line procedures are two extreme cases of batch algorithms. Batch algorithms collect a batch b of examples and then train on these examples; $b = 1$ corresponds to the on-line algorithm and $b = q$ corresponds to the off-line algorithm. It is interesting to study if any algorithm in between these two extreme cases would provide the accuracy and complexity of the networks obtained by the off-line algorithm at the computational costs of the on-line algorithm.

In addition we can extend the proposed algorithms in several other ways: We are currently working on adapting the algorithm for several other networks. This includes Gaussian mixture models trained with the EM algorithm and some unsupervised networks. We are also exploiting the possibility to use our algorithm for modular neural network design, and networks which use different types of hidden units.

Since the method proposed in this paper is quite general, it is quite easy to incorporate all these extensions without changing the general paradigm.

References

[1] H. H. Bülthoff and S. .Y. Edelman and M. Tarr, How are three-dimensional objects represented in the brain?, A. I. Memo No. 1479, C. B. C. L. Paper No. 96, Massachusetts Institute of Technology, April 1994.

[2] S.Chandrasekaran, B.S.Manjunath, Y.F. Wang, J. Winkler, and H. Zhang, An eigenspace update algorithm for image analysis, Technical report, TR CS 96-04, Dept. of Computer Science, Univ. of California, Santa Barbara.

[3] F. Girosi, M. Jones, and T. Poggio. Regularization theory and neural network architectures. *Neural Computation*, 7(2):219–269, 1995.

[4] B. Fritzke. Fast learning with incremental RBF networks. *Neural Processing Letters*, 1(1):2–5, 1994.

[5] Y. L. Le Cun and J. S. Denker and S. A. Solla. Optimal Brain Damage. *Advances in Neural Information Processing Systems 2.* Ed. D. S. Touretzky. San Mateo, CA: Morgan Kaufmann. pp. 598–605, 1988.

[6] B. Hassibi and D. G. Stork. Second Order Derivatives for network Pruning: Optimal Brain Surgeon. In Proceedings of NIPS 5. Ed. S. J. Hanson et *al.* Morgan Kaufmann. pp. 164–172, 1993.

[7] A. Leonardis and H. Bischof. Complexity Optimization of Adaptive RBF Networks. In *Proceedings of the ICPR'96*, Vol IV pages 654–658, IEEE Computer Society Press, 1996.

[8] A. Leonardis, A. Gupta, and R. Bajcsy. Segmentation of range images as the search for geometric parametric models. *International Journal of Computer Vision*, 14:253–277, 1995.

[9] D. Lowe. Adaptive radial basis function nonlinearities, and the problem of generalisation. In *1st IEE Conference on Artificial Neural Networks*, pages 171–175. London U.K., 1989.

[10] J. Moody and C. Darken. Fast learning in networks of locally tuned processing units. *Neural Computation*, 1(2):281–294, 1989.

[11] S. Mukherjee and S. K. Nayar. Automatic Generation of GRBF networks for visual learning. In *Proceedings of the ICCV'95*, pages 794–800, 1995.

[12] M. J. Orr. Regularization in the selection of basis function centers. *Neural Computation*, 7(3):606–623, 1995.

[13] T. Poggio and S. Edelman, A network that learns to recognize three-dimensional objects. Nature, 343:263–266, 1990.

[14] J. Ponce and A. Zisserman and M. Hebert (Eds.), Object Representation in Computer Vision II, ECCV'96 International Workshop, Cambridge, U.K., April 1996, Springer Verlag, LNCS-1144.

[15] A. Roy, S. Govil, and R. Miranda. An algorithm to generate radial basis function (RBF)-like nets for classification problems. *Neural Networks*, 8(2):179–201, 1995.

[16] L. Sardo and J. Kittler, Complexity analysis of RBF networks for pattern recognition, In *Proceedings of the CVPR'96*, pp. 574–579, 1996.

[17] M. Stricker and A. Leonardis. Figure ground segmentation using tabu search. In *Proceedings of the IEEE Int. Symposium on Computer Vision*, Coral Gables, Florida, November 1995.

[18] S. Ullman and R. Basri, Recognition by linear combination of models, IEEE Transactions on Pattern Analysis and Machine Intelligence, 13(10):992–1005, October 1991.

[19] T. Werner, R. D. Hersch, and V. Hlaváč, Rendering real-world objects using view interpolation. In Proceedings of the ICCV'95, pp. 957–962, Boston, USA, June 1995. IEEE Press.

A neural network model for blast furnace wall temperature pattern classification

Henrik Saxén[1] and Leif Lassus[2]

[1]Heat Engineering Laboratory, Åbo Akademi University, FIN-20500 Åbo, Finland
[2]TT Tieto Oy, Telecom, Kumpulantie 11, FIN-00521 Helsinki, Finland
E-mail: Henrik.Saxen@abo.fi, Leif.Lassus@tietogroup.com

Abstract

A model for classification, visualization and interpretation of temperature distributions from the wall of the ironmaking blast furnace is presented. The model classifies the patterns using a self-organizing map and depicts the evolution of the distributions on the feature map, which is used as an operation diagram. The model has been implemented at the blast furnaces of a Finnish steelmaking company to improve the alertness of the operators and to help them to take appropriate control actions. The generic features of the models make it possible to apply the proposed classification method to different furnaces with only minor overhead for model tuning. Use of the classifications in operation diagrams is finally discussed.

1   Introduction

The blast furnace is the most common unit for production of iron for primary steelmaking [1]. The furnace functions as a huge counter-current heat exchanger and chemical reactor, where the iron ores are reduced and smelted by hot combustion gases produced by burning coke at high temperature. Because of a hostile environment, it is very difficult to measure internal variables, which, however, should be known for the operators in their decision-making. Conventional mathematical models [2] and techniques from the field of artificial intelligence have been applied to tackle this problem, but the complexity of the process always constitutes an obstacle for proper control. However, several papers have reported progress in applying neural networks to pattern recognition, classification and control problems in the furnace [3-6].

The wall temperatures of the blast furnace can be interpreted in terms of several interesting variables that are related to the internal state of the furnace: They provide information about gas distribution, possible formation or "peeling" of accretions at the wall, changes in the high-temperature region and melting line, etc. Normally, the levels and trends of the temperatures are visually inspected to detect sudden or gradual changes, but the operators also have to consider hundreds of other important measurements simultaneously. A method that classifies the temperature patterns is therefore useful as a decision support.

The paper first briefly presents the classifier and then describes possible ways to interpret the results and to further develop the decision aid. Thermocouple signals from a Finnish blast furnace are used to illustrate the performance of the model. The method is shown to provide an illustrative visualization of the results on the feature map. The model is then further elaborated to allow for an interpretation of the classes on the feature map in terms of a set of important parameters characterizing the process.

2   Modeling and visualization

The wall temperature classification is based on Kohonen's self-organizing map [7]. A set of vertical temperature patterns collected during a longer period (here about 2 years of operation) is presented to the classifier (Fig. 1), which organizes its nodes to describe the features of the patterns. A typical result is shown in Fig. 2.

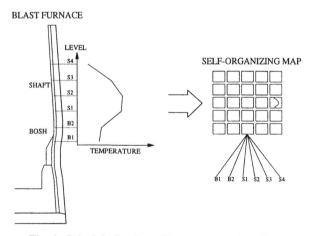

Fig. 1. Principle for the wall temperature classifier

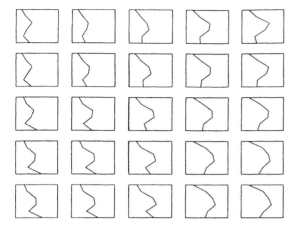

Fig. 2. Typical classes of the wall temperature classifier.

By depicting the evolution of the winning node on the map, it is possible to track changes in the state of the process with time. However, this makes the point jump from one class to another, which may distract the interpreter. If the coordinates of the winning node on the $N \times M$ map at time t are (N_t, M_t), the classification can be expressed as a two-dimensional array, $\mathbf{x} = (x_1, x_2)^T$,

$$x_{1,t} = N_t / N ; \quad x_{2,t} = M_t / M \quad (1)$$

in the unit interval. After this transformation, it is possible to filter the time evolution of the operation point, e.g., by taking moving averages (of l points)

$$\overline{x}_{i,t} = \frac{1}{l} \sum_{j=t-l+1}^{t} x_{i,j} ; \quad i = 1,2 \quad (2)$$

Figure 3 below shows an example of the evolution of the operation point ($\overline{\mathbf{x}}$) on the map, starting from S and ending in E.

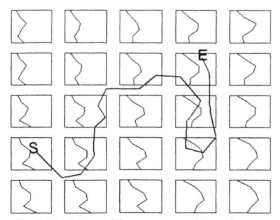

Fig. 3. Evolution of the operation point on the map.

3 Interpretation

A strength of the unsupervised classification algorithm applied is that no a priori information about the classes is required: The network simply creates its classes and the user only has to determine a suitable network dimension. Therefore, also retraining and -tuning of the model are simple and straightforward. However, a drawback of an unsupervised classification method is that the results have to be interpreted in order to make them usable in practice. This often implies that the network's interpretation has to be studied carefully (over a longer period of time) to be able to draw sensible conclusions from the results.

In order to help the blast furnace operator in this task, an effort was made to correlate the classifier's results with other important variables characterizing the process. The (smoothened) location of the winning node, $\overline{\mathbf{x}}$, was therefore used to model other important process variables, y (first subjected to a similar smoothing procedure to make a comparison sensible). For the sake of simplicity, polynomial models

$$\hat{y}_t = \sum_{i=0}^{n} \sum_{j=0}^{n} a_{ij} \overline{x}_{1,t}^i \overline{x}_{2,t}^j \quad (3)$$

were applied, and the coefficients, a_{ij}, were determined by linear regression. After this, contour lines, i.e., lines corresponding to constant values of $\hat{y}$, can be depicted on the map to reveal general features of the studied regime.

In the following, a few examples of the contour plots will be presented for the case $n=2$ in eq. (3), but where the cross-terms $(x_i x_j)$ were disregarded.

Figure 4 shows the relation between the classes and the hot metal carbon content, which is an important variable that characterizes the quality of the pig iron produced. It also reflects the contact between the molten pig iron and the coke bed in the lower part of the furnace. The highest values are obtained in the lower part of the diagram, where the heat load on the lowest level (bosh region, cf. Fig. 1) is considerable.

Figure 5 shows the effect of the top gas utilization degree of carbon monoxide – a variable that reflects the efficiency of the operation (of the upper furnace). Also this relation exhibits a saddle point, indicating that the lowest gas utilization corresponds to cases where there is a pronounced heat load (and loss) in the lower shaft. The highest values of η_{CO} are found on the lower edge of the diagram, in agreement with the findings presented above for the hot metal carbon content.

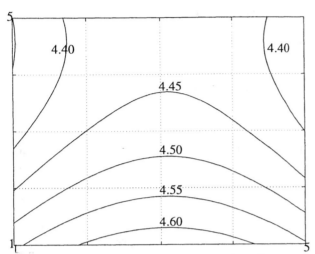

Fig. 4. Contour plots of hot metal carbon, C̲.

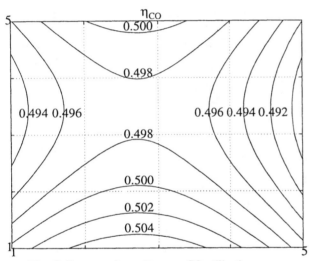

Fig. 5. Contour plots of top gas CO utilization, η_{CO}.

4   Conclusions

A method for classification of wall temperatures in an industrial furnace has been presented. An unsupervised learning algorithm forms the basis of the model, which has been augmented by including (regression) models mapping the classification to central process variables. The model can thus be seen as a combination of unsupervised and supervised learning techniques. The results indicate the existence of complex relations between the wall temperatures, which are affected by the way in which the furnace is charged and cooled, and production indices. Future work will study these interesting relations in more detail, and apply the technique in practice for interpretation of the state of the blast furnace.

References

[1] Biswas, A.K.: Principles of Blast Furnace Ironmaking. Brisbane: Cootha Publishing House 1981.
[2] Omori, Y., et al.: Blast Furnace Phenomena and Modeling. London: Elsevier Applied Science 1987.
[3] Hirata T., et al.:Blast Furnace Operation System using Neural Networks and Knowledge-Base. Proc. 6th International Iron and Steel Congress, Nagoya, Japan, pp. 23-27, 1990.
[4] Takada H., et al.: A Design Environment for Neural Networks and Its Applications. Control Engineering Practice, Vol. 2, pp. 123-128, 1994.
[5] Bulsari, A.B., Saxén H.: Classification of blast furnace probe temperatures using neural networks. Steel Research, Vol. 66, pp. 231-236, 1995.
[6] Otsuka Y., et al.: Application of Neural Network Systems to Pattern Recognition of Blast Furnace Operation Data. Kobelco Technology Review, Vol. 15, pp.12-16, 1992.
[7] Kohonen, T.: Self-Organization and Associative Memory. Berlin: Springer-Verlag 1989.

Monitoring an Industrial Plastic Injection Moulding Machine Using Neural Networks

N. Costa[1], A. Cunha[2], B. Ribeiro[1],

[1]CISUC – Centro de Informática e Sistemas da Universidade de Coimbra
Dep. de Eng. Informática da Univ. de Coimbra, Pinhal de Marrocos, Polo II, P-3030 Coimbra, Portugal
Tel: (+351) 39 790000 Fax: (+351) 39 701266
[2]Polymer Engineering Department, Univ. do Minho, Azurém, 4800 Guimarães, Portugal
E-mail: {nfilipe, bribeiro}@dei.uc.pt, amcunha@eng.uminho.pt

Abstract

Plastics are nowadays one of the most used materials in several industries. From domestic tools up to the automotive industry, there is an enormous number of possible applications. The pressures imposed by the market have led to the development of new strategies capable of answering the required demands. Neural networks have revealed high potential in a wide range of situations and have been successfully applied in fault detection and diagnosis systems. In this paper we intend to clarify, in part, the different diagnostic methodologies and, on the other hand, we suggest a neural network approach for monitoring the plastic injection moulding process. Future work will use the developed neural monitoring scheme for process fault diagnosis with the aim of industrial quality management.

1 Introduction

Plastics are doubtless part of our life. The wide range of applications, the lower production costs associated and the ease with which they can be processed, make them one of the most used materials in nowadays. As the quality and strength of materials improved, plastics began to replace steel, aluminium and other metals. Although the automotive industry leads in the amount of plastic used in their products, many other industries like agriculture, computer, appliance, medical, furniture, lawn and garden tools, are now turning to plastics. From all the existing plastic processing methods, one of the most common is injection moulding. New methodologies are being investigated, not only to increase production, but also to increase the quality of the final products. Under this perspective, new interest has arisen in on Fault Detection and Diagnosis (FDD) mechanisms. Generally, an FDD system is composed by a model used for comparison with the system's response. Faults are detected if the residual signal, ε, generated from the difference between real measurements and their estimates using

the model, exceeds a predefined threshold. If this happens, then the residual is analysed and the proper measures taken in order to stabilise the parameter values and, consequently, reduce the anomalies (see Figure 1). However, this is not an easy task as the dynamics of the process seem to be rather complex. The strong non-linear relationships between the process parameters and their reciprocal effects as well as the unpredictable behaviour of the material under temperature-pressure fluctuations, makes not only the mathematical analysis of its behaviour, but also the quality control of the moulded part difficult. Additionally, the lack of consistent relationships between the process parameters, the material properties and the moulded part quality as well as the environmental fluctuations, introduces several sources of variation which prevent the attainment of an ideal state of constant quality [1].

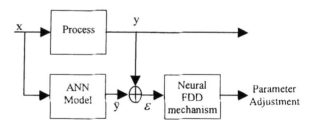

Fig. 1. Fault Detection and Diagnosis general proposed scheme.

So far, Statistical Process Control (SPC) has been used to monitor the injection process in order to prevent defects in the moulded parts. This technique, however, is time consuming and inefficient, since the evaluation of the whole production is based on samples. These disadvantages, together with the high equation development costs in conventional modelling, due to the complex dynamics of the injection moulding process, imply the need for new alternative methods. Because of their capability of learning from examples without needing the process analytical description, neural networks are an attractive approach not only in process modelling but also in fault diagnosis. Furthermore, it is believed that it is possible to increase the final product quality by reducing the anomalies and by acting, whenever necessary, in the whole process behaviour by predicting its future state using an Artificial Neural Network (ANN) model.

In section 2, a description of the process is given. Section 3 intends to clarify the current state of fault diagnosis methods. The need to understand the process behaviour and how it responds to changes in the environment, implies the analysis of process data, performed in section 4. To this end, two different statistical methods have been used: common statistics and Principal Component Analysis (PCA). Finally, in section 5, a neural network is derived and the obtained results displayed.

2 Process Description

In principle, injection moulding is a simple process. A thermoplastic, in the form of granules, passes from a feed hopper into a barrel where it is heated and melted. It is then injected into a cold mould which is clamped tightly closed. When the plastic has had enough time to return to a solid state the mould opens and the produced part is extracted. All details of the mould are reproduced in the finished part.

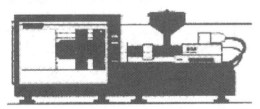

Fig. 2. Injection moulding machine.

Seven process variables are monitored: cycle time, dosage time, melt temperature at the nozzle, injection time, melt temperature before the nozzle, cushion and injection velocity.

Although the injection moulding process can be applied to a large variety of polymers, the manufacture of only one particular part, from the automotive industry, is considered here. It was moulded in a DN502 mould using the Hostacom D M2 T06 material.

3 Current State of Fault Diagnosis Methods

Over the past two decades, several diagnostic methods have been suggested, although their classification is still being ambiguous and subjected to discussion. There are two main approaches for fault diagnosis: the parametric and the non-parametric approach. The former is based on analysing an estimate of the feature vector, whereas the latter is based on analysing a signal, denoted as residual [2]. The parametric approach, in spite of not being implementable in real time, is based on estimating the coefficients of the process transfer function and on computing the physical parameter from the estimate. The non-parametric approach can still be subdivided into two main groups: (i) the model-based methods and (ii) the artificial intelligence based ones.

Model-based techniques can be further classified into estimation methods (either of state variables or of parameters) and pattern recognition methods [3]. Estimation of state variables is based on the generation of residuals in one of several ways: (a) parity space, which generates the residual by projecting the mathematical relationships linking process variables to eliminate the unknown state variables, (b) dedicated observers, which estimate the state variables in order to compute the outputs which are then compared to the real ones, and (c) fault detection filters. The estimation of parameters uses the inputs and the outputs of the process to estimate the structural parameters and their evolution. Pattern recognition methods cover three phases: measurement, feature extraction, which removes redundancy and generates a pattern strictly related to the actual mode of operation of the system, and classification, where the actual pattern is associated with one of the operation modes of the system. Pattern recognition techniques are applied when dynamics are negligible [4]. Model-based methods require, however, an exact model of the system which is not always possible as the process dynamics is, in most cases, unknown or partially known. A possible solution would be linear approximation. However, this means that some dynamic characteristics would be left out, leading to an unsuitable model. A more plausible solution is to use artificial intelligence techniques, such as knowledge-based systems, fuzzy logic sets or neural networks, since they do not need mathematical models. The choice of the correct method depends, in a great extent, on the desired accuracy and on the complexity of the problem. It is well known that in complex processes, like those found in the plastics processing industry, it is extremely difficult to develop an accurate mathematical or physical model due to the existence of process non-linearities. In this sense, there is almost never an exact agreement between the process and its respective estimated model. The necessary simplifications and assumptions leave out some relevant information, increasing the degree of imprecision of the model, which may lead to incorrect decisions by the FDD mechanism, affecting the whole system's response and leading to performance degradation.

Neural networks have remarkable advantages for heuristic modelling due to their capability to generalise by inductive learning from examples. Neural networks require little or no *a priori* knowledge of systems, and provide an effective tool for dealing with non-linearity [5]. However, whichever modelling method is chosen, data analysis is still a crucial step as it gives us the fundamental insight knowledge of the process, allowing the development of more effective models.

4 Principal Component Analysis on Process Data

In a modern injection-moulding machine, 10 to 20 process variables are usually provided during each production cycle, but only a small number has a significant influence on the final quality of the product. By examining the way the system reacts to variations in the process variables, it is possible to identify the most significant parameters related to product quality and detect the causes for process anomalies. Although the relationships between the process state and the product quality are generally unknown and not easily identified, they can be statistically described either by conventional approaches or PCA.

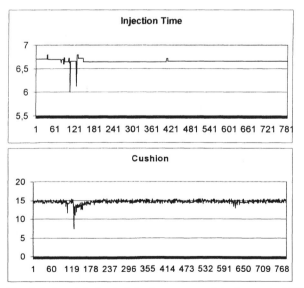

Fig. 3. Industrial plastic injection process data for a set of 782 samples.

Linear relationships between the variables may be found by examining the covariance matrices or the correlation coefficients. High coefficients clearly show strong correlation between the parameters. However, in the presence of time dependencies or non-linear relationships, the analysis will fail, since it only works for linear time-independent processes. Nevertheless, although the injection moulding process is known to be extremely non-linear, some variables, like Injection Time and Cushion, seem to be related, as indicated in Figure 3 and by a relatively high calculated correlation coefficient of 0.9148.

Alternatively, Principal Component Analysis allows clusters of linked variables to be derived. One may think of PCA as a technique to find the directions in which a cloud of data points is most stretched. One way of finding the Principal Components of a data set is to calculate the eigenvectors of the data correlation matrix. The first eigenvector corresponding to the largest eigenvalue is, by definition, the direction in the space defined by the columns of the data matrix that describes the maximum amount of variation or spread in the samples. In this way, it is possible to extract the main relationships between the data as well as to detect its abnormal behaviour. When the process is moving out of control variables indicative of the cause of the deviation need to be identified. Examining their contribution to the calculated scores, it is possible to identify the set of original variables whose contribution may be reflective of the non-conforming behaviour. As for early warning, variables that indicate a malfunction may have little impact on the first or on the second principal components but may be highlighted in the lower order components. The important components are those where particular variables are seen to make the largest contribution to an individual principal component. These variables are identified as the main indicators of the process moving away from the nominal operating region. Therefore, the lower components are of critical importance in the identification of malfunctions.

PCA has proved to be quite effective in distinguishing normal from abnormal behaviour in the injection-moulding problem. The process variables indicated in section 2 were monitored and an experimental data set with 188 samples obtained, from which 116 form a nominal data subset of good production. The other 72 correspond to a phase during which the process was subjected to various malfunctions.

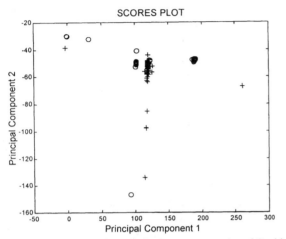

Fig. 4. Scores plot for principal components 1 and 2 with process data.

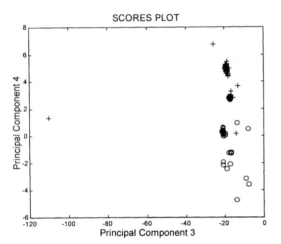

Fig. 5. Scores plot for principal components 3 and 4 with process data.

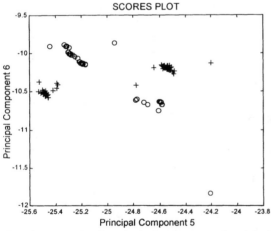

Fig. 6. Scores plot for principal components 5 and 6 with process data.

A principal component analysis was performed on the data set. The first two principal components explain approximately 96% of the variability in the process variables. Bivariate monitoring charts were constructed using the calculated principal components. Figures 4, 5 and 6 show the scores plots for principal components one and two, three and four and principal components five and six, respectively ('+' define the nominal data, whilst 'o' define abnormal data). As can be seen, the evidence of a problem is most clearly highlighted by principal component four. The identification of the combination of variables, which contribute to the particular principal component of interest (PC4), can be done by analysing the loadings. Variables six (cushion) and even four (injection time) exhibit the highest values. Interpreting the effect of these variables in the process behaviour would allow the identification of the most probable causes of the faults.

5 The Neural Network Process Monitoring Approach

In this paper, a method for process monitoring is suggested in the setting of the injection moulding process problem. The main objective is the detection of defects of the produced parts as well as the identification of their causes. The parts are manufactured for the automotive industry, very demanding in terms of product quality. Parts with burning marks, spots, streaks or anomalous dimensions should be detected and isolated. Once the faults are known and their causes identified, one can easily train a neural network to act on the system's responses in order to detect anomalies and identify its causes.

Table 1. Data measurement ranges.

	Minimum value	Maximum value	Average value
Cycle Time (s) 28; 38; 48	34.000	245.000	37.975
Dosage Time (s) 7.5; 12.5; 17.5	8.940	9.9000	9.341
Injection time (s) -0.5999; 4.4; 9.4	6.010	6.790	6.654
Cushion (mm) 12.8; 17.8; 22.8	7.400	15.400	14.622

As has been shown in previous work, neural networks have provided better results than other methods [1]. Their ability to learn from examples makes them suitable for modelling and predicting the process behaviour without detailed knowledge of the process (as is the case in injection moulding).

From the seven process variables indicated in section 2, cycle time, dosage time, injection time and cushion were chosen as inputs for the network. Table 1 shows the minimum, maximum and the average values for each one of the process variables used. The other three have been discarded from the training set since they exhibited constant behaviour and therefore do not contribute to the solution of the problem. The purpose was to train a network that could predict, one step in advance, the process behaviour in terms of the injection time and cushion parameters (Figure 7), since these were the ones that better identified the faults (see section 4). All data points were previously scaled according to the activation function of the hidden layer.

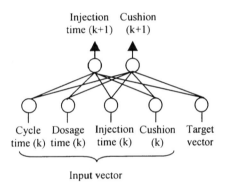

Fig. 7. Neural network general scheme.

To choose the best network, several neural network models were built and trained on 782 data samples using the backpropagation (BP) algorithm. Feedforward multi-layer perceptron (MLP) and Elman networks, with different complexity, were used and tested on a validation set formed by 784 independent samples. Table 2 shows the obtained results. Although both networks have similar performances, the Elman recurrent network, with 10 hidden neurons and tan-sigmoidal activation function, exhibits lower training times, converging more rapidly to the desired error value. This can partially be explained by the existence of a short-term dynamic memory responsible for the ability to learn to recognise and generate temporal patterns, as well as spatial patterns [6]. This memory, which is simply a copy of the hidden layer, acts as an extension to the input layer, holding the values of the hidden units from the previous time step and using them as inputs for the current time step.

Table 2. Neural network performance comparison.

Neural Networks	Elman	MLP
Hidden neurons	10	10
Sum Squared Error	0.0477	0.0499
Training epochs	4845	7000

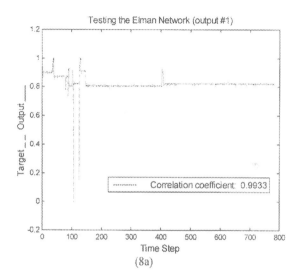

(8a)

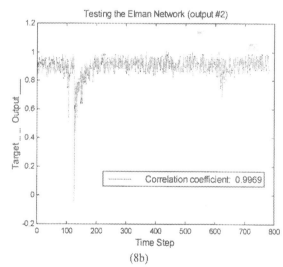

(8b)

Fig. 8. Neural network behaviour using the training set: a) injection time; b) cushion.

By using the above neural network, a process identification of the injection-moulding machine was performed. Figures 8 and 9 show the behaviour of the real injection time and cushion compared to the network's prediction. It is clearly seen that the network performs quite well.

94

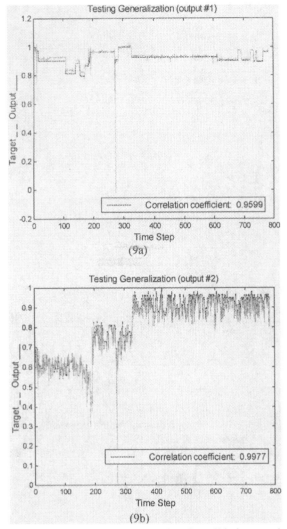

Fig. 9. Neural network behaviour using the validation set: a) injection time; b) cushion.

6 Conclusions

Neural networks are well known for their capacity to deal with non-linearities. An increase of accuracy of the process model and, implicitly, of the whole process, can be expected due to the identification of complex processes like the injection moulding process, characterised by delays, time variances and material properties.

There are several issues that remain to be addressed. For example, the number of process parameters used could be increased during training. Future work will include the results presented here for fault diagnosis of the injection moulding process with the aim of on-site quality management. Furthermore, it is expected that they will be used for process control in order to obtain the machine set-up parameters with the ultimate goal of reducing the production costs while increasing the overall production.

Acknowledgements

The authors would like to thank PLASFIL - Plásticos da Figueira, Lda. for the availability of the mould and for making the moulding data available. The authors would also like to express their gratitude to PRAXIS XXI, Sub-Programa Ciência e Tecnologia do 2º Quadro Comunitário de Apoio, which partially supports this project.

References

[1]   Tatiana Petrova and David Kazmer, *Development of a Hybrid Network for Quality Control of Injection Molding*. University of Massachusetts Amherst, 1997.

[2]   Ali Alawi and R. Doraiswami, *Real time fault diagnosis*. IFAC Conference SAFEPROCESS'97, vol. 1, pp 432-437, Hull, UK, 1997.

[3]   T. Sorsa, H. N. Koivo and H. Koivisto, *Neural networks in process fault diagnosis*. IEEE Transactions on Systems, Man and Cybernetics, vol. 21, no. 4, pp. 815-825, 1991.

[4]   H. N. Koivo, *Artificial neural networks in fault diagnosis and control*. Control Engineering Practice, vol. 2, no. 1, pp. 89-101, 1994.

[5]   J. Zhou and S. Bennett, *Dynamic System Fault Diagnosis Based on Neural Network Modelling*. IFAC Conference SAFEPROCESS'97, vol. 1, pp. 54-59, Hull, UK, 1997.

[6]   J. L. Elman, *Finding structure in time*. Cognitive Science, 14, pp. 179-211, 1990.

Rates of Learning in Gradient and Genetic Training of Recurrent Neural Networks

Ricardo Riaza, Pedro J. Zufiria
Grupo de Redes Neuronales
Dept. Matemática Aplicada a las Tecnologías de la Información
Escuela Técnica Superior de Ingenieros de Telecomunicación
Universidad Politécnica de Madrid
Ciudad Universitaria s/n. 28040 Madrid, Spain
Email: {rrr,pzz}@mat.upm.es

Abstract

In this paper, gradient descent and genetic techniques are used for on-line training of recurrent neural networks. A singular perturbation model for gradient learning of fixed points introduces the problem of the *rate of learning*, formulated as the relative speed of evolution of the network and the adaptation process, and motivates an analogous study when genetic training is used. The existence of bounds for the rate of learning in order to guarantee convergence is obtained in both gradient and genetic training. Some computer simulations confirm theoretical predictions.

1 Introduction

Fixed point learning in recurrent neural networks may be modeled as a problem of parameter adaptation in a dynamical system. In the supervised case, this adaptation is usually performed by means of an optimization technique, which attempts to minimize a criterion based on the distance to a prefixed objective or set of objectives.

Gradient descent techniques, which are pervasive in the training of feedforward neural networks, may be naturally extended to recurrent systems, as shown in [1, 5]. It has been proved that certain assumptions made in these works may be relaxed through the use of a singular perturbation model for gradient schemes in recurrent contexts [6]. This model introduces the problem of the *rate of learning*, based on the relative speed of evolution of the network and the gradient process, and states the existence of lower bounds for their ratio to guarantee convergence of the scheme.

The goal of this work is to introduce and analyze the question of the rate of learning when a genetic scheme is used for the adaptation of parameters in recurrent networks. The stochastic nature of genetic algorithms obstructs a direct application of the analytical results valid for the (deterministic) gradient case. A heuristic approach, however, predicts a similar behavior in the genetic setting. Some computer simulations confirm these predictions.

The document is structured as follows. Section 2 is devoted to the theoretical study of both gradient and genetic training of recurrent networks. A singular perturbation model for gradient learning of fixed points is presented in subsection 2.1. Single pattern and multiple pattern problems are considered. This model predicts the existence of bounds in the ratio between the speeds of the network and the adaptation process to guarantee convergence, and motivates an analogous study for a genetic training scheme which is proposed in 2.2. Some ideas for the formalization of the genetic process through a deterministic continuous-time system might help to rigorously state this phenomenon and are presented in this subsection. Section 3 summarizes some numerical experiments which confirm theoretical predictions. Finally, concluding remarks are compiled in section 4.

2 On-line Learning of Fixed Points in Recurrent Neural Networks

The problem of supervised fixed point learning in recurrent neural networks may be formalized as an optimization one, as it is the case in feedforward neural networks. The recurrent setting leads naturally to a formulation in a dynamical system context [1], where the network is defined by a system

$$\dot{u} = F(u, w) \tag{1}$$

being $u \in \mathbb{R}^n$ the state vector, while $w \in \mathbb{R}^m$ represents the set of parameters or weights in the network, and $F \in C^1(\mathbb{R}^n \times \mathbb{R}^m, \mathbb{R}^n)$. Particularization of (1) to get different neural models may be seen at [1]. The learning goal may then be described as the determination of a parameter set w^* such that

$$\dot{u} = F(u, w^*) \tag{2}$$

has an asymptotically stable equilibrium point at a prefixed point u^* or set $\{u_1^*, \ldots, u_p^*\}$. These two cases will be referred as *single pattern* and *multiple pattern* problems, respectively. When there is no significant difference between these two situations, the first one may be considered as a particular case of the second with $p = 1$.

Although off-line techniques may be employed to tackle this problem, we will focus our attention in on-line methods, that is, techniques that adapt the parameters w while (1) is evolving in time. In these methods, the problem of the *rate of learning*, relating the speeds of the network and the adaptation process, arises naturally. The study of this question is the main goal of the present paper.

All along the work, the existence of (at least) p asymptotically stable equilibrium points for any fixed value of w will be assumed. These equilibrium points will be denoted as $u_1^f(w), \ldots, u_p^f(w)$. The learning problem may then be formulated as the minimization of the following function $E : \mathbb{R}^m \to \mathbb{R}$:

$$E(w) = \sum_{i=1}^{p} \frac{1}{2} \|u_i^f(w) - u_i^*\|^2 \qquad (3)$$

where $\| \ \|$ denotes the euclidean norm in $\mathbb{R}^n$. This minimization problem will be addressed making use of gradient descent techniques (subsection 2.1) and genetic algorithms (subsection 2.2).

2.1 Gradient Descent Techniques

A general perspective of gradient techniques for recurrent neural network training may be found in [1]. This work covers different training techniques as particular cases, and is based on the following adaptation process:

$$\dot{w} = -E_w(u^f(w)) \qquad (4)$$

where E_w denotes the gradient $\partial E / \partial w$ and a single pattern problem is assumed. The computation of the gradient is carried out through the chain rule, which makes

$$E_w(u^f(w)) = E_{u^f}(u^f(w))u_w^f(w). \qquad (5)$$

Assuming that the Jacobian matrix $F_u(u^f(w), w)$ is regular, the derivative $u_w^f(w)$ may be computed by means of the application of the implicit function theorem to $F(u^f(w), w) = 0$, to get

$$u_w^f = -(F_u)^{-1}F_w. \qquad (6)$$

The substitution of (5) and (6) in (4) leads to

$$\dot{w} = E_{u^f}(F_u)^{-1}F_w, \qquad (7)$$

expression which must be evaluated in $(u^f(w), w)$. This process assumes that the evolution of the network to the point $u^f(w)$ is instantaneous for every w, making the computation of the quantities in (7) feasible. In [6], it is shown that this hypothesis may be relaxed making use of a singular perturbation model, which is presented below.

Single Pattern Model. The pair of equations (1)-(7), with the assumption of instantaneous convergence of (1) to $u^f(w)$, may be considered as the particularization to $\varepsilon = 0$ of the following singularly perturbed system:

$$\varepsilon \dot{u} = F(u, w) \qquad (8)$$
$$\dot{w} = E_u(u)F_u^{-1}(u, w)F_w(u, w) \qquad (9)$$

where the function E must now be understood as $E(u) = \frac{1}{2}\|u - u^*\|^2$. The point u^* will be used as initial value for (8) all along the paper. The case $\varepsilon = 0$ could be loosely interpreted as making (8) to evolve instantaneously to $u^f(w)$, and the substitution $u = u^f(w)$ in (9) leads to the gradient descent formula (7). However, in [6] it is shown that, under the assumption of exponential stability in both (1), with fixed w, and (7), there exists a range $(0, \varepsilon^*)$ of values of ε for which the system (8)-(9) has (u^*, w^*) as an asymptotically stable equilibrium point.

From a local point of view, this means that the assumption of infinite speed in the network is not necessary. Indeed, a ratio longer than $1/\varepsilon^*$ between the rates of the network (1) and the adaptation process (9) is sufficient to guarantee convergence. However, the explicit determination of ε^* requires the knowledge of adequate Lyapunov functions [2, 7], which are not available in most practical cases.

The question of the rate of learning will also be addressed for the genetic case in 2.2. Before that, a generalization to the multiple pattern problem of the model here presented is proposed.

Multiple Pattern Model. When $p > 1$ fixed points $u_1^*, \ldots, u_p^*$ must be learned by the network, the existence of p asymptotically stable equilibrium points $u_1^f(w), \ldots, u_p^f(w)$ must be assumed for every w. Furthermore, if $u_1^*, \ldots, u_p^*$ are used as initial points in the integration, convergence of each of these points to $u_1^f(w), \ldots, u_p^f(w)$, respectively, is necessary for every value of w. The on-line model previously presented for the single pattern case may then be extended to p patterns assuming that p systems evolve in parallel, that is, the state vector u

leads now to p vectors $u_1, \ldots, u_p$. The resulting singularly perturbed system is then

$$\varepsilon_1 \dot{u}_1 = F(u_1, w)$$
$$\vdots$$
$$\varepsilon_p \dot{u}_p = F(u_p, w)$$
$$\dot{w} = \sum_{i=1}^{p} E_{u_i}(u_i) F_u^{-1}(u_i, w) F_w(u_i, w)$$

with $E = \sum_{i=1}^{p} \frac{1}{2} \|u_i - u_i^*\|^2$. Now, there will exist p bounds $\varepsilon_1^*, \ldots, \varepsilon_p^*$ and one can choose $\varepsilon^* = \min\{\varepsilon_1^*, \ldots, \varepsilon_p^*\}$ to make the p networks evolve at the same rate. As mentioned before, the assumptions on the evolution in the p networks are stronger than in the single pattern case, making this problem considerably more difficult in practical situations.

2.2 Genetic Techniques

Genetic algorithms have been used for the optimization of different aspects of artificial neural networks, mainly in a feedforward context [8]. Some applications have also been developed using genetic techniques in recurrent neural networks [9]. Our aim in this paper is to address the question of the rate of learning, previously presented for gradient training, when a genetic algorithm is used for the adaptation of parameters in a recurrent network.

The genetic scheme employed is based on the use of a population of q (real-coded) individuals, each of them representing a complete set of parameters w. These individuals are initialized according to a uniform random distribution, and the selection of individuals for recombination is accomplished by means of the *integral roulette* method, where each old chromosome receives the integral part of its expected number of children. One-point crossover is used, and mutation is performed through the addition of a uniformly distributed variable. All of the individuals of a given population are replaced by their children, with the only exception of the one with the better fitness value, according to an *elitism* scheme.

The fitness criterion used is the key aspect of the learning process, and requires a more detailed explanation. The learning algorithm is based on the discretization of (1) making use of an Euler method, leading to the following discrete dynamical system:

$$u^{j+1} = u^j + \eta F(u^j, w). \tag{10}$$

This system is iterated J times for the individual w, from the initial point u^* (assuming a single pattern problem; the extension to a multiple pattern

case is entirely analogous to that of the gradient setting). The network reaches a state $u^J = \hat{u}(w)$, which may be considered as an approximation of the equilibrium point $u^f(w)$. The fitness value of this individual is then

$$E(w) = \frac{1}{2} \|\hat{u}(w) - u^*\|^2, \tag{11}$$

which is designed to approximate the error function

$$E(u^f(w)) = \frac{1}{2} \|u^f(w) - u^*\|^2.$$

The same operation is done for the next individual, integrating (1) from u^J (to keep the parallelism with the gradient method) and for the rest of the population. The process is repeated until a fitness value smaller than a prefixed E^* is reached.

Formally, this scheme may be described as follows. Let W stand for the population $(w_1, \ldots, w_q)$, and assume that the stochastic operators of selection, crossover and mutation are represented by a single stochastic process v^k. The genetic adaptation process may be then be described through a map G in the following way:

$$W^{k+1} = G(W^k, v^k) \tag{12}$$

being each child w_i obtained as

$$w_i^{k+1} = G_i(W^k, v^k).$$

The coupled evolution of the network and the genetic process is therefore given by

$$u^{j+1} = u^j + \eta F(u^j, w_1^k)$$
$$\vdots$$
$$u^{j+1} = u^j + \eta F(u^j, w_q^k)$$
$$W^{k+1} = G(W^k, v^k)$$

where J iterations are consecutively applied in the network defined by every w_i before applying a single step of the genetic process.

The fitness criterion employed naturally raises the problem of the rate of learning, in a similar manner to that of gradient schemes. If J is large enough for $\hat{u}(w)$ to be a good approximation of $u^f(w)$ (equivalently, if the network is fast enough with respect to the genetic process), the scheme may be expected to converge. On the contrary, if generation replacement is done with an excessively large rate, the value $\hat{u}(w)$ will not be representative and the scheme will not converge. In this case, the fitness value might fall below E^* but, since $\hat{u}(w)$ is not a good approximation of $u^f(w)$, the *real* error $E(w) = \frac{1}{2}\|u^f(w) - u^*\|^2$ may be large.

Therefore, the existence of a rate bound seems to be present also in the genetic context. Secondly, the existence of this bound raises the question of its dependence with respect to the concrete realizations of the stochastic process defined by the genetic algorithm. These topics will be addressed, for a specific case of study, in the following section.

It is worth mentioning that the use of a stochastic model for real-coded genetic learning could allow a rigorous characterization of this phenomenon which, by the moment, may only be heuristically asserted. For doing so, it might be interesting to analyze the possibility of rewriting equation (12) in the following form:

$$W^{k+1} = W^k + \mu^k h(W^k, v^k), \quad \mu^k > 0 \quad (13)$$

where μ is a decaying sequence and h is a deterministic function.

The analysis of this type of models has been traditionally performed via a fundamental theorem for the study of stochastic approximation algorithms, relating the stochastic model with a deterministic continuous-time (DCT) system [3, 4]. Several studies can be found in the literature employing such deterministic continuous-time formulations in order to indirectly interpret the adaptation process in the context of average behavior. Nevertheless, for doing so, (13) must satisfy a set of sufficient conditions concerning the rate of decay on μ, boundedness on W^k, and differentiability of h. Under such assumptions, the average operator could be applied, leading to a deterministic continuous-time system

$$\dot{W} = H(W),$$

where H is related to h in an average sense. If asymptotic stability is proved in such context, convergence in probability of the original discrete-time stochastic model is also guaranteed. Furthermore, the deterministic setting could allow the application of singular perturbation models to characterize the rate of learning of the resulting process, and the translation of these results to the stochastic context might introduce a random characterization of this rate.

Note that the study of the convergence of W^k strongly depends on some assumed properties for defining (13). A careful study should be performed for analyzing the following aspects: firstly, W^k must be assumed to vary slowly enough, so that it can be considered stationary with respect to short-time

statistics. Secondly, independence between v^k and W^k is to be considered. Finally, differentiability of h seems to be the stronger requirement whose relaxation might be addressed via approximation and limiting procedures.

3 Numerical Experiments

A summary of the computer simulation results for a simple example is presented in this section, in order to illustrate the discussion above. This example is defined by the following linear network:

$$\dot{u}_1 = w_0 u_1 + w_1 u_2 - w_2$$
$$\dot{u}_2 = w_3 u_1 + w_4 u_2 - w_5$$

where the point $u^* = (-1, -1)$ is to be learned. The problem of the rate of learning is analyzed in the following subsections, for both the gradient case (3.1) and the genetic one (3.2).

3.1 Gradient Training

When F takes the form of the network described above, the system (8)-(9) results

$$\varepsilon \dot{u}_1 = w_0 u_1 + w_1 u_2 - w_2$$
$$\varepsilon \dot{u}_2 = w_3 u_1 + w_4 u_2 - w_5$$
$$\dot{w}_0 = \frac{(u_1+1)w_4 - (u_2+1)w_3}{w_0 w_4 - w_1 w_3} u_1$$
$$\dot{w}_1 = \frac{(u_1+1)w_4 - (u_2+1)w_3}{w_0 w_4 - w_1 w_3} u_2$$
$$\dot{w}_2 = -\frac{(u_1+1)w_4 - (u_2+1)w_3}{w_0 w_4 - w_1 w_3}$$
$$\dot{w}_3 = \frac{-(u_1+1)w_1 + (u_2+1)w_0}{w_0 w_4 - w_1 w_3} u_1$$
$$\dot{w}_4 = \frac{-(u_1+1)w_1 + (u_2+1)w_0}{w_0 w_4 - w_1 w_3} u_2$$
$$\dot{w}_5 = -\frac{-(u_1+1)w_1 + (u_2+1)w_0}{w_0 w_4 - w_1 w_3}$$

Both the network and the adaptation process have been discretized in computer simulations using the Euler method with stepsize 0.001. The effect of ε has been analyzed through the parameter J, which corresponds to the number of iterations of the network before a single step of the adaptation process is applied. For instance, to simulate a value $\varepsilon = 0.2$ ($1/\varepsilon = 5$), the network must be iterated $J = 5$ times before each step of the gradient process.

Computer simulations have shown that a large number of initial guesses in the parameters leads to nonconvergence of the scheme, meaning that there exists a considerable domain in the space of parameters not included in the attraction domain of any acceptable w^*. However, in the cases in which the scheme converges, a clear dependence on the number J of iterations is displayed. Low values of J

(approximately smaller than 10), lead to nonconvergence of the algorithm, while larger values make it converge. These results suggest that the rate of learning, understood as the ratio between the speeds of the network and the adaptation process, must be greater than 10, at least in the cases here considered.

3.2 Genetic Training

In the genetic case, the network has been discretized with stepsize 0.001, as before, while the gradient process has been substituted by the algorithm described in 2.2. The dependence on the initial guesses of the parameters has been shown to be less critical than in the gradient case, possibly due to the random search performed by the genetic algorithm.

With regard to the rate of learning, the simulations have shown that there exists a gradual increase in the number of generations needed to achieve a minimum as the rate of the network (namely, the number of iterations J) decreases. In all the cases, a bound has been obtained for J, below which the scheme does not converge. However, this bound has been proved to be strongly dependent on the specific realization of the stochastic process associated with the genetic algorithm.

As indicated in 2.2, a stochastic model for genetic training could help to base the phenomenon here observed, as well as justify the dependence of the bounds obtained with respect to the realizations of the genetic algorithm. The random nature of ε^* could be clarified within this framework.

4 Concluding Remarks

The problem of the relation between the speeds of the network and the adaptation process in recurrent neural network training has been addressed in this paper. A singular perturbation model, which has been detailed for single pattern and multiple pattern problems, predicts the existence of a bound in the ratio between these speeds when gradient training is used, and motivates an analogous study in genetic training. A heuristic approach to the genetic case predicts a similar behavior, which has been observed in some computer simulations.

The use of a stochastic model for genetic training might allow a deeper characterization of the behavior here reported. Some ideas for the development of such a model have been presented, with the aim of linking the genetic process to a deterministic continuous-time system. It is worth mentioning that the ideas presented in this paper could be extended to different learning paradigms; in particular, for hybrid techniques combining gradient and genetic methods this extension seems to be straightforward.

Acknowledgements

This work has been financially supported by Proyecto Multidisciplinar de Investigación y Desarrollo 14908 of the Universidad Politécnica de Madrid and Project PB97-0566-C02-01 of the Programa Sectorial de PGC of Dirección General de Enseñanza Superior e Investigación Científica in the MEC. Ricardo Riaza wants also to thank the support of a graduate fellowship from the Universidad Politécnica de Madrid.

References

[1] Baldi, P.: Gradient Descent Learning Algorithm Overview: A General Dynamical Systems Perspective, IEEE Trans. on Neural Networks 6, 182-195 (1995).

[2] Kokotovic, P. V., Khalil, H. K., O'Reilly J.: Singular Perturbation Methods in Control: Analysis and Design. Academic Press 1986.

[3] Kushner, H. J., Clark, D. S.: Stochastic Approximation Methods for Constrained and Unconstrained Systems. Springer-Verlag 1978.

[4] Ljung, L.: Analysis of recursive stochastic algorithms, IEEE Tr. Aut. Cont. 22, 551-575 (1977).

[5] Pineda, F. J.: Generalization of Back-Propagation to Recurrent Neural Networks, Physical Review Letters 59, 2229-2232 (1987).

[6] Riaza, R., Zufiria, P. J.: A singular perturbation approach to fixed point learning in dynamical systems and neural networks, Proc. 2nd World Multiconf. Systemics, Cybernetics and Informatics, Orlando, USA, 1, 616-623 (1998).

[7] Saberi, A., Khalil, H., Quadratic-type Lyapunov functions for singularly perturbed systems, IEEE Tr. Aut. Cont. 29, 542-550 (1984).

[8] Whitley, D., Genetic Algorithms and Neural Networks, in "Genetic Algorithms in Engineering and Computer Science", Winter, G., Périaux, J., Galán, M., Cuesta, P., eds. John Wiley & Sons, 203-216 (1995).

[9] Wieland, A. P., Evolving Neural Networks Controllers for Unstable Systems, IEEE Intl. J. Conf. on Neural Networks, Seattle, USA, 2, 667-673 (1991).

Influence of the Learning Gain on the Dynamics of Oja's Neural Network

Pedro J. Zufiria

Grupo de Redes Neuronales

Depto. Matemática Aplicada a las Tecnologías de la Información

Escuela Técnica Superior de Ingenieros de Telecomunicación

Universidad Politécnica de Madrid

Ciudad Universitaria s/n. 28040 Madrid, Spain

Email: pzz@mat.upm.es

Abstract

In this paper, the dynamical behavior of Oja's neural network [7] is analyzed. Oja's net has been traditionally studied in the continuous-time context via some simplification procedures, some of them concerning the asymptotic behavior of the learning gain. The contribution of the paper is the study of a deterministic discrete-time (DDT) version, preserving the discrete-time form of the original network and allowing a more realistic behavior of the learning gain. As a consequence, the discrete-time nature of the new model leads to results which are drastically different to the ones known for the continuous-time formulation. Simulation examples support the presented results.

1 Introduction

Oja' neural net is a hebbian architecture widely employed the determination of the Principal Components of the correlation matrix associated with a given stochastic vector. At the same time, it also performs the Karhunen-Loeve transform of input data. The analysis of Oja's net has been traditionally performed via a fundamental theorem for the study of stochastic approximation algorithms [10], relating the stochastic model with a deterministic continuous-time (DCT) system [3, 6]. Hence, several studies can be found in the literature employing such deterministic continuous-time formulations in order to indirectly interpret Oja's net dynamics in the context of average behavior. In [15] a Lyapunov function is proposed for globally characterizing Oja's DCT model with a single neuron. Another single neuron generalized version of Oja's DCT net is studied in [14] by explicitly solving the system of equations. The global behavior of a several neuron

Oja's DCT net is determined in [13] by explicitly solving the equations of the model, whereas [9] addresses a qualitative analysis of generalized forms of this DCT network.

In this paper, we first reformulate some of the results from the point of view of indirect Lyapunov qualitative dynamical system analysis. The contribution of this work is the study of a deterministic discrete-time (DDT) version, preserving the discrete-time form of the original network and modeling the learning gain evolution in a realistic manner. This model approaches in certain sense the above mentioned continuous-time systems, but the discrete formulation leads to a drastically different behavior.

In order to illustrate concepts in a simplified manner, a single neuron Oja's model is considered, which provides the first Principal Component of the above mentioned correlation matrix.

The paper is structured as follows. Section 2 presents the single neuron Oja's neural net original model. Section 3 shows the stochastic formulation of the net and its continuous-time approximation. The deterministic discrete-time version is presented in Section 4. An example illustrating the behavior of the net in a very simple case is shown in section 5. Concluding remarks are presented in section 6

2 Oja's Neural Network

As mentioned above, Oja's neural network is usually employed in the context of determining Principal Components of second order moments matrices and the implementation of the Karhunen-Loeve transform.

The network is represented by a linear model which is trained via unsupervised learning. Therefore, its evolution can be modeled by the following set of stochastic difference equations:

$$y(n) = w^T(n)x(n),$$
$$w(n+1) = w(n) + \eta(n)y(n)[x(n) - y(n)w(n)],$$
$$\eta(n) > 0. \quad (1)$$

The input of the net $x(n) \in \mathbb{R}^{p+1}$, $n = 0, 1, \ldots$ can be modeled as a discrete-time stochastic process. Such process is constructed from a set $x(1), x(2), \ldots$ of independent and identically distributed samples upon a distribution of a random variable X. Therefore, the output evolution of the net will have to be considered as another stochastic process $y(n) \in \mathbb{R}$, $n = 0, 1, \ldots$, which is a function of the input and the parameters of the net. Specifically, such input/output relationship is determined by a weight vector $w(n) \in \mathbb{R}^{p+1}$, $n = 0, 1, \ldots$ which is updated as a function of the input and the output in each instant of time. Finally, the learning gain sequence $\eta(n)$ is selected according to the desired behavior, as commented later.

The network so designed, provides with w and y^2 an estimate of the the maximum eigenpair associated with the second order moments matrix of X –sometimes called the correlation matrix of process $x(n)$–.

The study of the convergence of $w(n)$ –in different senses– strongly depends on the assumed properties of $x(n)$. Such study can be framed within the so called *stochastic approximation algorithms* [3, 6, 10], and is partially addressed in the next section.

3 DCT Approximation of Stochastic Formulation

3.1 Fundamental Results of Continuous-Time Formulation

Following [4], the fundamental stochastic approximation theorem [3, 6] is applied to Oja's model defined in (1). The theorem is based in the following assumptions:

- $w(n)$ varies slowly enough and it can be considered as stationary with respect to short-time statistics:

$$E[w(n+1)/w(n)] = w(n) + \Delta w(n).$$

- $x(n)$ is an stationary stochastic process whose correlation matrix has distinct eigenvalues.

- $x(n)$ and $w(n)$ are independent. This assumption, usual in the context of convergence of linear adaptive filters, is true if $x(n)$ is independent from the previous values $x(0), x(1), \ldots, x(n-1)$. In general, it is obvious that $w(n) \in \sigma(x(0), x(1), \ldots, x(n-1))$, the sigma-field generated by previous values of x.

- There exists the average asymptotic value $q_0 = \lim_{n \to \infty} E[w(n)]$.

Then, the stochastic approximation theorem (based on the application of the expectation operator $E[\cdot]$ to (1)) leads to the study of the deterministic continuous-time system:

$$\frac{d}{dt}w(t) = Rw(t) - w(t)[w^T(t)Rw(t)], \quad (2)$$

where $R = E[x(n)x^T(n)]$ is the matrix of second moments of process $x(n)$. Finally, it is important to note that the condition $\eta(n) \to 0$ (usually, $\eta(n) = \frac{1}{n}$) is necessary for applying the theorem.

System (2) has been studied in [4] via Lyapunov stability theory, where a summary of the fundamental results about behavior the system is presented. Generalized formulations of (2) have been analyzed in [9, 13, 14]. Their fundamental results are:

- The equilibria of the system are the eigenvectors of R with norm one, as well as the origin of coordinates. This result can be directly obtained from the equilibrium point condition $0 = Rq - q(q^T Rq)$ to be satisfied by 0 and any value $q_0, \ldots, q_p$, eigenvectors of norm 1 of R.

- All of the equilibria are unstable, except for $\pm q_0$, eigenvectors associated with the largest eigenvalue λ_0.

- Such pair $\pm q_0$ defines an invariant globally asymptotically stable (except for a set with zero measure).

Results about (2) are employed in [11] to derive properties about system (1). For instance, there exists a compact set $A = \{w(n)/\|w(n)\| \leq a\}$ in $\mathbb{R}^{p+1}$ so that if $w(n) \in A$ then $\|w(n+1)\| < \|w(n)\|$ w. p. 1. Hence, the study concludes that $w(n) \to q_0$ w. p. 1.

3.2 Relevance of a Lower Bound on Step η

As mentioned earlier, due to computational round-off limitations, whenever n increases, small but constant values of η are employed in practice. This means that all the previous analyses are not directly applicable. As a matter of fact, this modification may lead to a crucial change of behavior in the network. Is is known that continuous time formulations approximated with fixed step discrete ones may lead to discrete systems with very different dynamics.

This fact motivates the direct study of the discrete-time formulation in the next section.

4 Deterministic Discrete-Time Formulation

In this section, a deterministic discrete system characterizing the evolution of deterministic Oja's network is considered. Such system can be obtained applying the expectation operator to (1):

$$w(n+1) = w(n) + \eta[Rw(n) - w(n)w(n)^T Rw(n)], \quad (3)$$

and it also can be obtained as the numerical integration of (2) via Euler's method with step η.

This formulation reflects in a very appropriate manner the lower bound limitation on η.

4.1 Local Study

Here, we begin considering some implications of η with respect to local dynamical aspects such as the stability of equilibria. The fundamental result is presented, where, for simplifying purposes in the exposition, each pair $\pm q_j$ of equilibria is considered as a single invariant.

Theorem: *Let us consider the deterministic discrete system characterizing Oja's net behavior (3). Fixed points $0, q_j$, $j = 1, \ldots p$ are unstable; q_0 is an asymptotically stable fixed point if*

$$\eta < \frac{1}{\lambda_0}.$$

Proof:
Let us consider first the origin $w = 0$. Following Lyapunov indirect method we have

$$\frac{\partial G}{\partial w}\Big|_0 = I + \eta R = J_0.$$

Since R is a positive definite matrix with eigenvalues $\lambda_i > 0$, $i = 1, \ldots p$, then J_0 has eigenvalues $1 + \lambda_i > 1$, $i = 1, \ldots p$, associated with the same corresponding eigenvectors, and instability of $w = 0$ is concluded.

Since the rest of equilibria are $w(n) = e_j$, normalized eigenvectors of R, associated with eigenvalues λ_j, then

$$\frac{\partial G}{\partial w}\Big|_{e_j} = I + \eta[R - \lambda_j I - 2\lambda_j e_j e_j^T] = I + \eta B_j = J_j,$$

where the stability of e_j will depend on the spectral analysis of J_j. Note that λ is eigenvalue of J_j iff

$$|J_j - \lambda I| = 0 \Leftrightarrow |I + \eta B_j - \lambda I| = 0$$
$$\Leftrightarrow \eta|B_j + \frac{1}{\eta}I - \frac{\lambda}{\eta}I| = 0 \Leftrightarrow |B_j - \frac{\lambda - 1}{\eta}I| = 0,$$

which is equivalent to $\frac{\lambda-1}{\eta} = \beta$ being eigenvalue of B_j, associated to the same eigenvector –i.e., $\lambda = 1 + \eta\beta$–. In addition, since $B_j = R - \lambda_j I - 2\lambda_j e_j e_j^T$ has β as an eigenvalue then

$$|B_j - \beta I| = 0 \Leftrightarrow |R - 2\lambda_j e_j e_j^T - (\beta + \lambda_j)I| = 0,$$

so that $\gamma = \beta + \lambda_j$ is eigenvalue of $R - 2\lambda_j e_j e_j^T$, with the same associated eigenvector as B_j. The relation $\beta = \gamma - \lambda_j$ reduces the original spectral analysis to the calculation of γ. Note that eigenvalues of $R - 2\lambda_j e_j e_j^T$ can be directly computed since such matrix has the same eigenvectors as matrix R. If we multiply the matrix by such eigenvectors we obtain that its eigenvalues are

- $\gamma_i = \lambda_i$, eigenvalue of R, if $i \neq j$,

- $\gamma_j = -\lambda_j$, eigenvalue of R with alternate sign.

Grouping terms we have that eigenvalues of J_j for fixed point e_j are:

- $\lambda = 1 + \eta\beta = 1 + \eta(\gamma - \lambda_j) = 1 + \eta(\lambda_i - \lambda_j)$ if $i \neq j$,

- $\lambda = 1 + \eta\beta = 1 + \eta(\gamma - \lambda_j) = 1 + \eta(-\lambda_j - \lambda_j) = 1 - 2\eta\lambda_j$.

From Lyapunov theory, the asymptotic stability condition is $|\lambda| < 1, \forall \lambda$ of J_j. Here, we must distinguish the case $j = 0$ corresponding to fixed point $e_0 = q_0$ associated with the largest eigenvalue λ_0.

- $j \neq 0$. Let us consider first the eigenvalues for $i \neq j$:

$$|1 + \eta(\lambda_i - \lambda_j)| < 1 \Leftrightarrow -1 \leq 1 + \eta(\lambda_i - \lambda_j) < 1$$
$$\Leftrightarrow \lambda_i < \lambda_j \quad \& \quad \eta < \frac{2}{\lambda_j - \lambda_i}.$$

It is obvious that for $i = 0$ we get a contradiction since $\lambda_0 > \lambda_j$, $\forall j \neq 0$. Therefore, point e_j is unstable $\forall j \neq 0$. Though it is not needed for stability analysis, we consider eigenvalue $i = j$ for completeness:

$$|1 - 2\eta\lambda_j| < 1 \Leftrightarrow -1 \leq 1 - 2\eta\lambda_j < 1$$
$$\Leftrightarrow \lambda_j > 0 \quad \& \quad \eta < \frac{1}{\lambda_j}.$$

- $j = 0$. Let us also consider first eigenvalues for $i \neq j = 0$:

$$|1 + \eta(\lambda_i - \lambda_0)| < 1 \Leftrightarrow -1 \leq 1 + \eta(\lambda_i - \lambda_0) < 1$$
$$\Leftrightarrow \lambda_i < \lambda_0 \ \& \ \eta < \frac{2}{\lambda_0 - \lambda_i}.$$

where $\forall i$ stability is preserved if $\eta < \frac{2}{\lambda_0 - \lambda_i}$. Considering now the eigenvalue $i = j = 0$:

$$|1 - 2\eta\lambda_0| < 1 \ \Leftrightarrow \ -1 \leq 1 - 2\eta\lambda_0 < 1$$
$$\Leftrightarrow \ \lambda_0 > 0 \ \& \ \eta < \frac{1}{\lambda_0}.$$

Note that $\frac{1}{\lambda_0} < \frac{1}{\lambda_0 - \lambda_j} < \frac{2}{\lambda_0 - \lambda_j}, \forall j \neq 0$. Hence, all the derived conditions are satisfied and, consequently, asymptotic stability of $e_0 = q_0$ is preserved if

$$\eta < \frac{1}{\lambda_0}.$$

e. q. d.

4.2 Global Behavior

As it has been shown in section 3.1, the continuous-time formulation leads to beneficial global results concerning the domains of attraction associated with the pair q_0. Here, some limitations concerning the size of these domains of attraction in the discrete case are indicated in the following theorem.

Theorem: *Let us consider the deterministic discrete system characterizing Oja's net behavior (3). In such system if $\|w(n)\| \geq \sqrt{(1 + \frac{2}{\eta\lambda_p})\frac{\lambda_0}{\lambda_p}}$ then $\|w(n + 1)\| > \|w(n)\|$.*
Proof:
The characterization of the evolution of $w(n)$ can be analyzed using the basis of eigenvectors $\{e_j\}$, $j = 1, \ldots p$:

$$w(n) = \sum_{j=1}^{p} z_j(n)e_j.$$

Plugging this expression into (3):

$$w(n+1) = \sum_{j=1}^{p}[1 + \eta\lambda_j(1 - \sum_{k=1}^{p}\frac{\lambda_k}{\lambda_j}z_k^2(n))]z_j(n), e_j \quad (4)$$

which can be written as follows

$$z_j(n+1) = [1 + \eta\lambda_j(1 - \sum_{k=1}^{p}\frac{\lambda_k}{\lambda_j}z_k^2(n))]z_j(n),$$
$$\forall j = 1, \ldots p,$$

where a sufficient condition of increasing modulus is:

$$\|w(n)\|^2 \geq (1 + \frac{2}{\eta\lambda_p})\frac{\lambda_0}{\lambda_p}$$
$$\Leftrightarrow \sum_{k=1}^{p}\frac{\lambda_p}{\lambda_0}z_k^2(n) \geq 1 + \frac{2}{\eta\lambda_p}$$
$$\Rightarrow \sum_{k=1}^{p}\frac{\lambda_k}{\lambda_0}z_k^2(n) > 1 + \frac{2}{\eta\lambda_p}$$
$$\Rightarrow \sum_{k=1}^{p}\frac{\lambda_k}{\lambda_j}z_k^2(n) > 1 + \frac{2}{\eta\lambda_p}, \ \forall j = 1, \ldots p$$
$$\Rightarrow \sum_{k=1}^{p}\frac{\lambda_k}{\lambda_j}z_k^2(n) > 1 + \frac{2}{\eta\lambda_j}, \ \forall j = 1, \ldots p$$
$$\Rightarrow [1 + \eta\lambda_j(1 - \sum_{k=1}^{p}\frac{\lambda_k}{\lambda_j}z_k^2(n))] < -1, \ \forall j = 1, \ldots p$$
$$\Rightarrow \|w(n+1)\| > \|w(n)\|.$$

e. q. d.

The fundamental new aspect of the discrete system is that, even in cases of stability of q_0, the globality of its domain of attraction is not guaranteed. From the previous theorem we conclude that such domain of attraction is a subset of:

$$A^* = \{w / \|w\|^2 \leq (1 + \frac{2}{\eta\lambda_p})\frac{\lambda_0}{\lambda_p}\}, \quad (5)$$

so that if $w(n)$ is not inside A^*, the trajectory will tend to infinity.

Obviously, if the system is studied under the assumption $\eta(n) \to 0$, the set $A^*(n)$ would depend on n and $A^*(n) \uparrow R^{p+1}$, the study of global stability being very complicated.

4.3 Invariant Subspaces

In this section we consider the dynamics of the system in the invariant subspaces defined by the eigenvector of R, e_j. If $w(n) = z(n)e_j$, $z(n) \in \mathbb{R}$, then equation (4) takes the form:

$$w(n+1) = z(n)e_j + \lambda_j\eta z(n)[1 - z(n)^2]e_j,$$

so that state vector is still proportional to e_j – remains within the invariant manifold– and its evolution can be studied in terms of $z(n)$

$$z(n+1) = z(n)\{1 + \mu[1 - z(n)^2]\} \quad (6)$$

where $\mu = \eta\lambda_j$. The fixed points of system (6) are, as expected, $z_1^* = 0$ and $z_{2,3}^* = \pm 1$. The Jacobian in such points is $J_1 = 1 + \eta\lambda_j$, $J_{2,3} = 1 - 2\eta\lambda_j$ respectively, so that unstability of $z_1^* = 0$ and stability (along the subspace e_j) of $z_{2,3}^* = \pm 1$ is guaranteed whenever $\eta < \frac{1}{\lambda_j}$.

A necessary and sufficient condition of increasing modulus along e_j is

$$|1 + \eta\lambda_j[1 - z(n)^2]| > 1,$$

and the points of the region $z(n)| > 1$ which satisfy such increasing condition are:

$$\|w(n)\|^2 > 1 + \frac{2}{\eta\lambda_j},$$

which matches with previous results.

Chaotic Behavior in Invariant Subspaces

As mentioned earlier, the parameter $\mu \geq 0$ determines the dynamics of the system along the invariant manifold defined by e_j. System (6) has been studied for $\mu \leq 0$ (see [1, 12]). Here, we perform an structural stability analysis of (6) for $\mu \geq 0$.

If $z(0) / z(0)^2 > 1 + \frac{2}{\mu}$ then $z(n) \to \infty$. Therefore, we will study the set of initial conditions $z(0) / z(0)^2 < 1 + \frac{2}{\mu}$. As it is shown below, analytical and computational studies support the existence of chaotic dynamics.

Analytical studies of system (6) can be performed based on the properties of the cubic map $f(z(n)) = z(n)\{1 + \mu[1 - z(n)^2]\}$. Among several results, the following can be underlined from a practical point of view:

- The roots of $f(z) = 0$ which also satisfy $z^* \geq 0$ are $z_1^* = 0$ and $z_2^* = \sqrt{\frac{1+\mu}{\mu}} > 1$.

- If we restrict our attention to positive values of z, then f is contractive within $I_1 = (\frac{1}{\sqrt{3}}, \sqrt{\frac{2+\mu}{3\mu}}) \subset I = [0, \sqrt{\frac{1+\mu}{\mu}}]$.

- If $\mu < \frac{3\sqrt{3}}{2} - 1$ then $I = [0, \sqrt{\frac{1+\mu}{\mu}}]$ is invariant under f (i.e., $f(I) \subset I$). This means that if the dynamics reach such region, they get trapped into it.

- Let $0 < \mu^* < 1$ be a root of $4\mu^3 + 12\mu^2 + 3\mu - 14 = 0$. If $\mu \in (0, \mu^*)$ then if we define $I_2 = [\frac{1}{\sqrt{3}}, \sqrt{\frac{2+\mu}{3\mu}}]$ we have that $f(I_2) \subset I_2$. Note that I_2 is invariant and it is a subset of the contraction region I_1. This implies that there must be an asymptotically stable fixed point in such region; since $\mu \in (0, 1)$ we know that such point is $z^* = 1$.

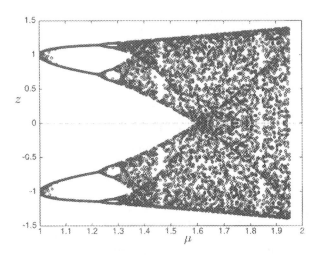

Fig. 1: Bifurcation diagram of z^* as a function of μ.

In addition, a computational study of system invariants has been also performed. Fig. 1 shows the bifurcation diagram corresponding to invariants of system (6), whose main aspects can be summarized as follows:

- $0 < \eta\lambda_j < 1$: system has two fixed points which are asymptotically stable $z^* = \pm 1$ –as shown earlier–.

- $1 \leq \eta\lambda_j < 1.3$: There is a period doubling cascade of bifurcations leading to chaotic behavior for $\eta\lambda_j \approx 1.3$. The are two strange attractor symmetric with respect to the origin.

  - $\eta\lambda_j = 1$: bifurcation point where fixed points become unstable and periodic solutions show up.

  - $1 < \eta\lambda_j < 1.25$: For initial conditions within $[-0.3, 0.3]$ the system tends to one of the two pairs of solutions.

  - $\eta\lambda_j \approx 1.25$: We have a new period doubling bifurcation.

- $1.3 < \eta\lambda_j < 1.6$: The attractors grow in size and they fuse for $\eta\lambda_j \approx 1.6$.

- $1.6 < \eta\lambda_j < 2$: The attractor keeps growing until $\eta\lambda_j = 2$ where it reaches region $z / z^2 > 1 + \frac{2}{\eta\lambda_j} = 2$. In such case, $\forall z(0)$ we have that $|z(n)| \to \infty$, and there is no chaotic behavior any more.

Note that these results support the idea of the existence of complicated dynamics in the discrete case, as well as the existence of unbounded solutions whenever η is not small enough.

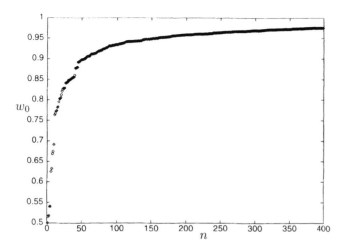

Fig. 2: Evolution of w_0 versus number of iterations. Case 1.

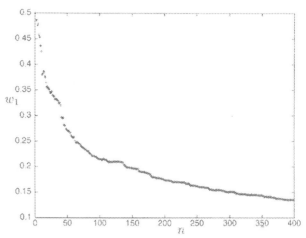

Fig. 3: Evolution of w_1 versus number of iterations. Case 1.

5 A Simple Example

In this section, an example is presented with a two-dimensional input vector whose components $[x_0, x_1]$ are considered to be zero-mean gaussian and independent. Hence, the correlation matrix is diagonal, its eigenvalues $\lambda_0 = \sigma_0^2$, $\lambda_1 = \sigma_1^2$ representing the variances of the corresponding random inputs x_i, $i = 0, 1$. Note that, if $\sigma_0 > \sigma_1$, the eigenvectors are $q_0 = [1, 0]$, $q_1 = [0, 1]$; hence, it is expected that $w(n)$ should converge to q_0.

Here we simply illustrate the relevance of the smallest eigenvalue λ_1 and the quotient $\frac{\lambda_0}{\lambda_1}$, both terms showing up in equation (5) that characterizes region A^*. For that purpose, two simulations are presented, where different values of σ_1 are considered. In both simulations, a decreasing learning gain is employed following the expression $\eta(n) = \frac{0.2}{n}$.

In the first simulation, input values following a gaussian distribution with $\sigma_0 = 2$ and $\sigma_1 = 0.5$ are employed. Fig. 2 and 3 show an evolution sample for $w(n)$ with starting point $w(0) = [0.5, 0.5]$. The evolution of the weights in this case suggests convergence to q_0 as expected, in a satisfactory manner.

In the second simulation, input values following a gaussian distribution with $\sigma_0 = 2$ and $\sigma_1 = 1.9$ are employed. Note that the corresponding regions $A^*(n)$ are considerably smaller in this case, and wandering iterations are more likely to occur. Fig. 4 and 5 show an evolution sample for $w(n)$ with the same starting point as in the previous case, $w(0) = [0.5, 0.5]$. The evolution of the weights in this case does not suggest convergence to q_0 and it can be considered as unsatisfactory.

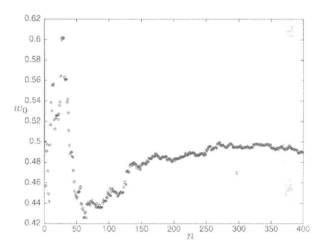

Fig. 4: Evolution of w_0 versus number of iterations. Case 2.

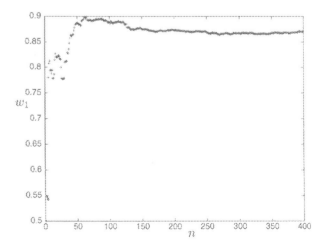

Fig. 5: Evolution of w_1 versus number of iterations. Case 2.

In general, simulations have shown that as λ_0 increases (e.g. $\lambda_0 = 4$) and λ_1 gets close to it (e.g. $\lambda_1 = 3$), $w(n)$ may increase in an unbounded manner aborting the simulations. The fact that large and very close eigenvalues may jeopardize the convergence properties of the model supports the theoretical results presented in this paper.

6 Concluding Remarks

Oja's neural network has been traditionally studied under the fundamental theorem of stochastic approximation theory. In this context, some beneficial local and global behavior properties of the network can be proved. Nevertheless, the fundamental supporting theorem is based on some restrictive assumptions concerning the learning gain limiting behavior. If such assumptions are not satisfied, a discrete formulation must be considered, where a drastic change in the local and global dynamic behavior of the network can be demonstrated. In addition, chaotic behavior shows up in some invariant subspaces. Simulations with a simple example support the presented results. Hence, practical applications are likely to encounter these limitations.

Acknowledgments

This work has been financially supported by Proyecto Multidisciplinar de Investigación y Desarrollo 14908 of the Universidad Politécnica de Madrid and Project PB97-0566-C02-01 of the Programa Sectorial de PGC of Dirección General de Enseñanza Superior e Investigación Científica in the MEC.

References

[1] F. T. Arecchi, R. Badii and A. Politi, *Low-frequency phenomena in dynamical systems with many attractors*, Phys. Rev. A, Vol. 29, NO. 2, pp. 1006-1009, 1984.

[2] J. A. Berzal, P. J. Zufiria and L. Rodríguez, *Implementing the Karhunen-Loeve transform via improved neural networks*, Solving Engineering Problems with Neural Networks. International Conference on Engineering Applications of Neural Networks, EANN'96. pp. 375-378, London 1996. ISBN 952-90-7517-0.

[3] H. J. Kushner & D. S. Clark, *Stochastic Approximation Methods for Constrained and Unconstrained Systems*, New York, Springer-Verlag, 1978.

[4] S. Haykin, *Neural Networks. A comprehensive foundation*, Macmillan Publishing Company, 1994.

[5] R.-W. Liu, Y.-F. Huang, and X.-T. Ling *A Novel Approach to the Convergence of Neural Networks for Signal Processing*. IEEE Transactions on Circuits and Systems, Vol. 42, NO. 3, 187-188, 1995.

[6] L. Ljung, *Analysis of recursive stochastic algorithms*, IEEE Transactions on Automatic Control AC-22, 551-575, 1977.

[7] E. Oja, *A Simplified Neuron Model as a Principal Component Analyzer*, Journal of Mathematical Biology, 15, 267-273, 1982.

[8] F. Peper and H. Noda, *A Symmetric Linear Neural Network That Learns Principal Components and Their Variances*, IEEE Transactions on Neural Networks Vol. 7, NO. 4, 1042-1047, 1996.

[9] M. D. Plumbley, *Lyapunov Functions for Convergence of Principal Component Algorithms*, Neural Networks, Vol. 8, NO. 1, 11-23, 1995.

[10] H. Robbins and S. Monro, *A Stochastic Approximation Method*, Ann. Math. Stat. 22, 400-407, 1951.

[11] T. D. Sanger, *Optimal unsupervised learning in a single-layer linear feedforward neural network*, Neural Networks, vol. 2, pp. 459-473, 1989.

[12] J. Testa and G. A. Help, *Study of a one-dimensional map with multiple basins*, Phys. Rev. A, Vol. 28, NO. 3, pp. 3085-3089, 1983.

[13] W.-Y. Yan, U. Helmke, and J.B. Moore, *Global Analysis of Oja's Flow for Neural Networks*, IEEE Transactions on Neural Networks Vol. 5, NO. 5, 674-683, 1994.

[14] Q. Zhang, and Z. Bao, *Dynamical System for Computing the Eigenvectors Associated with the Largest Eigenvalue of a Positive Definite Matrix*, IEEE Transactions on Neural Networks Vol. 6, NO. 3, 790-791, 1995.

[15] Q. Zhang, and Y. W. Leung, *Energy function for One-Unit Oja Algorithm*, IEEE Transactions on Neural Networks Vol. 6, NO. 5, 1291-1293, 1995.

ATR's Artificial Brain Project: CAM-Brain Machine (CBM) and Robot Kitten (Robokoneko) Issues

Michael KORKIN [(1)], Hugo de GARIS [(2)], N. Eiji NAWA [(2)], William Dee RIEKEN [(2)]

[(1)] Genobyte Inc., 1503 Spruce St., Suite 3, Boulder, CO 80302, USA

http://www.genobyte.com, korkin@genobyte.com

[(2)] Brain Builder Group, Evolutionary Systems Dept., ATR Research Labs,

2-2 Hikaridai, Seika-cho, Soraku-gun, Kyoto 619-0288, Japan

http://www.hip.atr.co.jp/~{degaris, xnawa, wrieken}

{degaris, xnawa, wrieken}@hip.atr.co.jp

Abstract— **This paper presents some ongoing issues concerning ATR's Artificial Brain (CAM-Brain) Project. The CAM-Brain Project evolves 3D cellular automata (CA) based neural networks directly in FPGA electronics at electronic speeds in special hardware called a CAM-Brain Machine (CBM). The CBM updates the CA cells at a rate of 150 Billion a second, and can perform a full run of a genetic algorithm (GA) in about 1 second. 32K of these evolved circuits (modules) are then assembled (in a large RAM space updated in real-time by the CBM) into humanly defined architectures to make an artificial brain to control a robot kitten ("Robokoneko"). The paper presents and discusses the latest design decisions for the CBM and the kitten robot, and maps out future plans aimed at having an artificial-brain-controlled robot kitten playing in the ATR labs by early 2001. A world wide, many membered, internet-videophone and electronic pen based, brain-architectural design and brainstorming group will be essential for distributing the design and evolution of the 32K modules. Designing and building an artificial brain within the next three years will be a major conceptual and managerial challenge.**

1. Introduction

This paper discusses ongoing issues concerning ATR's Artificial Brain (CAM-Brain) Project [1], specifically, the latest design decisions concerning the CAM-Brain Machine (CBM) [3] and the robot kitten ("Robokoneko"). We begin with a general introduction to the CAM-Brain Project as a whole, then describe the CBM design parameters in reasonable detail.The CAM-Brain Machine (CAM stands for Cellular Automata Machine) is a research tool for the simulation of artificial brains. An original set of ideas for the CAM-Brain project was developed by Dr. Hugo de Garis at the Evolutionary Systems Department of ATR HIP (Kyoto, Japan), and is currently being implemented as a dedicated research tool by Genobyte, Inc. (Boulder, Colorado).

An artificial brain, supported by the CBM, consists of up to 32,768 neural modules, each module populated with up to 1,152 neurons, a total of 37.7 million neurons. Within each neural module, neurons are densely interconnected with branching dendritic and axonic trees in a three-dimensional space, forming an arbitrarily complex interconnection topology. A neural module can receive efferent axons from up to 92 other modules of the brain, with each axon being capable of multiple branching in three dimensions, forming hundreds of connections with dendritic branches inside the module. Each module sends afferent axon branches to up to 32,768 other modules.

A critical part of the CBM approach is that neural modules are not "manually designed" or "engineered" to perform a specific brain function, but rather evolved directly in hardware, using genetic algorithms.

Genetic algorithms operate on a population of chromosomes, which represent neural networks of different topologies and functionalities. Better performers for a particular function are selected and further reproduced using chromosome recombination and mutation. After hundreds of generations, this approach produces very complex neural networks with a desired functionality. The evolutionary approach can create a complex functionality without any a priori knowledge about how to achieve it, as long as the desired input/output function is known.

2. CoDi Neural Model

The CBM implements a so called "CoDi" (i.e. Collect and Distribute) [2] neural model. It is a simplified cellular automata based neural network model developed at ATR HIP (Kyoto, Japan) in the summer of 1996 with two goals in mind. One was to make neural network functioning much simpler and more compact compared to the original ATR HIP model, to achieve considerably faster evolution runs on the CAM-8 (Cellular Automata Machine), a dedicated hardware tool developed at Massachusetts Institute of Technology in 1989.

In order to evolve one neural module, a population of 30-100 modules is run through a genetic algorithm for 200-600 generations, resulting in up to 60,000 different module evaluations. Each module evaluation consists of - firstly, growing a new set of axonic and dendritic trees, guided by the module's chromosome. These trees interconnect several hundred neurons in the 3D cellular automata space of 13,824 cells (24*24*24). Evaluation is continued by sending spiketrains to the module through its efferent axons (external connections) to evaluate its performance (fitness) by looking at the outgoing spiketrains. This typically requires up to 1000 update cycles for all the cells in the module.

On the MIT CAM-8 machine, it takes up to 69 minutes to go through 829 billion cell updates needed to evolve a single neural module, as described above. A simple "insect-like" artificial brain has hundreds of thousands of neurons arranged into ten thousand modules. It would take 500 days (running 24 hours a day) to finish the computations.

Another limitation was apparent in the full brain simulation mode, involving thousands of modules interconnected together. For a 10,000-module brain, the CAM-8 is capable of updating every module at the rate of one update cycle 1.4 times a second. However, for real time control of a robotic device, an update rate of 50-100 cycles per module, 10-20 times a second is needed. So, the second goal was to have a model which would be portable into electronic hardware to eventually design a machine capable of accelerating both brain evolution and brain simulation by a factor of 500 compared to CAM-8.

The CoDi model operates as a 3D cellular automata (CA). Each cell is a cube which has six neighbor cells, one for each of its faces. By loading a different phenotype code into a cell, it can be reconfigured to function as a neuron, an axon, or a dendrite. Neurons are configurable on a coarser grid, namely one per block of 2*2*3 CA cells. Cells are interconnected with bidirectional 1-bit buses and assembled into 3D modules of 13,824 cells (24*24*24).

Modules are further interconnected with 92 1-bit connections to function together as an artificial brain. Each module can receive signals from up to 92 other modules and send its output signals to up to 32,768 modules. These intermodular connections are virtual and implemented as a cross-reference list in a module interconnection memory (see below).

In a neuron cell, five (of its six) connections are dendritic inputs, and one is an axonic output. A 4-bit accumulator sums incoming signals and fires an output signal when a threshold is exceeded. Each of the inputs can perform an inhibitory or an excitatory function (depending on the neuron's chromosome) and either adds to or subtracts from the accumulator. The neuron cell's output can be oriented in 6 different ways in the 3D space. A dendrite cell also has five inputs and one output, to collect signals from other cells. The incoming signals are passed to the output with an 5-bit XOR function. An axon cell is the opposite of a dendrite. It has 1 input and 5 outputs, and distributes signals to its neighbors. The "Collect and Distribute" mechanism of this neural model is reflected in its name "CoDi". Blank cells perform no function in an evolved neural network. They are used to grow new sets of dendritic and axonic trees during the evolution mode.

Before the growth begins, the module space consists of blank cells. Each cell is seeded with a 6-bit chromosome. The chromosome will guide the local direction of the dendritic and axonic tree growth. Six bits serve as a mask to encode different growth instructions, such as grow straight, turn left, split into three branches, block growth, T- split up and down etc. Before the growth phase starts, some cells are seeded as neurons at random locations. As the growth starts, each neuron continuously sends growth signals to the surrounding blank cells, alternating between "grow dendrite" (sent in the direction of future dendritic inputs) and "grow axon" (sent towards the future axonic output). A blank cell which receives a growth signal be- comes a dendrite cell, or an axon cell, and further propagates the growth signal, being continuously sent by the root neuron, to other blank cells. The direction of the propagation is guided by the 6-bit growth instruction, described above. This mechanism grows a complex 3D system of branching dendritic and axonic trees, with each tree having one neuron cell associated with it. The trees can conduct signals between the neurons to perform complex spatio-temporal functions. The end-product of the growth phase is a phenotype bitstring which encodes the type and spatial orientation of each cell.

3. THE CBM

This section briefly describes the hardware implementation of the above CoDi-1Bit model, allowing CoDi neural net modules to be grown in hardware.

The CAM-Brain Machine (CBM) was especially designed to support the growth and signaling of neural networks built by the CoDi model. The CBM should fulfill the needs for high speeds, when simulating large-scale binary neural networks, a necessary condition when one is concerned with performing real-time control. The hardware core is implemented in XC6264 FPGA chips, in which the neural networks will actually grow. A host machine will provide the necessary interface to interact with the hardware core. It is planned that the CBM will be used to grow 32,000 neural networks modules, each with approximately 1000 cells. The modules will be organized in architectures defined in advance, so several neural network modules will be interconnected to form a functional unity. For a complete description of the CBM, refer to [3].

4. CHOOSING A REPRESENTATION FOR THE CoDi-1BIT SIGNALING

The constraints imposed by state-of-the-art programmable (evolvable) FPGAs in 1998 were such that the CA based model (the CoDi model) had to be very simple in order to be implementable within those constraints. Consequently, the signaling states in the model were made to contain only 1 bit of information (as happens in nature's "binary" spike trains). The problem then arose as to interpretation. How were we to assign meaning to the binary pulse streams (i.e. the clocked sequences of 0s and 1s which are a neural net module's inputs and outputs? Ultimately we chose a representation which convolves the binary pulse string with the convolution function shown in Fig.2. We call this representation "SIIC" (Spike Interval Information Coding) which was inspired by [5]. This representation delivers a real valued output at each clock tick, thus converting a binary pulse string into an analog time dependent signal. Our team has already published several papers on the results of this convolution representation work [4]. Figs. 3, 4, 5 and 6 show some results of CoDi modules which were evolved to output oscillatory signals (using the convolutionary interpretation). Fig. 7 shows the evolved output for a random target analog signal. We thought the results were good enough to settle on this representation. The CBM will implement this representation in the FPGAs when measuring fitness values at electronic speeds.

Fig. 1. Reproduction of the Robot kitten (*robokoneko*, in Japanese)

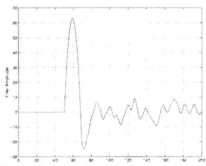

Fig. 2. Decoding filter for the spike trains.

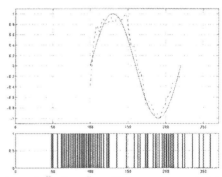

Fig. 3. Single period of a sinusoidal wave generated by the CoDi model and SIIC method. The lower part of the figure show the actual spikes that generated the waveforms.

5. THE ROBOT KITTEN ("ROBOKONEKO") AND RELATED ISSUES

An artificial brain with nothing to control is pointless, so we chose a controllable object that we thought would attract a lot of media attention, i.e. a cute life-size robot kitten that we call "Robokoneko" (which is Japanese for "robo-child-cat") [1]. We did this partly for political and

[1]For up-to-date data and images on the robot kitten (and the CBM, etc) see the web sites http://www.genobyte.com and http://www.hip.atr.co.jp/~degaris

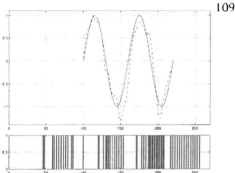

Fig. 4. Two periods of a sinusoidal wave generated by the CoDi model and SIIC method. The lower part of the figure show the actual spikes that generated the waveforms.

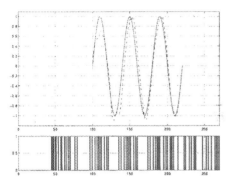

Fig. 5. Three periods of a sinusoidal wave generated by the CoDi model and SIIC method. The lower part of the figure show the actual spikes that generated the waveforms.

strategic reasons. Brain building is still very much in the "proof of concept" phase, so we wanted to show the world something (that is controlled by an artificial brain) that would not require a PhD to understand what it is doing. If the kitten robot can perform lots of interesting actions, this will be obvious to anyone simply by observation. The more media attention the kitten robot gets, the more likely our brain building work will be funded beyond 2000 (the end of our current research project).

Fig. 1 shows the mechanical design our team has chosen for the kitten robot. Its total length is about 25 cms, hence roughly life size. Its torso has two components, joined with 2 degrees of freedom (DoF) articulation. The back legs

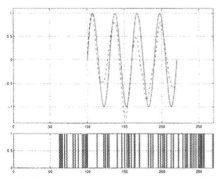

Fig. 6. Four periods of a sinusoidal wave generated by the CoDi model and SIIC method. The lower part of the figure show the actual spikes that generated the waveforms.

110

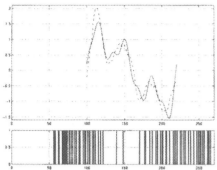

Fig. 7. Sum of sines and cosines generated by the CoDi model and SIIC method.

have 1 DoF at the ankle and the knee, and 2 DoF at the hip. All 4 feet are spring loaded between the heel and toe pad. The front legs have 1 DoF at the knee, and 2 DoF at the hip. With one mechanical motor per DoF, that makes 14 motors for the legs. 2 motors are required for the connection between the back and front torso, 3 for the neck, 1 to open and close the mouth, 2 for the tail, 1 for camera zooming, giving a total of 23 motors.

In order to evolve modules which can control the motions of the robot kitten, we thought it would be a good idea to feed back the state of each motor (i.e. a spiketrain generated from the pulse width modulation PWM output value of the motor) into the controlling module. Since each module can have up to 92 inputs (actually, 32 inputs repeated 3 times and distributed over three of the input surfaces of the module (minus 4 inputs positions reserved for the 4 output slots) feeding in these 23 motor state values will not be difficult. We are thinking we may install accelerometers and/or gyroscopes which may add another 6 or more inputs to each motion control module. It can thus be seen that the mechanical design of the kitten robot has implications on the design of the CBM modules. There need to be sufficient numbers of inputs for example.

The motion control modules will not be evolved directly using the mechanical robot kitten. This would be hopelessly too slow. Mechanical fitness measurement is impractical for our purposes. Instead we will soon be simulating the kitten's motions using an elaborate commercial simulation software package called "Working Model - 3D". This software will allow input from an evolving module to control the simulated motors of the simulated kitten. But, does not this approach rather destroy the whole philosophy of the CAM-Brain Machine and the CAM-Brain Project? It is a compromise, certainly, but in practice, the proportion of modules concerned with motion control will be very small compared to the total. Potentially, we have 32K modules to play with. Probably most of them will be concerned with pattern recognition, vision, audition, etc.

6. Future Plans and Challenges

Immediate plans are to use the latest specifications of the CBM to evolve a sample of single modules to show off the CBM's evolvability (using software simulation until the CBM is delivered). We will use the fitness definition type

(i.e. spiketrain comparisons) that will be implemented in the CBM. Once we get a feel for what is evolvable (and we already have quite a lot of experience in evolving CoD modules in simulation) we will be in a stronger position to start designing and evolving multi module systems. We need to specify a set of behaviors for the kitten robot and then evolve their motion control modules (in simulation) We need to specify what pattern recognition capacities we want. (We have the luxury of 32K modules, so we car afford to be ambitious, provided that the multimodule systems work as well as we hope they will). The first CBM should be delivered to ATR by the end of 1998 (delayed by a year due to a delay by Xilinx in supplying the XC6264 chips).

Section 4 showed how we convert a spike train into an analog signal. We may need to do the reverse, e.g. when a sensor sends an analog signal output voltage (potentiometer output) to an A/D (analog to digital) converter on the kitten, to the kitten's antenna, and is received by the CBM's antenna, which then goes through some converter which generates a spike train needed for the CBM modules. (CBM modules input and output spike trains). One idea is to evolve a module which takes an 8 bit input stream (a series of byte signals resulting from the A/D converter) and delivers the corresponding pulse train (i.e. if we convoluted it as in section 4, we would end up with the original analog signal). This may be difficult to evolve.

References

[1] DE GARIS, H. An artificial brain : ATR's cam-brain project aims to build/evolve an artificial brain with a million neural net modules inside a trillion cell cellular automata machine. *New Generation Computing Journal 12*, 2 (July 1994).

[2] GERS, F., DE GARIS, H., AND KORKIN, M. Codi-1 Bit: A simplified cellular automata based neuron model. In *Proceedings of AE97, Artificial Evolution Conference* (October 1997).

[3] KORKIN, M., DE GARIS, H., GERS, F., AND HEMMI, H. CBM (CAM-Brain Machine): A hardware tool which evolves a neural net module in a fraction of a second and runs a million neuron artificial brain in real time. In *Genetic Programming 1997: Proceedings of the Second Annual Conference* (July 1997), J. R. Koza, K. Deb, M. Dorigo, D. B. Fogel, M. Garzon, H. Iba, and R. L. Riolo, Eds.

[4] KORKIN, M., NAWA, N. E., AND DE GARIS, H. A 'spike interval information coding' representation for ATR's CAM-brain machine (CBM). In *Proceedings of the Second International Conference on Evolvable Systems: From Biology to Hardware (ICES'98)* (September 1998), Springer-Verlag.

[5] RIEKE, F., WARLAND, D., DE RUYTER VAN STEVENINCK, R., AND BIALEK, W. *Spikes: exploring the neural code.* MIT Press/Bradford Books, Cambridge, MA, 1997.

A Speaker Independent Arabic Isolated Spoken Digits Recognition System Using Fuzzy Kohonen Clustering Network

J. Elmalek IEEE Member, R. Tourki
Faculty Of Sciences, Route De Kairouan, 5000, Monastir, Tunisie. Fax : 00 216 3 500 278
Electronics And Micro-Electronic Laboratory
Email: Mehdi.Tekari@gnet.tn

Abstract

A Fuzzy Kohonen Network, which is capable of recognizing isolated Arabic spoken number speaker independently is described. The Fuzzy Kohonen Clustering Network (FKCN) algorithm, is based on the integration of Fuzzy C-Means (FCM) and Kohonen Clustering Network (KCN). FKCN is unsupervised, non-sequential, and uses fuzzy membership values from FCM as learning rates. Simulation results clearly indicate the superiority in recognition accuracy performance of FKCN when compared to that obtained for FCM, KCN and the conventional LBG (Linde-Buzo-Gray).

1 Introduction

Speech is one of the most direct means used by man to exchange information. Progress recorded in signal processing domain, computer development; equipment and software and contribution of the artificial intelligence, allow the use of speech for communication with a machine[1]. Several works intended for the realization of automatic speech recognition systems, have been achieved, However Research work on Arabic speech recognition, although lagging in comparison with other languages, is becoming more intensive than before and several researches have been published in the last few years.

Among the speech recognition system that recognize isolated Arabic Spoken Word, Several methods can be found in reference [1], [2], [3], [4], [5].

An Arabic phoneme recognition system, using supervised classifier LVQ (Learning Vector Quantization, LVQ is a variant of Kohonen network), is presented in [1], also a supervised learning algorithms, Beta Fuzzy Neural Network is proposed in [2]. However, in all, of these references, few words were used in the learning and testing processes.

The Soft Competitive Self-Organizing Map (SOM) algorithm has been extensively studied and applied successfully for different applications [8], [9].

The SOM was inspired by the way in which various human Sensory impressions are neurologically mapped into the brain such that spatial or other relations among stimuli correspond to spatial relations among the neurons. Kohonen's algorithm is a simple algorithm for the formation of such a mapping. A Sequence of inputs is randomly presented to the network, whose synaptic weights are then updated so that they eventually reproduce the input probability distribution as closely as possible. We investigate the use of the Kohonen method for simultaneous clustering and dimensionality reduction. Despite its many successes in practical applications, KCN suffers from some major deficiencies, many of which are highlighted in [9], KCN are heuristic procedures. So termination is not based on optimizing any model of the process or its data [12]. The final weight vectors are affected by the order of the input sequence and the initial conditions [12].

Several parameters of the KCN algorithms, such as the learning rate, the size of update neighborhood, and the strategy to alter these two parameters during learning, must be varied from one data set to another to achieve "useful" results[12]. Topology preserving mapping is not guaranteed [11].

To ameliorate problems suffered by KCN, several clustering algorithms have been proposed in recent years. Among these, the Fuzzy Kohonen Clustering Network (FKCN) [12]. This allows FKCN to employ less user defined parameters than KCN by computing the learning rate and the size of the update neighborhood directly from the data as a function of a weighting exponent m(t), where t is defined as the number of iterations.

In this paper, we used Arabic Spoken Digits (numbers "one" through "ten") to improve the effectiveness of the FKCN in speech recognition, and results are compared with KCN, FCM and the standard LBG (Linde-Buzo-Gray) algorithm.

The remainder of this paper is organized as follows. In the next section, since the learning rate we use is based on fuzzy membership values from FCM, FCM is briefly described. The structures of KCN are also presented. In section 3, we present the details of the FKCN model. In section 4, the principal operations of word recognition system is presented. In section 5, a comparison between FKCN, KCN, FCM and LBG is reported. Finally, we conclude, this work in section 6.

112

2 Fuzzy C-Means And Kohonen Clustering Network

Consider a set of n vectors $X = \{x_1, x_2, \ldots\ldots x_n\}$ to be clustered into c groups($1 < c < n$) of similar data. Each $x_i \in R^P$ is a feature vector consisting of p real-valued measurements describing the features (cepstrum coefficients) of the object (spoken number) represented by x_i

The objective function of the FCM algorithm to be minimized is written as follow [12]:

$$J_m(U,V) = \sum_{i=1}^{c} \sum_{k=1}^{n} (u_{i,k})^m \|x_k - v_i\|^2 , \qquad (1)$$

where u_{ik} is interpreted as the membership of x_k in the i th partitioning subset (cluster) of X; c is the total number of clusters, n is the total number of input patterns.

$\{u_{ik} = u_i(x_k), 1 \le k \le n, 1 \le i \le c\}$ satisfy three conditions ((2a), (2b), (2c)):

$$\forall\, i, k\;;\; 0 \le u_{ik} \le 1, \qquad (2a)$$

$$\forall k; \sum_{i=1}^{c} u_{ik} = 1, \qquad (2b)$$

$$\forall i; \sum_{i=1}^{c} u_{ik} > 0, \qquad (2c)$$

if all of the u_{ik} are in [0, 1], $U = [u_{ik}]$ is a conventional (crisp, hard) C-partition of X. and the objective function for clustering is defined as:

$$J_m(U,V) = \sum_{i=1}^{c} \sum_{k=1}^{n} (u_{i,k}) \|x_k - v_i\|^2 , \qquad (3)$$

2.1 Fuzzy C-Mean (FCM) Theorem [14]

Assume $\|x_k - v_j\|^2, \forall j, k$, at each iteration of (4); (U,V)

may minimize $\sum_{i=1}^{c} \sum_{k=1}^{n} (u_{i,k})^m \|x_k - v_i\|^2$ for m>1 only if

$$\forall i, k; u_{ik} = \frac{(\dfrac{1}{\|x_k - V_i\|^2})^{\frac{1}{m-1}}}{\sum_{j=1}^{c}(\dfrac{1}{\|x_k - V_j\|^2})^{\frac{1}{m-1}}}, \qquad (4a)$$

$$\forall i, v_i = \frac{\sum_{k=1}^{n} (u_{ik})^m x_k}{\sum_{k=1}^{n} (u_{ik})^m} , \qquad (4b)$$

A brief specification of FCM is given below.

2.2 Fuzzy C-Mean (FCM) Algorithm [14]

1) Fix : c; m and ε>0 some small positive constant.
2) Initialize network weight vector
$$v_0 = (v_{1,0}, v_{2,0}\ldots, v_{c,0}) \in R^{c\,p}.$$
3) For t=1,2,........., t_{max}.
   a. Update all memberships $\{u_{ik,t}\}$, $1 \le k \le n$, $1 \le i \le c$ with (4a)
   b. Update all weight vectors $\{v_{i,t}\}$, $1 < i < c$ with (4b)
   c. Calcul
   $$E_t = \sum_{i=1}^{c} \|v_{i,t} - v_{i,t-1}\|^2$$
   d. If $E_t \le \varepsilon$ stop; else next t.

The weighting exponent m, termed "amount of fuzziness"[12], is a user-defined resolution parameter belonging to range $(1, \infty)$; as $m \longrightarrow \infty$, $u_{ik,t} \longrightarrow 1/c$ The performance of FCM is quite dependent on a good way to choose m [15, 16]. FCM is non-sequential algorithms, the weights $\{v_{i,t}\}$ are updated after each pass through X. thus, the iterate sequence $\{v_{i,t}\}$ is independent of the data labels.

2.3 Kohonen Clustering Network

The structure of KCN consists of two layers an input (input neurons) layer and an output (output nodes) as shown in Fig1.

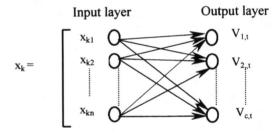

Fig. 1. The structure of a Kohonen clustering network

Each output node has a feature vector of the same dimension as the pattern to be classified. Initially the weight vectors of all the neural units are given small random values.

Given an input vector, the nodes in the output layer compete among themselves and the node on the map with the feature vector closest to that pattern is identified. That feature vector and the vectors belonging to nodes in the neighborhood of its associated node are modified slightly to bring them closer to the training pattern. The process is repeated until the weight vectors "stabilize". In this method, a learning rate must be defined which decreases with time in order to force termination, the size of the neighborhood must be defined and is also reduced with each training iteration.

Kohonen's feature mapping algorithm can be summarized as follows:

1) Fix c, $\varepsilon>0$ some small positive constant; Initialize the components of the feature vectors $v_0 = (v_{1,0}, v_{2,0}, \ldots v_{C,0})$ to small random values

2) For t =1,2,......,t_{max}.

   a. Present randomly a new input pattern x(t) to the input neurons.

   b. Measure the Euclidean distance d_j between the input pattern x(t) and each of the feature vectors $v_{j,t}$, where

$$d_j^2 = \left\| x(t) - v_{j,t} \right\|^2$$

Select the output node corresponding to the smallest d_j. Label it the "winner"

   c. Adjust the selected weight vector as

$$v_{s,t+1} = v_{s,t} + \alpha(t)(x(t) - v_{s,t}).$$

and the weight vectors of all the units in the topological neighborhood of the winner as

$$v_{j,t+1} = v_{j, t} + \alpha(t)(x(t) - v_{j,t}).$$

Where $\alpha(t)$ is the learning rate

3) Compute

$$Et = \sum_i \left\| v_{i,t} - v_{i,t-1} \right\|^2$$

4) If Et $<\varepsilon$ Stop; Else Next t.

KCN clustering is a sequential method, heuristic procedure, so termination is not guaranteed and the final weight vectors are dependent on the sequence of the input training vectors. Also Problems exist when a node is initialized to be far away from the data, and if it never wins, its weights will never be updated, Which decreases the performance of the system.

3 Fuzzy Kohonen Clustering Network

FKCN combines the on-line Kohonen weight adaptation rule with the fuzzy set membership function proposed by the batch fuzzy c-means (FCM) algorithm [12], [17]. This allows FKCN to employ less user defined parameters than KCN by computing the learning rate directly from the data as a function of a weighting exponent mt(t) The learning rate is defined as follow:

$$\alpha_{ik;t} = (u_{ik,,t})^{mt} \qquad (5)$$

where u_{ik} is calculated with (4a), mt is given by:

$$mt = m_0 - t\Delta m, \quad \Delta m = (m_0-1)/t_{max} \qquad (6)$$

A brief specification of FKCN is given below:

1) Fix c, $\varepsilon>0$ some small positive constant; Initialize network weight vector $v_0 = (v_{1,0}, v_{2,0}\ldots,v_{c,0}) \in R^{c\,p}$.

2) For t =1,2,......,t_{max}.

   a. compute $\alpha_{ik,t}$, $1\le k \le n$, $1\le i \le c$ with (4a), (5) and (6)

   b. Update all weight vectors $\{v_{i,t}\}$, $1<i<c$ with:

$$v_{i,t} = v_{i,t-1} + \frac{\left(\sum_{k=1}^{n} \alpha_{ik,t}(x_k - v_{i,t-1}) \right)}{\left(\sum_{s=1}^{n} \alpha_{is,t} \right)}$$

   c. Calcul

$$Et = \sum_i \left\| v_{i,t} - v_{i,t-1} \right\|^2$$

   d. If Et $<\varepsilon$ Stop; Else Next t.

KCN is sequential, different sequences of feeding the data did alter the final results. However, FKCN is non-sequential, unsupervised and use membership values from FCM as learning rate

4 System Operation

A bloc diagram indicating the principal operations of a word recognition system is presented in Fig.2. This system consists of five stages :Pre-treatment, segmentation, feature extraction, clustering and recognition.

The input analog signals are digitized at a 16kHz sampling rate, the digitized speech (spoken word) is then scanned forward from the beginning of the recording interval and backward from the end to determine the beginning and the end of the actual word, preemphasized by a 1st order digital network with a transfer function,

$$H(z) = 1-0.95*z^{-1}; \qquad (7)$$

Then a 20ms Hamming window is applied to the speech pattern every 5ms, and the first thirteen order cepstrum coefficients (features) are extracted from each of these frames. Then, an utterance (spoken word) can be represented by those sequential feature vectors.

The feature vector sequences is considered as speech pattern. We use FKCN algorithms to classify the speech pattern.

5 Experimental Results

FCM, KCN, FKCN and LBG algorithms have been tested on an Arabic database. The speech material consists of the spoken Arabic digits (numbers "one" through "ten"). The material was recorded from two male and two female speakers, each speaker recorded each word 10 times, so each speaker recorded 10*10 words = 100, in total we have 4*100 words =400. The first 200 utterances were used as training data to generate 10 clusters (numbers "one" through "ten"), the second 200 utterances were used for testing and evaluating the system

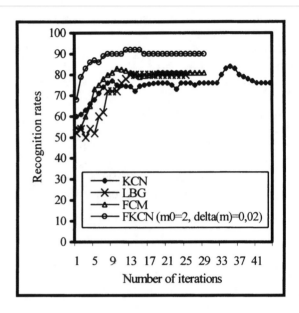

Fig. 3. Recognition rates as functions of time for KCN, LBG, FCM, FKCN.

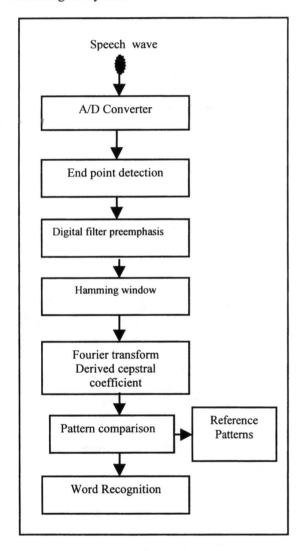

Fig. 2. Block diagram of pattern recognition Model for isolated word recognition

By testing the performance of FKCN and comparing it with FCM, KCN and LBG on an Arabic speech data base, (Fig. 3), the results indicate that the FKCN achieve the best recognition rate.

We also studied the stability of FKCN to m_0, Fig. 4 shows FKCN recognition rate evolution with time for fixed $\Delta m = 0.02$, it is evident that we have the best solution for $m_0 = 2$.

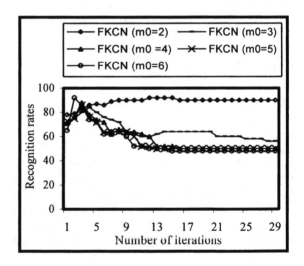

Fig. 4. FKCN Recognition rates as functions of time with $\Delta m = 0.02$.

6 Conclusion

In this paper we prove the efficiency of the implementation of the FKCN on an Arabic Isolated spoken digits recognition system, the overall performance of the recognition should be 93 %. A comparative study between FKCN and LBG, FCM, KCN is included.

Extension of the system to recognize continuous speech is in progress.

Performance Improvement using fuzzy integration of multiple classifier systems results is also considered.

References

[1] Mouria, B. F., Brahem, R., Nejma, A. M.: Syrapa: An automatic Arabic phoneme recognition system. 2nd IMACS International Multiconference CESA'98, Nabeul-Hammamet, Tunisia, vol. 4, pp. 196-200, 1998

[2] Jamaa, M., Alimi, A. M.: Recognition of spoken isolated Arabic words with the Beta Fuzzy Neural Network. 2nd IMACS International Multiconference CESA'98, Nabeul-Hammamet, Tunisia, vol. 4, pp. 348-351, 1998

[3] Belhaj, N., Boukadida, F., Ellouze, N.: Sapia :Expert system in Automatic segmentation and labelling of Arabic speech. 2nd IMACS International Multiconference CESA'98, Nabeul-Hammamet, Tunisia, vol. 4, pp. 201-203, 1998

[4] Sayoud, H., Selmane, M. K.: inter and intra-speaker variability of some phonetic parameters in standard Arabic. Classification in speaker recognition. 2nd IMACS International Multiconference CESA'98, Nabeul-Hammamet, Tunisia, vol. 4, pp. 216-219, 1998

[5] Oweiss, K., Abdel, A. O.: Performance evaluation of the time-delay fuzzy neural networks for isolated word recognition. International Symposium on Neuro-Fuzzy Systems, pp. 203-211, 1996.

[6] Alotaibi, Y. A., Shahsavri, M. M.: Speech recognition. IEEE Potentials, pp. 23-28, Feb-March 1998.

[7] Kohonen, T.: Self-organization and associative memory. Springer-verlag, 1988.

[8] Kohonen, T.: Self-organization map. Proc. of IEEE, vol. 78, no. 9, pp. 1464-1480, 1990.

[9] Kohonen, T.: Self-organizing maps. Berlin, Springer-verlag, 1995

[10] Fritzke, B.: Some competitive learning methods. Draft documend, available http:/www.neuroinformatik.ruhur-uni-bochum.de/ini/VDM/research/gsn/demogng.

[11] Martinetz, T. M., Schulten, K. J.: Topology representing networks. Neural Networks, vol. 7, no.3, pp. 507-522, 1994.

[12] Tsao, E. C., Bezdek, J. C., Pal, N. R.: Fuzzy Kohonen clustering network. Pattern Recognition, vol. 27, no. 5, pp. 757-764, 1994.

[13] Erwin, E., Obermayer, K., Schulten, K.: Self organizing maps: Ordering, convergence properties, and energy functions. Biol, Cybern, vol. 67, pp. 47-55, 1992.

[14] Bezdek, J. C.: Pattern Recognition with fuzzy objective function algorithms. Plenum, New York 1981.

[15] Mcbratney, A. B., Moore, A. W.: Application of fuzzy sets to climatic classification, Agric. Forest Meteor. Vol.35, pp. 165-185, 1985.

[16] Choc, H., Jordan, J.: On the optimal choice of parameters in a fuzzy c-means algorithms. Proc. First IEEE conf. on Fuzzy systems, san Diego 1992.

[17] Bezdek, J. C., Pal, N. R.: two soft relative of learning vector quantization.: Neural Networks, vol. 8, no. 5, pp. 729-743, 1995.

[18] Huntsberger, T., Ajjimarangsee, P.: Parallel self-organizing feature maps for unsupervised pattern recognition, Int. J. Gen. Syst, 16, pp. 357-372, 1989.

[19] Bezdek, J. C., Pal, N. R.: Generalization clustering networks and Kohonen's self-organizing scheme. IEEE. Trans. Neural Networks, vol. 4, pp. 549-557, 1993.

[20] Karayiannis, N. B., Bezdek, J. C., Pal, N. R., Hathaway, R. J., Pai, P.: Repair to GLVQ : A new family of competitive learning schemes. IEEE Trans. Neural Networks. Vol. 7, pp. 1062-1071, 1996.

Control of Complex of Parallel Operations Using Neural Network

D. Orski

Institute of Control and Systems Engineering,
Wroclaw University of Technology
Wyspianskiego 27, 50-370 Wroclaw, POLAND
Email: donat@ists.pwr.wroc.pl

Abstract

The paper presents a concept of the application of neural network to the control of a class of static operation systems. Two learning algorithms based on the known ideas of adaptation are investigated. Results of simulations obtained for the control of simple operation system using neuron-like control algorithms are presented.

1 Introduction

A great variety of problems in the field of control using neural networks have been studied over the last ten years (e.g., [1], [2], [7], [8]). This paper is concerned with a specific class of control systems known as operation systems, [3], [4]. The system under consideration consists of operations, i.e. some activities requiring an execution time, and the structure of the system is based on the time relationships between the separate operations. The control of such a system consists in the proper distribution of the global size of task (to be processed in the system) between the separate operations. Very often the models of such systems differ from the traditional ones and the determination of the exact control algorithm may be difficult except for some simple cases. Thus the application of the approaches and methods from the area of artificial intelligence, e.g. neural networks, may be justified and very useful. There are some concepts of application of neuron-like algorithms to the control of operation systems and there are also some algorithms that have been suggested in [5], [6], but till now have not been tested in any practical situation, i.e. in a real-life operation system. Before that the recommendation should be given based on the results of simulations. The purpose of this paper is to present simulation results obtained for the typical case of operation system – the system consisting of parallel operations (complex of parallel operations). The simulations were performed to investigate the properties of the proposed algorithms.

In Sec.2 the control problem for the system under consideration is formulated and solved in an analytical way. The concept of the application of neuron-like algorithms to the control of complex of parallel operations and two learning algorithms are presented in Sec.3, while in Sec.4 selected results of simulations are shown.

2 Control of the Complex of Parallel Operations

Let us consider the complex of m parallel operations with static models

$$T_i = \varphi_i(v_i, c_i) \qquad i \in \overline{1, m} \qquad (1)$$

where T_i is the execution time, v_i is the size of task and c_i is the parameter of the i-th operation. The control problem consists in finding the optimal distribution of the global size of the task V, i.e. the values $v_1^*, v_2^*, \ldots, v_m^*$ minimizing the execution time of the complex of operations

$$T = \max_i \varphi_i(v_i, c_i), \qquad (2)$$

subject to constraints

$$\bigwedge_{i \in \overline{1,m}} v_i \geq 0 \quad, \qquad \sum_{i=1}^{m} v_i = V. \qquad (3)$$

To find the solution of the control problem it is necessary to determine the inverse functions of (1):

$$v_i = \varphi_i^{-1}(T_i, c_i), \qquad (4)$$

to solve then the equation

$$\sum_{i=1}^{m} \varphi_i^{-1}(T, c_i) = V$$

117

with respect to T , and to obtain the control decisions by substituting the solution, i.e. the optimal time T^* into (4)

$$v_i^* = \varphi_i^{-1}(T^*, c_i) \triangleq F_i(c)$$

where $c = [c_1, c_2, \ldots, c_m]^T$.

The control consists in the determination of optimal task distribution based on the current values of operations parameters c_i measured at each stage of control. For the simple forms of the models (1) the control algorithm $F_i(c)$ may be obtained in an analytical way, for more complicated forms a numerical procedure should be applied. In both cases we will call it the *exact* control algorithm in contrast to the *approximate* neuron-like control algorithm.

For the typical case with the models

$$T_i = c_i v_i^{\frac{1}{\alpha}} , \quad \alpha > 0 , \quad c_i > 0 , \quad i \in \overline{1, m} \quad (5)$$

it is easy to obtain the algorithm $F_i(c)$ in an analytical form

$$v_i^* = \frac{c_i^{-\alpha}}{c_1^{-\alpha} + c_2^{-\alpha} + \ldots + c_m^{-\alpha}} . \quad (6)$$

3 Control with Neuron-Like Algorithms

In the case when the exact control algorithm is very complicated it may be useful and sufficient to approximate it using neuron-like algorithm

$$y_i = f_i(w_i^T x_i + b_i) \triangleq \Phi_i(x_i, a_i) \quad (7)$$

where y_i is one dimensional output, $x_i = [x_{i1}, x_{i2}, \ldots, x_{ik}]^T$ is the input vector, $w_i = [w_{i1}, w_{i2}, \ldots, w_{ik}]^T$ – the vector of weight coefficients, b_i – the threshold parameter and $a_i = [w_i^T \ b_i]^T$ – the vector of parameters in (7). The function f_i may have the commonly used forms (e.g. linear or sigmoid function). In the complex algorithm (i.e. the neural network) the outputs of the neuron-like elements from the first (input) layer are the inputs to the elements in the second (hidden or output) layer.

3.1 Concept of control

For the control of complex of parallel operations one-layer neural network with one neuron-like element for

each of $m-1$ operations was proposed, [6]. Based on that concept the approximation of the optimal decision v_i^* ($i \in \overline{1, m-1}$) is proposed as $v_i = y_i^{-1}$ where

$$y_i = | w_{i1} x_{i1} + w_{i2} x_{i2} + \ldots + w_{i,m-1} x_{i,m-1} + 1 | \quad (8)$$

and

$$v_m = | V - (v_1 + v_2 + \ldots + v_{m-1}) | , \quad (9)$$

$$x_{i1} = \frac{c_i}{c_1} , \quad x_{i2} = \frac{c_i}{c_2} , \ldots , x_{i,i-1} = \frac{c_i}{c_{i-1}} ,$$
$$x_{i,i} = \frac{c_i}{c_{i+1}} , \ldots , x_{i,m-1} = \frac{c_i}{c_m} . \quad (10)$$

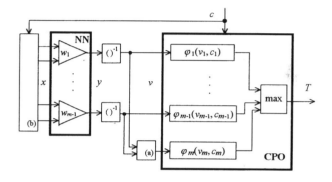

Fig.1. Control system.
NN – Neural Network, CPO – Complex of Parallel Operations, blocks (a), (b) correspond to formulae (9), (10)

The presented suggestion is based on the form (6) for the typical models (5) which may be considered as a good approximation of the real-life operations. The different form for v_m is suggested to ensure that the second condition in (3) is satisfied when $v_1 + v_2 + \ldots + v_{m-1} \leq V$.

3.2 Learning algorithms

Two learning algorithms corresponding to the algorithms presented in [5], [6] for operation systems are proposed here for the case of task distribution between the parallel operations:

Version 1
According to this version the correction of the weights is based on the difference between y_i (8) and the exact output $y_i^* = (v_i^*)^{-1}$ obtained from the analytical formula F_i, from the corresponding numerical procedure or given by a trainer (*supervised learning*):

$$w_i(n+1) = w_i(n) - \gamma_i(n) x_i(n)[y_i(n) - y_i^*(n)] \quad (11)$$

where $\gamma_i(n)$ satisfies the conditions

$$\bigwedge_n \gamma_i(n) \geq 0 , \quad \sum_{n=0}^{\infty} \gamma_i(n) = \infty , \quad \sum_{n=0}^{\infty} [\gamma_i(n)]^2 \leq \infty \quad (12)$$

or $\gamma_i(n) = const$ during the learning process.

Version 2

This version concerns a situation when we do not have a trainer and do not know the exact control algorithm. Thus it is not possible to evaluate the optimal control decisions. Then the model (1) of the system to be controlled is engaged in the learning process, which may be considered as step by step correction in adaptive control system. The parameters $w_i(n)$ are changed in the searching process, minimizing the execution time T. The learning algorithm (*unsupervised learning*) has the form

$$w_i(n+1) = w_i(n) - \gamma_i(n) d_i(n) \quad (13)$$

where the j-th component of the vector $d_i(n)$ is the following

$$d_{ij}(n) = \frac{Q[w_i(n)] - Q[w_i(n) - \overline{w}_i(n)]}{\delta_{ij}} ,$$

$$j = 1, 2, \dots, m-1 \quad (14)$$

and $\overline{w}_i(n)$ is a vector with components equal to zero except the j-th component, which is equal to δ_{ij} (the value of testing step). The coefficients $\gamma_i(n)$ should satisfy the conditions (12). The quality index

$$Q[w_i(n)] = T(n)$$

where

$$T(n) = \max_i \quad \varphi_i[v_i(n) , c_i(n)]$$

and $v_i(n) = [y_i(n)]^{-1}$, $y_i(n)$ is determined according to (8) for $w_i(n)$ and $x_i(n)$ (10).

4 Simulations

4.1 Concept of simulations

Simulations have been performed for the operation system consisting of $m = 3$ parallel operations. For the Version 1 the typical models of operations (5) were assumed. Other assumptions: the global size of task $V = 1$, initial values of the weight coefficients $w_{ij}(0) = 1$ $(i = 1, 2; j = 1, 2)$, $c_k(n)$ chosen randomly

from $[c_k - \Delta c , c_k + \Delta c]$ with uniform probability distribution $(k = 1, 2, 3)$. For the current evaluation of the result of learning the following error value was considered (mean relative error for the last β steps):

$$E(n) = \frac{1}{\beta} \sum_{l=n-\beta+1}^{n} \frac{T(l) - T^*(l)}{T^*(l)} \cdot 100\% . \quad (15)$$

The learning process was stopped if:

$$E(n) < \varepsilon \quad \text{or} \quad n > N_{max} \quad (16)$$

where ε was given accuracy and N_{max} – given maximal number of steps. For (16) the following performance index (taking into account a speed of convergence and an error value) was introduced to evaluate the quality of learning

$$P = N \cdot \min_{n \in \beta-1, N} E(n) \quad (17)$$

where $N \leq N_{max}$ is a number of steps in the learning process. The smaller value of P, the better performance of a learning algorithm. The objective of the simulations was to investigate the influence of α, Δc and $r_c = \frac{c_3}{c_2} = \frac{c_2}{c_1}$ on performance index P.

4.2 Results of simulations

Version 1

For $\beta = 10$, $\varepsilon = 2\%$, $N_{max} = 2000$ the results are presented in Fig.2 and Fig.3 for the case when $\gamma_i(n) = \frac{1}{n+1}$, and in Fig.4 and Fig.5 for $\gamma_i(n) = 1$. It should be emphasized that in both cases P-axis is in logarithmic scale.

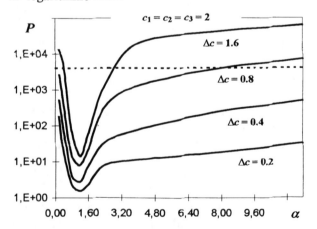

Fig.2. Version 1: decreasing γ , P vs. α for different Δc

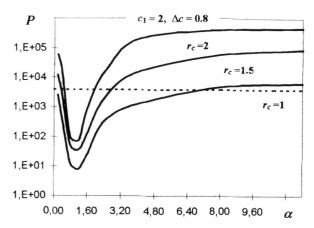

Fig.3. Version 1: decreasing γ, P vs. α for different r_c

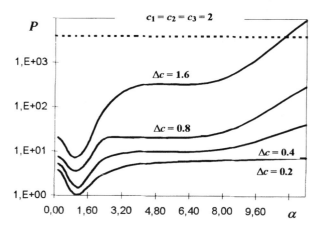

Fig.4. Version 1: constant γ, P vs. α for different Δc

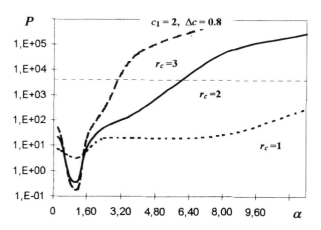

Fig.5. Version 1: constant γ, P vs. α for different r_c

In the presented figures the dashed line marks the value $P = 4000$, which for (17) and for the stop condition (16) with the assumed values of ε and N_{max} may be considered an upper limit of a good

performance. The error value $\varepsilon = 2\%$ is typical in the control systems and $N_{max} = 2000$ was determined after series of testing runs as a number of steps sufficient to ensure the convergence of the output value T to T^* with accuracy ε in most cases. The figures show a strong influence of the values of α, Δc and r_c on the performance of the learning algorithms and thus on the quality of control. A better performance (i.e. the smaller value of P) was achieved with constant coefficient $\gamma_i(n) = 1$ and in this case the performance is satisfactorily good for a wide range of the values of α, Δc and r_c. The best results are obtained for $\alpha = 1$, because in this case the neuron-like algorithms (8) with $w_{ij} = 1$ produce the optimal control decisions.

Version 2

Fig.6 and Fig. 7 show the results of analogous simulations obtained for the Version 2 of a learning algorithm with $\delta_{ij} = 0.1$ in (14).

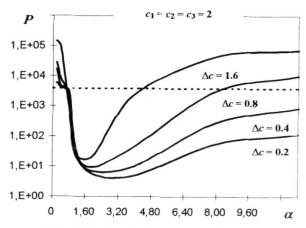

Fig.6. Version 2: P vs. α for different Δc

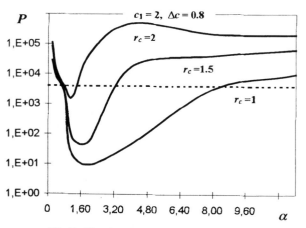

Fig.7. Version 2: P vs. α for different r_c

Since in this version of learning we assume that values of $T^*(n)$ are unknown, (15) was evaluated for $\frac{T_1(n) + T_2(n) + T_3(n)}{3}$ (with assumption that we can measure T_i) in place of the unknown $T^*(n)$. As was expected the performance of the Version 2 of a learning algorithm was a little worse than of the Version 1, but it is an advantageous approach when the *exact* control algorithm is not known.

In case when $\frac{\alpha_2}{\alpha_1} = \frac{\alpha_3}{\alpha_2} = r_\alpha$ and $r_\alpha \neq 1$ it is not possible to obtain the optimal control decisions v_1^*, v_2^*, v_3^* in an analytical way, and thus it is impossible to apply the Version 1 of a learning algorithm. For this situation the influence of r_α on the performance of learning was investigated with the Version 2 of a learning algorithm (Fig.8).

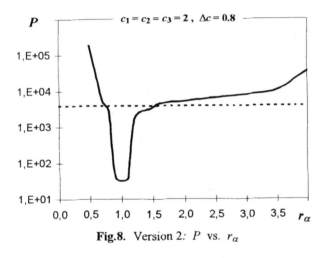

Fig.8. Version 2: P vs. r_α

A good performance was achieved for $r_\alpha \in [\,0.7\,,\,1.5\,]$ and a considerably good performance – for a quite wide range of r_α, i.e. $r_\alpha \in [\,0.6\,,\,3.5\,]$.

5 Conclusions and Final Remarks

The presented results of simulations show that the proposed algorithms may be successfully applied to the control of complexes of parallel operations and allow a recommendation to be given for the application of these algorithms to the control of a class of real-life static operation systems. The presented approach may be useful for the control of manufacturing systems when the mathematical models are not precisely defined and/or the determination of the exact control decisions is connected with great computational difficulties.

Further tests and computer simulations may concern a class of systems under consideration consisting of more than three parallel operations. The purpose of the simulations would be to investigate the influence of a number of operations on the performance of learning and the quality of control.

The suggested approach may be probably extended for more complicated cases of operation systems and for more complex structures of a neural network. This could be the main subject for further studies and simulations.

The other important direction to be followed in the future works is to test more sophisticated learning algorithms with particular stress on examining their convergence and behavior.

References

[1] Carelli, R., Camacho, E.F., Patino, D.: A neural network based feedforward adaptive controller for robots. IEEE Trans. on Systems, Man and Cybernetics 25 (1995).

[2] Chen, F.C., Khalil, H.K.: Adaptive control of a class of nonlinear discrete-time systems using neural networks. IEEE Trans. on Automatic Control 40 (1995).

[3] Bubnicki, Z.: Identification of Control Plants. Amsterdam, Oxford, New York: Elsevier 1980.

[4] Bubnicki, Z.: Optimal models of complex operation systems. Proc. of VI International Congress of Cybernetics and Systems, AFCET, Paris, 1984.

[5] Bubnicki, Z.: Application of neural networks to complex operation systems. In: Modelling, Identification and Control. Zurich: Acta Press 1994.

[6] Bubnicki, Z.: Adaptive control of operation systems using neural networks. Proc. of XI International Conference on Systems Engineering, Las Vegas, 1996.

[7] IEEE Contr. Syst. Mag., Special Issues on Neural Networks for Control Systems *1988-1990*.

[8] Narendra, K.S., Parthasarathy, K.: Identification and control of dynamical systems using neural networks. IEEE Trans. on Neural Networks *1* (1990).

Using Smoothing Splines in Time Series Prediction with Neural Networks

Uroš Lotrič[1] and Andrej Dobnikar[2]

[1] R&D Institute, Sava d.d., Škofjeloška 6, Kranj, Slovenia

[2] Faculty of Computer and Information Science, University of Ljubljana, Tržaška 25, Slovenia

E-mail: uros.lotric@sava.si, andrej.dobnikar@fri.uni-lj.si

Abstract

The smoothing spline based neural network is used for prediction of a trend from complex and noisy time series. First, the time series is smoothed by a cubic spline and then multilayered feedforward neural networks are applied to predict the parameters of the spline and by this the next values of the smoothed time series. The level of smoothing can be chosen by the smoothing parameter. We show that in the case of a complex time series like the bike tire sale, prediction of a trend with the smoothing spline based neural network gives us more reliable information than a classical prediction with the multilayered feedforward neural network.

1   Introduction

The prediction with the neural networks is based on their ability to extract relevant information from the time series themselves. However, this prediction can fail for complex and noisy time series. In such cases, smoothing the original time series might improve prediction. By smoothing, the information about sudden changes is lost, and therefore only the trend can be predicted from the smoothed time series. Prediction of the trend for the next sampling time is hardly reasonable – usually the values at few sampling times in advance are required.

In this paper, we propose a method which combines smoothing of a time series with cubic spline and prediction with neural network. Splines are often used together with the neural networks and fuzzy logic in modern process techniques, including modeling and regulation [1, 2]. There, the neural network output is a linear combination of a set of basis functions called B-splines. In our method however, the splines are used to transform a time series into a form more appropriate for prediction.

After giving a brief background of smoothing the time series in the next section, we present the method in the third section, and give some results on real time series in the fourth. The main conclu-

sions are drawn in the last section.

2   Smoothing the Time Series

The time series can be smoothed in various ways, for example by moving averaging or filtering [3]. These two simple methods can only be used, if the basic properties of the time series are known. Since this is usually not the case, we shall use statistical methods which enable us to calculate the noise level [4].

Consider the following time series

$$x_i = g(t_i) + \varepsilon_i \ , \quad i = 1, \ldots, N \ , \tag{1}$$

where x_i is the value of the time series in time t_i, g is an unknown smooth function, and ε_i is normally distributed noise with the standard deviation σ_i. The function f is considered the best approximation of g, if it minimizes the functional [4]

$$F(f) = \sum_{i=1}^{N} \left(\frac{x_i - f(t_i)}{\sigma_i} \right)^2 + \frac{\lambda}{2} \int_{t_1}^{t_N} dt \, f''^2(t) \ , \tag{2}$$

where f'' is the second time derivative of f. Function f is continuous with smooth first and second derivatives. The Euler-Lagrange smoothing parameter λ is chosen in such a way, that it minimizes the normalized mean squared error

$$E(\lambda) = \frac{1}{N} \sum_{i=1}^{N} \left(\frac{g(t_i) - f(t_i)}{\sigma_i} \right)^2 \ . \tag{3}$$

The smoothing parameter determines the smoothing level. For $\lambda = 0$, $f(t_i)$ equals the value of the time series at that time, x_i. The greater the smoothing parameter the greater the difference between $f(t_i)$ and x_i. The extremal function f which satisfies conditions (2) and (3) is composed of cubic polynomials [4]

$$f(t) = p_i(t) = a_i + b_i(t - t_i) + \\ + c_i(t - t_i)^2 + d_i(t - t_i)^3 \ , \tag{4}$$

where each polynomial p_i is defined only for the interval $[t_i, t_{i+1}]$, $i = 1, \ldots, N - 1$. The endpoints of polynomials are joined in such a way that f, f' and f'' are continuous. Hence, the solution is a cubic spline (Figure 1).

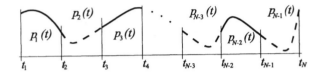

Fig. 1. A cubic spline composed of cubic polynomials.

The parameters of polynomials p_i are linear combinations of the values of the original time series [4]. Parameters $\mathbf{a} = (a_1, \ldots, a_N)^T$ for example, are obtained as

$$\mathbf{a} = \mathbf{A}(\lambda)\mathbf{x} \ , \tag{5}$$

where $\mathbf{x} = (x_1, \ldots, x_N)$, and $\mathbf{A}(\lambda)$ is a $N \times N$ square matrix.

In real time series, usually neither the function g nor the standard deviations σ_i are known, therefore $E(\lambda)$ cannot be calculated according to equation (3). However, in the case when $f(t_i)$ is a linear combination of x_i, the generalized cross-validation function [5, 6, 7]

$$GCV(\lambda) = \frac{\frac{1}{N}\|(\mathbf{I} - \mathbf{A}(\lambda))\mathbf{x}\|^2}{(\frac{1}{N}Tr(\mathbf{I} - \mathbf{A}(\lambda)))^2} \tag{6}$$

can be used instead.

Let us illustrate the spline smoothing on a simple sinus function with additive normally distributed noise

$$x_i = \sin(2\pi\nu t_i) + \varepsilon_i \ , \tag{7}$$

with $\nu = 0.02$ Hz and $\sigma = 0.5$. By minimizing the generalized cross-validation function $\lambda = 420$ was obtained. the thick solid line while the original time series is plotted with the thin solid line. For orientation, the sinus function without noise is plotted with the thick dashed line. The smoothed function follows the sinusoidal shape although the amplitude slightly varies.

3 The Smoothed Spline Based Neural Network

The smoothing procedure described in the previous section results in a spline which consists of cubic

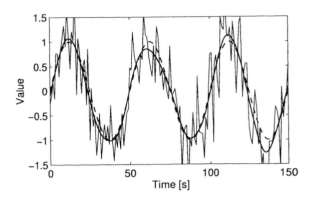

Fig. 2. Noisy sinus function smoothed with splines with smoothing parameter $\lambda = 420$.

polynomials. The time series is thus transformed into a 4-dimensional space of cubic polynomial parameters. By predicting the next values in this space, the smoothed time series can be predicted. In the case of a noisy time series, such a prediction might be more reliable than the prediction of the original time series. Before a spline can be used together with neural networks, some basic problems have to be solved – namely, the number of parameters, the parameterization of cubic polynomials, and missing parameters at the end of a time series.

For a time series with N samples, the spline is composed of $N - 1$ cubic polynomials, where each of them is described with four parameters, giving the total of $4(N-1)$ parameters. Thus, the number of parameters was increased from N to $4(N-1)$. In general, increasing the number of parameters does not lead to an improvement of prediction [8]. To give the relevant information on the future trend, the smoothed time series should be predicted for a few sampling times forward. If these values are simultaneously predicted the increase of the number of parameters can be avoided. T future values can be calculated simultaneously by joining T polynomials from spline into one approximating cubic polynomial

$$\begin{aligned} P_i(t) = A_i + B_i(t - t_i) + \\ + C_i(t - t_i)^2 + D_i(t - t_i)^3 \ , \end{aligned} \tag{8}$$

where $t \in [t_i, t_{i+T}]$. The approximating polynomial is constructed in such a way that in its starting point t_i, its value and the first derivative coincide with the value and the first derivative of the first polynomial in the spline, which is to be approximated, i.e.

$$P_i(t_i) \ = \ p_i(t_i) \tag{9}$$

$$\left.\frac{dP_i(t)}{dt}\right|_{t=t_i} = \left.\frac{dp_i(t)}{dt}\right|_{t=t_i} \quad . \qquad (10)$$

From the above conditions follows that $A_i = a_i$ and $B_i = b_i$. Parameters C_i and D_i are obtained by minimizing the square difference between the approximating polynomial and polynomials in the spline

$$\mathcal{E}_i = \sum_{j=0}^{T-1} \int_{t_{i+j}}^{t_{i+j+1}} dt \; (P_i(t) - p_{i+j}(t))^2 \quad . \qquad (11)$$

In Figure 3, the construction of one approximating polynomial is presented. We can see that

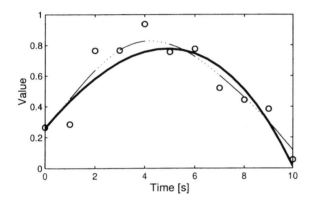

Fig. 3. Construction of the approximating polynomial from polynomials in the spline for $T = 10$. Polynomials in the spline are alternately denoted with the dotted and the thin solid line. The approximating polynomial is represented by the thick solid line.

the approximating polynomial is well adjusted to the spline at the beginning, but describes spline less accurately at the end. We can increase T as long as polynomials in the spline are well approximated with approximating polynomials over the entire time series. In some cases, T is determined by the application, but otherwise it should be selected in such a way that the square errors $\mathcal{E}_i$, $i = 1, \ldots, N-T$, are small enough and the approximation at the end of the approximating polynomial is acceptable.

In the training procedure each sampling point in the time series is used as a starting point of the approximating polynomial. All approximating polynomials can be described by a single approximating polynomial, whose parameters change in time. The prediction of the smoothed time series is achieved by predicting the next values of all polynomial parameters. The value of each parameter is predicted

from its previous values. Thus, by predicting the four parameters, the prediction of T future values is achieved. If we would have used the polynomials from the spline, only one future value would be obtained from the four parameters.

In the presentation of the approximating polynomial by equation (8), the parameters are related to its value at the starting point t_i and the first three derivatives at this point. Another representation, where the parameters are related to the values and the first derivatives in both, starting and ending point, has proved to be more appropriate for the parameter prediction with neural networks. By de Casteljeau parameterization [9], a polynomial can be transformed into a form

$$\begin{aligned} B_i(t) = & (1-u)^3 b_i^0 + 3(1-u)^2 u b_i^1 + \\ & + 3(1-u)u^2 b_i^2 + u^3 b_i^3 \end{aligned} \qquad , \qquad (12)$$

where $u = (t - t_i)/(t_{i+T} - t_i)$. Parameters $b_i^0 = B_i(t_i)$ and $b_i^3 = B_i(t_{i+T})$ represent values of the approximating polynomial at the beginning and at the end of each interval, respectively. The other two parameters are related to the first derivative at boundary points

$$b_i^1 = b_i^0 + \frac{t_{i+T} - t_i}{3} \left.\frac{dB(t)}{dt}\right|_{t=t_i} \quad , \qquad (13)$$

$$b_i^2 = b_i^3 - \frac{t_{i+T} - t_i}{3} \left.\frac{dB(t)}{dt}\right|_{t=t_{i+T}} \quad . \qquad (14)$$

In this notation, a polynomial is bounded by a polygon formed by points, obtained from parameters $b_i^0, \ldots, b_i^3$, as can be seen in Figure 4.

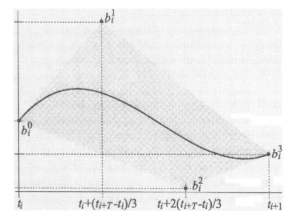

Fig. 4. Graphical representation of De Casteljeau parameterization of cubic polynomial.

From the known time series of length N, $N - T$ approximating polynomials can be generated (Figure 5). To predict T unknown values, the approximating polynomial with a starting point in the last known sampling point t_N of the time series is needed. However, the last known approximating polynomial starts in the sampling point t_{N-T}. If we look at the time series of parameter values, we have parameter values for up to $(N - T)$-th value to predict the N-th value. Usually, the next value of the time series depends quite strongly on a few recent values. In our case, T recent values are not available. We can obtain them by extending the original time series for T values using the mean value of the known time series. Consequently, the time series of approximating polynomial parameters are also extended for T samples, $b_{N-T+1}^k, \ldots, b_N^k$, where $k = 0, \ldots, 3$. This extension of the time series can be regarded as a very primitive prediction, which is then corrected by neural network.

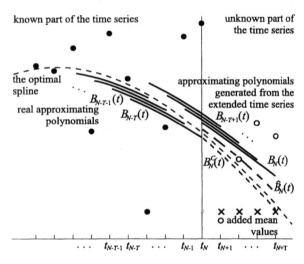

Fig. 5. Approximating polynomials at the end of the time series. Approximating polynomials are slightly shifted in direction of y-axis for easier review.

At each step of the prediction, the known time series is first extended by the primitive prediction for T sample points ahead and then the obtained time series is smoothed. The smoothing parameter λ can either be arbitrary – determined by the user – or obtained from the minimization of the generalized cross validation function (equation (6)). To get the prediction of T future values, T cubic polynomials from the spline are approximated by a single approximating polynomial (equation (12)). The time sequences of the approximating polynomial parameters generate four new time series. To predict the next values of these time series, four multilayered feedforward neural networks of equal structure are

used, one for each parameter.

The weights of the neural networks were set during the training procedure. 80% of the data in the training set are used for the adaptation of weights in neural networks, and 20% for validation purposes. The parameters of several last known approximating polynomials are the inputs to the neural networks. The output of each neural network, namely the predicted value of the corresponding parameter, is compared with the parameter obtained from the complete known time series. According to the difference between them, the weights of corresponding neural network are changed. The training procedure is stopped, when the average error between the predicted $\hat{B}_i(t)$ and the correct $B_i^C(t)$ approximating polynomials on the validation set

$$\mathcal{E}_P = \frac{1}{N_{va}} \sum_i \int_{t_i}^{t_{i+T}} dt \, (\hat{B}_i(t) - B_i^C(t))^2 \qquad (15)$$

reaches the minimum. By $\sum_i$ the sum over all N_{va} samples in the validation set is indicated. With this training criteria, all four neural networks are integrated into a single system – the smoothing spline based neural network.

4 Predicting the Sale of Bike Tires

The production of bike tires is usually planned one month or several months ahead. When preparing the production plan, the estimation of the future sales is of crucial importance. The time series describing the weekly sale of bike tires is presented by the dotted line in Figure 6. Within the time series, a yearly period consisting of spring and autumn season can be observed [10, 11]. The applied time series with $N = 143$ successive samples has the mean value of 45500 and the standard deviation $\sigma = 25300$.

The smoothing spline based neural network was compared to the classical feedforward neural network. For the both methods various neural network configurations were tested and the one with the smallest configuration and the smallest prediction error was selected. The first 80% of the samples in the time series were used for training ($N_{tr} = 114$) and the last 20% for testing ($N_{te} = 29$).

The prediction with the smoothing spline neural network gave the best results if multilayered feedforward neural networks with 12 inputs, one hidden and one output neuron were used. Smoothing parameter $\lambda = 127$ was obtained from the generalized cross validation function. The sale of bike tires was given in a weekly manner, therefore $T = 4$ was set for one-month trend prediction. The results are presented in Figure 6a. Furthermore, they were com-

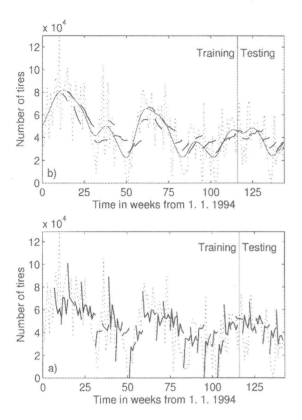

Fig. 6. Prediction with a) smoothing spline based neural network and b) multilayered feedforward neural network for $T = 4$ weeks in advance. Time series is plotted with the dotted line, predictions with the thick solid line. The optimal spline on the figure a) is plotted with the thin solid line. Only every fourth prediction is plotted.

pared with the prediction obtained from the original time series using the classical threelayered feedforward neural network (FNN). To obtain predictions for more sampling times ahead, this method was used iteratively. The neural network with 8 inputs, 6 hidden neurons and one output neuron gave the best results. This predictions are graphically presented in Figure 6b. Comparing the configurations of neural networks, the smoothing spline based neural networks are much smaller due to the preceding smoothing.

The errors of both methods are summarized in Table 1. From the errors, calculated considering the original time series, we can see that for the smoothing spline based neural network all errors were 3% to 7% smaller than for the threelayered feedforward

Table 1. Comparison of normalized roots of mean squared error in the prediction of the sale of bike tires with the threelayered feedforward neural network and the smoothing spline based neural network. The values in the table are normalized with the standard deviation of the time series, $\sigma = 25300$.

				Weeks
	1	2	3	4
FNN				
Training set	0.832	0.854	0.888	0.898
Testing set	0.762	0.791	0.803	0.799
SSBNN according to the original time series				
Training set	0.810	0.808	0.824	0.869
Testing set	0.715	0.751	0.749	0.761
SSBNN according to the smoothed time series				
Training set	0.270	0.320	0.364	0.407
Testing set	0.234	0.278	0.316	0.369

neural network.

However, the smoothing spline based neural network was trained according to the smoothed spline. Small errors in this aspect indicate how the original time series was simplified with smoothing. Smaller errors in the prediction of the smoothed time series also emphasize that in complex time series, like in the sale of bike tires, the prediction of the trend can be trusted more than the prediction of exact values of the original time series.

The effect of the smoothing parameter λ on the prediction error was also studied [10]. For different values of λ, namely 1, 10, 127, 1000 and 100000, the optimal configuration of the neural networks was found. Figure 7 presents the normalized square-root

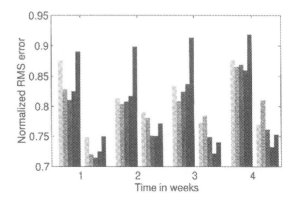

Fig. 7. Comparison of normalized root mean squared (RMS) errors gained by 4 week prediction for different smoothing parameters $\lambda = 1, 10, 127, 1000$ and 1000. For each week errors on training and testing set are presented respectively.

errors for four weeks on the training and the testing set, respectively. In most cases, the prediction with the smoothing parameter $\lambda = 127$ was close to the best prediction found. Thus, we may conclude that the generalized cross validation gives an almost optimal parameter not only for smoothing, but also for prediction.

5 Conclusion

Although the classical threelayered feedforward neural network is sometimes able to predict sudden changes in the time series, this predictions are rather unreliable. In many cases, especially for the prediction of monthly sale of bike tires, a reliable prediction of the trend is more useful for the user. Therefore, a method for trend prediction was presented in this paper.

First, the time series is smoothed by a cubic spline. Then several cubic polynomials that form the spline are described by an approximating cubic polynomial. In this way, four new time series are generated, each describing the time evolution of one of the polynomial parameters. The next value of each parameter is predicted by a neural network, and from the obtained parameters several future values are calculated.

Smoothing the time series is the fundamental procedure of the proposed method and there is still some room for improvements. We are currently using a statistical method for choosing the smoothing parameter, whereas a choice based on the prediction efficiency would be more appropriate.

We have observed that the predictions of the trend by the smoothing spline based neural network might be sometimes more reliable than the predictions obtained by a commonly used universal model. However, the main drawback of the purposed method is the time consuming computing algorithm.

Acknowledgements

This work is supported by Sava, Rubber & Chemical Industry, d.d. and the Slovenian Ministry of Science and Technology.

References

[1] Guarnieri, S., Piazza, F., Uncini, A.: Multilayered Neural Networks with Adaptive Spline-based Activation Functions, Proceeding of the 1995 World Congress on Neural Networks, 1 (1995).

[2] Harris, C. J., Moore, C. G., Brown, M.: Intelligent Control. Aspects of Fuzzy Logic and Neural Networks. World Scientific, 1994.

[3] Masters, T.: Neural, Novel and Hybrid Algorithms for Time Series Prediction, John Wiley & Sons, 1995.

[4] Reinsch, C. H.: Smoothing by Spline Functions. Numer. Math., 177-183, 10 (1967).

[5] Craven, P., Wahba, G.: Smoothing Noisy Data with Spline Functions. Numer. Math., 377-403, 31 (1979).

[6] Hutchinson, M. F., de Hoog, F. R.: Smoothing Noisy Data with Spline Functions. Numer. Math., 99-106, 47 (1985).

[7] Golub, G. H., Heath, M., Wahba, G.: Generalized Cross-Validation as a Method for Choosing a Good Ridge Parameter. Technometrics, 215-223, 21 (1979).

[8] Haykin, S.: Neural Networks, A Comprehensive Foundation. Macmillan College, New York, 1994.

[9] Farin, G.: Curves and Surfaces for Computer Aided Geometric Design. Third Edition, Academic, San Diego, 1993.

[10] Lotric, U.: Time Series Prediction with Smoothing Spline Based Neural Network. Masters Thesis, University of Ljubljana, Slovenia, 1997.

[11] Lotric, U., Dobnikar, A.: Functional Neural Gas Network versus Smoothing Spline Based Neural Network. International ICSC Symposium, Engineering of Intelligent Systems, Tenerife, ICSC Academic Press, 132-135 (1998).

Evolvable Hardware Chips for Neural Network Applications

I. Kajitani[1], M. Murakawa[2], N. Kajihara[3], M. Iwata[4], H. Sakanashi[4], T. Higuchi[4]

[1] University of Tsukuba, 1-1-1, Tenoudai, Tsukuba, Ibaraki, Japan
[2] University of Tokyo, Hongou, Tokyo, Japan
[3] RWCP Adaptive Devices NEC Laboratory, Kawasaki, Kanagawa, Japan
[4] Electrotechnical Laboratory, 1-1-4, Umezono, Tsukuba, Ibaraki, Japan

Abstract

This paper introduces two Evolvable Hardware LSIs for neural network applications. They are developed as part of MITI's Real World Computing Project. One is self-reconfigurable neural network chip for ontogenic neural network processing, having the processing capability equivalent to 10 Pentium II chips. The other LSI is for the pattern recognition for myoelectric artificial hand control.

1 Introduction

In contrast to conventional hardware where the structure is irreversibly fixed in the design process, Evolvable Hardware (EHW) is designed to adapt, like the chameleon changing its color to blend in with the environment, to changes in task requirements or changes in the environment, through its ability to reconfigure its own hardware structure dynamically and autonomously. This capacity for adaptation, achieved by employing efficient search algorithms known as genetic algorithms, has great potential for the development of innovative industrial applications.

Currently, five EHW chips are being developed as part of MITI's Real-World Computing Project. Although the concept of EHW is relatively new, some EHW chips are already being evaluated for their commercial value. In this paper, we introduce two EHW chips that can be used for neural network applications; a neural network EHW chip capable of autonomous reconfiguration and an adaptive control EHW chip for use in prosthetic hands and robot navigation.

This paper is organized as follows. In section 2, the basic idea of EHW is described. Sections 3 discusses the requirements for neural network processing in industrial applications. In section 4, a neural network EHW chip is described. Section 5 describes another EHW chip used for the pattern recognition in myoelectric hand control. Section 6 concludes this paper.

2 Evolvable Hardware: Basic Concepts

Evolvable hardware is based on the idea of combining reconfigurable hardware device with genetic algorithms to execute reconfiguration autonomously[5][4].

The structure of reconfigurable hardware devices can be changed any number of times by downloading into the device a software bit string called configuration bits. FPGA (Field Programmable Gate Array) and PLD (Programmable Logic Devices) are typical examples of reconfigurable hardware devices, for which there is already a market worth more than 2 Billion US dollars and growing at 23% per year. However it should be noted that the reconfiguration must still be executed manually by hardware designers.

A genetic algorithm (GA) is a robust search algorithm loosely based on population genetics[1]. It effectively seeks solutions from a vast search space at reasonable computation costs. Before a GA starts, a set of candidate solutions, represented as binary bit strings, are prepared. This set is referred to as a population, and each candidate solution within the set as a chromosome. A fitness function is also defined which represents the problem to be solved in terms of criteria to be optimized. The chromosomes then undergo a process of evaluation, selection, and reproduction. In the evaluation stage, the chromosomes are tested according to the fitness function. The results of this evaluation are then used to weight the random selection of chromosome in favor of the fitter ones for the final stage of reproduction. In this final stage, a new generation of the chromosomes are "evolved" through genetic operations which attempt to pass on better characteristics to the next generation. Through this process, which can be repeated as many times as required, less fit chromosomes are gradually expelled from a population and the fitter chromosomes become more likely to emerge as the final solution.

The basic concept behind the combination of

these two elements in EHW is to regard the configuration bits for reconfigurable hardware devices as chromosomes for genetic algorithms (See Figure 1). If a fitness function is properly designed for a task, then the genetic algorithms can autonomouly find the best hardware configuration in terms of chromosomes (i.e. configuration bits).

For example, in data compression with EHW, we use a prediction function. Optimal prediction functions vary greatly according to the different kinds of data to be compressed. It is, therefore, not possible to design in advance a prediction hardware function. Instead of specifying a detailed hardware design, we define a fitness function. In the case of data compression, the data compression rate is used as a fitness function. Accordingly, a circuit of prediction function with a higher data compression rate is likely to remain in a population. When a good chromosome is obtained, it is immediately downloaded into the reconfigurable device.

If the prediction performance of a given EHW is reduced due to changes in the nature of the data to be compressed, then the GA process is invoked and the search for a better hardware structure of prediction is initiated. In this way, EHW is capable of both autonomous and dynamic hardware reconfigurations.

3 Requirements for Neural Network Processing

Most of the industrial neural network (NN) applications are limited to neural networks with *off-line* learning where learning phase and the execution (recognition) phase are separate. Such a neural network never changes during its execution and therefore lacks the flexibility needed. We contend that in order to use neural networks in a broader range of practical applications, they have to be capable of *on-line* learning. On-line learning allows neural networks to adapt dynamically to changing problems. In this section we examine this in more detail.

3.1 Ontogenic Neural Network

The advantage of a neural network is the ability to adapt to problems by changing interconnection weights on-line. However, it is very difficult to determine the topology of neural network (i.e. the number of hidden layers and units per layer) in advance of the execution. If hidden layer nodes are less than required, it is impossible for the network to learn the problem. On the other hand, if hidden layer nodes are more than required, the network over-learn the problem and hence the poor generalization capability. As a result, trial and error while determining the topology are inevitable to obtain optimal performance. Some neural networks called *ontogenic neural networks* try to solve this problem by letting the network *autonomously* determine the best topology and interconnection weights for a given problem at the execution phase[2].

Ontogenic neural networks are promising for on-line learning and are necessary for practical applications.

3.2 Lack of Dynamically Reconfigurable Neural Network Hardware

Even though an ontogenic neural network allows the dynamic adaptation to a problem, conventional neural network hardware systems have not provided the reconfiguration capability of the network structure that is needed during execution. To our knowledge, digital neural hardware so far have been developed mainly for acceleration of neural computation.

Optimal performance of neural network hardware is obtained when the logical NN structure matches the physical NN hardware structure. Therefore, dynamically reconfigurable NN hardware is a key to applications of ontogenic neural networks. In addition, NN hardware systems for industrial applications should be compact for embedded systems. Stand-alone NN hardware is preferred to back-end NN hardware like SIMD neural machines.

The GRD chip is a building block for the configuration of scalable NN hardware systems. A system built with the GRD chips can dynamically reconfigure its hardware structure to be tailored to the optimal topology without a host machine.

3.3 Learning Time

Long learning time is one of the obstacles when neural networks are used to industrial applications. This is especially significant in multilayer perceptrons (MLPs) with back propagation (BP) learning. The MLP uses the sigmoid function as a node function.

To improve the learning speed, radial basis function (RBF) networks may be an appealing choice. In RBF networks, the node function is a Gaussian function where the response to inputs is more localized compared with the sigmoid function. This leads to faster convergence of up to three orders of

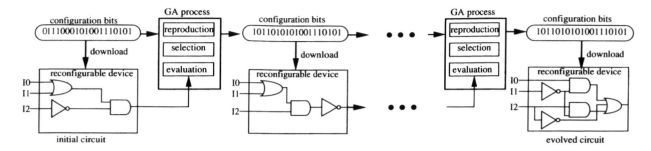

Fig. 1. Basic concept of Evolvable Hardware

magnitude compared with MLP.

However, compared with MLP, RBF networks requires large numbers of hidden layer nodes, particularly for high-dimensional input/output spaces (the "curse of dimensionality"). This becomes an obstacle to compact implementation. Thus, a trade off exists between learning time and size of the neural network. It was therefore our idea to mix the use of RBFs and sigmoid functions within a single architecture. Further, rather than specifying *a priori* how the two functions should be combined, we developed a method of using a GA to tailor the node functions in a network to a given problem adaptively. To reduce the learning time, a steepest descent method is first applied as a local learning algorithm to bring the network weights to a reasonable level. This local learning is performed in parallel by the hardware.

4 The GRD Neural Network Chip

4.1 Overview

The GRD (Genetic Reconfiguration of DSPs) chip is evolvable hardware designed for neural network applications[6]. Both the topology and the hidden layer node functions of a neural network mapped on the GRD chips are dynamically reconfigured using genetic algorithms. This *on-line* learning capability allows neural networks to adapt dynamically to changing problems.

In neural network applications, optimal performance for a given problem is obtained by creating a neural network with the most suitable topology and the most appropriate node functions (e.g. sigmoid function or Gaussian). Furthermore, in order to meet the time constraint imposed by real-time applications, neural network hardware systems need to be 'tailored' to an ideal network size for a problem. In general, it is very difficult to design an optimal neural network and process it with scalable parallel hardware.

With GRD chip, however, the GA program on the RISC processor continues to reconfigure the neural network topology and node functions in order to maintain the optimal performance.

The GRD chip consists of a 32-bit RISC processor and a binary-tree network of 15 DSPs. Each DSP can execute one node function. Using the binary-tree network, multiple GRD chips can be easily connected to configure a scalable neural network hardware.

Also, because a RISC processor is incorporated within the GRD chip, it does not need the host machine control for these tasks. This is desirable for embedded systems in practical industrial applications, together with the fast *on-line* learning capability.

The GRD chip was just manufactured in April 1998. The results of a simulation of an adaptive equalizer in digital mobile communication have showed that execution with a single GRD chip took 2.51 seconds, whereas execution on Sun Ultra2 200 MHz takes 36.87 seconds. The planned use of the GRD chip include applications that vary over time and have real-time constraints, such as CATV modems.

4.2 Genetic Learning

The neural network considered here is aimed at industrial applications which need a neural network for the approximation of non-linear functions.

The genetic learning determines the network topology (e.g. the number of nodes) and the choice of node functions (e.g. Gaussian or sigmoid function) adaptively for a given application. Initial values of the weights and the parameters of the node

Chromosome

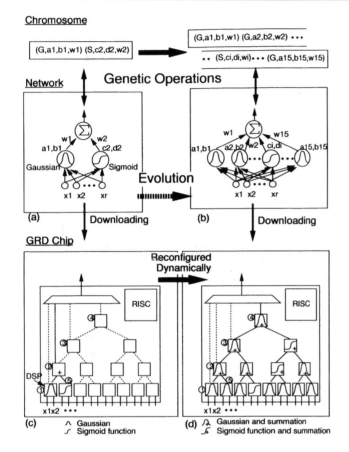

(a)

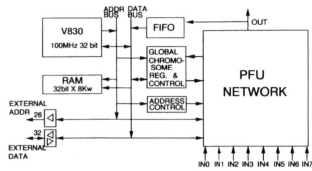

Fig. 3. Overview of the GRD chip

Fig. 2. Genetic learning with GRD chip

functions are also determined by the GA and then tuned by local learning with the steepest descent method.

Figure 2 illustrates this genetic learning. A chromosome of the GA represents one network. The network is evolved by applying the genetic operators to the chromosome. For example, Figure 2 shows how a network with two hidden layer nodes (a) is evolved to have 15 nodes (b).

Now we show how the network obtained by the GA is mapped on the GRD chips and how they are reconfigured dynamically with the example of Figure 2. The network structure obtained by the GA is immediately mapped on the GRD chips. For example, the network having a Gaussian function and a sigmoid function in Figure 2(a) can be mapped onto the GRD chip in Figure 2(c).

The functions and tree height of the GRD chips are dynamically controlled by rewriting the chromosome on the chips. For example, in Figure 2(c), the output of the GRD chip is connected to the out-

put of the DSP No.2. After the evolution, in Figure 2(d), the output of the GRD chip is reconfigured to be connected to the output of the DSP No.4. Also, in Figure 2(c),the DSP No.1 calculates the sigmoid function. After the evolution, in Figure 2(d), this DSP is reconfigured to calculate the Gaussian.

Binary tree connections are very useful when the summation of outputs of nodes is calculated. All the DSPs in Figure 2(d) are configured to conduct the summation in parallel. For example, the DSP No.2 calculates the Gaussian first and then add the result and the output of the DSP No.1 and No. 5.

The above implementation is for the fastest computation. For slower applications, one GRD chip suffices for processing more than 15 nodes. A DSP in the GRD chip can process up to 84 neurons in time division multiplexing.

4.3 The GRD Architecture and the Performance

The GRD chip consists of a 100 Mhz 32-bit RISC processor and 15 DSPs. Figure 3 gives the overall structure of the GRD chip. The RISC processor is the NEC V830 which is designed for multi-media applications. The DSP, a 33 Mhz 16-bit fixed point processor, is called a PFU (Programmable Function Unit). 15 PFUs are connected in a reconfigurable network of a binary tree shape. The GRD chip accepts eight 16-bit inputs and generates a 16-bit output with the MIMD parallel processing of 15 PFUs. The tree shape interconnection is powerful especially when the summation of neuron outputs is calculated.

The GRD chip includes the V830 RISC processor in order to perform genetic reconfiguration of PFUs. This means that the GRD chip can reconfigure itself. In the PFU, the content addressable memory (CAM) is employed to accelerate the computation of non-linear functions (e.g. sigmoid function and Gaussian). Consequently the GRD chip attains 319 MCPS (Mega Connection Per Second) performance in MLP.

The GRD chip can process up to 1260 neurons (84 neurons per PFU). To configure a scalable hardware system easily, GRD chips can be directly connected each other via FIFO buffers inside the PFU. For example, a 19-inch rack implementation of 16 VME triple-height boards (9 GRD chips on a board) can realize the performance of 46 GCPS (Giga Connection Per Second) in MLP.

Simulation results on two applications (the chaotic time series prediction and the adaptive equalization in digital mobile communication) show that the perforamnce of a GRD chip is almost ten times faster than Pentium II processor of 400 Mhz.

5 The EHW Chip for Adaptive Control

5.1 Overview

This chip was developed in April 1998 to serve as an off-the-shelf device for implementing adaptive control logic[9]. Currently, this chip is applied to two applications: an autonomous mobile robot and a myoelectric artificial hand.

The usage of this chip in artificial hand control is similar to neural network. The chip accepts noisy myoelectric signals and recognizes the desirable hand actions.

The hand can be controlled by myoelectric signals, which are the muscular control signals. However, myoelectric signals vary from individual person to person. Accordingly, anybody who has wanted to use a conventional myoelectric hand has to adapt to it through a long period of training (almost one month). To overcome this problem, research is carried out on controllers that can adapt themselves to the characterestics of an individual person's myoelectric signals. Most of this research is using neural networks with back propagation (BP) learning. However, learning with back propagation also needs a great deal of time.

EHW can perform the recognition tasks from noisy or incomplete inputs, which the human brain can easily perform in a flexible and robust manner. The combinatorial logic inside EHW can reconfigure itself to implement a noise-insensitive functions. Because the EHW chip can adapt itself quickly, the

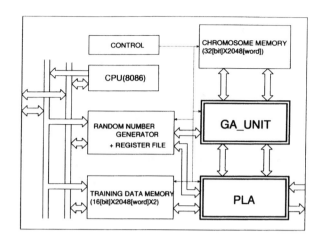

Fig. 4. A block diagram of the EHW chip

learning time for artificial hand can be reduced to a few minutes.

5.2 The Chip Architecture

Basically, EHW consists of genetic algorithms and the reconfigurable logic. In most research on EHW, genetic algorithms are executed with software on personal computers or workstations. This makes it difficult to use EHW in situations that need circuits to be as small and light as possible. For example, a prosthetic hand should be the same size as a human hand and weight less than 700 gram. Similar restrictions exist for autonomous mobile robots with EHW controllers. One answer to these problems is to integrate both the genetic algorithm and the reconfigurable logic into a single LSI chip.

This has been done with the EHW chip for adaptive control, which consists of three components,(1) PLA (Programmable Logic Array), (2) the GA (Genetic Algorithm) hardware, and (3) a 16-bit CPU core (NEC V30). Arbitrary logic circuits can be dynamically reconfigured on the PLA component according to the chromosomes obtained by the GA hardware. Figure 4 is the block diagram of the chip. The CPU core interfaces with the chips's environment, as well as supporting fitness calculations when necessary. The size of the GA hardware is almost one tenth of a 32bit CPU in gate size. However, genetic operations by this chip are 62 times faster than Sun Ultra2. The details of the GA hardware is described in the next subsection.

The chip is being used for a control circuit in a myoelectric artificial hand (See Figure 5). The

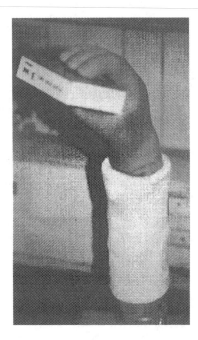

Fig. 5. A myoelectric artificial hand

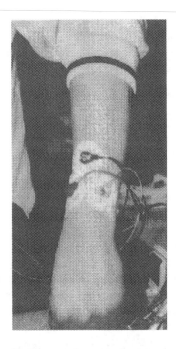

Fig. 6. Sensors for myoelectric signals

sensors for myoelectric signals is shown in Figure 6. The performance in the recognition of myoelectric signals is almost the same as the case with neural network (run on Pentium-Pro 200Mhz).

In general, the noise-insensitive recognition by EHW is very fast, but the precision of the recognition is sometimes a little bit worse than neural network as described in [8].

Another application is an adaptive navigation task for a real world mobile robot that must track a moving colored ball while avoiding obstacles (See Figure 7)[10]. Because the robot moves in an unknown and unpredictable environment, the robot is required to to change its behaviour adaptively. The robot, called *Evolver*, has two camera eyes and sensors (collision and proximity), but no a-priori knowledge of the shapes and the positions of obstacles. The contol logic on the EHW is continuously reconfigured towards improved behaviour. For example, even if one of the sensors becomes broken, the robot autonoumously reconfigures its control logic on the EHW, within a few minutes, to continue the tracking using the other functioning sensors. This adaptation speed is two orders of magnitude faster than with classical approaches. In this robot, EHW is used as a control logic of the robot which is continuously updated to adapt to the changing environment; the usage is completely dif-

ferent from the usage as an noise insensitive recognizer (just like neural network) in myoelectric hand.

5.3 The GA Hardware in the Chip

Implementation problems
Here, we present three problems for the effective implementation of genetic operators.

1. Size of the memory for chromosomes
   The size of the memory needed for chromosomes increases in proportion to the number of individuals. Although the number of individuals should be small to impelment them on a fixed size LSI, a larger population is better for effective genetic search.

2. Selection strategies
   Generational selection strategies which are used in simple GAs need extra memory to temporarily preserve selected individuals. Therefore, the selection strategy used should be a steady state one.

3. Random number generation for crossover points
   In the case of one point or multi-point crossover operations, we have to decide either one or a number of crossover points using random num-

Fig. 7. Autonomous mobile robot EVOLVER

bers. Although the maximum number of random numbers needed depends on the length of the chromosome, hardware implementation of random number generator can make only random bit strings in the form 2N (where N is the length of the bit string).

Based on these considerations, we adopt the following three criteria for the design of the GA hardware;
1. An effective search method using a small population,
2. Steady state GAs,
3. A crossover operation, which does not require random numbers in the form 2N.

Suitable genetic operators for hardware implementation

We propose linking Elitist Recombination and a uniform crossover, as an ideal combination capable of meeting the criteria identified in the previous section.

- Elitist recombination
  The workflow for Elitist Recombination is as follow;

  1. Randomly select two individuals parents (Parent A, Parent B).

  2. Operate crossover and mutation to Parent A and Parent B, to make two children

(Child A, Child B).

  3. Evaluate the two children.

  4. Select the two fittest individuals from amongst Parent A, Parent B, Child A, Child B.

This method is basically an elite-strategy in which the fittest individual always survives. However, here selection only occurs within a family (i.e., the two parents and two children) and diversity can be easily kept in the population. This method can therefore search effectively, even with a small number of individuals and thus meets the first criteria. Furthermore, because this method uses steady state GAs, it also satisfies the second.

- Uniform crossover
  In a uniform crossover, there is no need to select crossover points, unlike with one or multi-point crossovers. Each allele will exchange its information with a fixed probability of 0.5. This method, therefore, meets the final criteria stated in the previous section. By combining these two genetic operators it is possible to meet all three of the criteria identified in the previous section.

The GA hardware is developed according to above design considerations. In the block diagram of Figure 4, components related to the GA hardware are the followings.

- Random number generator
  A parallel random number generator using celluar automata was selected for implementation on the EHW chip, because this is very popular for GA hardware and can produce 576 bits of random bit string every clock cycle.

- Chromosome memory
  A memory for the chromosomes of all the individuals. The maximum length for each chromosome is 2048 bits, and the maximum population is 32. Therefore, the size of this memory is 32[bit] × 2048[word]. This memory has two input/output ports, and two chromosomes can be read in units of 32 bits from the GA UNIT.

- GA unit
  This block reads two chromosomes from the chromosome memory in units of 32 bits, then operates uniform crossover and mutation on them to make two segments (32 bits) of the chromosomes. Uniform crossover is operated

using a random 32bit string. If a location in the random bit string has a value of '1', then the information corresponding to the same location in the two segments the chromosomes is exchanged. The two new segments of the chromosomes are then read from PLA.

6 Conclusion

EHW is a key technology to developing new applications that have not been fully realized due to the lack of *autonoumously* reconfigurable hardware. Althogh FPGA is spreading rapidly and the usefulness of reconfigurable hardware is being recognized, the reconfiguration of FPGAs is not autonomous. EHW applications tend to be time-variant in nature and to have real-time constraints.

In addition to the two EHW chips described in this paper, other EHW chips are being developed for data compression[3], cellular phone[7], etc. The EHW concept is so generic that it can be applied for a wide variety of applications.

References

[1] D.E.Goldberg, "Genetic Algorithms in Search, Optimization, and Machine Learning", Addison-Wesley, 1989.

[2] E. Fiesler "Comparative bibliography of ontogenic neural networks", Proc. of Intl. Conf. on Artificial Neural Networks, 793-796, Springer Verlag, 1994.

[3] H. Sakanashi, M. Salami, M. Iwata, S. Nakaya, T. Yamauchi, T. Inuo, N. Kajihara, and T. Higuchi "Evolvable Hardware Chip for High Precision Printer Image Compression", The 15th National Conference on Artificial Intelligence (AAAI-98), Madison, Wisconsin, July, 26-30, 1998.

[4] X. Yao and T. Higuchi "Promises and Challenges of Evolvable Hardware", IEEE Trans. on Systems, Man, and Cybernetics, 1998.

[5] T. Higuchi, M. Iwata, and W. Liu (editors), Proc. of First International Conference on Evolvable Systems, Lecture Notes on Computer Science, 1259, Springer Verlag, 1996.

[6] M. Murakawa, S. Yoshizawa, I. Kajitani, and T. Higuchi "Evolvable hardware for Generalized Neural Networks", Proc. of the Fifteenth International Joint Conference on Artificial Intelligence (IJCAI-97), pp.1146-1151, Morgan Kaufmann Publishers, 1997.

[7] M. Murakawa, T. Yoshizawa, T. Adachi, S. Suzuki, K. Takasuka, D. Keymeulen, and T. Higuchi, Analogue EHW chip for Intermediate Frequency Filter, International Conference on Evolvable Systems, Sep. 1998.

[8] M. Iwata, I.Kajitani, H.Yamada, H.Iba and T. Higuchi, A Pattern Recognition System Using Evolvable Hardware, Proc. of Parallel Problem Solving from Nature (PPSN IV), Lecture Notes in Computer Science 1141, pp76-770, Springer Verlag, 1996.

[9] I. Kajitani, T.Hoshino, D. Nishikawa, H. Yokoi, S. Nakaya, T. Yamauchi, T. Inuo, N. Kajihara, M. Iwata, D. Keymeulen, and T. Higuchi, A gate-level EHW chip: Implementing GA operations and reconfigurable hardware on a single LSI, International Conference on Evolvable Systems, Sep. 1998.

[10] D. Keymeulen, M. Iwata, Y. Kuniyoshi and T. Higuchi, "Comparison between an Off-line Model-free and an On-line Model-based Evolution applied to a Robotics Navigation System using Evolvable Hardware". In Artificial Life VI: Proceedings of the Sixth International Conference on Artificial Life, C. Adami, R. K. Belew, H. Kitano and C.E. Taylor, eds.. MIT Press 1998.

Direct Inverse Control of Sensors by Neural Networks for Static/Low Frequency Applications

N. Steele, E. Gaura, R.J. Rider

Nonlinear Systems Design Group, Coventry University, Priory Street, Coventry, UK
tel: +44-(0)1203-838825, fax: +44-(0)1203-838949
e-mail: E.Gaura@coventry.ac.uk

Abstract

This paper addresses the issue of direct inverse control for two types of nonlinear transducer systems characterised by:
- piecewise linear input-output transfer function;
- hysteresis occurring in the input-output transfer function;

with the aim of establishing whether some relationship exists between the severity of different nonlinearities and the complexity of the network required to control such nonlinearities in static/low-frequency sensor applications.

The compensation is performed using an artificial neural networks approach. The networks chosen were a static MLP and a tap-delayed line MLP, both trained by an improved BKP method which included a form of dynamic learning management.

1 Introduction

The field of neural networks for control continues to flourish following the enormous world-wide resurgence of activity in the late 1980's. Apart from plant and process control, where the neural networks approach has been studied extensively, attention is now being directed towards the development of smart sensors.

The demand for intelligent sensor systems has increased considerably due to the growing number of applications depending on accurate measurement of physical quantities (e.g. monitoring systems, navigation, defence, medical applications, etc.). One important class of transducers is mechanical sensors for measuring quantities such as acceleration, velocity and position. Much research and development effort is being applied to improve existing designs for such sensors and to find novel control approaches, one aspect of which is the use of micromachined sensors.

In this study, we have considered two manufacturing problems associated with micromachined sensors having a capacitive type of pick-off. It has been shown, both by mathematical modelling and by measurements that these devices are inherently nonlinear [1]. Typical nonlinear effects which must be addressed are:

- the offset of the seismic mass from the central position between the plates. The offset arises due to the accumulation of manufacturing tolerances during the assembly process.
- a piecewise linear input-output characteristic;
- squeeze film damping (the gaseous medium between the electrodes and the seismic mass produces nonlinear damping due to the small physical dimensions of the micromachined product). This results in a dynamic nonlinearity.

In order to provide an overall linear measurement, the nonlinear characteristic of the sensor needs to be compensated. This paper addresses the issue of direct inverse control for two types of nonlinear transducer systems characterised by:

- piecewise linear input-output transfer function;
- hysteresis occurring in the input-output transfer function;

The compensation is performed using an artificial neural networks approach. The networks chosen were a static MLP and a tap-delayed line MLP, both trained by an improved BKP method which included a form of dynamic learning management [2].

2 Neural Networks Direct Inverse Control

Artificial neural networks can be used as a representation framework for modelling and controlling nonlinear dynamical systems. In the literature on neural networks architectures for control, a large number of control structures have been proposed and used [3, 4]. one of the simplest being direct inverse control.

The direct inverse control technique utilises an inverse system model. If the model of the sensor is invertable, then the inverse of the sensor can be approximated. This model is then used as the controller. The inverse model is simply cascaded with the controlled system in order that the composed system results in an identity mapping between the desired response (i.e. the network input) and the controlled system output. Thus, the network acts directly as a controller in such configurations [3].

Training a neural network using input-output data from a nonlinear system can be considered as a nonlinear functional approximation problem [5]. A number of results have been published showing that a feed-forward network of the MLP type can approximate arbitrary well a continuous function. However, what is needed now is an indication of the number of layers/units required to achieve a specific degree of accuracy for the function being approximated. Reported results show that networks with two hidden layers appear to provide higher accuracy and better generalisation than a single hidden layer network, and at a lower cost (fewer total processing units) [6]. Unfortunately, there are no constructive results which define the type and the structure of a suitable network for a given problem.

Our aim here is to determine whether some kind of relationship exists between the severity of different nonlinearities and the complexity of the network required to control such nonlinearities.

3 Case A: Correction of a Piece-Wise Linear Input-Output Transfer Function

Figure 1 shows a piece-wise linear input-output transfer function of a type commonly found in micromachined sensors.

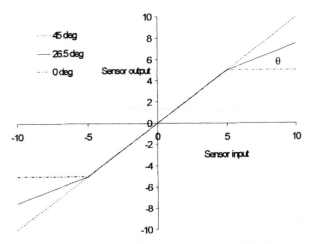

Fig. 1. Piecewise linear input-output transfer function

When $\theta = 45°$, the characteristic is linear over the whole dynamic range of the sensor and compensation is therefore unnecessary. When $\theta = 0°$, the sensor saturates and compensation becomes impossible. For intermediate values of θ, compensation is possible, but the ability to design a suitable neural network architecture may be expected to become increasingly more difficult as θ decreases.

The calibration task selected was to reconstruct a sine wave of fixed frequency after it had been distorted by the nonlinearity. An example is shown in figure 2, where the input sine wave is shown by a solid line and the distorted output of the sensor (at $\theta = 11.3°$) by a dashed line.

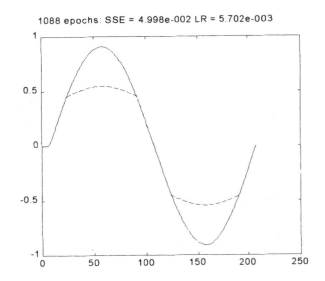

Fig. 2. Sensor input and output signals

The sine wave was scaled to have an amplitude of just less than 1 in order to facilitate reconstruction using a neural network containing sigmoid activation functions. The training set was constructed by uniformly sampling both the distorted and the original signals at 207 points/period, over a single period of the sine wave. These values were used as the network input and output respectively. The test set also contained 207 points taken over one period of the input and output signals, with each test point placed at the midpoint between two adjacent training points.

The dynamic training algorithm included both a variable learning rate and a momentum term. The learning rate is decreased if the sum squared error (SSE) at the network output tends to increase, in order to obtain a finer search over the network's state space. If the SSE decreases, the learning rate is increased in order to speed up the updating of the weights in the direction of the error's minimum.

The figures associated with the training process are presented in Table 1.

Table 1. Training process parameters

Parameter	Value
learning rate	0.04
l_inc	1.04
l_dec	0.99
momentum_term	0.5
lr_min	1E-09
lr_max	0.5
SSE_ratio	1.04

The feasibility of the approach was initially tested by selecting a mild nonlinearity and successfully training several neural networks architectures. Thereafter, five values of θ representing varying degrees of difficulty for the compensation task were chosen: θ = 26.5°; 11.3°; 5.7°; 3.4°; 1.1°, with the aim to discover suitable network configurations, trainable in a maximum of 100000 epochs.

In order to determine the minimum network configuration, an automated training program was designed to test a predefined set of network architectures. The key points are:

- for a given value of θ, up to 4 attempts were permitted to train a network with a given architecture from random initial conditions;
- after a successful configuration had been discovered, - defined by a training error of 0.05/207 per sample - the network architecture was reduced and the search restarted.

This procedure continued until all the architectures in the predefined set were exhausted. In order to abort unsuccessful training attempts without excessive loss of time, a number of milestones were placed in the program. Checks were placed on the rate of error reduction and if at a milestone the error had not reduced by a pre-set amount, that run was halted.

The program has proved of benefit since it has been able to test a very large variety of configurations without the need for human intervention and as a result, successful configurations have been found that would have only been chanced upon using manual restarting of the training process. As an example for θ = 26.5°, the program has found a minimum network architecture of 1-3-1, which trained in approximately 26000 epochs to an error of 0.05/207 per sample. Figure 3. shows the approximation performed by the network on the training set, where the network output is plotted as a dashed line. The generalisation abilities of the network subjected to the test set are presented in figure 4.

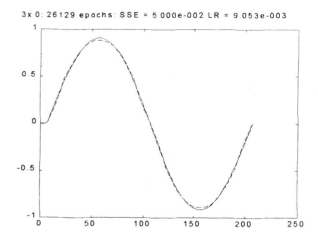

3x 0: 26129 epochs: SSE = 5.000e-002 LR = 9.053e-003

Fig. 3. Network approximation on the training set

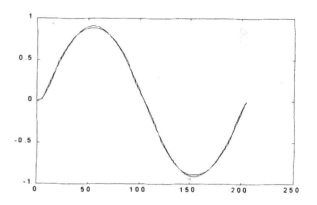

Fig. 4. Network approximation on the test set

The configurations of all of the networks which trained to the predefined output error and the corresponding number of epochs are shown in figure 5.

These results are very scattered, due to the use of random initial conditions for each new configuration. A general trend is, however, apparent in that results for the mildest nonlinearities are grouped into the lower left hand corner of the graph, whilst results for the severest nonlinearities tend towards the top right-hand corner. In an effort to eliminate the problems of the random restarts, it was decided to concentrate on the minimum network configuration that produced a trained result for each value of θ. Figure 6 shows the minimum network configurations found at each of the values of θ tested by the program.

The network configuration is defined here as:

$$L_1 + L_2 + L_1 x L_2,$$

where L_1 and L_2 are the number of neurons in the first and second hidden layers respectively. For single hidden layer networks, the configuration is calculated as $2 \times L_1$.

138

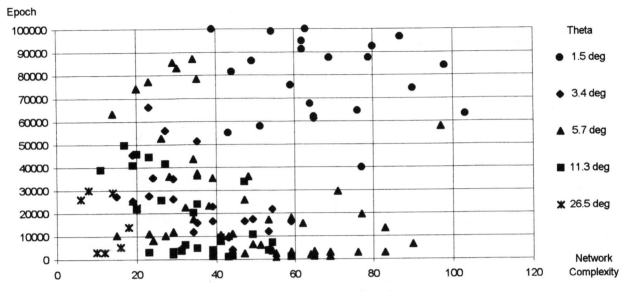

Fig 5. Trainable network configurations

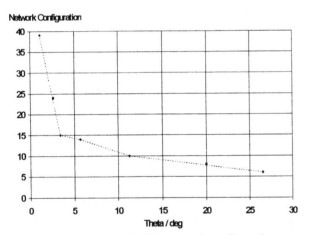

Fig. 6. Minimum trainable network configuration

It should be noted that the dashed line is merely illustrative and should not be interpreted as precisely defining the states between the experimental data points. However, in order to test the validity of this graph, two intermediate points were selected (θ = 2.6° & 20°) and corresponding network configurations ($L_1 = 4, L_2 = 4$ & $L_1 = 4, L_2 = 0$) were successfully trained using the same training regime as before.

The graph shows that for low values of θ, a small decrease in angle will lead to a large increase in the network complexity. On the other hand, for large values of θ, a small change in angle requires little or no change in network complexity. The inherent nonlinearity of the sensor and the range of

manufacturing tolerances will therefore have a significant influence on the required complexity.

4 Case B: Correction of Hysteresis

In addition to piecewise nonlinearities, hysteresis is also common in micromachined sensors. Figure 7 shows a set of four hysteresis loops considered for calibration. The area of each loop was chosen as a measure of the severity of nonlinearity.

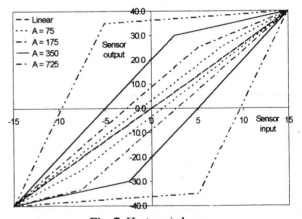

Fig. 7. Hysteresis loops

As for Case A, the calibration task was to reconstruct a sine wave of fixed frequency after it has been distorted by the nonlinearity. An example is shown in figure 8, where the input sine wave is shown by a solid line and the distorted output of the sensor (for a hysteresis area A = 350) by a dashed line.

12x 8: 100000 epochs: SSE = 1.440e-001 LR = 5.162e-004

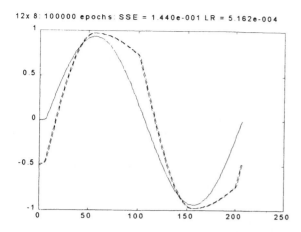

Fig. 8. Sensor input and output signals

Because of the time history dependency present in this type of nonlinearity, a tap-delayed line network was chosen to model the inverse of the sensor characteristic. A program was designed to generate the training sets for the networks, allowing any number of delayed inputs to be specified. After some preliminary investigations, it was decided that 2-input networks (the current and 1-unit delayed signals) should be suitable for this particular application. The training and test sets were constructed in a similar manner to Case A. The training algorithm and the learning rates and momentum parameters also remained the same.

An example of the approximation performed by a 2-9-6-1 network for an inverse loop area index of 0.006, on the training set (71300 epochs) and on the test set is shown in figures 9 and 10 respectively.

9x 6: 71302 epochs: SSE = 5.000e-002 LR = 2.683e-003

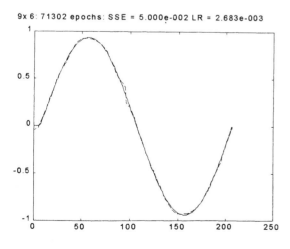

Fig. 9. Network approximation on the training set

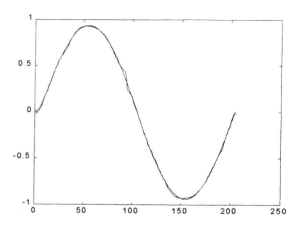

Fig. 10. Network approximation on the test set

The minimum network configurations found for different values of loop area are shown in figure 11. The network configuration was calculated in this case as $2 \times L_1 + L_1 \times L_2 + L_2$ for two-hidden layers networks and $3 \times L_1$ for single hidden layer networks.

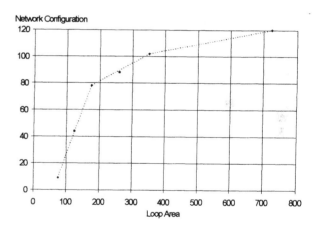

Fig. 11: Minimum trainable network configurations

Bearing in mind the illustrative, rather than precise, definition of the graph between the experimental data points, two intermediate test points were selected (A = 125 & 260) in order to check the validity of this graph. Corresponding network configurations ($L_1 = 9$, $L_2 = 4$ & $L_1 = 11$, $L_2 = 6$) were successfully trained using the same training regime as before.

The graphs in figures 6 and 11 indicate that as the sensor departs from linear behaviour, the required compensating neural network configuration changes much more rapidly for hysteresis than piece-wise linear transfer characteristics. Hence with a fixed hardware control configuration, only small variations in the parameters affecting hysteresis can be accepted against large variations permissible in those affecting the piece-wise linear characteristic.

5 Conclusions

In the context of sensor control for static/low frequency applications, the aim of this work was to establish a relationship between the degree and nature of nonlinearity in the sensor output and the complexity of a neural network required for control. Two types of nonlinearities were considered, namely piecewise linear and hysteresis input-output transfer characteristics. Except for the mildest nonlinearities, it was found that a single hidden layer of neurons was insufficient for the control task. By considering the minimum configurations which trained for each nonlinear case, useful graphs (figures 6 and 11) were produced which successfully provided a guide to network architectures required for intermediate cases tested.

In the case of manufactured devices, the span of manufacturing tolerances and thus the span of nonlinearities would be known. Such graphs could therefore be used to select appropriate architectures without either over or under specifying the capacity of the hardware associated with neural network implementations.

References

1. Kraft, M. : "Closed-loop accelerometer employing oversampling conversion", PhD Thesis, Coventry University, 1997.
2. Gaura E., Burian A., "A dedicated medium for the synthesis of BKP networks", Romanian J. of Biophysics, Vol. 5, No. 15, 1995, Bucharest, Romania.
3. Irwin, G.W., Warwick, K., Hunt, K.J., "Neural networks applications in control", IEE Control Engineering Series 53, Short Run Press Ltd., UK, 1995
4. Godjevac, J., Steele, N. "Fuzzy systems and neural networks", Autosoft J. Intelligent Automation and Soft Computing, 1995
5. Poopalasindam, S., "Neural network based digital compensation schemes for industrial pressure sensors", Ph.D. Thesis, Coventry University, Sep 1995
6. Trifa, R. Munteanu, E. Gaura, "Neural Network based modelling and simulation of PM-Hybrid Stepping Motor Drives", Proc. International Aegean Conference on Electrical Machines and Power Electronics, Vol. 2/2, pp. 460-464, Kusadasi, Turkey.

Centre and Variance Selection for Gaussian Radial Basis Function Artificial Neural Networks

F Scheibel[1], N C Steele[2] & R Low[2]

[1]Fachhochschule Darmstadt, Haardtring 100, 64295 Darmstadt, Germany
[2]Coventry University, Priory Street, Coventry, CV1 5FB, UK.
email: N.Steele@coventry.ac.uk

Abstract

The quality of the response of a RBF neural network depends strongly on the calculation method of the centres and the variance matrices. This paper describes an algorithm which combines the calculation of the centres and variances of the Gaussian nodes to improve the response of a RBF neural network. The selection of the centres is made using a modified version of the K-means algorithm and the variances are based on the sample variance-covariance matrices of the input values associated with the centres. Applications to classification and function approximation problems are considered.

1 Introduction

In a radial-basis function artificial neural network (RBF) the activation function for the units in the single hidden layer are often chosen to be Gaussian functions of the form

$$\phi_l(\mathbf{x}) = \exp\left\{-(\mathbf{x} - \mathbf{c}_l)^T \boldsymbol{\Sigma}_l^{-1}(\mathbf{x} - \mathbf{c}_l)\right\}, l = 1, \ldots, k \tag{1}$$

where k is the number of units in the hidden layer and $\mathbf{c}_l, l = 1\ldots, k$ is the centre vector associated with unit l. Here $\boldsymbol{\Sigma}_l^{-1} = \boldsymbol{\Lambda}_l^{-1}$ is a diagonal matrix, with non-zero entries corresponding to the Gaussian radii. Given this type of internal architecture, the RBF network may be viewed as constructing a Gaussian mixture model of the training data. The model constructed takes the form

$$y = \sum_{l=1}^{k} a_l \phi_l(\mathbf{x})$$

where $a_l, l = 1, \ldots k$ are parameters which have to be determined along with those of the Gaussian basis functions $\phi_l, l = 1, \ldots k$. A possibly related approach to this parameter optimisation problem, based on the expectation-maximisation (EM) algorithm [1], is discussed by Bishop [2].

There have been many successful applications of RBF's to classification problems with $\boldsymbol{\Lambda}_l^{-1}$ chosen as a simple mutiple of the identity matrix. In other situations, notably in applications to fuzzy logic, then $\boldsymbol{\Lambda}_l^{-1}$ has been selected as a diagonal matrix with positive entries, [3]. This latter case can be shown to be equivalent to the use of weights on the connections from the input layer to the hidden layer.

In this paper we consider alternative choices for $\boldsymbol{\Sigma}_l^{-1}$ based on an examination of the training data and its distribution. In the proposed method for centre selection, we also take into account the known classification of the training data.

2 The Sample Variance-Covariance Matrix

Let $\mathbf{X}$ be a set of n points $\mathbf{x}_1, \mathbf{x}_2, \ldots, \mathbf{x}_n$, with $\mathbf{x}_i \in \mathbb{R}^p$ and let $\mathbf{c}$ be the mean vector of set $\mathbf{X}$, which is defined by its components as

$$c_j = \frac{1}{n} \sum_{i=1}^{n} x_{ij}, 1 \leq j \leq p$$

The sample variance-covariance matrix S is then defined by the following the equation:

$$s_{jr} = \frac{1}{n-1} \sum_{i=1}^{n} (x_{ij} - c_j) * (x_{ir} - c_r)$$

Then using the inverse of S as the variance matrix for the Gaussian function

$$f(\mathbf{x}) = \exp -(\mathbf{x} - \mathbf{c})^T \mathbf{S}^{-1}(\mathbf{x} - \mathbf{c})$$

we can achieve a good representation of clustered data in terms of its location and the orientation of the cluster. When $\mathbf{S}^{-1}$ is a diagonal matrix the diagonal norm is induced which generates hyperellipsoidal clusters. The symmetric matrix $\mathbf{S}^{-1}$ induces a Mahalanobis norm which also generates hyperellipsoidal clusters. When $\mathbf{S}^{-1}$ is a diagonal matrix, the axes of these hyperellipsoids are constrained to be parallel to the coordinate axes but in the symmetric case, the orientation is arbitrary. Figure 1 demonstrates the effect of using the symmetric variance-covariance matrix within a two dimensional Gaussian function representing clustered data.

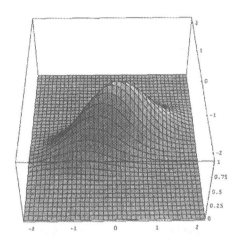

Figure 1. A Gaussian representation of clustered data.

3 The Centre-Variance Algorithm

After considering the sample variance-covariance matrix, we now come to the Centre-Variance Algorithm. For classification problems we assume that the training data for the RBF network is divided into m classes or categories. The idea behind this algorithm is that we use the knowledge as to which of the m classes each of the individual input vectors $\mathbf{x}_i$, $i = 1, \ldots, n$ belongs and we determine the centre $\mathbf{c}^l, l = 1, \ldots, m$ as the centre of gravity of each of these classes which we will refer to as clusters. Thus based on the classification of the training data we obtain m centres as the mean vectors of each class, and we calculate the Gaussian variance matrix for each of the m clusters. It is often the case that the use of just these m centres is insufficient for the classification task, and additional centres are required. In our approach, additional centres are found using the K-means algorithm and Gaussian variance matrices are calculated for these also. The Centre-Variance Algorithm is described in detail below.

- Step 1: Assign a data point $\mathbf{x}_i$, $i \in [1, n]$ to cluster C_l if $\mathbf{x}_i$ belongs to class l, $1 \leq l \leq m$, where m is the number of classes into which the training data has been separated.

- Step 2: Calculate the centre of gravity $\mathbf{c}^l$ for each of the m clusters.

- Step 3: Calculate the Variance-Covariance Matrix $\mathbf{S}_l$ for each of the m clusters using the equation:

$$s^l_{jr} = \frac{1}{n_l - 1} \sum_{i=1}^{n_l} (x^l_{ij} - c^l_j) * (x^l_{ij} - c^l_r), \; j, r = 1, \ldots, p$$

where c^l_j is the jth component of the centre vector of cluster l, x^l_{ij} is the jth component of the ith input vector of cluster l.

- Step 4: Set

$$\boldsymbol{\Sigma}_l = f * \frac{(n_l - 1) * \delta_l}{\max_l(s_{ll}) * n_l} * \mathbf{S}_l,$$

with f a constant scaling factor, δ_l as the square of the maximum distance between the cluster centre $\mathbf{c}^l$ and the input vectors $\mathbf{x}^l_i$, and $\max_l(s_{ll})$ is the maximum of the main-diagonal elements of $\mathbf{S}_l$. n_l is the number of training vectors which are assigned to group l, $l = 1, \ldots, m$

- Step 5: If the total number of centres to be used by the RBF network is k, $k > m$, calculate the extra $k - m$ centres using the K-means algorithm on the entire input data set and repeat Steps 3 and 4 for these new centres.

4 Results

We demonstrate the success of the algorithm on both classification and function approximation problems. The first classification task considered concerned the well known iris data and the second, data on seagulls. The centre selection procedure also extended well into function approximation problems.

4.1 The iris data

This data set contains data on three different species of iris (setrosa, virginia and versicolor) with the four measurements of sepal length, sepal width, petal length and petal width defining the input vectors in $\mathbb{R}^4$. Before we start the calculation we have to fix the total number of centres k and the scaling factor f. In this case the number of classes into which the training data is divided is $m = 3$ and we set the total number of centres as $k = 5$. The scaling factor f was set to 28.

Figure 2 shows the first two components of the input vectors, the 3 centre vectors selected on training data classification, and the 0.5 level of the Gaussian functions. We can see that the Gaussian functions achieve a good representation of the distribution of the different groups in these two dimensions. Fig-

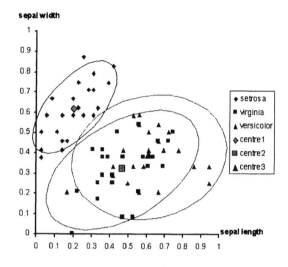

Figure 2. The 3 centres and associated ellipses for the training data.

ure 3 shows all 5 centre vectors and the 0.5 level of the Gaussian functions after the algorithm has finished. An RBF neural network with a bias on the

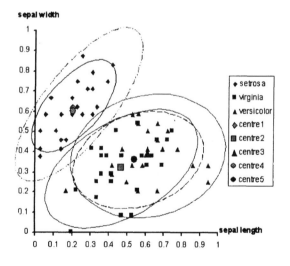

Figure 3. The 5 centres and associated ellipses for the training data.

output nodes, four input nodes, five Gaussian nodes and three output nodes, was trained using the centres and variances from above with data on 75 iris flowers (25 of each type). A testing set of the same size was also used and the results shown in Table 1 were achieved. Figure 4 illustrates the testing data

Table 1. Training and testing results

	Training Data	Testing Data
Number correct	75	73
Percentage correct	100	97.33

with the 5 centres and associated ellipses. Again, the structure and orientation of the data clusters are well represented.

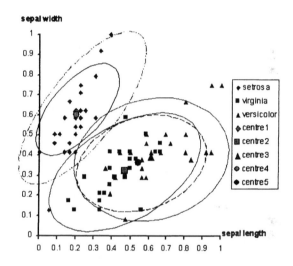

Figure 4. The 5 centres and the associated ellipses superimposed on the test-data.

The scaling parameter f and the number of centres are both important factors for the quality of the results, and the optimal value for f will depend on the nature of the input space. Another possibilty to influence the size of the ellipses shown in Figures 2 and 3 might be to require an inclusion of, say, 95% of the cluster members.

Table 2 compares the response for an RBF network with a bias at the output node for the iris data, using different methods for centre and Gaussian variance calculation. In the Table, 'CV' denotes the Centre-Variance algorithm with $f = 27$, 'KM' denotes the K-means algorithm and 'KNN' denotes the K nearest-neighbour algorithm, used for finding the widths as in Moody and Darken [4]. Whilst performance on the training data can be a misleading indicator, it can be seen from Table 2 that in

Table 2. Results for the iris data

Centre algorithm	Variance algorithm	k	training data	test data
CV	-	4	75	69
KM	KNN, k=2	4	67	65
KM	KNN, k=3	4	67	66
KM	constant radius, r=0.2	4	69	65
KM	constant radius, r=0.3	4	69	65
CV	-	5	75	73
KM	KNN, k=3	5	73	70
KM	KNN, k=4	5	73	70
KM	constant radius, r=0.2	5	72	69
KM	constant radius, r=0.3	5	73	70
CV	-	6	75	69
KM	KNN, k=3	6	71	71
KM	KNN, k=4	6	72	70
KM	constant radius, r=0.2	6	72	69
KM	constant radius, r=0.3	6	72	68
CV	-	7	75	69
KM	KNN, k=4	7	72	71
KM	KNN, k=5	7	72	69
KM	constant radius, r=0.2	7	72	70
KM	constant radius, r=0.3	7	73	68

all cases the CV algorithm achieved 100% performance on this set. The performance on the test set is indicative of generalisation performance, and for lower numbers of centres, the CV algorithm is the best performer. Indeed, the best performance of all is achieved by the CV algorithm with $k = 5$. It is noticeable, and to be expected, that the degree of generalisation reduces as the number of centres is increased.

4.2 The seagull data

The data for this problem concerned four measurements on seagulls. Based on these data the seagulls were identified as members of one of two species. Table 3 shows the response of an RBF network with a bias at the output node trained on this data. The same notation is used as in Table 2 for the centre and variance calculations, however this time, in the CV algorithm, f was set to 15 in each case. Both training and test sets contain 100 input values. The input space dimension is 4 and the output space dimension is 2. The results in Table 3 serve to confirm the excellent performance of the CV algorithm both in terms of success in the classification of the training set and in generalisation performance. Again the performance for the lower values of k is superior to that of the other algorithms tested. There is also

again further evidence of a peak performance being reached at $k = 6$, and after this, although generalisation ability declines at $k = 7$, the CV algorithm still returns (equal) best performance.

4.3 Function Approximation

A considerable amount of work has been done recently on applications of the Tageki-Sugeno fuzzy controller. When the fuzzy sets involved have Gaussian membership functions, this controller can be implemented using a modified RBF of the type considered here. The details of the modifications can be found in detail in [5], but briefly they are:-

1. An additional node is introduced which has as its inputs the unweighted output of the hidden nodes and gives as its output o_{nn} their sum. This is fed to the output nodes which now, as well as summing their weighted inputs, perform division by o_{nn}.

2. The network weights, which only occur between the hidden layer and the output node, are replaced by a function of the input to the network, often a linear function, so that

$$w_i = \mathbf{a}_i . \mathbf{x} + b_i, \ i = 1, \ldots, n_w$$

where n_w is the number of weights in the original network.

Table 3. Results for the Seagull data

Centre algorithm	Variance algorithm	k	training data	test data
CV	-	4	96	97
KM	KNN, k=2	4	94	93
KM	KNN, k=3	4	94	93
KM	constant radius, r=0.2	4	94	93
KM	constant radius, r=0.3	4	94	93
CV	-	5	97	98
KM	KNN, k=3	5	94	93
KM	KNN, k=4	5	94	93
KM	constant radius, r=0.2	5	95	93
KM	constant radius, r=0.3	5	95	93
CV	-	6	97	99
KM	KNN, k=3	6	94	93
KM	KNN, k=4	6	94	93
KM	constant radius, r=0.2	6	97	97
KM	constant radius, r=0.3	6	97	97
CV	-	7	97	98
KM	KNN, k=4	7	94	93
KM	KNN, k=5	7	95	91
KM	constant radius, r=0.2	7	96	98
KM	constant radius, r=0.3	7	97	97

The result of these modifications is that for a network with a single output y this is now given by

$$y = \frac{\sum_{l=1}^{k} \phi_i(\mathbf{x})\,(\mathbf{a}_i.\mathbf{x} + b_i)}{\sum_{l=1}^{k} \phi_i(\mathbf{x})},$$

where k is the number of units in the hidden layer, and the ϕ_l are given by (1). Notice that if the variances of all the Gaussians are small, so that for any input vector $\mathbf{x}$, only one, $\phi_m(\mathbf{x})$, say is significant, then

$$y \approx \mathbf{a}_m.\mathbf{x} + b_m.$$

Thus, a locally linear approximation is generated, however, in view of the form of the approximation, it is differentiable everywhere.

The values of the parameters $\mathbf{a}_i$ and b_i $i = 1,\ldots,n_w$ can be learnt from data, using either backpropagation or by an SVD based solution of a system of linear equations. Alternatively, we can make use of a bi-product of the new clustering procedure, namely knowledge of the orientation of clusters. Figure 5 shows noisy data distributed about an underlying non-linear signal. This signal was that used by Babuska [6] in his discussion of fuzzy clustering algorithms, and is given by

$$y = \frac{x^3 \sin(0.001 x^2)}{10^4} + \epsilon,$$

where $\epsilon \sim N(0,25)$ is normally distributed random noise.

For approximation purposes, the data was categorised into 5 notional classes, based on the x value and in such a way that the 5 elliptic clusters shown were generated. The equations of the major axes of these ellipses were used as the local linear linear approximations between the points of intersection with the major axes of adjacent ellipses. The variance of the Gaussian associated with either the extreme left or right hand ellipse (which now become a function of the single variable x), is set as a small multiple of twice the difference between the x coordinate of its centre and that of the intersection of its major axis with that of its neighbour. In order to produce satisfactory behaviour at the join of adjacent local linear approximations it is necessary to impose the condition that adjacent Gaussians take the same value at the points of intersection of adjacent major axes. Suppose that $\phi_m(x)$ and $\phi_{m-1}(x)$ are associ-

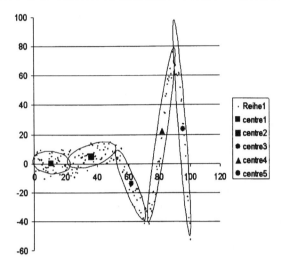

Figure 5. Noisy data

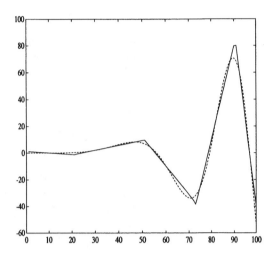

Figure 6. The approximation generated.

ated with ellipses L_m and L_{m-1}, with L_{m-1} to the left of L_m. Then the variances of these Gaussians are related by

$$\sigma_m^2 = \frac{(x_{m-1} - c_m)^2}{(x_{m-1} - c_{m-1})^2}\sigma_{m-1}^2,$$

where x_{m-1} is the x-coordinate of the intersection of the major axes and c_p, $p = m-1, m$ are the Gaussian centres.

The results of this procedure are shown in Figure 6, and a reasonable representation of the underlying function has been obtained. The approximation is least satisfactory in its overshooting behaviour at the turning points. A possible solution to this problem is the location of carefully placed additional small clusters, and ideas on how to automate this location process, and indeed the others, will be developed, drawing on ideas from the field of fuzzy clustering.

5 Conclusions

The CV algorithm has been tested in two classification tasks, and results indicating superior performance over the other algorithms tested have been obtained. This superiority appears in terms of training set performance and, more important, in terms of generalisation capability. In each case, optimum performance was achieved with a low number of centres, which is an important factor in managing network complexity. The results obtained here must be considered as preliminary in statistical terms, and a programme of cross-validation studies is required in order to obtain confirmation. This will form part of the further study. There have been many advances in the field of classification in recent years, and there is a rich literature on the topic. For example, Albrecht and Werner [7] also employed a Mahalanobis distance measure in their successful work on handwritten character recognition.

An approach to function approximation/signal extraction has also been examined, based on an RBF neural network implementation of the Takegi-Sugeno fuzzy controller. A locally linear, differentiable approximation to a signal embedded in noise was constructed, using as line segments, the major axes of the induced data clusters. Again further work is required to refine the technique, both in terms of performance near turning points, and in the general automation of the process. Nevertheless, results are encouraging.

The CV algorithm described here has been developed primarily for detecting 'crisp' clusters and their centres for use with the RBF artificial neural network. It is interesting to note the similarities, particularly in terms of the distance metrics used, between this and Gustafson and Kessel's extension of the 'fuzzy c-means' algorithm, [8]. Thus, some of the techniques already developed in the extraction and manipulation of fuzzy clusters may assist in our future work.

References

[1] Dempster A P, Laird N M and Rubin D B: Maximum Liklehood from Incomplete Data via the EM Algorithm. Journal of the Royal Statistical Society B 39, 1 pp 1-38, 1977.

[2] Bishop C M: Neural Networks for Pattern Recognition. Oxford University Press, Oxford, 1995.

[3] Godjevac J: Neuro-Fuzzy Controllers, Design and Application, Presses Polytechniques et Universitaires Romandes, Lausanne, 1997

[4] Moody J and Darken C J: Fast Learning in Networks of Locally-tuned Processing Units. Neural Computation 1,2, pp 281-294, 1989.

[5] Steele N, Godjevac J: Adaptive Radial Basis Fumction Neural Networks and Fuzzy Systems. Proc CESA'96 Symposium on Discrete Events and Manufacturing Systems, Lille, France, pp 143-148, 1996.

[6] Babuska R: Fuzzy Modeling for Control, Kluwer Academic Publishers, Boston, 1998.

[7] Albrecht R, Werner W: Ein Verfahren zur Identifizierung von Zeichen, deren Wiedergabe stationaeren statistischen Stoerungen unterworfen ist. Computing, 1,1, pp 1-7, 1966.

[8] Gustafson D, Kessel W: Fuzzy Clustering with a Fuzzy Covariance Matrix. Proc IEEE CDC, San Diego, CA, USA, pp 761-766, 1979.

Minimum Square-Error Modeling of the Probability Density Function

M. Kokol, I. Grabec

University of Ljubljana, Faculty of Mechanical Engineering,
Aškerčeva 6, SI-1000 Ljubljana, Slovenia
E-mail: miran.kokol@fs.uni-lj.si, igor.grabec@fs.uni-lj.si

Abstract

Training of normalized radial basis function neural networks can be considered as a probability density function estimation of the experimental data. A new unsupervised method of probability density function estimation is proposed. The method is applied to a multivariate Gaussian mixture model. Batch-mode learning equations are derived and some simple examples are given. Training method is called a *minimum square-error modeling of the probability density function*. It is similar to the maximum-likelihood method but is numerically less demanding.

1 Introduction

Information about natural phenomena is obtained by a measurements. Experimental data are most generally described by the probability density function (p.d.f.). The fundamental problem in modeling of natural phenomena is related to an estimation of the p.d.f. A new method for estimation of the parameters in the parametric p.d.f. is proposed in this paper. The method is compared with the maximum likelihood approach. Estimated p.d.f. is applied for the construction of the normalized radial basis function neural network. Generalization properties of the network are demonstrated on simple regression problems.

Let us consider an independent variable $x \in \Re^n$ and dependent variable $y \in \Re^m$ describing a phenomenon we observe. Let they both be normalized to the unity cubes. In order to simplify notations in further work, we introduce a joint vector $z = x \otimes y \in \Re^{n+m}$. In N measurements we obtain the set $\{z_1, z_2, ..., z_N\}$ of empirical data. The empirical probability density function estimator defined by these data is [1]

$$p_e(x, y) = p_e(z) = \frac{1}{N} \sum_{i=1}^{N} \delta(z - z_i) \qquad (1)$$

If z is a continuous variable then singular estimator must properly be transformed into a smooth function. With this aim we approximate the delta functions by a continuous window function

$$p_e(z) = \frac{1}{N} \sum_{i=1}^{N} \delta_\sigma(z - z_i, \sigma) \qquad (2)$$

This representation of p.d.f. is known as a *Parzen window estimator* [2] and describes a smooth *empirical probability density function*. The smoothing parameter σ depends on the number of samples and tends to zero when N grows to infinity. Detailed discussion on selecting the optimal smoothing parameter can be found in [3].

In practice, we are usually dealing with a large number of multi-component empirical data. This can cause a saturation of memory resources or make the processing time unacceptable long. We overcome this problems by limiting the number of basis functions to a proper fixed value K which is generally much less than N. The estimator (2) is then represented by a mixture of localized component densities $p(z|\mu_i, \theta_i)$

$$p_r(x, y) = p_r(z) = \frac{1}{K} \sum_{i=i}^{K} p(z|\mu_i, \theta_i) \qquad (3)$$

In this parametric description of the p.d.f. the parameters μ_i denote centers and θ_i determine receptive field shapes of component densities in the joint vector space. We call $p_r(z)$ a *representative probability density function*.

Various methods for estimation of parameters μ and θ are known. The most popular unsupervised method is based on the maximum-likelihood approach in which the following log-likelihood measure is to be optimized

$$\mathcal{L} = -\frac{1}{N} \sum_{i=1}^{N} \ln p_r(z_i). \qquad (4)$$

Optimization of the measure (4) can be achieved in different ways. In this work we have used the expectation-maximization (EM) algorithm for comparison with our algorithm [4].

In numerical experiments we found that optimization of the log-likelihood suffers from some difficulties, e.g. slow convergence, numerous local minima, numerically demanding calculation of posterior probabilities. Consequently, we developed an alternative strategy for unsupervised training of the p.d.f. It is based on the comparison of empirical p.d.f. and its mixture model representation via the square-error criterion. As a special case a Gaussian mixture model is applied. Error measure is optimized and obtained training equations are explored and demonstrated in one and two dimensions.

1.1 Application to neural networks

Accurate estimate of the p.d.f. is desired in many scientific and technical areas. We use it to construct the *normalized radial basis function neural network* (NRBF NN). This type of artificial neural network was introduced almost a decade ago [5, 6, 7, 8], but has still not attracted sufficient attention in literature [4, 9, 10]. The NRBF NN can be interpreted either as an approximation to the nonparametric kernel regression estimators or can be derived as an optimal regression estimator when joint p.d.f. is represented by the mixture model. In the second case the optimal estimator is written in the form [1]

$$\widehat{\boldsymbol{y}}(\boldsymbol{x}) = \frac{\int \boldsymbol{y} p(\boldsymbol{x}, \boldsymbol{y}) \, d^m y}{\int p(\boldsymbol{x}, \boldsymbol{y}) \, d^m y}. \tag{5}$$

Numerator in this expression is a statistical average of the dependent variable while the denumerator represents the marginal p.d.f. of the independent variable. A specific form for the regression function $\widehat{\boldsymbol{y}}(\boldsymbol{x})$ depends on the selected mixture component densities. A tractable choice for the individual mixture density is a multivariate radial Gaussian function[1]

$$p(\boldsymbol{z}|i) = \frac{1}{(2\pi\sigma_i^2)^{(n+m)/2}} \exp\left[-\frac{\|\boldsymbol{x} - \boldsymbol{\mu}_i\|^2}{2\sigma_i^2}\right] \tag{6}$$

where the parameter σ_i represents a width and $\boldsymbol{\mu}_i$ a center of i-th basis function in the joint space. In this case the optimal regression function of variable $\boldsymbol{y}$ given variable $\boldsymbol{x}$ is

$$\widehat{\boldsymbol{y}}(\boldsymbol{x}) = \sum_k^K \boldsymbol{w}_k \frac{\frac{1}{\sigma_k^n} \exp\left[-\frac{\|\boldsymbol{x} - \boldsymbol{q}_k\|^2}{2\sigma_k^2}\right]}{\sum_j^K \frac{1}{\sigma_j^n} \exp\left[-\frac{\|\boldsymbol{x} - \boldsymbol{q}_j\|^2}{2\sigma_j^2}\right]}. \tag{7}$$

---

[1]More compact notation for basis functions is applied: $p(\boldsymbol{z}|\boldsymbol{\mu}_i, \sigma_i) = p(\boldsymbol{z}|i)$.

Centers of basis functions in the input space $\boldsymbol{q}_i$ and output weights $\boldsymbol{w}_i$ are obtained from the joint centers $\boldsymbol{\mu}_i$ by splitting them to first n and remaining m components

$$\boldsymbol{\mu}_i = \|\boldsymbol{q}_i, \boldsymbol{w}_i\|.$$

We note that each mixture model component density corresponds to a separate basis function representing the response function of hidden neuron.

We have applied the p.d.f. estimate to the function approximation problem. In the strict approach to this problem, we should obtain all parameters by optimizing the error measure between desired and actual output. In the case of NRBF NN this is not an easy optimization problem. We realize that our approach is only sub-optimal, because the method optimally fits the p.d.f. and not the final error measure. This way we certainly can not achieve the optimal function approximation, but instead we gain some other desirable properties of the model: easier optimization problem, possibility of applying various methods for p.d.f. estimation, robust model, strict statistical interpretation. On the other hand, the p.d.f. is more general model of stochastic phenomenon than $\widehat{\boldsymbol{y}}(\boldsymbol{x})$ and has a great importance in various other applications [1].

2   Square-error minimization

The representative and empirical probability density functions are both continuous and differentiable and can be compared using an appropriate probabilistic distance measure. Many of them have been used [11, p.258] of which the most known is Kullback-Leibler distance leading to the log-likelihood approach. Here we utilize the Patrick-Fisher probabilistic distance measure taking the form of the square-error criterion

$$\mathcal{N} = \int_{-\infty}^{\infty} \left(p_e(\boldsymbol{z}) - p_r(\boldsymbol{z})\right)^2 d^{n+m}z. \tag{8}$$

After squaring and omitting the first term which does not depend on unknown parameters we obtain

$$\mathcal{N} = \int_{-\infty}^{\infty} p_r^2(\boldsymbol{z}) \, d^n x - 2 \int_{-\infty}^{\infty} p_e(\boldsymbol{z}) p_r(\boldsymbol{x}) \, d^{n+m}z. \tag{9}$$

By applying the equation (2) and assuming that number of samples is sufficiently large, so that $\sigma \to 0$, we express the error-function in a form suitable for optimization purpose

$$\mathcal{N} \approx \int_{-\infty}^{\infty} p_r^2(\boldsymbol{z}) \, d^{n+m}z - \frac{2}{N} \sum_{i=1}^{N} p_r(\boldsymbol{z}_i). \tag{10}$$

The integrals in the latest expression have an interesting interpretation. Let us rewrite equation (10) in the form

$$\mathcal{N} = \frac{1}{K^2} \sum_{i=1}^{K} \sum_{j=1}^{K} \int_{-\infty}^{\infty} p(z|i)p(z|j)\, d^{n+m}z$$

$$- \frac{2}{NK} \sum_{i=1}^{N} \sum_{j=1}^{K} p(z_i|j)$$

The integral $\int_{-\infty}^{\infty} p(z|i)p(z|j)\, d^{n+m}z$ can be interpreted as an overlapping of two basis functions. Therefore, the first term on the right-hand side describes the average mutual overlapping of the basis functions. The second term describes an average overlapping of basis functions with experimental data.

Minimization of the measure (10) yields the system of optimizing equations

$$\frac{\partial \mathcal{N}}{\partial \boldsymbol{\mu}_l} = 2\int_{-\infty}^{\infty} p_r(z)\frac{\partial p_r(z)}{\partial \boldsymbol{\mu}_l}\, d^{n+m}z - \frac{2}{N}\sum_{i=1}^{N} \frac{\partial p_r(z_i)}{\partial \boldsymbol{\mu}_l}$$

$$\frac{\partial \mathcal{N}}{\partial \boldsymbol{\theta}_l} = 2\int_{-\infty}^{\infty} p_r(z)\frac{\partial p_r(z)}{\partial \boldsymbol{\theta}_l}\, d^{n+m}z - \frac{2}{N}\sum_{i=1}^{N} \frac{\partial p_r(z_i)}{\partial \boldsymbol{\theta}_l} \tag{11}$$

This is a nonlinear system of gradients which must be optimized by a proper iterative method. The resulting training procedure is called *the minimum square-error learning of the p.d.f.* The explicit form of equations depends on a selected probabilistic mixture model. It seems that the major problem in practical application of the system (11) is calculation of the required integrals. An analytically expressible case is the radial Gaussian probabilistic mixture model which is treated in the next section.

3 Gaussian mixture model

Radial normal mixture model was defined by the expressions (3) and (6). It is inserted in the error measure (10) and in the system (11). After several arithmetic steps we obtain the explicit expression for square-error and the system of gradient equations for optimization of the parameters $\boldsymbol{\mu}_i$ and σ_i

$$\mathcal{N} = \frac{1}{K^2} \sum_{i=1}^{K} \sum_{j=1}^{K} h(i,j) - \frac{2}{NK} \sum_{i=1}^{N} \sum_{j=1}^{K} p(z_i|j) \tag{12}$$

$$\frac{\partial \mathcal{N}}{\partial \boldsymbol{\mu}_l} = \frac{2}{K} \sum_{k}^{K} \frac{(\boldsymbol{\mu}_k - \boldsymbol{\mu}_l)}{\sigma_k^2 + \sigma_l^2} h(k,l)$$

$$- \frac{2}{N} \sum_{i}^{N} \frac{(z_i - \boldsymbol{\mu}_l)}{\sigma_l^2} p(z_i|l)$$

$$\frac{\partial \mathcal{N}}{\partial \sigma_l^2} = -\frac{1}{K} \sum_{k}^{K} \frac{n}{\sigma_k^2 + \sigma_l^2} \left(1 - \frac{\|\boldsymbol{\mu}_k - \boldsymbol{\mu}_l\|^2}{\sigma_k^2 + \sigma_l^2}\right) h(k,l)$$

$$+ \frac{1}{N} \sum_{i}^{N} \frac{(n\sigma_i^2 - \|z_i - \boldsymbol{\mu}_k\|^2)}{\sigma_i^4} p(z_i|l) \tag{13}$$

where the following interacting function is introduced

$$h(i,j) = \frac{1}{(2\pi(\sigma_i^2 + \sigma_j^2))^{n/2}} \exp\left[-\frac{\|\boldsymbol{\mu}_i - \boldsymbol{\mu}_j\|^2}{2(\sigma_i^2 + \sigma_j^2)}\right]. \tag{14}$$

This function has the form of p.d.f. and depends on the parameters of network only. It can be interpreted as a measure of interaction between neurons i and j. System (13) may be used in an appropriate nonlinear optimization method [12, 4] but, as an alternative approach, we found the following iterative scheme to be convergent

$$\boldsymbol{\mu}_l(t+1) = \frac{1}{\frac{1}{N}\sum_i p(z_i|l)} \left(\frac{1}{N}\sum_i^{N} z_i p(z_i|l)\right.$$
$$\left. - \frac{1}{K}\sum_k^{K} \frac{\sigma_l^2}{\sigma_k^2 + \sigma_l^2}(\boldsymbol{\mu}_k - \boldsymbol{\mu}_l)\, h(k,l)\right)$$

$$\sigma_l^2(t+1) = \frac{1}{\frac{1}{N}\sum_i p(z_i|l)} \left(\frac{1}{N}\sum_i^{N} \frac{\|z_i - \boldsymbol{\mu}_l\|^2}{n} p(z_i|l)\right.$$
$$\left. + \frac{1}{K}\sum_k^{K} \frac{\sigma_l^4}{\sigma_k^2 + \sigma_l^2}\left(1 - \frac{\|\boldsymbol{\mu}_k - \boldsymbol{\mu}_l\|^2}{\sigma_k^2 + \sigma_l^2}\right) h(k,l)\right) \tag{15}$$

Only the first term on the right-hand side in each equation depends on the experimental data z_i. The second term is determined by K parameters of network only and is for $K \ll N$ computationally nondemanding. It is important to note, that iterative system do not include a normalization of Gaussian functions what is in opposition to the equations for log-likelihood optimization [4] where calculation of the posterior probabilities takes a lot of processing time. System (15) represents a method for self-organization of neurons resembling the one introduced in [13].

In the next section we examine the training equations (15) on the p.d.f. estimation and regression problems.

Table 1. Estimation of the test normal mixture density.

Likelihood method		Square-error method	
q_k	σ_k	q_k	σ_k
-0.008	0.043	-0.005	0.044
0.270	0.160	0.273	0.158
0.971	0.487	0.999	0.469
Elapsed time: 1.0		Elapsed time: 0.65	

4 Examples

We have examined minimum square-error learning of the p.d.f. in one and two dimensions so far. First type of simulations explores the abilities of derived method for p.d.f. estimation. Second type of numerical experiments is referred to the approximation of continuous functions.

4.1 Estimation of p.d.f.

The accurate estimation of the mixture of normal modes is expected when the parametric Gaussian mixture model is used. In the first example we tested the algorithms by estimating the three component normal mixture density

$$p_r(x) = \frac{1}{3}p(x|0, 0.05) + \frac{1}{3}p(x|0.3, 0.15) + \frac{1}{3}p(x|1, 0.5)$$

This is an easy problem for minimum square-error and maximum likelihood method. We used 500 training points and perform 50 iterations from random starting values. The results are presented in Table 1. Values in Table are random variables depending on training set, but they don't variate significantly for different training sets. From elapsed time values being given in a normalized unit, we note much faster convergence of minimum square-error method.

In the second simulation the uniform distribution from interval [0..1] is estimated. 1000 training points is used and 100 iterations is made. Representative mixture model includes 5 normal components. Left Figure represents the result from the maximum likelihood method, while the right one is the result obtained by the system of equations (15). Minimum square-error method is more than 50% faster than maximum likelihood method while both p.d.f. estimates are very similar. It is not very naturally to estimate the uniform distribution with a Gaussian mixture model. Thus, we can not expect flat p.d.f. estimates by using small number of mixture components. This fact is demonstrated

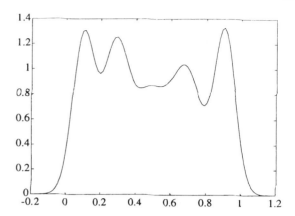

Fig. 1. Estimation of the uniform distribution by maximum likelihood method. Elapsed time in normalized unit is 1.0.

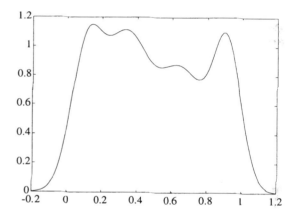

Fig. 2. Estimation of the uniform distribution by minimum square-error method. Elapsed time in normalized unit is 0.47.

in the next example. We simulated two-modal normal distribution by 1000 random points. Mixture model includes 6 components and 100 batch iterations of maximum likelihood and minimum square-error methods is evaluated. Initial state was randomly chosen. Maximum likelihood method results in a more bumpy p.d.f. estimate than minimum square-error method. This seems to be a general valid property for both methods. Half of the components converges to the first mode while remaining components describe the second mode. It is obvious that 6 component is too small number to represent underlying distribution accurately.

152

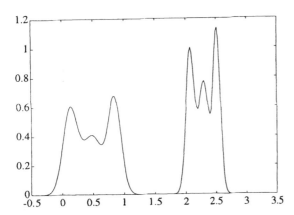

Fig. 3. Estimation of the two-modal uniform distribution by the maximum likelihood method. Elapsed time in normalized unit is 1.0.

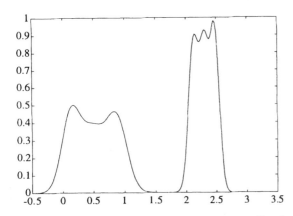

Fig. 4. Estimation of the two-modal uniform distribution by minimum square-error method. Elapsed time in normalized unit is 0.56.

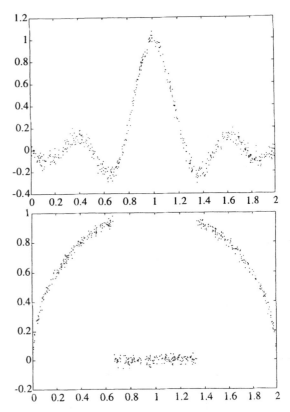

Fig. 5. Simulated hat and hole functions. Each function is presented by 500 points by normal noise $N(0, 0.05)$ added. Distribution of independent variable is chosen to be uniform from the interval $[0..1]$.

4.2 Function approximation

Numerous neural network applications deals with function approximation problems. This is a fundamental task in modeling of the natural phenomena. Pattern recognition problems may also often be treated as a discrete function approximation. In the following simple examples we illustrate how information about p.d.f. can be used in NRBF NN for function approximation purpose. We have chosen two-dimensional examples which are presented in the Figure 5.

Each function is represented by a sample of 500 noisy points. In order to construct the estimator we must select the number of mixture components. This number is not unique for all examples. We tried various structures and selected the acceptable

one. The joint p.d.f. was trained by using maximum likelihood and minimum square-error criterion. Resulting mappings are calculated by using the estimator (7) and are presented in Figures 6 and 7. Non-smoothness of the regression functions is a consequence of the sub-optimal approach we use. There is a systematic error in the applied method, because the radial basis functions are used. Probability that should follow the samples is spread into the neighboring area, therefore the conditional average is over- or under-estimated between the basis functions. This problem of the radial estimator is eliminated by increasing the number of basis functions, or by using the more general basis functions.

More realistic example is presented in Figures 8 and 9 by the discontinuous hole function. Parameters of training are given in the captions of figures.

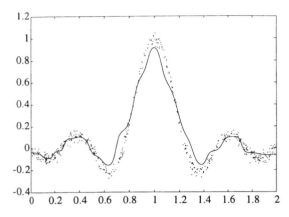

Fig. 6. NN estimation of the hat function when using maximum likelihood training method. Number of data points is 500, number of basis functions is 15, number of iterations is 100, elapsed time is 1.0.

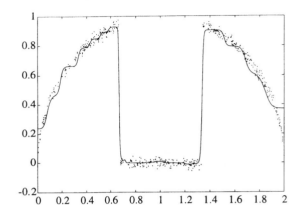

Fig. 8. NN estimation of the hole function when using maximum likelihood training method. Number of data points is 500, number of basis functions is 20, number of iterations is 100, elapsed time is 1.0.

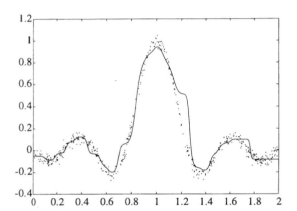

Fig. 7. NN estimation of the hat function when using minimum square-error training method. Number of data points is 500, number of basis functions is 15, number of iterations is 1000, elapsed time is 0.7.

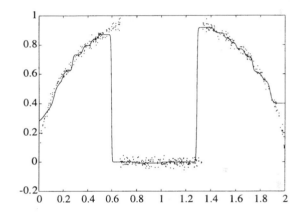

Fig. 9. NN estimation of the hole function when using minimum square-error training method. Number of data points is 500, number of basis functions is 20, number of iterations is 500, elapsed time is 0.35.

5 Conclusions

From the extensive literature on the theory of probability and pattern recognition a reader may get a feeling that only maximum likelihood and Bayesian estimations of unknown parameters in p.d.f.s are possible. In this work we have shown a very general approach in which the parametric and empirical p.d.f.s are compared through a probabilistic distance measure. By minimizing it a new method for p.d.f. estimation is obtained. Method based on the minimization of the Patrick-Fisher measure leads to the minimum square-error modeling of the p.d.f. From the simple examples that have been elaborate it is evident that minimum square-error method gives smoother estimates than maximum

likelihood method and takes considerably less time to converge. In other aspects the both methods are similar.

It is important to note that our approach can be extended easily in two ways: 1) by using various probabilistic distance measures, 2) by using various parametric models for p.d.f. Radial normal mixture model is not flexible enough to represent the underlying p.d.f. in all cases. Selection of more appropriate basis functions is still an open problem. We are trying to extend the radial model to the elliptic basis functions, but this introduces a large number of free parameters.

Another open problem in using the mixture probabilistic models is a number of mixture components. As a possible solution to that problem the approach based on the minimization of information redundancy is presented in an another article at the same conference.

Mixture of probability density components (3) can be given a neural network interpretation. The vectors q_i represent the centers of neuron responses, while the widths σ_i determines the receptive fields of neurons. Regression estimator (7) is interpreted as a recall rule for the normalized RBF NN. Parameters of the network are trained by using the system (15) which represents the self-organization method for presenting the empirical p.d.f. on a finite number of neurons.

References

[1] Grabec I., Sachse W.: Synergetics of Measurements, Prediction and Control. Springer-Verlag, Heidelberg, 1997.

[2] Duda R.O., Hart P.E.: Pattern Classification and Scene Analysis. Academic Press, New York, 1973.

[3] Scott D.W.: Multivariate Density Estimation. John Wiley & Sons, New York, 1992.

[4] Bishop C.M.: Neural Networks for Pattern Recognition. Oxford University Press, Oxford, 1996.

[5] Moody J., Darken C.J.: Fast Learning in Networks of Locally-Tuned Processing Units. Neural Computation, vol. 1, 281-294, 1989.

[6] Stokbro K., Umberger D.K., Hertz J.A.: Exploiting neurons with localized receptive fields to learn chaos. Complex Systems, vol. 4, 603-622, 1990.

[7] Grabec I.: Modeling of chaos by self-organizing neural network. Published in Kohonen T., Mäkisara K., Simula O., Kangas J. (eds.): Proceedings of International Conference on Artificial Neural Networks, Espoo, Finland, Elsevier Science Publishers, North-Holland, Amsterdam, 151-156, 1991.

[8] Schiøler H., Hartman U.: Mapping neural network derived from the Parzen window estimator. Neural Networks, vol. 5, 903-909, 1992.

[9] Ripley B.D.: Pattern Recognition and Neural Networks. Cambridge University Press, Cambridge, 1996.

[10] Cherkassky V., Mulier F.: Learning from Data. John Wiley & Sons, New York, 1998.

[11] Devijver P.A., Kittler J.: Pattern Recognition - A Statistical Approach. Prentice Hall International, Englewood Cliffs, New Jersey, 1982.

[12] Press W.H., Teukolsky S.A., Veterling W.T., Flannery B.P.: Numerical Recipes in C. 2nd edition, Cambridge University Press, Cambridge, New York, 1995.

[13] Grabec I.: Self-organization of neurons described by the maximum-entropy principle. Biological Cybernetics, vol. 63, 403-409, 1990.

Adaptation of NN Complexity to Empirical Information

Igor Grabec

Faculty of Mechanical Engineering, University of Ljubljana
Aškerčeva 6, POB 394, SI-1000 Ljubljana, Slovenia
Email: igor.grabec fs.uni-lj.si, Fax: +386 61 1253 135

Abstract

Instrumental observation of a natural phenomenon represents a transmission of information which is generally subject to random disturbances. In the article the scattering of empirical data provided by observation of a certain state is described by a probability distribution function. It is further applied at the estimation of probability distribution of a compound phenomenon which must be characterized by several possible states. The uncertainty of observation is commonly described by the information entropy. It is shown that the empirical information I_e, which is defined as the difference between the entropies of a compound phenomenon and a single state, characterizes a complexity of observed phenomenon. With an increasing number of observations N the value of I_e increases less quickly as $\log N$ and converges to a fixed value I_∞. A proper number of empirical samples sufficient to represent the phenomenon can be estimated as $K_r = \exp I_\infty$. The redundancy of empirical observations is therefore defined by the excess complexity $R = \log N - I_\infty$. When the phenomenon is modeled by a radial basis function NN a proper number of neurons can be described by the parameter K_r. An optimal NN structure can be obtained by minimizing the objective function which is comprised of the redundancy of the model and the information divergence between the representative and empirical distribution .

Keywords: information, complexity, redundancy, objective function, RBFNN

1 Introduction

Artificial neural networks (NN) can be considered as models of natural laws that are adapted to empirical information provided by repeated observation. [1] Efficient procedures for adaptation of a NN with a given structure have already been developed and profoundly explained in the existing literature while a proper formulation of network structure still represents an outstanding problem. [1, 2] This problem is tried to be solved mainly by minimizing an objective function in which the fitness of a given model to describe the empirical data is accounted together with a model complexity. [2,3] However, this does not answer a question how the NN structure itself could be properly formulated. The aim of this article is to proceed in this direction by the analysis of empirical modeling of natural laws. [1]

The models of natural laws are most frequently formulated analytically in terms of mathematical relations between variables representing the observable properties of the phenomenon. The formulation is usually based on intuition of scientists and at present no unique method formalizes this procedure. Consequently various models could be applied to describe the same phenomenon and there appears a question how one could generally avoid this arbitrariness. With this aim we propose here to consider any observation as a stochastic phenomenon which must be described by methods of probability and statistics. [1] The main problem is thus essentially reduced to formulation of a generally applicable method for empirical modeling of probability distribution.

A commonly accepted method at the analytical modeling of natural laws is to select from some appropriate class of functions an applicable one and to describe it in terms of parameters which are further adapted to the given empirical data. [4] Repeated observations are normally subject to stochastic variations which can be considered as "experimental errors". Consequently, the fitness of the model can generally be described by some statistical measure of discrepancy between experimental data and predictions of the model. [3,4] Quite intuitively, such a measure alone is not applicable as a proper basis for comparison of different models. For this purpose a measure is needed which beside the fitness also accounts the model complexity. [2,3] For practical application it is convenient if both characteristics are quantified in a similar way and joined in a common measure of model quality. [3] The aim of this article is to proceed to the definition of a corresponding objective function by describing the properties of empirical modeling of natural laws in terms of information theory. [1,5] Based on this theory the fitness as well as the model complexity can simply be described in terms of information entropy. In order to keep the treatment as general as possible we assume that probability distributions can most generally be modeled non-parametrically. [1] By this assumption the arbitrariness connected with a selection of parametric model is avoided.

2 Non-parametric Model of Probability Distribution

Any result of measurement can be quite generally treated as a realization x of some random variable X. Without essential lack of generality we further assume that its sample space S_X is continuous. In most experimental situations this space is limited because the ranges of measuring instruments cannot extend to infinity. The fundamental problem of empirical description is to estimate the probability distribution of variable X. [1,4] For this purpose we assume that one can always perform at least some limited number N of statistically independent, identically prepared observations which provide measured data that comprise the sample vector: $x = \{x_1, \ldots, x_N\}$. With these data the empirical model of probability density function (pdf) of variable X can be analytically described by: [1]

$$f(x) = \frac{1}{N} \sum_{n=1}^{N} \delta(x - x_n) \tag{1}$$

This non-parametric estimator is unbiased, but because of the singularity of the delta function it is not consistent. [1,4] The inconsistency is a consequence of the essential discrepancy between the discreteness of observation process and continuity of the sample space S_X. In fact, a finite number of observations cannot provide sufficient information to describe the properties of a continuous random variable exhaustively. Parzen has shown that one can surpass this discrepancy by introducing a continuous approximation of delta function such as the Gaussian function of certain width s. [4,6] This causes the estimator in Eq.1 to become biased, but if s is properly decreasing with increasing number of samples N, then the estimator is asymptotically unbiased as well as consistent. [4,6] In Parzen's estimator the empirical samples represent

natural parameters of the model, but in order to provide consistency, the asymptotic procedure $s \to 0$ requires their number to increase without limit. One can argue that such a model is too complex for practical purposes and there appears a problem how to avoid properly the asymptotic procedure. In this article we propose a possible solution of this problem by considering the uncertainty of instrumental observations of natural phenomena.

For the sake of simplicity we further consider a phenomenon that can be characterized by a scalar variable X because a generalization to a multivariate variable is straightforward. We assume that this variable is observed by an instrument whose accuracy is limited and that the scattering of experimental data acquired by observation of a particular state x is described by some given pdf function $g_o(x)$. Most frequently experimental scattering is described by normal pdf $g_o(x) = g(x - x_o, s_o)$ in which the values x_o and s_o represent the output of the instrument and its root mean square deviation respectively. When the value x_o is observed on the instrument output the value $dP = g_o(x)\, dx$ describes the probability that the true state takes place in the interval of width dx around x. Consider now a more general situation in which it is not certain that the phenomenon could be characterized by just a single state. In this case we talk about the compound phenomenon, or a mixture, that must generally be characterized by a set of different possible states. [4] Based on N observations of such a phenomenon the pdf of observed phenomenon can be expressed as:

$$f_e(x) = \frac{1}{N} \sum_{n=1}^{N} g(x - x_n, s_o) \tag{2}$$

This appears as Parzen's estimator, [6] but the width s_o has here an essentially different meaning. In our case its value is fixed and characterizes statistical scattering caused by experimental observation which is not exactly reproducible, while in the Parzen's estimator it describes an analytically introduced smoothing parameter whose value is asymptotically vanishing with the number of samples N. [4,6] Comparison of Eqs.1 and 2 reveals that $f_e(x)$ is a filtered version of the singular empirical pdf $f(x)$ with a filter kernel $g_o(x)$. [4]

3 Entropy of Empirical Information and Redundancy of Data

Our next goal is to describe the uncertainty related to observation of the selected phenomenon over given instrument. For this purpose we apply the entropy of information H introduced by Shannon.[5,7] For a discrete random variable, having the probability distribution $\{p_j, j = 1, \ldots, J\}$, it can absolutely be described as:

$$H = -\sum_{j=1}^{J} p_j \log p_j \qquad (3)$$

When all J states are equally probable: $p_j = 1/J$, we get the maximal possible value of entropy $H = \log J$, while in the case of only one possible state the entropy is 0. Therefore, the inequality $0 \leq H \leq \log J$ is generally fulfilled.

Contrary to above definition only a relative information entropy can properly be defined for continuous variables. [8] Without loss of generality we assume as a reference distribution a uniform one on arbitrary selected span of the instrument $[0, L]$. Thus the reference, or non-informative pdf is: $\varphi(x) = 1/L$ and the information entropy of a $f(x)$ relatively to $\varphi(x)$ is defined by: [5,8]

$$H = -\int_{S_X} \log\left[\frac{f(x)\,dx}{\varphi(x)\,dx}\right] f(x)\,dx =$$

$$= -\int_{S_X} \log\left[f(x)\right] f(x)\,dx - \log L \qquad (4)$$

Application of the reference pdf yields only an arbitrary additive constant $\log L$ which is not essential for further treatment.

Using Eq. 4 we define the uncertainty related to instrumental scattering of observations of a particular state inside $[0, L]$ as:

$$H_o = -\int_{S_X} \log\left[g_o(x)\right] g_o(x)dx - \log L \qquad (5)$$

For the Gaussian distribution with $s_o \ll L$ it follows:

$$H_o \approx -\log\left[\frac{L}{2s_o}\right] + \frac{1}{2}\left[1 + \log\left(\frac{\pi}{2}\right)\right] \qquad (6)$$

The leading term depends on the ratio $L/2s_o$ which describes the relative span of the instrumental observation. The greater it is, the smaller is the relative uncertainty of observation of a result on the instrument span. The value $\log[L/2s_o]$ describes how much

information can be obtained from the experiments when using the instrument having span L and effective width of observation scattering $2s_o$. The last, constant term is determined by the geometrical properties of Gaussian distribution and is not essential for the properties of H_o.

Similarly as the uncertainty of observation of a particular state we define the uncertainty of empirical observation of the compound phenomenon with respect to the span of the instrument as:

$$H_e = -\int_{S_X} \log\left[f_e(x)\right] f_e(x)dx - \log L \qquad (7)$$

Eq. 2 shows that the pdf obtained from observations of a compound phenomenon cannot be more concentrated as the density $g_o(x)$ describing scattering of observation on a particular state x. This means that there pertains to the observation of a compound phenomenon an uncertainty that is greater or equal to the uncertainty pertaining to observation of a particular state. The difference :

$$I_e = H_e - H_o = -\int_{S_X} \log\left[f_e(x)\right] f_e(x)dx +$$

$$+ \int_{S_X} \log\left[g_o(x)\right] g_o(x)\,dx \qquad (8)$$

is introduced to describe *the empirical information* about the phenomenon. In order to show that it is of fundamental importance for characterization of information provided by observations of continuous variables over instruments we describe its properties for several characteristic cases.

Consider first a non-compound phenomenon, which is characterized by arbitrary single value x_1. In this case the probability density $f_e(x)$ is compressed to the scattering density $g_1(x) = g(x - x_1, s_o)$, both integrals in Eq. 8 are equal, and $I_e = 0$. Consider next an observation of a compound phenomenon when all empirical samples $\{x_1, \ldots, x_N\}$ are mutually separated for much more than s_o so that the functions $g_n(x) = g(x - x_n, s_o)$ are practically non-overelapping. The first term of Eq. 8 can then be approximated as:

$$-\frac{1}{N}\sum_{n=1}^{N}\int_{S_X} g_n(x) \log\left[\frac{1}{N}\sum_{n=1}^{N} g_n(x)\right]dx \approx$$

$$\approx \log N - \int_{S_X} \log\left[g_o(x)\right] g_o(x)\,dx \qquad (9)$$

158

which yields $I_e \approx \log N$. In a more general case, when the samples are overlapping, but not concentrated on just one point, then the inequality $0 \le I_e \le \log N$ holds. We have thus obtained the same relation as is characteristic for the entropy of information of a discrete random variable. The empirical information can therefore be considered as a characteristic that in a continuous case has a similar absolute meaning as the entropy of information in a discrete one. It describes in entropy units how much information is provided about the phenomenon by N repeated experiments on the instrument having instrumental scattering specified by $g_o(x)$. In other words it measures the complexity of the observed phenomenon.

According to the above analysis N repeated experiments can maximally provide $I_{e,max} = \log N$ information, which happens when the functions $g_n(x)$ are completely non-overlapping. Generally we can expect that some overlapping takes place and consequently $I_e \le I_{e,max}$. This means that samples are not maximally informative, or that the complete experimental observation is to certain extent redundant. We quantify the redundancy of N-dimensional empirical sample with respect to given instrumental scattering by the excess complexity:

$$R(N, s_o) = I_{e,max} - I_e \qquad (10)$$

If the width s_o is decreased by improvement of instrument accuracy, the redundancy is reduced and tends to 0 with $s_o \to 0$. With an increasing number of samples N there appears ever more overlapping between functions $g_n(x)$ on average. This overlapping causes that I_e generally increases less quickly than $I_{e,max} = \log N$. Therefore the redundancy increases on average with the increasing number of samples. In accordance with this we can conjecture that empirical information I_e describes how many non-overlapping functions is needed to represent the phenomenon under observation. We define the corresponding number K_r of representative functions by the equation $K_r = \exp I_e$. Its value can be determined from empirical samples and scattering function g_o and generally depends on N. But with $N \to \infty$ it converges to some fixed value K_∞ that can be estimated rather accurately from a finite number of empirical samples. This value corresponds to a characteristic parameter that quite naturally represents the complexity of observation of given phenomenon over an instrument with limited accuracy. The estimated value K_r indicates when on average it is reasonable to stop repetition of the experimental observation. This happens when the number N essentially exceeds K_r.

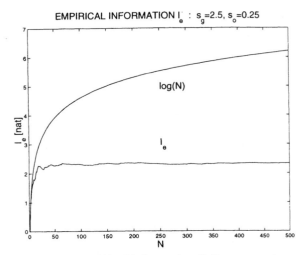

Fig. 1. Empirical information I_e for a normal distribution with standard deviation $s_g = 2.5$ and scattering width of instrumental observation $s_o = 0.25$.

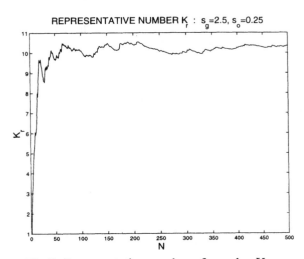

Fig. 2. Representative number of samples K_r as function of the sample number N.

Above mentioned properties are here demonstrated by numerical treatment in which the measurement procedure is simulated for an example of compound phenomenon with Gaussian distribution having standard deviation $s_g = 2.5$. Fig. 1 shows the dependence of the empirical information on the number of samples N for the case of normal scattering of measurements with $s_o = 0.25$. The convergence of empirical information to a fixed value is clearly visible. In the same Figure the maximal possible empirical information $I_{e,max} = \log N$ is represented by the upper curve. The difference between this curve and the lower one represents the redundancy of observation. Figure 2 shows the dependence of the representative number K_r on the number of samples N. The limiting value $K_\infty \approx 10$ is close to the ratio $s_g/s_o = 10$ which is an intuitively expectable result.

4 Compressed Representation of Empirical Model by NN

The fact that the value of K_r quickly converges to K_∞ with increasing N indicates that it is reasonable to utilize only a finite set of $K_r \approx K_\infty$ parameters $\boldsymbol{q} = \{q_1, \ldots, q_{K_r}\}$ to represent in a compressed way the information provided by observations. The problem is thus essentially reduced to the search of a generally applicable cost function whose minimization would lead to adaptation of parameter vector $\boldsymbol{q}$ and its dimension K_r to empirical data.[1] Our final goal is, however, to find such parameters q_k which could simply be interpreted as representative empirical samples because in this case the representation of pdf resembles the non-parametric model and can consequently be considered as a quite general basis for a compressed description of empirical information. We therefore assign to parameters q_k the same properties as are characteristic for the empirical samples, which means that q_k can be interpreted as a possible outcome of the experiment which is subject to scattering determined by the properties of the applied instrument. The representative model of the pdf is therefore described as:

$$f_r(x) = \frac{1}{K_r} \sum_{k=1}^{K_r} g(x - q_k, s_o) \qquad (11)$$

We require that during the adaptation of the model to experimental data this representative pdf will tend to empirically estimated one while the dimension of the representative sample vector $\boldsymbol{q}$ will converge to a fixed value with increasing number of observations N. With this in mind we first describe the discrepancy between the empirical and representative pdf by the Kullback information divergence: [1]

$$D_{re} = \int_{S_X} \left[f_r(x) - f_e(x) \right] \log\left[\frac{f_r(x)\,dx}{f_e(x)\,dx} \right] dx \qquad (12)$$

In accordance with the definition of the empirical information in Eq.8 we define also the information content of the representative model:

$$I_r = H_r - H_o = -\int_{S_X} \log\left[f_r(x) \right] f_r(x)\,dx +$$

$$+ \int_{S_X} \log\left[g_o(x) \right] g_o(x)\,dx \qquad (13)$$

Its maximal possible value for a representative vector of dimension K_r is: $I_{r,max} = \log K_r$ so that the redundancy of the representative model is:

$$R_r = I_{r,max} - I_r = \log K_r - I_r \qquad (14)$$

Our final goal is to proceed towards a model which would represent the empirical pdf with low divergence and also low inherent redundancy. Consequently we define the objective function that describes the discrepancy between both pdfs by the functional:

$$\mathcal{L} = D_{re} + R_r = D_{re} + \log K_r - I_r \qquad (15)$$

It turns out that $\mathcal{L} \geq 0$. An optimally compressed representation of empirical observations is further defined by the representative vector at which this functional has a minimum. At present there is still not known how an exact solution of this problem could be found, but we can obtain suggestions for a heuristic approach to the solution based on the following analysis of the properties of the functional $\mathcal{L}$.

Assume that we try to find the solution by variation. For this purpose we can first select $K_r = N$ and $\boldsymbol{q} = \{x_1, \ldots, x_N\}$. In this case $f_r(x) = f_e(x)$ which yields $\mathcal{L} = R_e$. If $I_e \ll \log N$ we can immediately argue that the complexity of the representative model is too high and try to reduce the dimension K_r of the representative vector. A judiciously selected value would be $K_r = \exp I_e$. This yields the redundancy $R_r \approx I_e - I_r \approx 0$ if the components of the representative vector are distributed so that $f_r(x) \approx f_e(x)$. In this case also D_{re} is small and a significant drop of $\mathcal{L}$ takes place. It is important that adaptation of $f_r(x)$ to $f_e(x)$ can be achieved at fixed K_r by minimizing just the information divergence D_{re}. For this purpose a standard calculus of variation could be applied. Unfortunately, the corresponding system of equations is non-linear and cannot be solved analytically. However, one can hope to find an approximately optimal solution by linearizing the problem similarly as has been shown in reference [1] at the description of self-organized adaptation of neurons of a radial basis function NN. Similarly as here, the problem of an optimal presentation of empirical probability distribution by a representative one has

160

been treated also there. The main difference between both treatments is that the number of prototype data was there considered to be given, while here we estimate it from the complexity of empirical data. Beside this we have previously described the discrepancy between both distributions by the mean square difference, which is after linearization of minor significance.

In relation to the treatment given in reference [1] we propose the following interpretation of an optimal model. The components of the representative vector q can be considered as contents of Gaussian neurons in a RBFNN. The instrumental scattering s_v thus represents the width of the receptive fields of neurons. With the increasing number of empirical samples the number of neurons can be adapted to the complexity of the observed phenomenon by the equation: $K_r = \exp I_e$ while the stored data can be adapted to the empirical distribution by the self-organized co-operation of neurons. [1]

5 Conclusions

The main result of our treatment is that the new definition of complexity of observation in terms of empirical information provides a simple estimation of a dimension of an optimal representative vector. This can further be considered as the number of neurons in a RBFNN that is optimally suited for storage of empirical data provided by instrumental observation. By following the changes of the representative number K_r caused by an increasing number of empirical samples N one can obtain suggestions when a creation or pruning of neurons in the process of adaptation is reasonable.

Assigning the same scattering width s_o to the empirical samples as well as to the representative parameters appears too much constrictive, although it provides simple interpretation. Therefore, procedure presented here is now developed further by considering representative distributions with adjustable widths. Simple examples indicate that such models could exhibit lower redundancies as the representative models with fixed scattering width.

References

[1] I. Grabec, W. Sachse, Synergetics of Measurement, Prediction and Control, Springer-Verlag, Berlin, 1997

[2] A. Leonardis, H. Bischof, An efficient MDL-based construction of RBF networks, Neural Networks, 11, 963- 973, (1998)

[3] J. Risanen, "Complexity of Models", Complexity, Entropy, and the Physics of Information, Vol VIII, Ed. W. H. Zurek, Addison-Wesley, 1990, p-p 117-125

[4] R. O. Duda, P. E. Hart, Pattern Classification and Scene Analysis, J. Wiley and Sons, New York, 1973, Ch. 4

[5] T. M. Cover, J. A. Thomas, Elements of Information Theory, J. Wiley & Sons, new York, 1991

[6] E. Parzen, "On Estimation of Probability Density Function and Mode", Ann. Math. Stat., 35, 1065-1076 (1962)

[7] C. E. Shannon, "Mathematical Theory of Information", Bell. Syst. Tech. J., 27, 379-423 (1948)

[8] A. N. Kolmogorov, On the Shannon Theory of Information Transmission in the Case of Continuous Signals", IEEE Trans. Inform. Theory, IT-2, 102-108 (1956)

The RBF Neural Network in Approximate Dynamic Programming

Branko Šter and Andrej Dobnikar

Faculty of Computer and Information Science, University of Ljubljana, Slovenia

e-mail: Branko.Ster@fri.uni-lj.si

Abstract

A radial basis function (RBF) neural network was applied to an optimal control problem. The role of an approximation architecture in the task of dynamic programming is emphasised. While it has been proved that dynamic programming works well for moderate discrete spaces, research is continuing on how to apply dynamic programming techniques to large discrete and continuous spaces. For continuous spaces there does not yet exist a universal approach, but it seems that a RBF network is able to solve the problem with a negligible amount of manual experimentation.

1 Introduction

Reinforcement learning [1], [2], [3], [4], [5], is a type of learning that needs no teacher. Only a scalar reinforcement signal after a certain period of time is available. Most kinds of reinforcement learning are connected to dynamic programming methods and principles [6], [7]. The task to solve is mostly optimal control, i.e., how to control a dynamic system while optimising some criterion. Optimal control is traditionally linked to the calculus of variation, which is analitically well founded but rather cumbersome in practice. Dynamic programming is the most widespread method for optimal control problems. In discrete environments dynamic programming is proved to work well, while in continuous and large discrete domains certain approximation architectures must be introduced. In this area there are many ill-conditioned cases, and problems are much more difficult.

We connected our work with the basics of classical incremental dynamic programming. We applied the radial basis function (RBF) neural network [8], [9] as the approximation architecture to perform incremental dynamic programming. As a test-bed we took a simplified version of a container crane with a simple goal to stabilise the rope. We show the results of the experiments and conclude the paper.

2 Dynamic Programming and Reinforcement Learning

Let us first introduce the basic assumptions of the problem of dynamic programming. We have a discrete dynamic system with n states. In state i an action $u \in U(i)$ is applied, where $U(i)$ is a finite set of actions; $p_{ij}(u)$ denotes the probability of the transiton from state i to state j when the control u is applied. Such a system can be described as a Markov decision process (MDP). Every transition has a certain reinforcement, in the dynamic programming context usually referred to as a cost. The cost of the transition from state i to state j with the control action u is denoted by $g(i, u, j)$.

The value function $V^\mu(i)$ is defined as the sum of all costs following some predetermined stationary policy μ

$$V^\mu(i) = \lim_{N \to \infty} E \left[\sum_{k=0}^{N-1} \alpha^k g(i_k, \mu_k(i_k), i_{k+1}) | i_0 = i \right].$$
(1)

The discount factor α, $0 < \alpha \leq 1$, gives less importance to more distant transitions. It is also a mathematical trick which assures the integrability of the sum.

The optimal value function of the state i is the sum of all costs following the optimal policy until the end.

$$V^*(i) = min_\mu V^\mu(i)$$
(2)

In the *shortest path problems* holds $\alpha = 1$, and the final state 0 is assumed, which is cost-free.

$$p_{00}(u) = 1, \qquad g(0, u, 0) = 0, \qquad \forall u \in U(0)$$
(3)

The basic equation of dynamic programming is the *Bellman equation*

$$V^*(i) = \min_{u \in U(i)} \sum_{j=1}^{n} p_{ij}(u)(g(i, u, j) + \alpha V^*(j)),$$
$$i = 1, ..., n,$$

In the case of a deterministic system it can be written as

$$V^*(i) = \min_{u \in U(i)} (g(i,u,j) + \alpha V^*(j)), \qquad i = 1, ..., n, \tag{4}$$

while for the shortest path problems we can write

$$V^*(i) = \min_{u \in U(i)} (g(i,u,j) + V^*(j)), \qquad i = 1, ..., n, \tag{5}$$

The simplest form when the cost g is constant is

$$V^*(i) = g + \min_{u \in U(i)} V^*(j), \qquad i = 1, ..., n, \tag{6}$$

The *value iteration* is one of the basic procedures for the incremental updating of the value function (also called cost-to-go function). It is similar to the Bellman equation, except that during the procedure the value function is not yet optimal.

$$V(i) = \min_{u \in U(i)} (g(i,u,j) + V(j)), \qquad i = 1, ..., n, \tag{7}$$

The value function begins at zero and is updated incrementally. It satisfies Bellman equation after a long enough period of time. There are two general ways to update the value function: following trajectories (on-policy) and with independent sampling (off-policy). On-policy assumes that the states for updating are taken from actual trajectories where the learning system is moving, while in the case of off-policy updating the states are taken independently all over the state space. When the existence of the model of the process is assumed, updating can be done off-policy.

3 The RBF Neural Network in Approximate Dynamic Programming

For discrete problems convergence has been proved. In cases of continuous or large discrete spaces, certain approximation architectures must be applied instead of look-up table representation:

$$V(i) \rightarrow \tilde{V}(i,w), \tag{8}$$

where w is the set of parameters of an approximation architecture. The radial basis function (RBF) neural network is a suitable architecture for representing the value function. Each radial basis function

$$f_k(i) = \phi(i, \mu_k, \sigma_k) \tag{9}$$

can be considered as a feature, while μ_k and σ_k are the centres and the widths of the RBFs. The approximation value for the value function V is

$$\tilde{V}(i,w) = \sum_{k=1}^{K} w_k \phi(i, \mu_k, \sigma_k). \tag{10}$$

This is a linear architecture with respect to the weights w_k. The approximation is therefore the dot product of the weight vector and the feature vector $F(i)$

$$\tilde{V}(i,w) = w^T F(i) = \sum_{k=1}^{K} w_k f_k(i), \qquad \forall i. \tag{11}$$

The error is defined as

$$E = \frac{1}{2}(V_d - V)^2 \tag{12}$$

where V_d is the desired value function $(g + V(j))$, and the weight change is defined as

$$\Delta w = -\eta \frac{\partial E}{\partial w}. \tag{13}$$

The partial derivative of the error E on the weight w_i can be written as

$$\frac{\partial E}{\partial w_i} = (V_d - V)(-\frac{\partial V}{\partial w_i}) = -(V_d - V)f_i, \tag{14}$$

and the weight change is

$$\Delta w_i = \eta(V_d - V)f_i. \tag{15}$$

Because of $\partial V / \partial w_i = f_i$, the change is proportional to the f_i. The output weights of the RBF network are therefore updated proportionally with respect to the outputs of the hidden neurons, i.e., dependent on the distance of the state to the centre of a basis function. The closest radial basis function (according to Euclidean distance) is updated to the largest extent, which is logical because the RBFs are typical local features. Gaussian basis functions are used as RBFs.

4 Experiments

We used a simplified model of a container crane. The original problem [12] has six degrees of freedom: x (the position of the trolley), $\dot{x}$ (the velocity of the trolley), l (the length of the rope), $\dot{l}$ (the change of

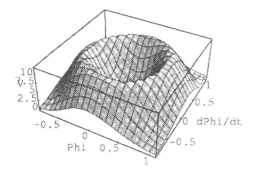

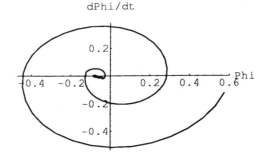

Fig. 1. The value function V at $\sigma = 0.1$

Fig. 2. The resulting trajectory at $\sigma = 0.1$

the length of the rope), φ (the angle of the rope) and $\dot{\varphi}$ (the angular velocity of the rope). The original problem is rather complex, especially concerning the goal; the original goal in experiments was first to swing the rope, then to move the crane to another horizontal position while simultaneously lifting it up and down and trying to stabilise the rope. The angle of the rope was expected to be inside some pre-specified area. The actual physical model was used to train volunteers for many hours, who later described how they obtained their skills and what kind of problems they encountered during learning.

We simplified the problem considerably. We defined the goal as stabilising the rope, starting from some arbitrary initial position. We maintained the length of the rope constant. Our goal was mainly to apply an approximate architecture such as the RBF network to a continuous optimal control problem. We took the model

$$\ddot{\varphi} = \frac{1}{lmM}(-F_x m \cos(\varphi) - gMm \sin \varphi) \quad (16)$$

where the mass of the load is $m = 17200kg$, the mass of the trolley is $M = 12800kg$, $g = 9.81m/s^2$ and the set of the possible actions is $F_x = \{-10000N, 0, 10000N\}$, where N stands for Newtons.

The state space is two-dimensional and we have $K = 11 \times 11 = 121$ radial basis functions with centres μ_k on a grid (i,j), $i = -1, -0.8, -0.6, ..., 0.8, 1$, $j = -1, -0.8, -0.6, ..., 0.8, 1$. This choice is arbitrary.

The final state is $(0, 0)$, where the angle φ and the angular velocity $\dot{\varphi}$ are equal to zero. Using the value iteration procedure, the value function $V(x, y)$ is built.

The value of the approximation is

$$\tilde{V}(x, y, w) = \sum_{i \in I} \sum_{j \in J} w_{ij}$$
$$exp(-\frac{(x - i)^2 + (y - j)^2}{2\sigma^2}),$$

where $I = J = \{-1, -0.8, ..., 0.8, 1\}$ and w_{ij} is the weight between the hidden neuron (i, j) and the output neuron. Therefore only the output weights are updated

$$w_{ij} = w_{ij} + \eta(g + \min V(x', y') - V(x, y))$$
$$exp(-\frac{(x - i)^2 + (y - j)^2}{2\sigma^2})$$

for all hidden neurons, where (x', y') is the next state.

If we draw the state vector on the $(\varphi, \dot{\varphi})$ plane, each circle, centered at $(0,0)$, defines the swinging of the rope with some specified energy. Our goal is to lower the energy as quickly as possible by applying a sequence of appropriate forces on the trolley. Thus the state vector approaches the $(0, 0)$ point in the position-velocity plane, i.e. the equilibrium point.

Figure 1 shows the value function V after 10000 iterations of the value iteration algorithm. Figure 2

164

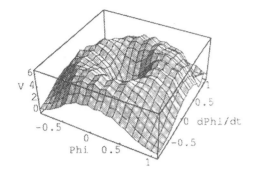

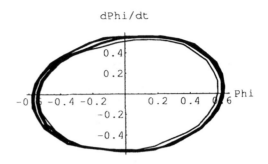

Fig. 3. The value function V at $\sigma = 0.08$

Fig. 4. The resulting trajectory at $\sigma = 0.08$

shows the trajectory obtained following the greedy policy with respect to the value function. (A greedy policy simply chooses that action which leads to the minimal accumulated cost or value function.) The trajectory is expected and logical. The axes are position and velocity. The energy of the system gradually decreases.

It is crucial that the width of the RBF is chosen carefully. Figures 3 and 4 illustrate the effect of a chosen width which is too small. It is obvious that the value function is distorted, and also that the resulting greedy policy is not appropriate. On the other hand, Figures 5 and 6 show the effect of a chosen σ which is too large.

The fact is that for a discrete space σ should be small enough, so that it doesn't interfere with the neighbouring state too much. In that case the convergence is proved ([13]), but it doesn't help much for a continuous space, where relative smoothness is a necessary condition for the greedy policy to be successful.

5 Conclusion

Since dynamic programming is a very successful method for discrete optimal control problems, in recent years there has been extensive research on how to apply approximate methods in cases of very large or continuous state spaces. Artificially constructed problems exist which show that approximate dynamic

programming cannot be applied blindly. In this paper we show that by using the RBF network it is possible to solve a typical optimal control problem successfully, provided some care is taken concerning the widths of the radial basis functions. We simply show what happens when different widths are chosen; we use no method to choose the widths, they are chosen arbitrarily. It is shown that the choice of the width of the RBFs can affect the problem drastically. How to choose the widths appropriately in some systematic way, can be the topic worth further exploring. This paper is a part of the PhD thesis (Branko Šter) on Faculty of Computer and Information Science, University of Ljubljana.

References

[1] Watkins, J.C.H., and Dayan, P.: Technical Note: Q-learning. *Machine Learning*, Vol. 8, pp. 279-292 (1992).

[2] Barto, A.G., Sutton, R.S., and Anderson, C.W.: Neuronlike adaptive elements that can solve difficult learning control problems. *IEEE Transactions on Systems, Man, and Cybernetics*, Vol. 13(5), pp. 834-846 (1983).

[3] Sutton, R.S.: Learning to predict by the methods of temporal differences. *Machine Learning* Vol. 3(1), pp. 9-44 (1988).

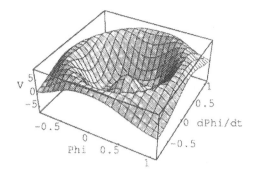

Fig. 5. The value function V at $\sigma = 0.2$

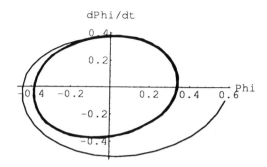

Fig. 6. The resulting trajectory at $\sigma = 0.2$

[4] Barto, A.G.: Reinforcement Learning and Adaptive Critic Methods. In Handbook of Intelligent Control: Neural, Fuzzy, and Adaptive Approaches. D. A. White and D. A. Sofge (Eds.). New York: Van Nostrand Reinhold, pp. 469-491 (1992).

[5] Kaelbling, L.P., Littman, M.L., and Moore, A.W.: Reinforcement Learning: A Survey, *Journal of Artificial Intelligence Research*, Vol. 4 (1996).

[6] Bellman, R.: *Dynamic Programming*. Princeton University Press, Princeton, NJ (1957).

[7] Bertsekas, Dimitri P., and Tsitsiklis, John N.: *Neuro-Dynamic Programming*. Athena Scientific, Belmont, Massachusetts (1996).

[8] Moody J. and Darken C.J.: Fast Learning in Networks of Locally Tuned Processing Units. *Neural Computation 1*, pp. 281-294 (1989).

[9] Haykin, S.: *Neural Networks*. Macmillan College Publishing Company (1994).

[10] Boyan, J.A., and Moore, A.W.: Generalization in Reinforcement Learning: Safely Approximating the Value Function, in G. Tesauro, D.S.Touretzky and T.K.Leen, eds., *Advances in Neural Information Processing Systems 7*, MIT Press, Cambridge MA (1995).

[11] Bradtke, S.: *Incremental Dynamic Programming for On-line Adaptive Optimal Control*. PhD Thesis. University of Massachusetts, Amherst (1994).

[12] Urbančič, T.: PhD Thesis, University of Ljubljana (1994).

[13] Tsitsiklis, J.N., and Van Roy, B.: Feature-Based Methods for Large Scale Dynamic Programming, *Machine Learning*, Vol. 22, pp. 59-94 (1996).

Adaptive Modularity and Time-Series Prediction

Mira Trebar, Andrej Dobnikar
Faculty of Computer and Information Science
University of Ljubljana
Trzaska 25, 1001 Ljubljana, SLOVENIA
mira.trebar@fri.uni-lj.si

Abstract

This paper focuses on the use of recorded time-series to estimate future values as a function of their past values. We study the local events in input space and apply them as classes of similar patterns to the problem of short-term prediction. The decomposition of the time-series into the patterns formed from d past values denoted as an input vector and the true future value in the observed time-series is performed. From the observation of past values we can conclude that similar input vectors often have similar predictive values. We assume that this principle can be expanded in the future. The time-series is based on a similarity measure partitioned into similar patterns grouped into classes. Each of these classes computes the predicted value. The final predicted value is then determined with only one class obtained by the classification from the present input vector.

We apply the concept of modularity in neural networks to perform the local computation of predicted values based on classes of similar patterns. The approach under discussion is a combination of classification and prediction problems. The classification of the input space defines the classes of similar patterns. The number of classes is adaptive and based on the length of the observation pattern for time-series. The predicted value in time-series is obtained by the class modular neural network, which generalises from the classes of similar patterns in training sets in combinations of classification input data into the class of similar patterns.

1 Introduction

In many areas of science and life there is a great deal of interest in looking into the future. We can say that short-term predictions are often possible. Neural networks, with their adaptive, non-linear properties, have been used in different applications: stock-market and exchange-rate prediction [5], control problems [2] and pattern classification [1,4]. Different types of neural network were also used in time-series prediction [1,7,8], where satisfactory results were obtained. At the hierarchical level of organisation, a modular neural network that relies on the combined use of supervised and unsupervised learning paradigms was proposed [1]. The structure consists of supervised modules called expert networks, and an integrating unit called a gating network, which mediates between the expert networks. They are all trained simultaneously and the learning algorithm used to train the modular neural network will model the probability distribution of the set of training patterns. The modules of the network are specialised in learning the different regions of the input space, and the form of competitive learning performs this specialisation naturally. From a practical point of view, modularity in neural networks means the decomposition of the complex task into simpler sub-tasks. In modular architecture it is possible to use different neural structures or different kinds of neural network algorithm or different output functions.

The class modular neural network based on the use of a selection module to define the predicted class was used in time-series prediction [7]. It consists of a selection module and class neural networks. The selection module precisely implements the classification rule derived from the similarity measure. In the proposed modular neural network, the number of class modules was defined by the input vector length and the resolution in a similarity measure, and was not adapted to the data of the observed time-series. Some of the classes were never used in training, and if the test input vector is fixed to that class the prediction gives very bad fits to the time-series data.

In this paper we will introduce an adaptive approach to modularity in neural networks. The described prediction method, based on classes of similar patterns, is divided into two parts, where the definition of the similarity measure and the classification of input patterns are first given in the pre-processing. In the second part, prediction with classes of similar patterns and the implementation of the class modular neural network are outlined. The training algorithm in the proposed neural network relies only on supervised learning paradigms made off-line. In the pre-processing the classification of similar patterns into classes based on the similarity measure is defined. Many similarity measures exist in input space classification [3]. We define a kind of decision-tree classification of observation patterns, where the number of classes is adaptive and depends on the length of the input pattern and the nature of the time-series. If the observed time-series can be locally decomposed into different parts representing similar patterns, the number of these parts affects the number of classes. Each class of similar patterns will give us one predicted value for the present input vector. To define the predicted value of the observed time-series, only one class must be selected. We therefore define the class prediction function that gives us the class connected to the present input vector. Our goal was to find the optimal number of classes from which the best predicted results could be obtained.

We first introduce the basic idea of time-series prediction based on classes of similar patterns in terms of input vector classification, followed by the implementation of the class modular neural network (CMNN) with the supervised learning of class neural networks (CNN) and class prediction neural networks (CPNN). The prediction experiments with a class modular neural network for several time-series are shown at the end, and compared with the results obtained by using a well-known multi-layer perceptron (MLP) and a modular neural network (MNN) that refers to the adaptive mixtures of local experts [1]. The examples we work on are chaotic deterministic time-series and real time-series.

2 Prediction based on classes of similar patterns

We define a time-series prediction method based on classes of similar patterns. In pre-processing, the observation pattern and the similarity measure used in pattern classification are defined. The observation pattern $p=x_{t-d-1},...,x_{t-1},x_t,x_{t+1}$ consists of input vector $X=x_{t-d-1},...,x_{t-1},x_t$ defined in terms of d past values used for prediction in time t and the next future value x_{t+1}. The number of classes depends on the input pattern length d, and is adaptively associated with the time-series under observation. The similarity measure is chosen in advance by an expert. There are many measures to determine the similarity between input patterns, where a commonly used measure is Euclidean distance. We decided to observe the similarity in the meaning of the input pattern shape. The positions of two successive values of the observation pattern give us the criteria to classify two patterns in the same class. Two input patterns consisting of X_i, X_j input vectors of length $d=4$ and their next values $x_i(t+1)$ and $x_j(t+1)$ in Figure 1 a) are similar, and are classified in the same class. For the other two input patterns consisting of X_i, X_k input vectors and their next values $x_i(t+1)$ and $x_k(t+1)$ in Figure 1 b), we assess that they are not

similar, and the second input pattern is classified in a new class.

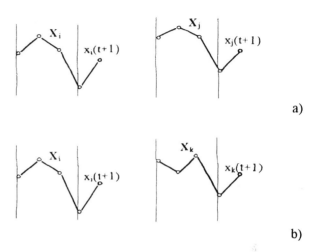

Fig. 1. a) Similar patterns; b) Non-similar patterns

We can formulate the similarity measure with a tree-structured classifier. It is constructed by repeated splits based on a comparison of two successive values in the observation pattern that lead to the terminal node defined as a class of similar patterns. The splits are formed by conditions on the data of the observation pattern $p=x_{t-d-1},...,x_{t-1},x_t,x_{t+1}$. For example, if we begin the tree construction with node N_1, split 1 could be of the form $N_2 = \{p; x_{t-d-1}-x_{t-d-2} < 0\}$ and $N_3 = \{p; x_{t-d-1}-x_{t-d-2} \geq 0\}$, and split d could be of the form $N_k = \{p; x_t-x_{t+1} < 0\}$ and $N_{k+1} = \{p; x_t-x_{t+1} \geq 0\}$. After d splits, the terminal node defines the class of the observation pattern. It is represented with a binary combination of bits obtained during the splits from the observation pattern. The number of terminal nodes is adaptive and is associated with the observed time-series and the input vector length d.

In this paper, the prediction method presented is divided into three basic functions: pre-processing, pattern classification and class prediction. In pre-processing, each input vector $X_i, i=1,...,n-d$ from the observed time-series is classified in exactly one class of similar patterns with regard to the next value $x_i(t+1)$. The number of classes m is defined from the observed time-series. For each class we construct the set of classification training samples. The sample in the classification set consists of an input vector X and the desired response, which is class vector $C = (c_1, ... ,c_m)$. If the input pattern is classified in class $c_i, i=1,2..,m$, it has the value 1; otherwise it is 0. The class vector C will define the class used in calculating the predicted value for the present input vector X.

168

In pattern classification, we predict the class for the input vector used in time-series prediction. The classification set defined in pre-processing is used to obtain the predicted class based on the input vector.

In class prediction, we calculate the predicted value for the present input vector. For each class of similar patterns we have a set of training examples to calculate the predicted value/prediction set. Each example in a prediction set consists of an input-output pair: an input vector X_i, and the corresponding desired response, which is the next value x_i $(t+1)$.

The defined prediction sets and the classification set are used for the definition of the prediction model that is able to calculate the actual predicted value for each class of similar patterns from the present input vector with regard to the class obtained from the classification module. The basic idea of the proposed modularity with classes of similar patterns lies in the adaptive dimensionality of the input vector length and the number of classes for different time-series defined in pre-processing and used in class modular neural network implementation.

3 Class modular neural network

The implementation of the prediction method based on classes of similar patterns requires m prediction modules to calculate the prediction values for each class, and the classification module to predict the class of input vector. We will present a class modular neural network (CMNN) where neural networks are used for the prediction modules and the classification module. For every time-series we can build several CMNNs with different numbers of class neural networks defined by input vector length. The described prediction algorithm consists of three steps for choosing the best class modular neural network used for the time-series under observation.

In pre-processing, we begin with input vector length $d = 1$ where, for the observation pattern p used in the classification tree, the classes of similar patterns are defined. The number of classes $1 \leq m \leq 2^d$ is adaptive and depends on the time-series being described. We write down the classification set and the prediction sets for the defined classes. We continue with the training of the neural networks off-line. A class prediction neural network for input vector classification training is defined and trained with the classification set obtained in pre-processing from the observed time-series data. For every class c_i, we build a class neural network CNN_i $i=1,...,m$ and train them with the defined prediction set of similar patterns. In the evaluation phase at the end, we combine all described

neural networks within the class modular neural network. The output of the class modular neural network is the predicted value and is compared with the next future value obtained in time $t=1,2,....$ to analyse the performance of the proposed neural network. For every input vector the class prediction neural network matches the class of similar patterns used for class neural network selection at output.

The length of input vector d is updated, and we stop the process if the results of the prediction evaluation begin to increase or the number of classes is not acceptable. The proposed prediction method can be described as

$d= 1$
REPEAT
    Pre-processing: *classification,*
                        number of classes definition
    Training: *class prediction neural network,*
                class neural networks
    Prediction: *testing set evaluation*
    $d= d+1$
UNTIL   *{the number of classes is small*
            and the evaluation error is satisfied}

We use multi-layer perceptrons to build the class modular neural network. The class neural networks have only one output y_i, $i=1,...,m$, defined as a function of d past values $y_i = f(x_{t-d-1},...,x_{t-1},x_t)$ to predict the next values for the input vector. The class prediction neural network has as many outputs c_i, $i=1,2,...,m$ as number of classes of similar patterns used. The class modular neural network shown in Figure 2, which implements the prediction with classes of similar patterns, is trained separately off-line for the class neural networks and for the class prediction neural network.

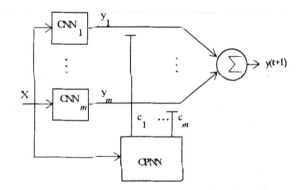

Fig. 2. Class modular neural network

Let the training examples for prediction modules CNN_i, $i=1,...,m$ be denoted by input vector X_i of dimension d and desired future value $x(t+1)$. Let y_i denote the output of the class neural networks and c_i denote the activation of the i-th output neuron of the class prediction neural network, and let $y(t+1)$ denote the output of the class modular neural network that represents the predicted value. The input vector X is applied to the class neural networks and the class prediction neural network. We define the output of the class modular neural network as

$$y(t+1) = \sum_{i=1}^{m} c_i y_i \, .$$

Only one output neuron of the class prediction neural network is active, and the actual predicted value $y(t+1)$ is thus obtained with the class neural network output y_i ,$i=1,2,...,m$ selected by the class prediction neural network.

4 Prediction experiments

The learning and testing results of the class modular neural network are given for several types of discrete time-series. For measuring the performance of the prediction, we use the average relative variance (ARV), which is defined by the normalisation of the mean squared error with the variance of the whole time-series. In CMNN, the training of a particular neural network is done off-line, and convergence is achieved for supervised modules using the back-propagation learning algorithm. The dimensionality of classes is selected in an experimental way in the pre-processing phase for each length of input vector for the observed time-series.

The training process was done for several lengths $d=1,2,3,...$ of input vector X, and was finished when the errors in the testing set began to increase. The class neural networks are multi-layer perceptrons containing one hidden layer of nodes between the input and output nodes, where sigmoid logistic non-linearity is used. Length d is the parameter that biases the number of classes. Our goal is to find the minimal number of classes in order for the observed time-series to achieve good prediction results. From the results obtained in the training process for parameter d, we use the smallest length for input vector X, and the number of classes obtained with these parameters gives us the number of CNNs. The classification of the input vector to obtain the predicted class in CMNN was also trained using a multi-layer perceptron of the same type as was used for the prediction problem, where we try to obtain a 100% classification rate on the testing set, or the best

match of input vector and prediction class. The training parameters used in multi-layer perceptrons have a learning rate of 0.8 and a momentum term of 0.6 to show the fastest and most stable convergence for the observed time-series. In a class modular neural network to implement class neural networks and the class prediction neural network, multi-layer perceptrons were used.

The first time-series used in our experiments with a class modular neural network is the logistic map [1], where only the previous sample x_{t-1} determines the value of the present sample x_t (qmap). This time-series is known to be chaotic at the interval {0,1}. In Figure 3 we present this time-series for 268 samples, where the first 214 samples were used in training and the last 54 samples were used for measuring the performance of the prediction.

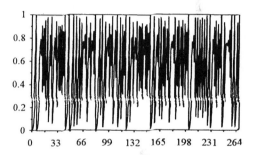

Fig. 3. Qmap time-series

In the second experiment, a real time-series (wbike), which included 144 samples of the weekly sales of bicycle tyres by a domestic producer, was used (see Figure 4). We had 115 samples in the training set; the other 29 were in the testing set.

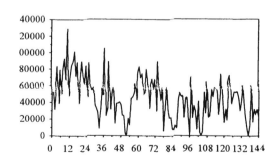

Fig. 4. Wbike time-series

The other three time-series used in the experiments illustrate a few examples of the wide range of behaviour that is observed in time-series. The sources from which the data sets are selected are listed [6].

170

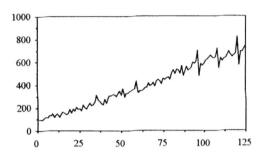

Fig. 5. Consumer time-series

The consumer time-series in Figure 5 includes a monthly index of shipments of consumer products using 125 pieces of data.

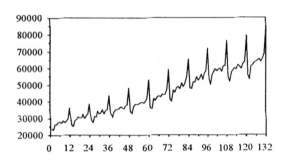

Fig. 6. Retsales time-series

In the retsales time-series in Figure 6, the monthly retail sales of nondurable goods stores in millions of dollars using 132 pieces of data is shown.

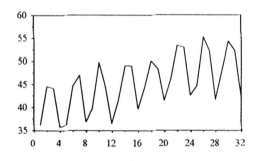

Fig. 7. Beer time-series

The last time-series in Figure 7 shows 32 pieces of data for quarterly US beer production in millions of barrels.

All the time-series described above were used for measuring the performance of prediction using the class modular neural network. As we can see, the time-series used in the experiments have trends, seasonal

cycles and added noise. The input data is first pre-processed using the normalisation function and then scaled into the (0,1) range. We present the prediction results obtained using a class modular neural network (CMNN), a multi-layer perceptron (MLP) and an associative Gaussian mixture model of a modular neural network (MNN).

Table 1 presents the results obtained using a class modular neural network. These results are compared with the prediction results obtained using a well-known multi-layer perceptron (Table 2) and a modular neural network (Table 3). A class modular neural network with two class prediction modules gives better results for qmap time-series.

Table 1. Prediction results (CMNN)

Time-series	d	CMNN ARV Train	CMNN ARV Test
Qmap	1	0.0008	0.0007
Wbike	1	0.41	0.68
Consume	1	0.02	0.096
Retsales	5	0.013	0.073
Beer	5	0.048	0.035

Table 2. Prediction results (MLP)

Time-series	d	MLP ARV Train	MLP ARV Test
Qmap	1	0.0047	0.0046
Wbike	3	0.78	0.57
Consume	3	0.26	0.13
Retsales	4	0.05	0.22
Beer	4	0.049	0.037

Table 3. Prediction results (MNN)

Time-series	d	MNN ARV Train	MNN ARV Test
Qmap	1	0.012	0.011
Wbike	2	0.42	0.45
Consume	5	0.63	0.89
Retsales	7	0.54	0.89
Beer	4	0.09	0.07

Each CNN module is a multi-layer perceptron with one input node, one hidden layer of two neurons, and one output node. The class prediction neural network is also a multi-layer perceptron with one input node and one hidden layer, where the number of nodes is the same as for the output layer, which is defined by the number of classes of similar patterns. The patterns were partitioned into two sub-sets and we obtained 100% classification accuracy with the class prediction neural network, and the prediction results are better than those obtained using the multi-layer perceptron with one input node, one hidden layer of three neurons and one output node (1-3-1). The actual and predicted outputs of the class modular neural network for qmap time-series provide a very close fit with the time-series under observation. The results in the modular neural network, where two expert modules were used for the qmap time-series, are not comparable with the other two neural networks.

The second time-series and their predictions on the testing data showed us that the class modular neural network was mostly able to predict average values, with the trend included in the wbike time-series data. We can say that we did not expect the results of the real wbike time-series to be better, owing to the lack of regularity of the data used in similar patterns. The results of the wbike time-series are slightly better with the multi-layer perceptron. We can conclude that time-series data very much affects the prediction results with a class modular neural network.

In the next two time-series we obtained much better results with CMNN, where the number of classes for the consumer time-series was 2 and for the retsales time-series 20. The multi-layer perceptrons used in CMNN have dimensionality connected to the number of input nodes for CNN and output nodes for the class prediction neural network.

The last time-series shows that the results for both neural networks are very similar. The beer time-series deals with a significant periodical component. The number of CNNs used in prediction was 8. For this type of data, very good predictive results can be obtained with a multi-layer perceptron, and the clustering of data into several classes slightly improves the prediction.

The results obtained by MNN in Table 3 are comparable only for the wbike time-series. For the other four time-series we could not achieve better results with a modular neural network, where only one expert module was used in time-series prediction.

5 Conclusion

A prediction method using classes of similar patterns and implemented using a class modular neural network defined in advance is presented here. The most important feature of the prediction method was the search for the optimal length of input vector d used in time-series prediction. The number of classes depends on the length of input vector used for the observed time-series, and on the similarity measure used in pre-processing to obtain the prediction sets and the class prediction set of training patterns. The class modular neural network architecture is adaptive and is connected with the observed time-series. We use supervised learning in class prediction and class neural networks to obtain the best prediction with a class modular neural network. The results are quite good if the observed time-series can be realised using a local structure of input-output mapping which is extended into the future. The prediction results also depend on the accuracy of the classification used to define the class of similar patterns for choosing the best prediction module.

References

[1] Haykin, S.: Neural Networks - A Comprehensive Foundation, Macmillan College Publishing Company, New York 1994

[2] Jacobs, R.A., Jordan, M.I.: Learning Piecewise Control Strategies in a Modular Neural Network Architecture, IEEE Trans. on Syst., Man, and Cyb., Vol.23, No.2, pp. 337-345, 1993

[3] Ronco, E., Golle, H., Gawthrop, P.: Modular Neural Networks and Self-Decomposition, Technical Report: CSC-96012, February 11, 1997

[4] Petridis, V., Kehagias, A.: Modular Neural Network for MAP Classification of Time-series and the Partition Algorithm, IEEE Trans. on Neural Networks, Vol. 7, No. 1, pp. 73-86, January 1996

[5] Pham, D.T., Xing, L.: Neural Networks for Identification, Prediction and Control, Springer Verlag, London, 1995

[6] Pole, A., West, M., Harrison, J.: Applied Bayesian Forecasting and Time Series Analysis, Chapman & Hall, New York, 1994

[7] Trebar, M., Dobnikar, A.: Time Series Prediction with Modular Neural Network Architecture, SOCO'97 Second International ICSC Symposium on Soft Computing, Nimes, France, pp. 152-155, 1996

[8] Weigend, A.S., Huberman, B.A., Rumelhart, D.E.: Predicting the Future: A Connectionist Approach. International Journal of Neural Systems, Vol. 1, N°3, pp.193-209, 1990

Some Comparisons Between Linear Approximation and Approximation by Neural Networks

M. Sanguineti [1], K. Hlaváčková–Schindler [2]

[1]Department of Communications, Computer and System Sciences

DIST - University of Genova, Via Opera Pia 13, 16145, Genova, Italy

E-mail: marce@dist.unige.it

[2] Institute of Computer Science, Academy of Sciences of the Czech Republic

Pod vodárenskou věží 2, 182 07, Praha 8, Czech Republic

E-mail: katka@uivt.cas.cz

Abstract

We present some comparisons between the approximation rates relevant to linear approximators and the rates relevant to neural networks, i.e., nonlinear approximators represented by sets of parametrized functions corresponding to a type of computational unit. Our analysis uses the concept of variation of a function with respect to a set. The comparison is made in terms of Kolmogorov n-width for linear spaces and a proper nonlinear n-width for the nonlinear context represented by neural networks.

The results of this paper contribute to the theoretical understanding of the superiority of neural networks with respect to linear approximators in complex tasks, as is confirmed by a wide variety of applications (recognition of handwritten characters and spoken numerals, approximate solution of functional optimization problems from control theory, etc.).

1 Introduction

Artificial neural networks have greatly outperformed with respect to linear approximators in complex applications such as recognition of handwritten characters and spoken numerals [3], stabilization of high-order strongly nonlinear dynamic systems [12], vocalization of text [15], etc. This performance brings forward the need for theoretical comparison of the approximation capabilities of linear and nonlinear approximation schemes.

The *universal approximation* property has been proved by various authors (see, for example, [4], [7], [13]), in many function spaces and for different types of architectures and activation functions of hidden units (e.g., radial-basis-functions and perceptrons). *Rates of approximation* express the relationship between the accuracy of approximation and the complexity of the approximators required to achieve such an accuracy. The complexity is usually expressed as the size of a properly defined parameter vector: for example, the degree of a polynomial or the number of knots of a fixed-knots spline in the linear case, the degree of a rational function or the number of hidden units of a neural network in the nonlinear context.

To theoretically understand the superior experimental performance of neural networks with respect to linear approximators, it is important to study the comparison of rates by linear and nonlinear approximation schemes in the same functional spaces. What makes this comparison difficult is the fact that each approximator proposed in the literature has been developed to approximate functions from different spaces, i.e., has been obtained under different assumptions on the functions to be approximated. It is then expected that each approximator performs better than the others if such assumptions are satisfied [5]. A better convergence rate for nonlinear approximators with respect to linear ones in the same functional space has been proved by Barron in [1]. However, this comparison is again made only for functions belonging to a particular space and for a specific class of nonlinear approximators.

In [9] Kůrková has defined a norm, called variation of a function with respect to a set of functions, which extends a concept introduced by Barron in [2]. Such a norm is assigned to a given class of networks and allows the comparison of rates of convergence within a common framework (see [9], [10]). In [6] we have used this norm as a tool for comparison of the optimal bounds on the approximation error achievable by nonlinear approximators in certain spaces of functions of finite-dimensional Hilbert spaces and the rates obtained in the same spaces by linear approximators.

In this paper, we further develop the comparisons started in [6]. Using the variation norm, we show that, in some functional spaces, lower bounds on the rates of linear approximation are greater than upper bounds on the rates achievable by nonlinear approximators represented by neural networks (i.e., sets of parametrized functions corresponding to a type of computational unit).

The organization of the paper is the following. Section 2 contains preliminary notations and definitions. Section 3 presents the comparison of the optimal bounds and some final remarks are concluded in Section 4.

2 Preliminary Notations and Definitions

The following notations and definitions will be used (see also [8] and [10]). We assume to work in a normed linear space $(\mathcal{X}, \|.\|)$; $\|.\|_2$ denotes the norm induced by the inner product in case of a Hilbert space.

The approximation is called *linear approximation*, when the approximating functions form a linear subspace of $(\mathcal{X}, \|.\|)$. On the contrary, the approximating functions can be members of unions of finite-dimensional subspaces generated by a given computational unit. In other words,
$\mathcal{G} = \{g(., \theta) : Y \to \mathcal{R}; \theta \in \Theta \subset \mathcal{R}^p\} \subset (\mathcal{X}, \|.\|)$,
$Y \subset \mathcal{R}^d$, is a parametrized set of functions corresponding to the computational unit represented by the (activation) function g. The set of all linear combinations of n elements of $\mathcal{G}$ is considered. This set, denoted by $span_n\mathcal{G}$, is the union of all linear subspaces formed (spanned) by n-tuples of elements of $\mathcal{G}$, i.e., $span_n\mathcal{G} := \{f \in \mathcal{X}; f = \sum_{i=1}^{n} w_i g_i; w_i \in \mathcal{R}, g_i \in \mathcal{G}\} = \bigcup\{span\{g_1, \ldots, g_n\}; g_i \in \mathcal{G}, i = 1, \ldots, n\}$. In this case, the approximation is called *nonlinear approximation*. Note that $span\, \mathcal{G} = \bigcup_{n \in \mathcal{N}} span_n\mathcal{G}$. $\mathcal{G}^0$ denotes the set of normalized element of a given set $\mathcal{G}$, i.e. $\mathcal{G}^0 = \{g^0 = \frac{g}{\|g\|}, g \in \mathcal{G}\}$.

Let $\mathcal{G}(b) := \{wg; w \in \mathcal{R}, |w| \leq b, g \in \mathcal{G}\}$. For a subset $\mathcal{G}$ of a normed linear space $(\mathcal{X}, \|.\|)$, $\mathcal{G}$-variation of $f \in \mathcal{X}$ is

$$\|f\|_\mathcal{G} := \inf\{b > 0; f \in cl\, conv\, \mathcal{G}(b)\}$$

where the notation is motivated by the fact that

$\mathcal{G}$-variation is a norm on $\{f \in \mathcal{X}; \|f\|_\mathcal{G} < \infty\}$ [9]. Although in general the concept of $\mathcal{G}$-variation depends on the choice of the norm, to simplify the notation we write $\|f\|_\mathcal{G}$ instead of $\|f\|_{(\mathcal{G}, \|.\|)}$ (note that when $\mathcal{X}$ is finite-dimensional, all norms on it are equivalent, hence in such a case $\mathcal{G}$-variation does not depend on $\|.\|$).

For a subset $\mathcal{S}$ of $(\mathcal{X}, \|.\|)$, the n−width in the sense of Kolmogorov (or the *Kolmogorov n-width*) of $\mathcal{S}$ in $\mathcal{X}$ [14] is

$$d_n(\mathcal{S}, \mathcal{X}) := \inf_{\mathcal{X}_n} d(\mathcal{S}, \mathcal{X}_n) = \inf_{\mathcal{X}_n} \sup_{f \in \mathcal{S}} \inf_{h \in \mathcal{X}_n} \|f - h\|$$

where the left-most infimum is taken over all n-dimensional subspaces $\mathcal{X}_n$ of $\mathcal{X}$ and

$$d(\mathcal{S}, \mathcal{Y}) := \sup_{f \in \mathcal{S}} \|f - \mathcal{Y}\|$$

Nonlinear n-width of $\mathcal{S}$ in $\mathcal{X}$ is defined as

$$\delta_n(\mathcal{S}, \mathcal{X}) :=$$

$$\inf_{\mathcal{G}} d(\mathcal{S}, span_n\mathcal{G}) = \inf_{\mathcal{G}} \sup_{f \in \mathcal{S}} \inf_{h \in span_n\mathcal{G}} \|f - h\|$$

where $\mathcal{G}$ is a member of a family of parametrized subsets of $\mathcal{X}$. Nonlinear n-width as the alternative to the Kolmogorov linear n−width for nonlinear approximation was first suggested in [8].

We finally denote

$$d_n(f, \mathcal{X}) := \inf_{\mathcal{X}_n} \inf_{h \in \mathcal{X}_n} \|f - h\|$$

and

$$d(f, span_n\mathcal{G}) := \inf_{h \in span_n\mathcal{G}} \|f - h\|$$

that correspond to $d_n(\mathcal{S}, \mathcal{X})$ and $d(\mathcal{S}, span_n\mathcal{G})$, respectively, for $\mathcal{S} = \{f\}$.

3 Comparison of Bounds on Approximation Rates

The following is a reformulation of Jones-Barron's theorem in terms of $\mathcal{G}^0$−variation:

Theorem 1 *[10] Let $(\mathcal{X}, \|.\|_2)$ be a Hilbert space, $\mathcal{G}$ be its subset. Then for every $f \in \mathcal{X}$ and for every positive integer n*

$$\|f - span_n\mathcal{G}\|_2^2 \leq \frac{\|f\|_{\mathcal{G}^0}^2 - \|f\|_2^2}{n}.$$

It is easy to see that $\|f\|_2 \leq \|f\|_{\mathcal{G}^0}$ holds for every $f \in (\mathcal{X}, \|.\|_2)$, i.e. the unit ball of $\mathcal{G}^0$-variation is contained in the unit ball of $\|.\|$. Then there exists a constant $c_{f,\mathcal{G}} \geq 1$ such that $c_{f,\mathcal{G}}\|f\|_2 = \|f\|_{\mathcal{G}^0}$. We then get the following corollary:

Corollary 1 *Let $(\mathcal{X}, \|.\|_2)$ be a Hilbert space and $\mathcal{G}$ its subset. Then for every $f \in \mathcal{X}$ there exists a constant $c_{f,\mathcal{G}} \geq 1$ such that for every positive integer n*

$$\|f - span_n\mathcal{G}\|_2^2 \leq \|f\|_2^2 \frac{c_{f,\mathcal{G}}^2 - 1}{n}.$$

It follows that Theorem 1 gives a "good" upper estimate for the class of functions $F_\epsilon := \{f : c_{f,\mathcal{G}}\|f\|_2 = \|f\|_{\mathcal{G}^0}, 1 \leq c_{f,\mathcal{G}} \leq 1 + \epsilon\}$, $\epsilon > 0$.

Let us now consider the following result from Pinkus:

Theorem 2 *[14] Let S_n be the unit ball of any $(n + 1)$-dimensional subspace $\mathcal{X}_{n+1}$ of a normed linear space $(\mathcal{X}, \|.\|)$. Then*

$$d_k(S_n, \mathcal{X}) = 1, \quad k = 0, 1, \ldots, n.$$

In the following, the unit ball of an $(n + 1)$-dimensional subspace $\mathcal{X}_{n+1}$ of a normed linear space $(\mathcal{X}, \|.\|)$ will be denoted by S_n. Let $B_r(\|.\|)$ be the ball of radius r in the metric $\|.\|$, i.e.

$$B_r(\|.\|) := \{f \in \mathcal{X}; \|f\| \leq r\}.$$

Then, $B_r(\|.\|_{\mathcal{G}^0})$ is the ball of radius r in $\mathcal{G}^0$-variation. It follows from Theorem 1 that the approximation error ϵ in the case of nonlinear approximation by the parametrized family $\mathcal{G}$ is

$$\epsilon^2 \leq \frac{\|f\|_{\mathcal{G}^0}^2 - \|f\|_2^2}{n}.$$

On the other hand if, for a given $r \in \Re^+$, there exists $n \in \mathcal{N}$ such that $B_r(\|.\|_{\mathcal{G}^0}) \supseteq S_n$, then from Theorem 2 we get

$$d_n(B_r(\|.\|_{\mathcal{G}^0}), \mathcal{X}) \geq 1$$

Since a sufficient condition for $\frac{\|f\|_{\mathcal{G}^0}^2 - \|f\|_2^2}{n} \leq 1$ is $r \leq \sqrt{n}$ in $B_r(\|.\|_{\mathcal{G}^0})$, we obtain the following proposition:

Proposition 1 *Let $(\mathcal{X}, \|.\|_2)$ be a Hilbert space, $\mathcal{G}$ its subset and $n \in \mathcal{N}$ such that $B_{\sqrt{n}}(\|.\|_{\mathcal{G}^0}) := \{f \in \mathcal{X}; \|f\|_{\mathcal{G}^0} \leq \sqrt{n}\} \supseteq S_n$. Then the upper bound on $d(B_{\sqrt{n}}(\|.\|_{\mathcal{G}^0}), span_n\mathcal{G})$ is less than the lower bound on $d_n(B_{\sqrt{n}}(\|.\|_{\mathcal{G}^0}), \mathcal{X})$.*

Now we focus on Kolmogorov n-width. We will use the following characterization of Kolmogorov n-width.

Theorem 3 *[14] If K is a closed, convex, centrally symmetric proper subset of an $(n + 1)$-dimensional subspace $\mathcal{X}_{n+1}$ of a normed linear space $(\mathcal{X}, \|.\|)$ and δK denotes the boundary of K, then*

$$d_n(K, \mathcal{X}) = \inf\{\|f\| : f \in \delta K\}.$$

Note that, based on the properties of the Kolmogorov n-width [14], if $\mathcal{K} \subset (\mathcal{X}, \|.\|)$, and K is a centrally symmetric set created from the closure of the convex hull of $\mathcal{K}$, then

$$d_n(\mathcal{K}, \mathcal{X}) = d_n(K, \mathcal{X}).$$

Given $f \in \mathcal{X}$, it follows from the definition of $\mathcal{G}$-variation that $f \in clconv(\mathcal{G}(\|f\|_{\mathcal{G}})) = B_{\|f\|_{\mathcal{G}}}(\|.\|_{\mathcal{G}})$. If we denote $b := \sup_{g \in \mathcal{G}} \|g\|$, $B_{\|f\|_{\mathcal{G}}}^{n+1}(\|.\|_{\mathcal{G}}) := B_{\|f\|_{\mathcal{G}}}(\|.\|_{\mathcal{G}}) \bigcap \mathcal{X}_{n+1}$ and apply Theorem 3 with $K = B_{\|f\|_{\mathcal{G}}}^{n+1}(\|.\|_{\mathcal{G}})$, we get

$$\|f\|_{\mathcal{G}^0} \leq b\|f\|_{\mathcal{G}} = b\, d_n(B_{\|f\|_{\mathcal{G}}}^{n+1}(\|.\|_{\mathcal{G}}), \mathcal{X})$$

and, using Theorem 1:

$$\|f - span_n\mathcal{G}\|_2 \leq \sqrt{\frac{b^2 d_n(B_{\|f\|_{\mathcal{G}}}^{n+1}(\|.\|_{\mathcal{G}}), \mathcal{X})^2 - \|f\|_2^2}{n}}$$

$$\leq b\frac{d_n(B_{\|f\|_{\mathcal{G}}}^{n+1}(\|.\|_{\mathcal{G}}), \mathcal{X})}{\sqrt{n}}.$$

This is concluded in the following proposition:

Proposition 2 *Let $(\mathcal{X}, \|.\|_2)$ be a Hilbert space, $\mathcal{G}$ its subset, $b := \sup_{g \in \mathcal{G}} \|g\|$ and $f \in \mathcal{X}$. Moreover, let $B_{\|f\|_{\mathcal{G}}}^{n+1}(\|.\|_{\mathcal{G}}) := B_{\|f\|_{\mathcal{G}}}(\|.\|_{\mathcal{G}}) \bigcap \mathcal{X}_{n+1}$, where $\mathcal{X}_{n+1}$ is an $(n + 1)$-dimensional subspace of $(\mathcal{X}, \|.\|_2)$. Then*

$$d(B_{\|f\|_{\mathcal{G}}}^{n+1}(\|.\|_{\mathcal{G}}), span_n\mathcal{G}) \leq b\frac{d_n(B_{\|f\|_{\mathcal{G}}}^{n+1}(\|.\|_{\mathcal{G}}), \mathcal{X})}{\sqrt{n}}.$$

In other words, the upper bound on nonlinear approximation by a parametrized set $\mathcal{G}$ of functions corresponding to a type of computational unit is better at least for a multiplicative factor $\frac{1}{\sqrt{n}}$ than the upper bound on linear approximation in the set $B_{\|f\|_{\mathcal{G}}}^{n+1}(\|.\|_{\mathcal{G}})$. It follows that linear approximation of functions in the intersection of $\mathcal{G}$-balls whith $(n+1)$-dimensional subspaces, is a weak tool in comparison to nonlinear approximation for $n \gg 1$.

Now, suppose that $(\mathcal{X}, \|.\|)$ is a normed functional space defined on $\mathcal{R}^d$ and that linear approximators in $B^{n+1}_{\|f\|_{\mathcal{G}}}(\|.\|_{\mathcal{G}})$ suffer of the *curse of dimensionality*, i.e. the number of parameters necessary to achieve a given accuracy increases exponentially with increasing dimension d. This is expressed by a factor of the form Cn^d in the approximation rate, where C is a constant with respect to n. Note that Proposition 2 does not a priori imply that neural networks corresponding to the parametrized set of functions $\mathcal{G}$ avoid this problem, since the multiplicative factor $\frac{1}{\sqrt{n}}$ can not cope with an exponential term.

For an orthonormal basis $\mathcal{A}$ of a finite-dimensional Hilbert space $(\mathcal{X}, \|.\|_2)$ we denote the l_1-*norm with respect to* $\mathcal{A}$ by $\|.\|_{1,\mathcal{A}}$, i.e. $\|f\|_{1,\mathcal{A}} = \sum_{i=1}^m |w_i|$, where $f = \sum_{i=1}^m w_i g_i$. It easy to verify that, for every $f \in \mathcal{X}$, $\|f\|_{\mathcal{A}} = \|f\|_{1,\mathcal{A}}$, i.e. $\mathcal{A}$-variation is the l_1-norm with respect to $\mathcal{A}$ [10]. The following theorem holds.

Theorem 4 *[10]* Let $(\mathcal{X}, \|.\|_2)$ be a finite-dimensional Hilbert space and $\mathcal{A}$ its orthonormal basis. Then for every $f \in \mathcal{X}$ and for every positive integer n there exists $f_n \in span_n \mathcal{A}$ such that

$$\|f - f_n\|_2 \leq \frac{\|f\|_{1,\mathcal{A}}}{2\sqrt{n}}.$$

This implies that $\forall S \subset \mathcal{X}$, $d(S, span_n\mathcal{A}) \leq \sup_{f \in S} \frac{\|f\|_{1,\mathcal{A}}}{2\sqrt{n}}$. If the only information available about f is the value of its $\mathcal{A}$-variation, then this upper bound can not be improved [10]. However, the upper bound in Theorem 4 can be improved if in addition to $\|f\|_{1,\mathcal{A}}$ also $\|f\|_2$ is known [10].

We are now interested in approximating functions from the unit ball S_n of an $(n+1)$-dimensional subspace $\mathcal{X}_{n+1}$ of a Hilbert space $(\mathcal{X}, \|.\|_2)$. It follows from Theorem 4 that

$$\sup_{\|f\|_2=1} \{\|f - span_n\mathcal{A}\|_2\} \leq \sup_{\|f\|_2=1} \left\{ \frac{\|f\|_{1,\mathcal{A}}}{2\sqrt{n}} \right\}.$$

If $\mathcal{A} = \mathcal{E}_l$ (the Euclidean basis of $\mathcal{X}_l$), where $l \geq n$, we get

$$\sup\{\|f\|_{1,\mathcal{E}_l} : \|f\|_2 = 1\} =$$
$$\max \{\|f\|_{1,\mathcal{E}_l} : \|f\|_2 = 1\} =$$
$$\max \left\{ \sum_{i=1}^l |f_i| : \|f\|_2 = 1 \right\} = \sqrt{l}.$$

Then

$$d(S_{l-1}, span_n\mathcal{E}_l) = \sup_{f \in S_n} \inf_{g \in span_n\mathcal{E}_l} \|f - g\|_2 \leq \frac{1}{2}\sqrt{\frac{l}{n}}.$$

If we use this for $l = n + 1$, we get

$$d(S_n, span_n\mathcal{E}_{n+1}) \leq \frac{1}{2}\sqrt{1 + \frac{1}{n}}.$$

On the other hand, we know from Theorem 2 that

$$d_n(S_n, \mathcal{X}) = \inf_{X_n} \sup_{f \in S_n} \inf_{g \in X_n} \|f - g\|_2 = 1.$$

The above results can be summarized in the following proposition.

Proposition 3 *Let $(\mathcal{X}, \|.\|_2)$ be a Hilbert space and $\mathcal{E}_{n+1}$ the Euclidean basis of its $(n+1)$-dimensional subspace $\mathcal{X}_{n+1}$. Then the upper bound on $d(S_n, span_n\mathcal{E}_{n+1})$ is less than $d_n(S_n, \mathcal{X})$.*

In other words, the upper bound on nonlinear approximation by $span_n\mathcal{E}_{n+1}$ of the unit ball S_n of $\mathcal{X}_{n+1}$ is better than the upper bound on linear approximation of the same ball for $n \geq 1$.

We finally make some remarks on the approximation of a single function. The above results on approximation are too general for this case and they need not provide the optimal rates for approximation of a single function. We compare them with the rates of approximation in Hilbert spaces achieved in [11].

Let $(\mathcal{X}, \|.\|_2)$ be a separable Hilbert space with $\mathcal{G}$ an orthogonal basis of $\mathcal{X}$. Then every $f \in \mathcal{X}$ can be written in the form $f = \sum_{k=1}^\infty a_k(f)g_k$, where the series converges in the norm of $\mathcal{X}$. Define $\mathcal{G} := \{g_k; k = 1, 2, \ldots\}$ and $S_{\mathcal{G}} = \{f \in \mathcal{X}; \sum_{k=1}^\infty |a_k(f)| \leq 1\}$. Let $\Lambda \subset Z_0^+$ (Z_0^+ represents the set of non-negative integers) and $\mathcal{U}_\Lambda := span\{g_k, k \in \Lambda\}$. Let $\mathcal{T}_\Lambda$ denote the projection operator on $\mathcal{U}_\Lambda$ and $\mathcal{C}_\Lambda(f) := \inf_{g \in \mathcal{U}_\Lambda} \|f - g\|, g \in \mathcal{G}$. It holds that $\mathcal{C}_\Lambda(f) = \|f - \mathcal{T}_\Lambda(f)\|$. Denote

$$\Delta_n(S_{\mathcal{G}}, \mathcal{X}) = \sup_{f \in S_{\mathcal{G}}} \inf_{\Lambda \subset Z_0^+, |\Lambda| \leq n} \mathcal{C}_\Lambda(f), \ n = 1, 2, \ldots.$$

We deal only with infinite dimensional Hilbert spaces here as for $\mathcal{X}$ having a finite dimension we get the rate $\Delta_n(S_{\mathcal{G}}, \mathcal{X}) = 0$. The following result holds [11]:

Theorem 5

$$\Delta_n(S_{\mathcal{G}}, \mathcal{X}) \leq \frac{1}{\sqrt{n+1}}, \ n = 1, 2, \ldots$$

Moreover, if $f \in S_G$ then there is a sequence$\{\rho_n\}$ of numbers such that $\rho_n \in (0, 2]$, $n = 1, 2, \ldots$, $\lim_{n \to \infty} \rho_n = 0$ and

$$\inf_{\Lambda \subset Z_0^+, |\Lambda| \leq n} \mathcal{C}_\Lambda(f) \leq \frac{\rho_n}{\sqrt{n}}, \quad n = 1, 2, \ldots$$

Then we get for $f \in S_G$:

- $\Delta_n(f, \mathcal{X}) = \inf_{\Lambda \subset Z_0^+, |\Lambda| \leq n} \mathcal{C}_\Lambda(f) = \inf_{\mathcal{X}_n} \inf_{g \in \mathcal{X}} \|f - g\| = d_n(f, \mathcal{X})$

- Linear approximation:
  $d_n(f, \mathcal{X}) = \inf_{\mathcal{X}_n} \inf_{g \in \mathcal{X}_n} \|f - g\|$, where $\mathcal{X}_n$ is an n-dimensional subspace of $\mathcal{X}$

- Nonlinear approximation: $d(f, span_n G) = \inf_{g \in span_n G} \|f - g\|$

As for every $\mathcal{X}_n \subset \mathcal{X}$, also $\mathcal{X}_n \subset span_n G$, $d(f, span_n G)$ is infimum over a 'bigger' set than in the case of $d_n(f, \mathcal{X})$. So it follows that, for an orthonormal basis G of $(\mathcal{X}, \|.\|_2)$, we have

$$d(f, span_n G) \leq d_n(f, \mathcal{X}) = \Delta_n(f, \mathcal{X}) \leq \frac{\rho_n}{\sqrt{n}}$$

and $\lim_{n \to \infty} \rho_n = 0$.

For functions $f \in S_n \bigcap S_G$, this is a better bound than the one given in Theorem 2, i.e. $d_n(f, \mathcal{X}) = 1$, especially for $n >> 1$.

4   Concluding Remarks

A theoretical framework for the comparison of linear approximators and nonlinear approximation schemes is necessary for understanding of the experimental outperformance of neural networks with respect to traditional linear approximators. In this paper, we have shown the superiority of neural network approximation in some functional spaces. The variation of a function with respect to a set and a proper nonlinear n-width, analogous to the Kolmogorov n-width for the linear case, play a key-role in this analysis.

The proposed results provide further theoretical understanding of the surprising success of neural networks in complex approximation tasks (e.g, vocalization of text, optimization problems from control theory, etc.). The analysis in more general functional spaces is of high importance and is still an open problem.

Acknowledgements

We are grateful to Věra Kůrková for valuable discussions and remarks.

Marcello Sanguineti was partially supported by grants CNR 96.02472.CT07 and CNR 97.00048.PF42. Kateřina Hlaváčková–Schindler was partially supported by grants CR 201/96/0917 and 201/99/0092.

References

[1] Barron, A.R.: Universal approximation bounds for superpositions of a sigmoidal function. IEEE Transactions on Information Theory 39, pp. 930-945, 1993.

[2] Barron, A. R.: Neural net approximation. Proc. 7th Yale Workshop on Adaptive and Learning Systems. K. Narendra Ed., Yale University Press, 1992.

[3] Burr, D.J.: Experiments on neural net recognition of spoken and written text. IEEE Trans. Acoust. Speech and Signal Processing 36, pp. 1162-1168, 1988.

[4] Cybenko, G.: Approximation by superposition of a sigmoidal function, Math. Control Signal Systems 2, pp. 303-314, 1989.

[5] Girosi, F., Jones, M. and Poggio, T.: Regularization theory and neural networks architectures. Neural Computation 7, pp. 219-269, 1995.

[6] Hlaváčková, K., Sanguineti, M.: On the rates of linear and nonlinear approximations. Proc. 3rd IEEE European Workshop on Computer-Intensive Methods in Control and Signal Processing (CMP), pp. 211-216, 1998.

[7] Hornik, K., Stinchcombe, M., White H.: Multilayer feedforward networks are universal approximators. Neural Networks 2, pp. 359-366, 1989.

[8] Kainen, P.C., Kůrková, V., Vogt, A.: Approximation by neural networks is not continuous. Submitted to Neurocomputing.

[9] Kůrková, V.: Dimension-independent rates of approximation by neural networks. Computer-intensive methods in Control and Signal Processing: Curse of Dimensionality (Eds. K. War-

wick, M. Kárný). Birkhäuser, Boston, pp. 261-270, 1997.

[10] Kůrková, V., Savický, P., Hlaváčková, K.: Representations and rates of approximation of real–valued Boolean functions by neural networks. Neural Networks 11, pp. 651-659, 1998.

[11] Mhaskar, H.N., Micchelli, C.A.: Dimension-independent bounds on the degree of approximation by neural networks. IBM Journal of Research and Development 38, pp. 277-284, 1994.

[12] Parisini, T., Sanguineti, M., Zoppoli, R.: Nonlinear stabilization by receding-horizon neural regulators. International Journal of Control 70, no.3, pp. 341-362, 1998.

[13] Park J., Sandberg, I. W.: Approximation and radial-basis-function networks. Neural Computation 5, pp. 305–316, 1993.

[14] Pinkus, A.: $N-$Widths in Approximation Theory. Springer-Verlag, New York, 1986.

[15] Sejnowski, T.J., Rosenberg, C.: Parallel networks that learn to pronounce English text. Complex Systems 1, pp. 145-168, 1987.

Transposition: A Biological-Inspired Mechanism to Use with Genetic Algorithms

A. Simões[1,2], E. Costa[2]

[1] Escola Superior de Educação, Instituto Politécnico de Leiria, Rua Dr. João Soares, 2400 Leiria, Portugal
[2] Centro de Informática e Sistemas, Universidade de Coimbra, Polo II, 3030 Coimbra, Portugal
E-mail: {abs, ernesto}@dei.uc.pt

Abstract

Genetic algorithms are biological inspired search procedures that have been used to solve different hard problems. They are based on the neo-Darwinian ideas of natural selection and reproduction. Since Holland proposals back in 1975, two main genetic operators, crossover and mutation, have been explored with success. Nevertheless, in nature there exist much more mechanisms for genetic recombination based in phenomena like gene insertion, duplication or movement. The goal of this paper is to study one of these mechanism, called transposition. Transposition is a context-sensitive operator that promotes the movement intra or inter chromosomes. In this preliminary work we empirically study the performance of the genetic algorithm where the traditional crossover operator was substituted by transposition. The results are very promising but must be confirmed by a more extensive empirical study and the correspondent theoretical justification.

1 Introduction

Genetic Algorithms (GA's) are a search paradigm that applies ideas from evolutionary biology (crossover, mutation, natural selection) in order to deal with intractable search spaces. The power and success of GA's is mostly achieved by the diversity of the individuals of a population which evolve, in parallel, following the principle of "the survival of the fittest". In the standard GA the diversity of the individuals is obtained and maintained using the genetic operators crossover and mutation which allow the GA to find more promising solutions and avoid premature convergence to a local maximum [7].

In order to find the most efficient ways of using GA's, many researchers have carried out extensive studies to understand several aspects such as the role of types of selection, space representation and how to apply the genetic operators. Several studies were made concerning the genetic operators crossover and mutation. For instance, Schaffer and Eshelman empirically compared mutation and crossover and concluded that mutation alone is not always sufficient.

Spears and De Jong analyse the role of crossover and mutation in terms of disruption theory, trying to understand the power of the two operators [14]. Later, De Jong and Spears in [3] present a formal study of the role of multipoint crossover in GA, in order to analyse their recombination potential and exploratory power. Their work provides a better understanding of when and how to use n-point and uniform crossover. Spears in [15] proposes an adaptive GA which decides between 2-point and uniform crossover as it runs. He concludes that this adaptive mechanism works well especially with larger populations.

Although the classical GA uses these two main genetic operators to achieve population diversity, in nature the diversity of the species' genetic material is obtained by several mechanisms which involve gene insertion, duplication or movement. With this respect, Mitchell and Forrest point out the importance of study other "mechanisms for rearranging genetic material (e.g., jumping genes, gene deletion and duplication, introns and exons)" to know if any of these is significant algorithmically [10].

Some authors proposed other biological inspired genetic operators besides crossover and mutation. Harvey and Smith suggest alternative genetic operators inspired in a bacterial form of recombination called conjugation [6], [12], [13]. This process involves the uni-directional transfer of genetic material by direct cellular contact between a donor bacterial cell and a recipient cell. Harvey suggests a type of conjugation based on tournament selection [6]. Parents are selected on a random basis, the two parents "fight" in a tournament. The winner of the tournament becomes the donor and the loser the recipient of the genetic material. Smith uses conjugation as a method of genetic recombination to solve hard satisfiability problems [12]. He constructed a simple model using a GA operating directly the phenotype (the satisfiability expressions) and using mutation operator. A random population is created and placed in a 15x15 matrix. The individuals are allowed to move in the matrix and to conjugate genetic material if placed in adjacent positions. Later, Smith proposes a simple conjugation operator involving two individuals randomly chosen [13]. Both authors achieved good results using those substitute genetic operators.

In this paper we will introduce a genetic operator alternative to crossover, inspired in real biology. This mechanism is known as transposition and consists in the presence of genetic mobile units called transposons, that are capable of relocating themselves, or transposing, onto the chromosome and subsequently jumping into new zones of the same or other chromosome.

We will compare the performance of the GA in finding an optimal solution to a given function using either crossover and transposition followed by mutation.

This paper is organised in the following manner. First, in section 2, we introduce the classical way to use the traditional GA. In section 3, we describe how transposition works in nature and discuss how we implement it. In section 4, we present our case study and we make an exhaustive comparison of the results obtained with transposition, 1-point, 2-point and uniform crossover. Finally we conclude with a discussion and direction for future research in this area.

2   The Classical Genetic Algorithm

The mechanics of a simple GA is very simple, involving nothing more complex than copying strings and swapping partial strings [4]. It starts with an initial population of individuals created at random. Then, this population is evolved through time by a string manipulation process based in three genetic operators: reproduction, crossover and mutation (see Figure 1).

```
1.  Generate population
2.  Do
    2.1. Evaluate population
    2.2. Reproduction (Select parents)
    2.3. Crossover
    2.4. Mutation
    2.5. Substitute old population
Until (DONE)
```

Fig. 1. - The classical GA

Reproduction is a process in which individuals are copied to a mating pool according to their fitness. Individuals with higher fitness have higher probability of generate offspring in next generation. The "goodness" of a solution is measured by the fitness function and is typically defined with the respect to the current population. This operator mimics the natural

selection process in which the fittest individuals are determined by their ability to survive predators, sickness and other obstacles.

After reproduction, crossover may proceed in two steps. First, members of the individuals in the mating pool are selected and mated at random. Second, each pair of strings is crossed-over, exchanging genetic material between them. There are several types of crossover operator, but the general idea of all of them is to swap genetic material between two strings. For instance, in 1-point crossover, a cut point is chosen at random and the genes are swapped according the cut point (see Figure 2):

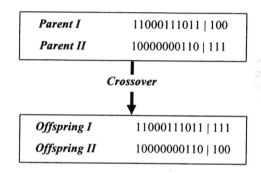

Fig. 2. Crossover operator

The power of the GA is mostly due to crossover. It is the most important operator to the GA. Diversity is indispensable to evolution. The populations' diversity is obtained and maintained by crossover. which allows the GA to find better solutions in the search space [9].

The offspring generated by reproduction and crossover can be affected by mutation. The effect of this operator is to change the value of a single gene (see Figure 3). Although mutation plays a secondary role in the operation of the GA, it is needed to avoid premature convergence of the GA to a local optima. Mutation is applied with a low rate and has the ability to "shake" the GA enabling it to continue evolving.

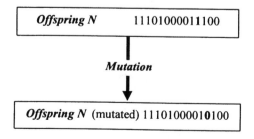

Fig. 3. Mutation operator

3 Transposition

In nature the genetic diversity of the individuals is preserved by several mechanisms that involve operations like gene insertion, duplication or movement. In each one of these categories there are several processes that in one way or another produce changes in the genome of the species enabling the genetic diversity , so important to the evolutionary process. For instance we can find phenomenon like transformation, transduction, conjugation, retroinsertion, etc. (involving gene insertion); break and fusion, unequal recombination, transposition, etc (involving either gene duplication or gene movement). Table 1 shows the categorisation of these mechanisms [5], [11].

In this paper we explore one of these mechanisms, the transposition.

Table 1. Biological mechanisms of changing genetic material

Mechanism	Possible Consequences
Transformation	New genes from a dead cell imported from surrounding medium incorporated into chromosome of bacterium.
Transduction	New genes accidentally picked up from previous host and imported into cell by a virus.
Conjugation	Genetic material from a donor bacteria is transferred to a recipient cell.
Lysogenic Insertion	Novel genes of temperate phage inserted into host genome.
Retroviral Insertion	cDNA copy of novel genes of retrovirus inserted into host genome.
Intron Insertion	Excised introns inserted into genome, mainly at exon-exon junctions in cDNA insertions.
Retroinsertion	cDNA copies of transcribed host DNA incorporated into genome providing duplicate copies of genes.
Breakage and Fusion	Part of one chromosome breaks off and fuses to the end of another during gamete formation; some gametes may obtain duplicate copies of genes on the broken fragment.
Unequal crossing over	Chromosomes may be misaligned during the crossover process; some gametes may obtain duplicate copies of some genes.
Transposition	Chromosomal DNA moved with genome, or both duplicated and moved.

3.1 Biological Transposition

Transposition characterises itself by the presence of mobile genetic units that move about in the genome, either removing themselves to new locations or by duplicating themselves for insertion elsewhere. These mobile units are called transposons.

Transposons (also known as jumping genes) can be formed by one or several genes or just a control unit, and can move in several ways, none of which is fully understand: some transposons move from one site on the chromosome to a new point of the same or to the other chromosome; others leave a copy behind, still others remain fixed but dispatch copies to other sites. In some cases, the transposon, before inserting in the target position, duplicates itself and the seek for another insertion point continues in the same way (see Figure 4).

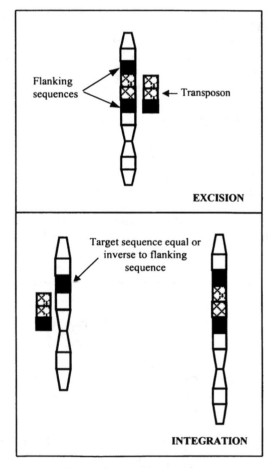

Fig. 4. Transposition mechanism

Transposition was first discovered by Barbara McClintock in the 50's (when the DNA structure was not completely understood). She proved that certain phenomena present in living beings exposed to UV radiation could not be the result of the normal recombination and mutation processes. She found that in corn certain genetic elements occasionally move producing kernels with unusual colours that could not have resulted from crossover or mutation. Transposons were for a long time considered as some sort of abnormality, but in 1983 when she was awarded with the Nobel Prize, many such transposons had been discovered and their possible role in evolution was beginning to be recognised. For instance, the genetic alterations caused by transposons are responsible for the growth of cancers in human or the resistance to antibiotics in bacteria [5], [11].

In order for a transposable element to transpose as a discrete entity it is necessary for its ends to be recognised. So, transposons within a chromosome are flanked by identical or inverse repeated sequences, some of which are actually part of transposon (see Figure 5).

> *Inverse Flanking Sequences*
> NNNNN<u>ATTGA</u> (Transposon) <u>AGTTA</u>NNNNNN
>
> *Identical Flanking Sequences*
> NNNNN<u>ATTGA</u> (Transposon) <u>ATTGA</u>NNNNNN

Fig. 5. Inverse or equal flanking sequences

When the transposon moves to another zone of the genome one of the sequences goes with it. The insertion point can be chosen at random, but there are transposons that show a regional preference when inserting into the same gene. Other method can be a correspondence (identical or inverse) in the new position with the flanking sequences. The last method is described in Figure 6.

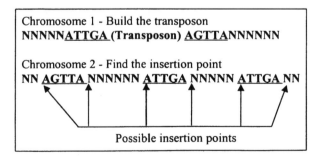

Chromosome 1 - Build the transposon
NNNNN**<u>ATTGA</u> (Transposon) <u>AGTTA</u>**NNNNNN

Chromosome 2 - Find the insertion point
NN **<u>AGTTA</u>** NNNNNN **<u>ATTGA</u>** NNNNN **<u>ATTGA</u>** NN

Possible insertion points

Fig. 6. Building the transposon and finding the insertion point

The sequence in into which the transposon is inserted requires no homology with the transposon. This is in

marked contrast to classical recombination, where relatively long sequences of DNA must share homology to permit a recombination event to occur (same cut point(s)). As a consequence, transposition is sometimes referred to as illegitimate recombination.

3.2 Computational Transposition

We implemented the transposition mechanism following the inspiration from biology. After selecting two parents for mating we look for the transposon in one of them. The insertion point will be found in the second parent. The same amount of genetic material is exchanged between the two chromosomes according to the found insertion point.

Now, we are going to describe how the transposon will be formed, how it will move in the genome, how to define the insertion point, how to define the flanking sequences' length and how the integration in the new position will take place.

Our case study uses chromosomes of fixed size. Suppose this size is CL (Chromosome Length). The transposition method will work as follow (see Figure 7):

- FSL is the length for the flanking sequences;
- Choose at random a gene (gene **T**) between **0** and **CL**, from which we will build the transposon;
- The **FSL** genes immediately before gene **T** will form the first flanking sequence;
- The second flanking sequence can be identical or inverse to the first one;
- Look in the chromosome, from bit **T**, for a possible second flanking sequence;
- The transposon will be formed with all the genes between gene **T** and the last gene of the second flanking sequence;
- The second flanking sequence always moves with the transposon.

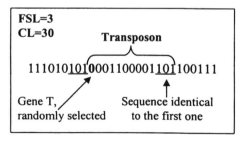

FSL=3
CL=30
Transposon

11101010<u>1</u>00011000011<u>01</u>100111

Gene T, randomly selected

Sequence identical to the first one

Fig. 7. Computational transposition: building the transposon

We will look in the second parent for a sequence of bits equal or inverse to the flanking sequences. The

182

insertion point will be the first gene after that sequence. After finding the insertion point the same number of genes, equal to the transposon's length, will be exchanged between the two parents.

All the process is exemplified in Figure 8.

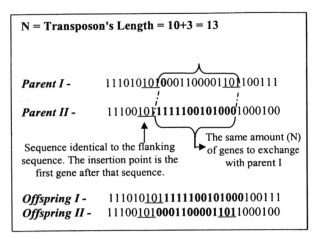

Fig. 8. Computational transposition

Some particular cases can occur:

- The search of the second flanking sequence is made between gene T+1 through the end of the chromosome. If no sequence is found the search starts at the beginning of the chromosome. In a limit situation the sequence found will be the first one. In this case there will be no transposition.
- If there is no equal or inverse sequence in the target chromosome, the insertion point is defined randomly.

4 A Case Study: Transposition versus Crossover

To study the performance of transposition we will compare it with the standard mechanism of crossover. We will use the function [8]:

$$f(x1, x2) = 21.5 + x1.\sin(4\Pi x1) + x2.\sin(20\Pi x2) \quad (1)$$

where $-3.0 \leq x1 \leq 12.1$ and $4.1 \leq x2 \leq 5.8$

The optimal solution is 38.87
This function has some interesting aspects as can be seen in Figure 9.

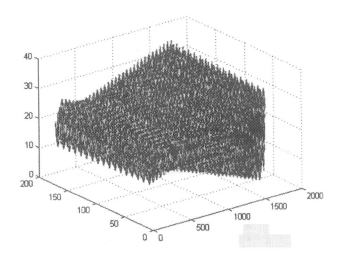

Fig. 9. Michalewicz's test function

x_1 and x_2 domains' length are, respectively, 15.1 and 1.7. We assume a four decimal cases precision. So, we need 18 bits to represent x_1 and 15 bits to represent x_2. The chromosome's length is 33.

In all experiments we used roulette wheel with elitism as the selection method. The elite size is 20% population size. Mutation rate used is 0.01 and crossover/transposition rate is 0.7. We made experiments involving 1-point, 2-point and uniform crossover. In each crossover type we used population's size of 50, 100 and 200 individuals. Transposition was tested with flanking sequences from 1 to 10 bits and, in each case, we used populations of 50, 100 and 200 specimens. We run each experiment 10 times. All the tests were run over 1000 generations. The results analysed are the average of best individual fitness obtained in the ten trials made with each experiment.

First we will analyse transposition results alone, showing how the flanking sequences' length can influence the performance of the GA.

Then, we will show the results obtained with transposition, 1-point crossover, 2-point crossover and uniform crossover. We will compare results obtained with transposition and with each one of the crossover methods to get a clear idea of the performance of the genetic operators used.

4.1 Transposition Performance

We analysed the results obtained with the mechanism of transposition using flanking sequences with lengths from 1 to 10 bits. In each case, we use populations with 50, 100 and 200 individuals.

Observing the average of the results got in the 10 simulations we conclude two main aspects:

1. With larger populations the results are better.
2. In general, with bigger flanking sequences the performance of the transposition becomes worst. (see Figure 10).

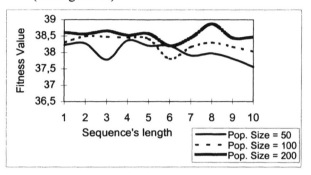

Fig. 10. The GA's performance changing sequence's length

The first conclusion seems obvious.

The second one must be justified by the fact that if the flanking sequence length is greater, then the transposon's length will be bigger. Thus, in most cases the transposition mechanism won't work because the second flanking sequence is never found. In practice, with bigger sequences the rate of transposition will decrease. Table 2 shows the average results of the transposon's size achieved in all simulations executed with the flanking sequences from 1 to 10 bits. We also show the percentage of times that the transposon had the same size of the chromosome, i.e., no transposition occurred.

Table 2. Variation of the transposon's length, increasing the size of flanking sequences.

Sequence length	Average of transposon's length	Transposon's length = 33 (%)
1	3	0%
2	6	6%
3	9	13%
4	15	19%
5	19	27%
6	22	33%
7	25	38%
8	27	49%
9	28	68%
10	29	73%

As we can see that, with larger sequences, the amount of genetic material exchanged is bigger. Besides this, it is harder to find the second flanking sequence so, the percentage of no occurring transposition is very high. This could lead to a loss of the population diversity and, subsequently, to the worst results achieved.

4.2 Transposition versus 1-point Crossover

In all the simulations made with transposition (with populations size of 50, 100, 200 individuals and with flanking sequence's length from 1 to 10) the results outperformed the results obtained with 1-point crossover with the same populations size.

An interesting result is that, using transposition, in most cases, with a population of 50 individuals the results were much better than 1-point crossover using 50, 100 or 200 individuals.

Only in the worst results of transposition (sequences' length of 9 and 10 bits) this is not true. But in these cases, transposition with a population of 50 individual outperforms 1-point crossover with 50 and 100 individuals. In all the other experiments a population of 50 individual got better results than 1-point crossover with 50, 100 or 200 individuals. Transposition using 100 or 200 individuals, indifferently of the flanking sequence's length, is always better than 1-point crossover with 50, 100 and 200 individuals in population.

For instance, in Figure 11, we show the results obtained with transposition (flanking sequence's length of 4 bits, 50 individuals) and with 1-point crossover (using 50, 100 and 200 individuals). We can see that with a smaller population, transposition gets much better results. In Figure 12, we present the worst results obtained with transposition using 50 individuals (flanking sequence's length = 10). As we can see the results are better than 1-point crossover with 50 and 100 population's size.

4.3 Transposition versus 2-point Crossover

The results obtained with transposition and 2-point crossover were very close. We still observe the same characteristics observed with 1-point crossover, i.e. , better results with transposition using smaller populations, but the results are not so obvious. We observed that transposition with a population size of 50 individuals rarely gets the same results of 2-point crossover with 200 individuals, but frequently gets better results than 2-point crossover with 50 or 100 individuals (except when the performance of transposition is worst). With populations size of 100 and 200 specimens, transposition is often better than 2-point crossover with the same population size.

To illustrate these results we show, in Figure 13, the best values obtained with transposition using 50 individuals (sequence length = 4) and, in Figure 14, the worst ones (sequence length =10).

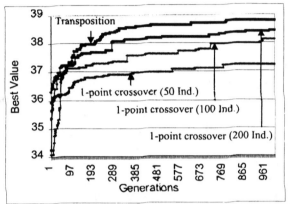

Fig. 11. - Transposition, 50 individuals, sequence length = 4; comparing results with 1-point crossover, 50, 100 and 200 individuals.

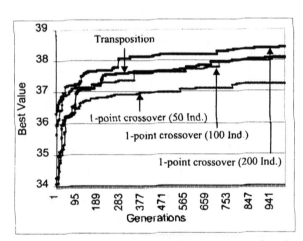

Fig. 12. - Transposition, 50 individuals, sequence length = 10; comparing results with 1-point crossover, 50, 100 and 200 individuals.

4.4 Transposition versus Uniform Crossover

Comparing the results achieved with transposition and uniform crossover we can get the same conclusions. In the best results transposition with smaller populations (50 individuals) exceed uniform crossover using 50, 100 or 200 individuals. If we analyse the worst results for transposition we conclude that with a population size of 50 individuals, the results are better than the ones achieved by uniform crossover with population of 50 and 100 individuals, but not enough to the results got with 200 individuals.

We show these results in Figures 15 and 16.

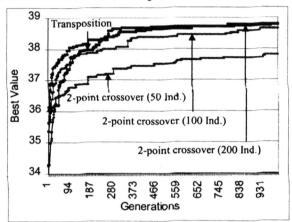

Fig. 13. - Transposition, 50 individuals, sequence length = 4; comparing results with 2-point crossover, 50, 100 and 200 individuals.

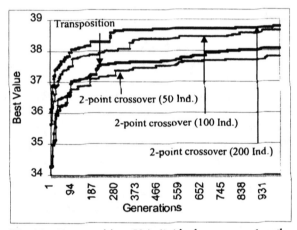

Fig. 14. - Transposition, 50 individuals, sequence length = 10; comparing results with 2-point crossover, 50, 100 and 200 individuals.

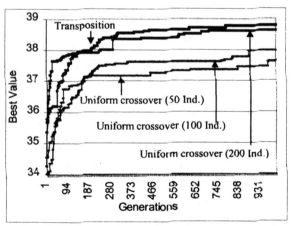

Fig. 15. - Transposition, 50 individuals, sequence length = 4; comparing results with uniform crossover, 50, 100 and 200 individuals.

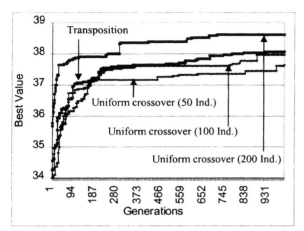

Fig. 16. - Transposition, 50 individuals, sequence length = 10; comparing results with uniform crossover, 50, 100 and 200 individuals.

Transposition using populations of 100 and 200 specimens most of the time overtake uniform crossover results using the same populations size.

If we compare transposition and 1-point crossover performances we can conclude that transposition, with a population size of 50 individuals, exceed 1-point crossover results for populations size of either 50, 100 or 200 individuals in all situations except the one presented in Figure 11.

Comparing transposition with 2-point and uniform crossover performances we observe that, for the same population size, then transposition's results always exceed 2-point and uniform crossover's results. However, in 2-point crossover the results were much closer to transposition results than in uniform crossover, we can say that, for both cases, if we choose with some care the flanking sequence's length we can obtain better results using transposition with smaller populations than 2-point crossover and uniform crossover.

5 Conclusions and Future Work

In this paper we presented a new biological-inspired genetic operator, alternative to the traditional crossover. This genetic operator is called transposition. We used both transposition and crossover (1-point, 2-point and uniform) to solve the same function optimisation problem. We analysed the average results of the best individuals in all the experiments made, changing parameters such as the population size and the flanking sequences' length.

We conclude that transposition performance is related with the flanking sequences size: bigger sequences implies worst results due to a loss of diversity. Comparing the results with crossover we saw that transposition is always better than traditional crossover when using the same populations size, and in most cases with smaller populations we can get better results with transposition.

Besides the good results obtained with this new genetic operator we intend to make more empirical work, namely, study the implications of changing other parameters besides population size and sequence's length, for instance the crossover and mutation rates. An exhaustive study will be made using other selection methods such as tournament selection and roulette wheel without elitism. We will extend our work to other functions. For instance, some preliminary work already done with the five De Jong Test Bed functions [2], was very promising but need to be completed in order to make well supported conclusions.

Also, we will extend our work to other domains than function optimisation, namely, applications which use based-order genetic operators. It will be important to make the correspondent theoretical justification.

We will also analyse another version of transposition for problems using variable length chromosomes and follow the suggestion made by [1] and see how this operator can be used in genetic programming.

6 Acknowledgements

This work was partially financed by the Portuguese Ministry of Science and Technology under the program Praxis XXI.

7 References

[1] Banzhaf, W., Nordin, P., Keller, R. E., Francone, F. D.: Genetic Programming - An Introduction - On the Automatic Evolution of Computer Programs and its Applications. Morgan Kaufmann 1998.

[2] De Jong, K. A.: Analysis of the Behavior of a Class of Genetic Adaptive Systems. Ph.D. Dissertation, Department of Computer and Communication Science, University of Michigan 1975.

[3] De Jong, K. A., Spears, W. M.: A Formal Analysis of the Role of Multi-Point Crossover in Genetic Algorithms. Annals of Mathematics and Artificial Intelligence, *5 (1)*, 1-26 (1992).

[4] Goldberg, D. E.: Genetic Algorithms in search, optimization and machine learning. Addison-Wesley Publishing Company, Inc 1989.

[5] Gould, J. L., Keeton, W. T.: Biological Science. W. W. Norton & Company 1996.

[6] Harvey, I.: The Microbial Genetic Algorithm. Submitted as a Letter to Evolutionary Computation. MIT Press 1996.

[7] Holland, J. H.: Adaptation in Natural and Artificial Systems: an introductory analysis with applications to biology, control and artificial intelligence. MIT Press 1992.

[8] Michalewicz, Z.: Genetic Algorithms + Data Structures = Evolution Programs. Springer-Verlag 1994.

[9] Mitchell, M.: An Introduction to Genetic Algorithms, MIT Press 1996.

[10] Mitchell, M., Forrest, S.: Genetic Algorithms and Artificial Life. Artificial Life, *1 (3)*, 267-289 (1994).

[11] Russell, P. J.: Genetics. 5th edition, Addison-Wesley 1998.

[12] Smith, P.: Finding Hard Satisfiability Problems Using Bacterial Conjugation. Presented at the Workshop on Evolutionary Computing, University of Sussex, April 1996.

[13] Smith, P.: Conjugation - A Bacterially Inspired Form of Genetic Recombination. Late Breaking Papers of the First International Conference on Genetic Programming. Stanford University, California, July 1996.

[14] Spears, W. M., De Jong, K. A: An Analysis of Multi-Point Crossover. Proceedings of the First Workshop on Foundations of Genetic Algorithms, pp. 301-315, Morgan Kaufmann 1991.

[15] Spears, W. M.: Adapting Crossover in a Genetic Algorithm. Technical Report AIC-92-025, Washington, DC, Naval Research Laboratory, Navy Centre for Applied Research on Artificial Intelligence, 1992.

[16] Spears, W. M.: Crossover or Mutation?. Proceedings of the Second Workshop on Foundations of Genetic Algorithms, pp. 221-238, Morgan Kaufmann 1993.

Genetic Algorithms For Decision Tree Induction

Z. Bandar, H. Al-Attar, K. Crockett

The Intelligent Systems Group, The Manchester Metropolitan University, Chester Street, Manchester, UK, M1 5GD.

Email : Z.Bandar@doc.mmu.ac.uk

Abstract

This paper presents a novel Genetic Algorithm (GA) [1], [2] based approach for decision tree induction.
Decision tree induction algorithms such as ID3 [3], [4] and CHAID [5], are based on a stepwise search procedure. This is essentially a heuristic search technique based on selecting the best local attribute/values split for each internal node. This is performed according to some given criteria, regardless of the impact on subsequent splits. Once a split is selected, these algorithms have no backtracking mechanism to enable them to change an attribute split. Thus a potential weakness of these algorithms is the lack of a globally optimal search strategy.
GAs perform a non-linear search for the optimal or near optimal solution in a pre-defined search space. The aim is to cover the entire search space without performing an exhaustive search of the domain. This paper presents GAs as an effective alternative to the step-wise search strategy employed by traditional decision tree induction algorithms. The new algorithm has been applied to two data sets and has shown a clear improvement in classification accuracy over ID3 generated by a more traditional methods.

1 Introduction

Decision tree induction is based on the premise that at least one decision tree model exists that would represent the whole domain by correctly classifying all, or at least a significant proportion, of its cases[3], [6]. The only certain way of finding the best model is through an exhaustive search. This involves the evaluation of every possible decision tree model and the selection of the model that would most accurately classify cases in the domain. Except for very small domains, the total number of different possible decision trees is very large, therefore building and evaluating all the possible decision tree models is beyond all practical time constraints. A good decision tree induction algorithm must avoid the need to search the whole domain, yet find the best decision tree model to represent that domain.

Attempts to build GA based decision tree induction systems have been very limited and focused on deterministic domains [7]. In this paper some success has been reported in replicating a decision tree model generated by ID3 on an artificially designed deterministic domain using 14 examples. This approach suffers from the same problems associated with the original ID3 algorithm, namely static branching strategy and no provision for pruning. Therefore this will rule out the application of such an approach in non-deterministic domains.

2 Genetic Algorithms

Genetic Algorithms (GA) perform non-linear searches for the optimal solution in a pre-defined search space using techniques based on the principles of reproduction and natural selection. GAs, originally developed by Holland [1], maintain a population of chromosomes which encode potential solutions to a specific problem. Once a generation of chromosomes has been initialised, the fitness of each member is evaluated and chromosomes with the best valuations are allowed to reproduce more often. These chromosomes are then recombined to form a new population of possible solutions to the problem. This process is repeated until either the required solution to the problem is found or the search stagnates, which occurs when the GA converges on a premature solution (local optimum). During this process a number of generic operators such as crossover and mutation may be applied to introduce diversity into a population [1], [8].

3 GA Based Decision Tree Induction

The task of decision tree induction in non-deterministic classification domains can be defined as a search for a set of attribute/value grouping splits organised in a tree structure that best represents the domain without overfitting the training sample. From this definition of the decision tree induction task, the following three distinct sub-search tasks can be identified which correspond to the process already identified in the conventional stepwise algorithms.

(i) Branching Strategy: search for the best grouping of attribute values for all potential split attributes in every internal node. For a binary branching, the search task is restricted to branch value membership.

(ii) Attribute Selection: search for the best attribute combination for all splits.

(iii) Pruning Criteria: search for the positions and decision values of terminal nodes.

Other searches, relating to the decision tree modelling language, can also be identified. However, the main thrust of this paper is focused around the attribute selection search. The GA driven attribute selection is combined with the other heuristics used by the stepwise approach for both branching and pruning.

3.1 Gene Representation

A fixed decision tree structure template, with binary branching, was used as the basic organisation primitive. Each internal node in this template can either be configured by an attribute/value grouping selection or replaced, together with its sub-structure, by a leaf node with a class value.

This fixed decision tree structure was represented as a single chromosome with a fixed number of genes G, given by:

$$G = 2^{depth} - 1$$

where depth is the maximum number of attribute split levels of the decision tree (variable parameter). G will be equal to the number of genes selected in the tree. A three level deep binary decision tree with a 7 gene chromosome and 8 leaves that has no repetition is shown in figure 1.

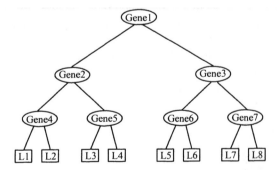

Fig. 1. Three levels binary decision tree chromosome

3.2 Interpreting the Genes

To evaluate the fitness of the decision tree model represented by a set of gene values, the decision tree model is built using heuristics adopted from the stepwise decision tree induction approach:

• attribute selection: for each node, the value returned by the gene specifies which attribute should be selected;

• branching strategy: the binary branching was applied to the attributes using the training set cases filtering down to each node. The choice of binary branching was adopted because it removes the need for a further lengthy search for the best number of splits, allows the use of a fixed number of genes and for its superiority over static and dynamic branches [9], [10].

3.3 The Fitness Function

The rate of correct classification is the most obvious candidate for the fitness function as it represents the measure which is used to judge the performance of the decision tree. The fitness function is given by:

$$f = \frac{c}{t} * 100$$

where c is the number of training cases correctly classified by the tree and t is the total number of training cases.

3.4 Parent selection

The biased roulette-wheel method [11] was used to select the two parents of new individuals.

4 Experiments

4.1 Data sets

The algorithm has been applied to two real world data sets Bank Loan and Cleveland Heart Disease. The Bank Loan data set represents records of the decision made by credit managers of a bank to grant or refuse requests for loans. the data set comprises of 434 records featuring 7 discrete and 6 continuous attributes (238 classed as reject and 196 classed as accept). The Cleveland Heart Disease data set represents the records of medical observation and measurement at a hospital and the subsequent diagnosis of the medical condition suffered by the patient. The data set consist of 303 records featuring 8 discrete and 5 continuous attributes (164 classed as negative and 139 classed as positive).

4.2 GA parameters

A series of experiments were under taken to determine a set of suitable GA parameters. The most appropriate parameters are given in table 1.

Table 1. GA parameters

Parameters	Default value
Number of generations	50
Individuals per generation	50
Crossover probability	0.6
Mutation probability	0.04

4.3 Experimental procedures

Given the relatively small size of the training sets, three training samples for each domain were used to check the effect of the sampling process on the results. The training and test sets were mutually exclusive and equal in size. The process was repeated with a pruning level for the most widely used Chi-square significance level values of 0.1%, 0.5%, 1.0%, 2.0% and 5%. Given the degree of randomness inherent in GAs and given that the various GA parameters were not exhaustively tested, each data set was run 3 times to test the algorithm's consistency. Therefor a total of 180 decision trees were generated. The classification accuracy of all 180 decision trees generated, for both the training and test sample, was measured.

The Maximum Tree Depth parameter was set to 6 levels. This level was found to be sufficient given the size of the training samples used. The maximum observed decision tree depth actually generated by the trials was 5.

5 Results

Tables 2 and 3 show the best results achieved from the above trials on both data sets. The result of the binary

Table 2 Bankloan Data

	%GACA (Tr)	%GACA (Tst)	ID3 (Tr)	ID3 (Tst)
Sample 1	82.0	74.2	80.6	74.2
Sample 2	77.4	76.0	76.0	74.7
Sample 3	80.6	75.6	82.9	74.1
Average	**80.0**	**75.3**	**79.8**	**74.3**

Table 3 Cleveland Heart Data

	%GACA (Tr)	%GACA (Tst)	ID3 (Tr)	ID3 (Tst)
Sample 1	83.6	80.1	88.2	78.8
Sample 2	87.5	83.4	84.9	76.2
Sample 3	88.2	82.1	90.1	80.1
Average	**86.4**	**81.9**	**87.7**	**78.4**

ID3 is also given for purposes of comparison. %GACA is the percentage average classification accuracy for the training set (Tr) and the test set (Tst) over the three samples for both data sets. The results in Tables 2 and 3 show clear improvement using the GA based decision tree induction over the classification accuracy of the stepwise generated binary ID3. The GA based decision tree performed consistently better than the stepwise binary ID3. An average improvement of %1 Bank Loan data and %3.5 for the Cleveland Heart Disease data was achieved.

6 Discussion and Conclusion

This paper introduced the use of GAs in rule induction and how they can be applied effectively to improve their classification accuracy over existing decision tree induction algorithms. The task of decision tree induction was re-defined in terms of searches to which GAs can be applied. A chromosome/gene structure was defined and related to the selection of attribute splits. The gene based attribute selection method was combined with two heuristics adopted from existing stepwise induction algorithms.

The over all improvement in classification accuracy achieved by the GA based algorithm is evidence that the lack of a globally optimal strategy in stepwise algorithms has a negative impact on their performance. The improvements in performance, is very much dependent on the nature of the data domain and its separability. If the training set contains records which are easily separable then a very simple tree will be capable of modelling the domain to a high degree of accuracy. In this case very little improvement can be expected from a GA generated decision tree. Conversely if data is complex and difficult to partition with large number of attributes finding a good decision tree to partition the data is very difficult if not impossible. For such data domains GA based algorithms are ideal.

7 References

[1]   Holland, J.: Adaptation in Natural and Artificial Systems. University of Michigan (1975).

[2]   Grefenstett, J.: Genetic Algorithms For Machine Learning. A Special Issue of Machine Learning. Machine Learning, Vol. 13, Nos 2-3. Kluwer Academic Publishers (1992).

[3]   Quinlan, J.: Induction of Decision Trees. Machine Learning 1, Kluwer Academic Press (1986).

[4]   Michie, D.: Personal Models of Rationality. Technical Report, Turing Institute Document No. TIRM-88-035 (1989).

[5]   Kass, G. V.: An exploratory Technique for Investigating Large Quantities of Categorical Data. Applied statistics, 29, No. 2 (1980).

[6]   Quinlan, J.: Simplifying Decision Trees. International Journal of Man-Machine studies (1987).

[7]   Koza, J.: Concept formation and Decision Tree Induction Using the Genetic Peogramming Paradigm. Proceeding of the 1st Workshop. Parallel Problem Solving from Nature. Dortmund, Germany. Springer-Verlag (1991).

[8]   Goldberg, D.: A Meditation on the Application of Genetic Algorithms. IlliGal Technical Report No. 98003 (1998).

[9]   Bratko, I., and Kononenko, I.: Learning Diagnostic Rules from Incomplete and Noisy Data. Technical Report.   Turing Institute Document No. ANT24362 (1987).

[10] Michie D., Spiegelhalter D.J., Taylor C.C.: Machine Learning, Neural and Statistical Classification, Ellis Hopwood Series in Artificial Intelligence, Ellis Hopwood, (1994).

[11] Davis, L.: Handbook of Genetic Algorithms. Van Nostrand, Rheinhold (1991).

Optimising Decision Classifications Using Genetic Algorithms

Keeley A Crockett[1], Zuhair Bandar[1], Akeel Al-Attar[2]

[1]The Intelligent Systems Group, The Manchester Metropolitan University, Chester Street, Manchester, UK, M15 GD.
[2]Attar Software Limited, Newlands House, Newlands Road, Leigh, Lancashire, WN7 4HN, UK
Email:{K.Crockett, Z.Bandar}@doc.mmu.ac.uk

Abstract

A difficult problem associated with traditional decision tree rule induction algorithms is how to achieve common currency between the actual classification accuracy and the distribution of this accuracy between the outcome classes. This paper introduces a novel method which successfully shows that it is possible to attain decision trees which exhibit both good performance and balance. The method first involves the induction of a decision tree, which is then fuzzified using a Genetic Algorithm (GA). Each solution generated by the GA produces a set of fuzzy regions which are mapped onto all nodes in the tree. The GA's fitness function consists of two components : classification accuracy and balance which are optimised concurrently. Three alternative functions are defined, each of which opposes different penalties on the classification accuracy depending on the selected weighting associated with the balance component. The method is applied to two real world data sets and is shown to achieve a high degree of common currency between accuracy and balance.

1 Introduction

When generating a decision tree for a particular application, the performance of the tree in terms of accuracy may not be the only concern for the user. If a relatively high correct classification rate is achieved but the outcomes are poorly distributed between the possible classes then the tree could be seen to be ineffective as a classifier. In [1][2], a new genetically optimised fuzzy inference algorithm (FIA) was proposed, which applied the principles of fuzzy theory [3][4] to the branches of an induced decision tree. A Genetic Algorithm [5..7] was used to determine a sufficient amount of fuzzification which would be needed to relax the strict partitions of attributes which are inherent in the generation of the tree. In these instances the fitness function maximised the percentage average classification accuracy of the decision tree. It did not take into consideration the distribution of correct classifications between the outcome classes. This paper proposes three new fitness functions. The purpose of each new function will be to improve the distribution of correct classifications whilst simultaneously maintaining a reasonably good performance. This will be achieved by using the average classification accuracy as the fundamental component of the fitness and using an imbalance function (as the second component) to reward or penalise this value depending on the balance achieved.

2 Genetic Algorithms

The Genetic Algorithm (GA) is a stochastic searching algorithm inspired by the process of natural evolution. Originally developed by Holland [7], GA's have been successfully used in problems requiring the optimisation of fuzzy membership functions[8..9] A GA maintains a population of chromosomes onto which potential solutions to a specific problem are encoded. Once such a population has been created, simulated evolution takes place where each chromosome within the population is evaluated in terms of it's fitness to reproduce. The chromosomes with the best evaluations are selected to reproduce more often and are then recombined to form a new population of possible solutions to the problem. The cycle of evaluation, selection and recombination is then repeated until either a satisfactory solution to the problem is found or the search is stagnated, which occurs when there is premature convergence on a local optimum. Throughout the cycle, generic operators such as crossover and mutation may be applied with some probability to introduce diversity into a population. A basic version of the algorithm can be described as follows :

1. Initialise a population of chromosomes containing M strings of Length, L
2. Repeat for all required generations
    2.1 Evaluate the *fitness* of each individual chromosome
    2.2 Repeat M times

2.2.1 *Select* the Parents

2.2.2 Apply ***crossover*** and ***mutation*** operators as parents reproduce

2.2.3 Place both new offspring in the population

2.3 Replace current generation with next generation

3. Until a ***near optimal*** solution found or required generation is reached.

3 Fuzzifying Trees

In [10] it was shown that the classification accuracy of a crisp decision tree can be improved by introducing fuzzification onto the branches of the tree. This was achieved by creating fuzzy regions around each tree node in order to soften the sharp decision thresholds. A fuzzy region is defined using a pair of linear membership functions for each decision node. This is illustrated in Fig. 1 for a tree node with a decision threshold of 3. Each linear membership function is defined by upper and lower bounds *dm* and *dn,* about the decision threshold (*dt*) of the attribute which is determined by the tree induction algorithm.

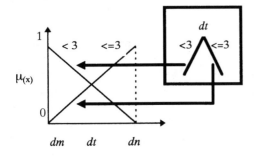

Fig. 1. Membership Function Representation

FIA first requires a tree to be created using an ID3 algorithm [11][12]. Each path from root to leaf is then converted into a series of fuzzy IF-THEN production rules. A case passing through the tree will result in all branches in the tree firing to some degree, which is determined by each specific attributes degree of membership in the corresponding fuzzy region. To determine the classification outcome of a case passing through the tree, the membership grades at all branches are combined using a pre-selected fuzzy inference technique.

3.1 Determining Sufficient Fuzziness

When fuzzifying a tree, it is essential to obtain a balance of fuzziness. Too much fuzzification leads to additional uncertainty in the tree, whilst too little has insignificant impact on the performance. To determine sufficient fuzziness for a given tree a GA is used. The membership functions are encoded onto a chromosome (Figure 2). Each gene will represent a real value n, used in the determination of one domain delimiter (dm_i or dn_i). Given M is a complete set of ordered domain delimiters in a decision tree consisting of i branches then

$$M = \{ \ (dm_1, dn_1), (dm_2, dn_2)....... (dm_i, dn_i) \ \}$$

where each pair represents the domain of a membership function at branch i and (dm_i, dn_i) is defined as

$$dm_i = dt_i - n_j\sigma_i \quad \text{and} \quad dn_i = dt_i + n_{j+1}\sigma_i$$

where M can be represented as a series of genes $g_1....g_n$ (Fig. 2)

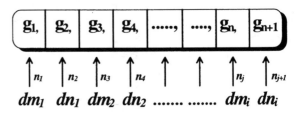

Fig. 2. Chromosome Representation

Each individual gene will represent a value of n used within the determination of a domain delimiter representing a possible solution. In order to focus the search, constraints can be applied in order to restrict the range of values a gene may take.

The GA used to optimise FIA was provided by the software package XpertRule [13]. Each chromosome was passed as a real number array. Once a population of chromosomes had been generated by XpertRule, each chromosome was passed into FIA in order for it's fitness to be evaluated. This involved the process of mapping each gene to a membership function domain delimiter and then generating one run to obtain the cost function to be returned to XpertRule for further optimisation.

3.2 Constraining the Genes

In [1] it was shown through extensive manual experiments that the best performance of FIA was achieved when the fuzzy regions created around each tree node were no more than two standard deviations around the decision threshold of the attribute. It was found that this restriction on the amount of fuzziness could be used to constrain the genes, therefore reducing the search space and hence the time required to seek a optimal or near optimal amount of fuzzification .

4 Variations of the Fitness Function

The original fitness function, F_1 was defined to maximise the percentage average classification accuracy of the Training set. However, in-order to determine the degree of common currency between accuracy and balance three variations on the original fitness function F_1 were investigated. The purpose of each new function will be to improve the distribution of correct classifications whilst simultaneously maintaining a reasonably good performance. This will be achieved by using the average classification accuracy as the fundamental component of the fitness and using an imbalance function (as the second component) to reward or penalise this value depending on the balance achieved. The three alternative fitness functions are defined as :

$$F_2 = AVG \bullet IB \rightarrow max \qquad (3.1)$$
$$F_3 = AVG + (1 \bullet IB) \rightarrow max \qquad (3.2)$$
$$F_4 = AVG + (2 \bullet IB) \rightarrow max \qquad (3.3)$$

where

max is the maximum value which is attained from the fitness function during the search.

IB is an imbalance function calculated as the fractional division of

smallest % outcome
largest % outcome

where outcome refers to the classification outcome from a binary data set.

AVG is the percentage average classification accuracy.

The fitness function F_2 can be seen to have a very strong imbalance component. A high inconsistency in the balance would oppose a strong penalty on AVG resulting in a low fitness. In comparison, the balance is the weaker component of F_3 contributing only 1% to the overall fitness. F_4 is also similar to F_3 except up to 2% of the final fitness is determined by the degree of imbalance. The three distinct imbalance component weightings were selected to investigate the degree by which accuracy may be compensated in the quest for establishing a balanced tree.

5 Experiments

5.1 Data Sets

Two real world data sets were considered in-order to investigate the three fitness functions. The first set known as Diabetes, provides diagnosis on whether a patient shows signs of diabetes according to World Health Organisation criteria. The population lives near Phoenix, Arizona, USA. The data set comprises of 768 records (500 Class 1 indicating that 'patient has tested positive for diabetes', 268 Class 2 indicating ' person shows no signs of the disease') and 9 continuos attributes. The second set known as Mortgage, contains details of applicants and the status of the Mortgage two years after acceptance ('Good' = not in arrears, 'Bad' = 3 or more months in arrears) The Mortgage data set is considerably larger than the Diabetes set, comprising of 8611 records (4306 Class 'Good' and 4305 Class 'Bad') with 11 discrete and 15 continuous attributes.

5.2 Methodology

Each data set was partitioned into two sets of randomly selected, mutually exclusive cases, which are referred to as the Training and Test sets. A restriction of the Training set, was that it contained an equal number of both class 1 and class 2 examples. From each Training set, a binary ID3 tree was created and then pruned with a significance level of 0.1%. The membership function domains were mapped onto a series of genes, with chromosome in the population representing one possible solution (Fig. 2). The results obtained from the crisp ID3 tree are shown in Table 1. For purposes of comparison, Table 2 shows the average results of three runs (300 generations) from using the GA to optimise fuzziness using F_1.

Table 1. ID3 Test Results

Data Set	%AVG Test	% Class1	%Class 2
Mortgage	67	70	64
Diabetes	70	89	52

Table 2. Original Fitness Function F_1

Data Set	%AVG Test	% Class1	%Class 2
Mortgage	70.2	71.2	69.2
Diabetes	74.1	81.6	66.7

5.3   Parameters

Table 3 shows a summary of parameters used in these experiments. Previous work [1] was undertaken to determine suitable values for a number of common generic operators [5]. Zadeh's Min-Max inference technique [4] was selected to combine membership grades to allow the focus of the investigation to be based upon the impact of the fitness functions F_2, F_3, and F_4 on the overall performance.

Table 3. Parameters

Data Sets	Mortgage Diabetes
Significance Level of trees	0.1%
Inference Technique	Min-Max
Domain delimiters dm and dn	Gene constrained {0,2}
Number of Generations	50 - 300 (varied)
Number of Individuals	50
Crossover Probability	0.5
Mutation Probability	0.05

6   Results

In order to examine the convergence rate of the GA, three runs were undertaken of each 50 and 300 generations per data set. Tables 4 - 9 show the percentage average classification accuracy (%Avg) of the unseen Test set obtained for each of the three fitness functions F_2, F_3, and F_4.

Table 4. Mortgage F_2

No' Gens	% Avg Test	%Good (Test)	%Bad (Test)
50	69.6	69.9	69.3
300	70.4	71.9	68.8

Table 5. Diabetes F_2

No' Gens	% Avg Test	%Class1 (Test)	%Class2 (Test)
50	73.5	71.1	75.9
300	73.8	72.0	75.6

Table 6. Mortgage F_3

No' Gens	% Avg Test	%Good (Test)	%Bad (Test)
50	70.2	74.7	65.7
300	70.1	74.1	66.1

Table 7. Diabetes F_3

No' Gens	% Avg Test	%Class1 (Test)	%Class2 (Test)
50	70.2	74.7	65.7
300	70.1	74.1	66.1

Table 8. Mortgage F_4

No' Gens	% Avg Test	%Good (Test)	%Bad (Test)
50	70.4	71.1	69.8
300	70.1	71.3	68.9

Table 9. Diabetes F_4

No' Gens	% Avg Test	%Class1 (Test)	%Class2 (Test)
50	73.4	82.9	63.8
300	71.2	73.7	68.7

7   Discussion

Tables 4 and 5 show that fitness function F_2 has resulted in a substantial improvement in the distribution of correct cases for both data sets compared with that obtained using only ID3 (Table 2). The accuracy of the fuzzified trees is also significantly better than the crisp trees, with an average 3% improvement on the Mortgage data and 3.7% improvement on the Diabetes data sets. However, the accuracy of the Diabetes data set is marginally lower (on average 0.3%) than that achieved with the original function F_1, indicating that a small sacrifice has been made in order to improve the balance. For the run of 300 generations, the modular difference between the outcome classes is 3.6%, compared with the 14% difference achieved through optimisation using F_1.

The performance of F_3 on both data sets was not as significant as that achieved by F_2 due to the balance only effecting the fitness by at most 1%. The results (Table 6) show that the Mortgage data has however maintained a 3% improvement in accuracy. In contrast the Diabetes data (Table 7) has suffered a substantial decline in performance, achieving similar results in terms of accuracy as the crisp tree. This is likely to be a result of the entirely continuous nature of this data set.

The application of F_1 to the Mortgage data showed a difference of 1.3% between the distribution of correct classifications was obtained with only 50 generations (Table 8). For the Diabetes data set, the balance which was obtained on the 50 generation run was only comparable to that achieved when the original fitness function was applied. The 300 generation runs produced a slightly better balance, but a relatively poor accuracy compared with all other GA runs.

8 Conclusion

This paper has introduced a new method which enables common currency to be obtained between classification accuracy and the distribution of the correct classifications in crisp decision trees. The method used a GA to optimise fuzzy regions at each node in a crisp tree using a previously developed Fuzzy Inference Algorithm (FIA). Three fitness functions were investigated where each imposed various penalties on the search with a biased towards balancing the distribution. Each function was applied to trees generated from two real world data sets. The results indicated that it was possible to achieve a significant improvement in balance compared with the crisp trees. It was found that the best results were obtained when the imbalance component had a significant impact on the fitness (i.e. F_2). The amount of improvement varied depending on the degree by which the imbalance contributed to the overall fitness and would be selected at the users discretion. The accuracy of the crisp trees were also improved on average by 3.4%. The factors influencing these improvements were the nature of the data set, the softening of sharp decision thresholds by introducing fuzzy regions and the combining of all potential information throughout the tree.

9 References

[1] Crockett. K, Bandar. Z, Al-Attar. A, A Genetically Optimised Fuzzy Inference Algorithm. International Conference on Artificial Intelligence and Soft Computing, Mexico, 1998.

[2] Crockett. K, Bandar. Z, Al-Attar. A, A Fuzzy Inference Framework For Induced Decision Trees. ECAI 98. 13th European Conference on Artificial Intelligence, Brighton, UK. Eds. H. Prade. John Wiley & Sons Ltd, 425-429, 1998

[3] Zadeh, L. Knowledge Representation In Fuzzy Logic. An Introduction To Fuzzy Logic Applications In Intelligent systems, edted by Yager, R and Zadeh, L, Kluwer Academic Publishers,1992.

[4] Klir, G. Folder, T. Fuzzy Sets, Uncertainty and Information. Prentice Hall, 1988.

[5] Grefenstette, J. Genetic Algorithms For Machine Learning, A Special Issue of Machine Learning. Machine Learning Vol. 13, Nos. 2-3. Kluwer Academic Publishers, 1993

[6] Goldberg, D. A Meditation on the Application of Genetic Algorithms. IlliGal Technical Report No. 8003. Solution: http://gal4.ge.uiuc.edu/illigal.home.html, Feb 1998

[7] Holland, J. Adaptation in Natural and Artificial Systems. University of Michigan, 1975.

[8] Karr, C. Freeman, L. Meredith, D. Improved Fuzzy Process Control Of Spacecraft Autonomous Rendezvous Using A GA, SPIE Vol 1196 Intelligent Control and Adaptive Systems, 1989

[9] Mahmood, A. Meashio, Y. Singer, G. A Genetic Algorithm Based Fuzzy Logic Controller For Non-Linear Systems, GALESIA-97 Genetic Algorithms in Engineering Systems : Innovations and Applications, Conference Publication 446, IEE, 30-35, 1997

[10] Crockett. K, Bandar. Z, Al-Attar. A, Fuzzy Rule Induction From Data Sets. Proceedings of The 10th International Florida Artificial Intelligence Research Symposium, 332-336, 1997

[11] Quinlan, J, R. C4.5 : Programs for Machine Learning. Morgan Kaufmann Publishers, 1993

[12] Quinlan, J, R. C5.0 An Informal Tutorial. http://www.rulequest.com/see5-unix.html#XVAL, 1997

[13] XpertRule, Attar Software, Newlands Road, Leigh, Lancashire, WN7 4HN, UK

Scheduling Tasks with Non-negligible Intertask Communication onto Multiprocessors by using Genetic Algorithms

G. Jezic [1], R. Kostelac [2], I. Lovrek [1], V. Sinkovic [1]

[1] Department of Telecommunications, Faculty of Electrical Engineering and Computing,
University of Zagreb, Unska 3, HR-10000 Zagreb, Croatia
[2] Croatian Academic and Research Network CARNet, Marohniceva bb, HR-10000 Zagreb, Croatia
Email: {gordan.jezic, ignac.lovrek, vjekoslav.sinkovic}@fer.hr, robert.kostelac@carnet.hr

Abstract

The paper deals with genetic algorithms for scheduling the tasks with non-negligible intertask communication. Three genetic algorithms for scheduling with the primary goal to minimise finishing time are reported. The basic genetic algorithm includes the reproduction, crossover and mutation operators. Its improved version has additional cloning operator that allows duplicated scheduling. The third algorithm is adaptive. The experiments describing influence of genetic operators' probabilities, population size and number of generations on the resulting schedules, comparison of algorithms and the results obtained for different task granulation are discussed.

1 Introduction

Generally, multiprocessor scheduling is the problem of scheduling tasks of the precedence constrained task graph representing a parallel program in the way that minimises the completion time by exploiting optimally the capabilities of the multiprocessor architecture. It means that the work must be specified for each processor separately, so that the entire set of tasks is processed within shortest possible time, satisfying all precedence relations, and using the available processor resources [1].

This paper presents scheduling with non-negligible intertask communication based on genetic algorithms with the primary goal to minimise the finishing time. Genetic algorithms belong to the robust search techniques rooted in the mechanism of evolution and natural genetics [2]. Those for scheduling described in the paper are studied for the purpose of analysing distributed parallel call and service processing in telecommunications. They are used for finishing time evaluation in a simulation based method for the analysis of distributed parallel processing systems [3-7]. An efficient simulation method uses the algorithms that can produce the results close to an optimum by using simple operations. Genetic algorithms meet these conditions, if a satisfactory schedule is derived by a small number of generations.

The paper is organised in the following way. Section 2 contains model description. Basic genetic algorithm, genetic algorithm with cloning operator and adaptive genetic algorithm are described in Section 3. Section 4 deals with comparison of algorithms, while different task granulation is presented in Section 5. The conclusions are given in Section 6.

2 Model Description

A network of M equal processors, $P = \{P_1, P_2, ..., P_M\}$, is supposed. Each processor has half-duplex communication channels for interconnection with other processors and has separate communication units that enable simultaneous processing and intertask communication. Consequently, the algorithm can exploit the schedule-holes generated in the processors and communication channels in order to produce better results. The number of channels is not restricted and the system is fully interconnected.

The set of tasks and precedence relations between tasks are represented by a directed acyclic graph, $G = (N, E)$. The set of tasks (nodes) is denoted by $N = \{n_1, n_2, ..., n_i, ..., n_Q\}$. E represents the set of directed edges e_{ij} between n_i and n_j implying that n_i must precede n_j. It is assumed that each task n_i can be executed on any processor P_k. The model is deterministic, i.e. precedence relations, execution time t_i of task n_i, as well as communication time c_{ij} between tasks n_i and n_j are specified. It also assumes the non-pre-emptive scheduling wherein the task cannot start before all its predecessors are completed. Each task n_i has a height value v_i depending on predecessors in the following way. If a task n_i has no predecessors, then:

$$v_i = 0,$$

if a task n_i has tasks $n_{j1}, n_{j2}, ..., n_{jk}$ as predecessors, then:

$$v_i = \max (v_{j1}, v_{j2}, ..., v_{jk}) + 1.$$

A task graph example is shown in Figure 1.

If two tasks n_i and n_j are allocated on different processors P_k and P_l, the task n_j cannot start before the current task n_i and the intertask communication c_{ij} are completed [8]. If the task n_i starts at some time t_x, the successor task will be finished at time $t_x + t_i + c_{ij} + t_j$. The finishing time will be reduced to $t_x + t_i + t_j$ when the successor task is allocated to the same processor as the current task. That is, the intertask communication is neglected, $c_{ij} = 0$. Communication delay will appear if a communication channel is busy when requested. In this case, the finishing time will be increased because the processors must wait till former communication is performed.

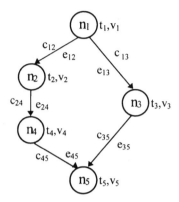

Fig. 1. Task graph

In order to include the intertask communication, the transformed task graph, $G_c = (N`, E`)$ is introduced. Each edge is transformed to a node and the corresponding intertask communication is represented as a communication task. Each communication task has the execution time equal to c_{ij} if n_i and n_j are allocated to different processors. If n_i and n_j are allocated to the same processor, the communication time equals 0. In this way, the analysis of the intertask communication is replaced by the analysis of the transformed task graph G_c with the communication and processing tasks. The communication tasks are being included with the execution time c_{ij} or 0, depending on the successor and predecessor processing task allocation.

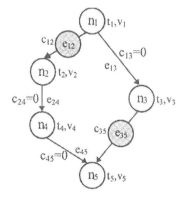

Fig. 2. Transformed task graph

An example showing the transformed graph from the Figure 1 with five processing and two

communication tasks, is presented in Figure 2. Processing tasks n_2, n_4 and n_5 belong to the first processor, and the tasks n_1 and n_3 to the second one. Because the tasks n_1 and n_2, as well as the tasks n_3 and n_5, are allocated to different processors, two communication tasks, e_{12} and e_{35}, are inserted.

3 Genetic Algorithms for Scheduling

The scheduling problem being complex, various approaches to their solving have been developed: different optimisation and approximation algorithms, state-space search algorithms, models based on neural networks, etc. By comparing different approaches, genetic algorithms have been found as very suitable for the problems with excessive search space as is scheduling [9-11]. Computational simplicity of genetic operators supports the applications in the simulation based methods for the analysis of distributed and parallel processing systems. Quick estimation of the finishing time for a defined set of partially ordered tasks is of great importance. So, we are interested in the genetic algorithms that offer nearly optimal result in a small number of generations [12].

3.1 Basic Genetic Algorithm

The genetic algorithm (GA) operates so that the members of an actual population are evaluated, the best members are selected and a new population is created by using genetic operators. The population members are different task schedules. The schedules are represented as the strings consisting of lists. A string has as many lists as processors available. The list describes tasks allocated on a processor. The string representing schedule for the task graph from Figure 1 onto processors P_1 and P_2 is shown in Figure 3.

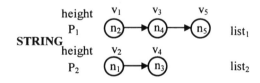

Fig. 3. String representation

Three tasks belong to one list and two tasks belong to another. The sequence of tasks inside the list must satisfy the precedence condition, i.e. the heights of tasks must not be in a decreasing order. In our example that is: $v_2 = 1$, $v_4 = 2$, $v_5 = 3$ and $v_1 = 0$, $v_3 = 1$.

The genetic operators created and accommodated to the scheduling problem produce new schedules in order to find the optimal one, i.e. the schedule with the minimum finishing time. The basic algorithm, Simple Genetic Algorithm (SGA) is based on the reproduction, crossover and mutation operators that follow the approach proposed in [9].

The initial population includes the starting task schedules onto processors. It is a set of the randomly generated schedules that satisfy two conditions: precedence relations between tasks and single appearance of each task in a schedule.

The fitness function is the objective function that has to be optimised. The finishing time of the schedule (string) S equals $FT(S) = max \{ft(P_k)\}$, where $ft(P_k)$ is the finishing time for the last task in the processor P_k. Because the reproduction operator tries to maximise the fitness function, the conversion of the finishing time to $f(FT) = T - FT(S)$, where $T = \sum t_i$ for a transformed task graph, is needed.

The reproduction operator creates a new string population (next generation), by selecting the strings in the actual population using the fitness value as a selection criterion. The strings with a higher fitness value should have a better chance of entering the next generation. In SGA, elitism is used and the best string is marked to be passed to the next generation. In this way, the best string can never be lost.

The crossover operator produces two new strings by combining the parts of original strings. It selects the crossover sites at the same height and exchanges the parts of the strings. Two strings are chosen and the crossover with a predefined probability p_c is performed.

Mutation produces a new string by exchanging two tasks from the old one. These tasks must have the same height. For each string, mutation is performed with a predefined probability p_m.

3.2 Genetic Algorithm with Cloning Operator

The cloning operator is based on duplication scheduling heuristic method. The operator randomly chooses a task from a list and performs cloning that duplicates the current task and includes a clone-task on the other list. The clone-task n_c is identical with the original task n_o; execution time $t_c = t_o$, height $v_c = v_o$ and precedence relations are the same. The application of the cloning operator is determined by cloning probability p_l. The most important goal of the operator is to minimise the intertask communication in order to avoid a potential communication delay [13].

The genetic algorithm with a cloning operator (SGA-C) performs the cloning in the following way:
1. (Pick a task) Randomly picks a task n_o with height v_o.

2. (Select a processor) Randomly generates a number, c, between 1 and M.
3. (Find a cloning site) Searches the task n_i in the processor c that has the first higher height $v_i > v_o$.
4. (Add clone-task) Forms a new string by inserting the clone-task n_c in the schedule, unless it already exists.

For the task graph from Figure 1 and the string from Figure 3, one of possible strings after cloning operation is shown in Figure 4. The clone-task, n_{c1}, allocated to the processor P_1 is identical to the task n_1 allocated to the processor P_2.

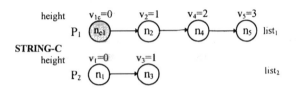

Fig. 4. String after cloning operation

Figure 5a shows the task schedule *STRING*, and Figure 5b shows the task schedule *STRING-C*. It is supposed that the execution times of all tasks (processing and communication) are the same and equal, $t_i = c_{ij} = 1$. *STRING-C* has better finishing time; such a schedule can be generated only by using the cloning operation.

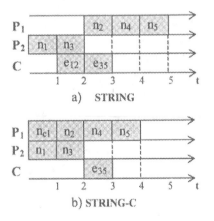

Fig. 5. Influence of cloning operation

3.3 Adaptive Genetic Algorithm

Adaptive genetic algorithm (AGA) is based on the principle of self adapting parameters during the performance of an algorithm (crossover probability p_c, mutation probability p_m, cloning probability p_l and population size N_p). In SGA and SGA-C, all parameters are set before starting the algorithm, and they remain unchanged while the algorithm is performed.

To improve the quality of an algorithm and to avoid circling around the local optimum, an adaptive setting of parameters is introduced, with the ultimate goal to free the user of setting the parameters by himself. Thus, it is possible to increase parameter values while the population approaches the local optimum, and to decrease the values while the population reaches the best schedule. By using this algorithm, better schedules have more chance to enter next generation, while the average and under-average ones are being eliminated.

The steady-state policy is used to selectively replace the schedules, and protect one or more schedules in order to enter next generation (no elitism is used). The key factor in adapting the parameters is a fitness function. The method of finding the algorithm convergence is applied by comparing the average fitness value f_{av} of a population with the highest value in a population f_{max}. The difference between the maximum and the average value is decreased if the population approaches the best solution. The expressions for calculating p_c, p_m and p_l follow:

$$p_{c,m,l} = A + B \cdot \frac{f_{max} - f}{f_{max} - f_{av}} \qquad (1)$$

where $A = 0.8$, $B = 0.2$ for p_c, $A = 0.05$, $B = 0.15$ for p_m and $A = 0.2$, $B = 0.1$ for p_l. A and B constants are determined experimentally. f represents the fitness value of a current schedule.

4 Comparison of Genetic Algorithms

The genetic algorithms SGA, SGA-C and AGA were tested by using different parameter values on different task graphs with 20 – 50 tasks.

The analysis presented in the paper was performed on a task graph with 35 tasks and the execution times of 1 or 2 intervals, and 41 edges with the same communication time, one interval each. Total execution time of all the tasks was 52 intervals. Considering the expression (1), the crossover probability p_c was between 0.8 and 1.0, mutation probability p_m between 0.05 and 0.2 and cloning probability p_l between 0.2 and 0.3. These had been the ranges within which the operators yielded satisfactory results. Lower values were applied in AGA for the best schedules and the highest ones for the schedules with the finishing time below the average.

The analysis was done in three steps: without a particular genetic operator, with the fixed value of the

parameter outside the range, with the fixed value of the parameter within the range or with the adaptive values. The following parameters were used: population size $N_p = 30$, number of processors $M = 5$ and number of generations $N_g = 200$. Only the best finishing times (*FT*) gained from the executions of algorithms are shown. The influence of the cloning operator with the fixed crossover and mutation probabilities $(p_c = 0.9, p_m = 0.1)$ is presented in Figure 6. The result obtained by using the adaptive genetic operators is shown in Figure 7. The influence of the population size was analysed by using AGA and the results are shown in Table 1.

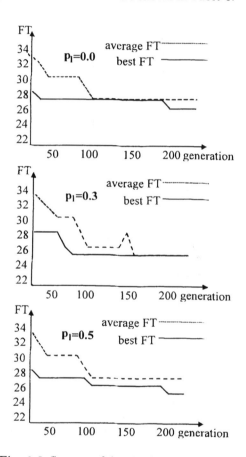

Fig. 6. Influence of the cloning probability

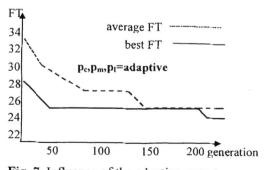

Fig. 7. Influence of the adaptive operators

Experiments show that SGA and SGA-C with the crossover probability p_c between 0.8 and 0.9, mutation probability p_m between 0.1 and 0.15 and SGA-C with the cloning probability p_l between 0.25 and 0.3 will produce the acceptable results in 30 - 50 generations, and with the population of 10 - 20 strings. Being complex AGA is not generally acceptable for our purposes. Its major role is to help determine the operator's probabilities and for detailed analysis of the specific task graphs.

Table 1. Influence of population size

population size N_p	FT	number of generations
10	26	200gen
20	25	245gen
30	24	216gen
40	24	279gen
50	24	290gen

The comparison of results for *SGA* and *SGA-C* with cloning probability $p_l = 0.25$ is performed. The results show that *SGA-C* generates better schedules, and, generally, unlike an algorithm without a cloning operator, a genetic algorithm with a cloning operator produces on an average about 10% better results. For instance, by using the parameters $p_c = 0.9$ and $p_m = 0.1$, SGA's finishing time will be within the range of 25-27, and the results for SGA-C will be 24-26.

The results show that the application of the genetic algorithms fulfils the most important requirement: quick estimation of the finishing time in a small number of generations. The result obtained in a small number of generations is not always the best one, but the deviation is 10% at the most. A quick estimation of the result, which is close to the optimum, as well as computational simplicity of the genetic operators, support the application of the algorithm in the simulation-based method for the analysis of a parallel call and service control.

5 Task Granulation

Each task graph can be modified in two ways (Figure 8).

The first one begins with a detection of task sequences without parallelism. The recognition of such a sequence provides the execution of the whole sequence on the same processor and facilitates the convergence. In order to exploit this, it is necessary to bind the area in which the crossover sites can be chosen. In this way, the tasks that make one sequence

can be moved from one processor to another only as one group. This method is called "rough granulation". The tasks are grouped in a "bigger" task, with an execution time equal to the sum of the execution times of all the tasks comprising a new one. The new task graph now contains less tasks than the original one. This approach can help to minimize an intertask communication [14].

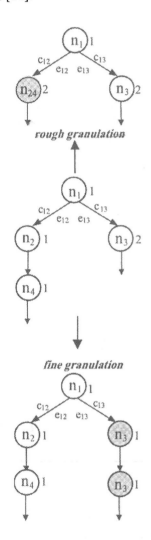

Fig. 8. Granulation sizes

The second method provides decomposition of the tasks into elementary tasks, each of equal execution time. This method is called "fine granulation". Each task with the execution time $t_i > 1$, is replaced by the sequence that contains as many tasks as the time intervals preceding fine granulation. In this way, it is possible to exploit the schedule-holes caused by dependent tasks, i.e. the tasks on a critical path or the task sequences with no parallelism [14].

The analyses and comparison of these two methods follow. The starting task graph has 35 tasks. After performing rough granulation, a graph with 19 tasks is created. New, modified graph has 10 new tasks and the execution time of the "biggest" task equals 6 intervals. After fine granulation, a task graph with 52 tasks is created, all with the same execution time (one interval).

The best results are obtained for the task graph with rough granulation. Therefore, it can be concluded that better schedules are produced if a task graph has a smaller number of tasks because of minimisation of intertask communication and potential communication delay. Fine granulation improves the possibility of exploiting the schedule-holes and generating better results if a task graph contains the tasks with bigger differences between the execution times.

6 Conclusions

Genetic algorithms for scheduling the tasks with non-negligible intertask communication onto multiprocessors are reported. The basic algorithm includes the reproduction, crossover and mutation operators. The improved version has an additional cloning operator. The adaptive algorithm applies changeable genetic operator probabilities. The experiments with operators' probabilities, population size and number of generations as well as the comparison of algorithms and the results obtained for different task granulation, show that the genetic algorithm with cloning operator produces the results quite suitable for the analysis of distributed parallel call and service control in telecommunications.

References

[1] El-Rewini, H., Lewis, T.G., Ali, H.H.: Task Scheduling in Parallel and Distributed Systems, Prentice Hall, 1994.

[2] Goldberg, D.E.: Genetic Algorithms in Search, Optimization & Machine Learning, Addison-Wesley, 1989.

[3] Sinkovic, V., Lovrek, I.: An Approach to Massively Parallel Call and Service Processing in Telecommunications, Proceedings MPCS'94 Conference on Massively Parallel Computing Systems (MPCS): the Challenges of General-Purpose and Special-Purpose Computing, Ischia, Italy, pp. 533 – 537, 1994.

[4] Sinkovic, V., Lovrek, I.: A Model Massively Parallel Call and Service Processing in Telecommunications, Journal of System Architecture - The EUROMICRO Journal, 43(1997), pp. 479-490.

[5] Nemeth, G., Lovrek, I., Sinkovic, V.: Scheduling Problems in Parallel Systems for Telecommunications, Computing, Vol. 58, No. 3, pp. 199-223, 1997.

[6] Sinkovic, V., Lovrek, I.: Load Balancing Potential of Distributed Parallel Call and Service Processing in Telecommunications, Proceedings Second International Conference on Massively Parallel Computer Systems MPCS 96, Ischia, Italy, pp. 366-373, 1996.

[7] Lovrek, I., V. Sinkovic, Jezic, G.: Analysis of Call and Service Control in Telecommunications by using Genetic Algorithms, Proceedings KES 98 2nd International Conference on Knowledge-Based Intelligent Electronic Systems, pp. 302-310, Adelaide, Australia, 1998.

[8] Selvakumar , C., Murthy, S.R.: Scheduling Precedence Constrained Task Graphs with Non-negligible Intertask Communication onto Multiprocessors, IEEE Transactions on Parallel and Distributed Systems, Vol. 5, No. 3, pp. 328 – 336, 1994.

[9] Hou, E.S.H., Ansari, N., Ren, H.: A Genetic Algorithm for Multiprocessor Scheduling, IEEE Transactions on Parallel and Distributed System, Vol. 5, No. 2, pp. 113-120, 1994.

[10] Lovrek, I., Jezic, G.: A Genetic Algorithm for Multiprocessor Scheduling with Non-negligible Intertask Communication, Proceedings MIPRO 96 Computers in Telecommunications, Rijeka, Croatia, 1996.

[11] Jezic, G., Kostelac, R., Lovrek, I., Sinkovic V.: Genetic Algorithms for Scheduling Tasks with Non-negligible Intertask Communication onto Multiprocessors, Proceedings GP 98 Third Annual Genetic Programmnig Conference: Symposium on Genetic Algorithms, pp. 518, Madison, USA, 1998.

[12] Sinkovic, V., Lovrek, I.: Performance of Genetic Algorithm Used for Analysis of Call and Service Processing in Telecommunications, Proceedings ICANNGA 95 International conference on Artificial neural networks and Genetic Algorithms, Ales, France, pp. 281-284, 1995.

[13] Jezic, G., Kostelac, R.: The Use of Task Duplication in the Genetic Algorithm for Multiprocessor Scheduling with Non-negligible Intertask Communication, Proceedings MIPRO 97 Computers in Telecommunications, Rijeka, Croatia, pp. 2-204-2.209, 1997, (in Croatian).

[14] Jezic, G., Kostelac, R.: The Influence of Different Granulation Sizes on the Task Scheduling in the Genetic Algorithm for Multiprocessor System, Proceedings SoftCOM 97, Split-Bari-Dubrovnik, Croatia-Italy, pp. 423-432, 1997, (in Croatian).

Design of Robust Networks of Specific Topologies using a Genetic Algorithm

A. Webb[1], B.C.H.Turton[1], J.M. Brown[2]

[1] Cardiff School of Engineering, Division of Electronic Engineering, Cardiff University, Queen's Buildings, PO Box 689, Newport Road, Cardiff, CF2 3TF, UK

[2] Magellan Business Networks, Northern Telecom House, Maidenhead, SL6 8XB, UK

Email: {Turton, WebbA}@cf.ac.uk, John.Brown.jmbrown@nortel.co.uk

Abstract

This paper deals with the design and optimisation of networks with specific (as opposed to arbitrary mesh) topologies. The problem is of relevance to the design of telecommunication networks, where robust and low-cost solutions need to be found within strict deadlines. Real design problems often have enormous search spaces that are difficult to search efficiently within reasonable time scales. Designing networks of specific topologies is a way of reducing the available search space and the time taken to find good solutions. This method also has the additional benefit of being able to guarantee networks that have a number of desirable topological properties. A method of encoding specific topologies is used that allows the GA to apply traditional genetic operators and fully explore the search space containing the topology of interest only. Tariffs from an industry-standard database are used to realistically cost networks.

1   Introduction

This paper presents a method of optimising networks that have specific, pre-defined topologies. Previous work involving GA-based network optimisation has been done using arbitrary ('random') mesh topologies on networks with a relatively small number of nodes [9, 11-13]. Other work in this area has concentrated on optimising the allocation of link capacities or link frequencies for an existing network [6-8, 10].

Modern enterprise networks might contain anything from 5 to 500 nodes. The problem of interconnecting these nodes using links that can have one of several discrete capacities can present us with a combinatorial optimisation problem with an enormous solution space. A GA-based network tool that optimises arbitrary mesh topologies will yield the optimal solution eventually, due to the mesh topology covering the whole available search space. However as the size of the network and hence the search space increases, optimal solutions become harder to find. In addition, each individual network ('chromosome') has to undergo a lengthy evaluation process to determine among other things, its true economic cost, overall delay, suitability for routing all traffic requirements and its ability to survive link failures. The amount of computation rises dramatically as the network size increases. Given that the time scale for designing a complete network is often quite short, with insufficient time to perform lengthy simulations and optimisations, the problem we seek to address is that of generating good, robust network solutions within a reasonable time scale. To limit the topology is one way of doing this.

For an n-node network with L available link sizes, the number of potential topologies is given by $(L+1)^{n(n-1)/2}$ [15]. For example, a 100-node network with 9 available link capacities will evaluate to 10^{4950} topologies - a very large solution space. Many of these topologies will be undesirable or even infeasible thus making the search for good networks difficult, even for a powerful search technique such as a GA. The search can be made easier by optimising a pre-specified architecture instead of an arbitrary network structure. By optimising pre-specified topologies we may also be able to determine the suitability of these architectures under various applications and customer requirements. Different points in the solution space may contain networks that have desirable properties for surviving link or node failures, for example. Other areas of the search space may have networks that are the most cost-effective when the traffic levels are low and there are no reliability requirements. A summary of the search spaces for each topology is shown in Table 1.

Table 1. A summary of search space sizes

Topol.	Ring	Torus	Mesh
S	$\frac{1}{2}(n-1)!L^n$	$\frac{1}{4}(n-1)!L^{2n}$	$(L+1)^{n(n+1)/2}$
n=8	1.08×10^{11}	2.33×10^{18}	1×10^{28}
n=16	1.21×10^{27}	1.12×10^{42}	1×10^{120}
n=32	1.41×10^{64}	2.42×10^{94}	1×10^{496}

In this table, n is the number of nodes in the network, S is the total search space size and L is the number of available link sizes.

For this paper we have compared the use of the arbitrary mesh topology with two pre-specified topologies: a self-healing ring and torus, for designing robust networks of sizes 8, 16 and 32 nodes respectively. The results obtained from optimising the ring and torus topologies are compared with those results obtained from the mesh network optimisations. The 3 topologies of interest are illustrated in Figure 1.

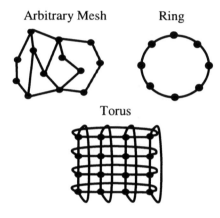

Fig. 1. The topologies used in the simulations

2   Arbitrary Mesh Representation

Given a network with n number of nodes, interconnected with any one of m discrete link capacities, an arbitrary ('random') mesh topology can be completely represented using the following chromosome structure:

$$C = \{C_0, C_1, C_2, ..., C_{p-1}\}$$

Where p is the total number of potential links and is equal to $n(n-1)/2$. A 16-node network, for example, would have a total of $16(16-1)/2 = 120$ possible links. C is the link capacity value, and can be one of nine discrete capacity values if a link is present, or zero if no link is present. The discrete link capacity values are 64 kbit/sec, 128 kbit/sec, 256 kbit/sec, 384 kbit/sec, 512 kbit/sec, 768 kbit/sec, 1024 kbit/sec, 1536 kbit/sec and 1920 kbit/sec respectively. [6, 9-11, 15] also use this method of encoding link capacities. To ensure the networks can survive any single link or node failure, additional heuristic repair techniques are applied to individual chromosomes to ensure bi-connectivity. Adding an appropriate number of randomly chosen

links will repair networks that violate the bi-connectivity requirement. The result is a network where there are 2 node-disjoint paths for every node pair. The reliability of a network also depends to a large extent on the link sizes. When a link fails the remaining link(s) must have sufficient extra bandwidth to cope with the inevitable increase in load.

3   Specific Topology Representation

Two chromosomes are used to encode a network of a specified topology: a node chromosome and a link chromosome. The node chromosome, N, is an ordering of node numbers of size n, where n is the number of nodes in the network.

$$N = \{N_0, N_1, N_2, ..., N_{n-1}\}$$

The link chromosome, M, is an array containing m number of elements. Each array element contains one of nine discrete link capacity values, ranging from between 64 kbps to 1920 kbps. The value of m is the number of links that the specific topology has - not the number of potential links. The link chromosome is encoded as follows:

$$M = \{M_0, M_1, M_2, ..., M_{m-1}\}$$

An example encoding of a 5-node ring topology is shown in Figure 2.

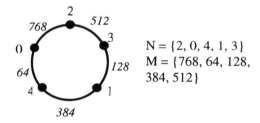

Fig. 2. An example encoding

Since the topology in this case is a ring the number of links used is equal to the number of nodes, i.e. m = n. Genetic operators crossover and mutation are applied to the link and node chromosomes. Their operation is described more fully in the next section.

3.1   Decoding Specific Topologies

In order that we may evaluate and cost each chromosome, we need to decode the node and link chromosomes so that we can determine how the nodes are actually connected together. A lookup table is used to convert the node ordering into a form of representation that can be evaluated. Each node chromosome array element contains lookup values that consist of pointers to neighbouring array elements. The contents of these neighbouring array elements are used to determine the set of adjacent nodes for each node. This is illustrated in Figure 3:

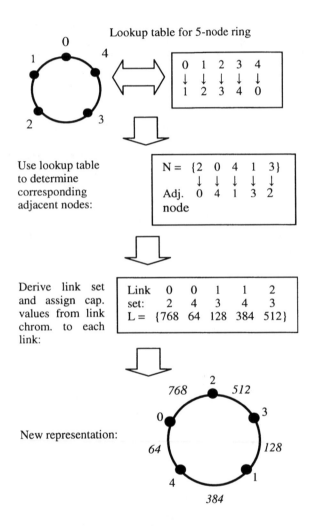

Use lookup table to determine corresponding adjacent nodes:

Derive link set and assign cap. values from link chrom. to each link:

New representation:

Fig. 3. Decoding the node and link chromosomes into a usable representation

By using the lookup table we can deduce that node 0 is connected to node 2 and node 4, node 1 is connected to node 3 and node 4 and node 2 is connected to node 3. The links connecting these nodes are assigned the capacity values given in the link chromosome. This method of determining the network connectivity can be applied to other standard topologies of varying sizes such as hypercubes, cube-connected cycles and torii.

This can be achieved by adjusting the chromosome sizes and lookup tables to suit the problem in hand.

4   Genetic Operators

Standard two-point crossover and three different mutation operators were applied to the arbitrary mesh chromosomes. The first mutation operator, alters a chosen link in the following way. If the link capacity is zero, an arbitrary link capacity of between 64 kbits/sec and 1920 kbits/sec is assigned to it. If the link has a value other than zero then it is assigned a value of zero. The second mutation operator increases or decreases a link by a discrete amount. If for example the link has a value of 384 kbits/sec then it is assigned a value of 256 kbits/sec or 512 kbits/sec at an equal probability. However, if a link has a capacity of 64 kbit/sec, the lowest available capacity, its value is incremented to 128 kbit/sec. Similarly, a link with a capacity of 1920 kbit/sec, the highest available capacity, is decremented to 1536 kbit/sec. The third mutation operator selects two links and swaps their capacity values. [6-8, 10] have previously used these mutation operators.

For the torus and ring networks Partially Matched Crossover (PMX) was applied to the node chromosomes. A full description of the PMX method can be found in [5]. A simple operator that swaps an arbitrary pair nodes is used as the mutation operator. The link chromosomes use the same genetic operators as those used for the arbitrary mesh chromosomes. The link chromosome size is equal to the number of actual links as opposed to the number of potential links. Since the number of actual links in a specific topology is known, a link chromosome contains no zero capacity values. The 'Roulette Wheel' method was the mechanism used to select new, fitter generations. The optimising parameter used by the objective fitness function was overall network cost. The lower the network cost, the higher the fitness value. Population sizes of 100 chromosomes were used and the crossover rate for all network types was 60%. The mutation probability for each separate mutation operator was set at 0.5 %.

5   Methodology and Evaluation

Simulation results for networks of sizes 8, 16 and 32 nodes were obtained. The emphasis was on the design of wide-area backbone networks and each node represents a major European city. The GA was run 10 times per network and a single GA run consisted of 100 generations. The best result obtained by the GA was stored every 10 generations. In all simulations a uniform traffic matrix was used. The total throughput requirement in all problems was 2480 kbps. The overall network cost was the sum total of all node and link costs. The link costs were extracted from the Tarifica™ manual [4] and are based on monthly connection and rental charges for using private digital network facilities. For example, when setting up a bi-directional link between the UK and Germany, British Telecom and Deutsche Telekom will both charge an amount for connecting and renting this link. The cost of a bi-directional link is the sum of these two amounts. A lookup table containing tariff values for 9 different link sizes in 32 different European countries was compiled giving a total of 9216 different values.

The routing algorithm used as part of the network evaluation is based on Nortel's connection-oriented routing system (PORS) and is essentially a modified version of Dijkstra's shortest path algorithm [1, 2]. The link 'lengths' used in the determination of the shortest path can be metrics based on either the available link bandwidth, link cost, link delay or simply the number of hops. Link delay metrics were used in the simulations such that the path chosen for a given traffic requirement between two nodes is that which has the least overall delay. If any of the network links have insufficient spare capacity with which to carry the present traffic requirement, then these links are 'repaired' by increasing their sizes to their next discrete values. This ensures to a reasonable degree that the GA-generated network topology is able to carry the traffic using routing methods used in industry.

The evaluation procedure also applies a robustness check to ensure that each network can survive single link failures. Each link is removed in turn and the traffic is run using the Dijkstra-based routing scheme described above. If there is insufficient spare capacity in any of the links to carry the current traffic requirement, the network is corrected in the manner described previously. This is repeated for all network links so that the network can survive any single link failure. To perform robustness checking on a 32-node torus network with 64 links requires performing 64 separate traffic simulations; one traffic simulation for every severed link. One traffic simulation involves routing 32(32-1)/2 = 496 traffic requirements for all source-destination node pairs. Thus, for a population size of 100 networks a total of 100 x 64 x 496 = 3,174,400 separate routings are required for each generation. Applying the robustness check is therefore computationally expensive, and for this reason the number of generations was limited to 100.

6 Results

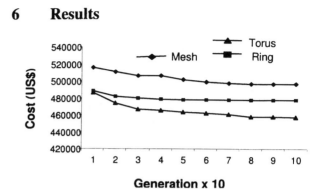

Fig. 4. Best overall results for 8-node networks

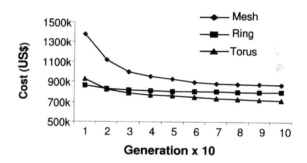

Fig. 5. Best overall results for 16-node networks

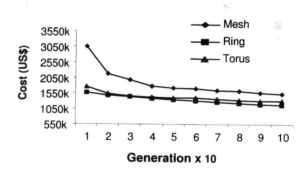

Fig. 6. Best overall results for 32-node networks

The graphs in Figures 4-6 show the best results obtained by the GA averaged over 10 runs, for each type of network. Each size network had the same total traffic requirement of 2480 kbit/sec. It can be seen that the torus and ring topologies both outperform the mesh topology for the first 100 generations, with the torus giving the best results overall in the 8-node and 16-node networks. The ring topology gave the best overall result for the 32-node networks.

Table 2. Network statistics obtained from the best overall computational results

	Cost	ALU	TNC	TNF	ALC	ALF	NL
n=8							
R	476	59	12.6	7.3	1584	924	8
T	451	50	8.57	4.5	714	381	12
M	464	50	10.1	5.1	1011	519	10
n=16							
R	783	55	24.9	13.6	1560	850	16
T	697	53	9.2	5.8	310	183	32
M	880	30	17.5	6.1	604	210	29
n=32							
R	1011	52	53.7	27.2	1680	852	32
T	1218	36	10.2	26.1	409	160	64
M	1474	34	26.8	8.1	478	144	56

The statistics shown in Table 2 show the best overall results obtained by the GA for each network. The parameters that are considered are as follows: cost (US$ x 1000), ALU (% average link utilisation), TNC (total network capacity in Mbps), TNF (total network flow in Mbps), ALC (average link capacity in kbps), ALF (average link flow in kbps) and NL (number of links). From other results obtained using the GA, it was seen that if the total traffic requirement was kept sufficiently low, the ring topology always gave the best overall results for both robust and non-robust networks. At low traffic levels there is enough bandwidth in the ring network links to accommodate the traffic even under link failure conditions. At these traffic levels there would be no additional benefit in increasing the number of links. However, when traffic levels are increased and the links become increasingly utilised, ring networks do not perform so well. This is because traffic in ring networks has to travel a larger number of hops on average than in more highly connected networks. This places a greater burden on a higher number of links, thus raising the overall average link flow. Despite this effect, the GA optimising mesh topologies was still not able to find better solutions within a reasonable time, due to the very large solution space.

7 Conclusions

This particular type of optimisation problem is made more difficult by the fact that the link sizes are quantised. Using discrete instead of continuous link sizes produces discontinuities in the fitness function, as the network traffic changes. For example, the network cost may remain constant for a number of traffic level increments and then increase sharply at the next slight increase in traffic. This causes the search space to become much less well behaved and prohibits the use of traditional optimisation techniques.

The torus has a number of advantages over the mesh and ring topologies, especially when we are concerning ourselves with robust networks. It's network connectivity, i.e., the number of nodes or links that must fail to partition the network, is higher than a ring, and unlike the arbitrary mesh topologies, is guaranteed. A problem for simple ring networks is their large network diameter. Network diameter is the maximum number of links that must be traversed to send a traffic requirement to any node along the shortest path. The torus has the advantage of having a relatively low network diameter (again guaranteed) so that the time taken to send traffic to distant nodes is shorter. In general the average path length for the torus will be much shorter than for a ring, so that more efficient use of the available bandwidth can be made. In addition, the bisection width for a torus is much higher than a ring. The bisection width is the minimum number of links that must be broken to partition the network into two separate halves, equal to within one node. Breaking just two links can partition a ring topology. A torus will require the failure of a far higher number of links to partition the network. The results also show that torii can be implemented at a lower cost than rings.

The very large solution space that a mesh topology presents us with is very difficult to search within a reasonable amount of time. When a full search of the solution space is infeasible an optimisation process must select which subset of solutions to evaluate. Consequently better solutions may be obtained in the limited time available by searching a subspace that contains a large number of good solutions but is not guaranteed to hold the optimal solution. In network topology design this can be achieved by specifying a particular type of topology that limits the search to a much smaller space, but is guaranteed to have certain useful properties.

It may be argued that if the network has a sufficient number of nodes there is always a point at which it is advantageous to search smaller spaces. The results show that a limited search is advantageous even for modestly sized design problems of 8 to 32 nodes. Optimising pre-specified networks has the distinct advantage of being able to guarantee a number of desirable topological properties, such as the degree, bisection width and diameter, which cannot be enforced in a mesh network. Only by manipulating the fitness function can we encourage, but not guarantee these qualities in a mesh network.

References

[1] Kershenbaum, A.: Telecommunications Network Design Algorithms: McGraw-Hill 1993.

[2] Standish, T.A.: Data Structures, Algorithms and Software Principles in C: Addison-Wesley 1995.

[3] Tanenbaum, A.S.: Computer Networks, 3rd ed.: Prentice/Hall Int., Inc. 1996.

[4] Tarifica Europe Manual 1996.

[5] Davis, L.: Job Shop Scheduling with Genetic Algorithms. Proc. Int. Conf. Genetic Algorithms and their Apps, pp.136-140, 1985.

[6] Davis, L., Orvosh, D., Cox, A., Qiu, Y.: A Genetic Algorithm for Survivable Network Design. Proc. Fifth Int. Conf. on Genetic Algorithms, pp. 405-415, 1993.

[7] Davis, L., Coombs, S.: Genetic Algorithms and Communication Link Speed Design: Theoretical Considerations. Proc. Second Int. Conf. on Genetic Algorithms, pp. 252-256, 1987.

[8] Davis, L., Coombs, S.: Optimizing Network Link Sizes with Genetic Algorithms. Modelling and Simulation Methodology, pp. 317-331, Elsevier Science Publishers, 1989.

[9] Hewitt, J., Soper, A., McKenzie, S.: CHARLEY: A Genetic Algorithm for the Design of Mesh Networks. Genetic Algorithms in Eng. Systems: Innovations and Applications, Conf. Pub. No. 414, 12-14 September 1995.

[10] Kapsalis, A, Rayward-Smith, V.J., Smith, G.D.: Using Genetic Algorithms to Solve the Radio Link Frequency Assignment Problem. Proc. of the Second Int. Conf. on Artificial Neural Networks and Genetic Algorithms, pp.37-40, 1995.

[11] Ko, K.T., Tang, K.S., Chan, C.Y., Man, K.F.: Packet Switched Communication Network Designs Using GA. Genetic Algorithms in Eng. Systems: Innovations and Applications, Conf. Pub. No. 446, 2-4 September 1997.

[12] Pierre, S., Legault, G.: An Evolutionary Approach for Configuring Economical Packet Switched Computer Networks. Artificial Intelligence in Engineering, vol. 10, pp. 127-134, 1996.

[13] Sinclair, M.C.: Minimum Cost Topology Optimisation of the COST 239 European Optical Network. Proc. of the Second Int. Conf. on Artificial Neural Networks and Genetic Algorithms, Alés, France, pp. 26-29, April 1995.

[14] Sinclair, M.C.: NOMaD: Applying a Genetic Algorithm/Heuristic Hybrid Approach to Optical Network Topology Design. Proc. of the IEE Colloquium on Multiwavelength Optical Networks: Devices, Systems and Network Implementations, London, June 1998.

[15] White, A.R.P.: Genetic Algorithms and Network Design, BNR (Nortel) Technical Report, S&SE Tech. Rep. No. 20832 (S&SE/GA/tr/20865/1.0), 24 July 1996.

The Effect of Degenerate Coding on Genetic Algorithms

Colin R Reeves and Ping Dai*
CTAC Computational Intelligence Group
School of Mathematical and Information Sciences
Coventry University
UK
Email: C.Reeves@coventry.ac.uk

Abstract

The choice of a suitable coding for the application of a GA to optimization problems is often critical to the effectiveness of the GA in finding good solutions. The phenomenon of *degeneracy* has been pointed out by Radcliffe as one that may be detrimental to GA performance [1, 2]. This problem is characteristic of applications of GAs to such cases as the travelling salesman problem (TSP), neural network design and training, and system identification in control. Previous experimental work by Hancock [3] found that the problem in practice appears less detrimental than expected. In this paper we examine a simple probability model for the occurrence of degenerate crossover that explains why degeneracy causes difficulties in some cases and not others. Experimental results in the case of system identification verify these expectations.

1 Introduction

In many combinatorial and other optimization problems, there is an inherent difficulty in finding a suitable coding for a genetic algorithm (GA)—the presence of what is variously described in the literature as redundancy or competing conventions [4], the permutation problem [3]. Radcliffe and Surry, in a carefully argued analysis [5] prefer the term *degeneracy*, and that is the term we shall use in this paper. The problem arises when the string that encodes the solution can be divided into a number of independent *units*.

For example, there is widespread interest in

designing and training neural nets using GAs, where typically each neuron (unit) is encoded by means of a string of binary or other symbols denoting its connectivity and/or weights. The problem is that the labelling of the units is arbitrary, so that a given phenotype can be represented by many genotypes. To put it mathematically, the genotype-phenotype mapping is not injective, and this can lead to difficulties in the application of genetic recombination operators. This example has been investigated by Hancock [3], who found (somewhat to his surprise) that the difficulties were not as great as had been imagined.

Another example is the problem of modelling a high order plant in applications of control theory (the system identification problem). As an approximation, it is often convenient to use a combination of lower (usually 2nd) order components (units), owing to the relatively simple computation and realisation in hardware of the approximate model.

In this paper the investigation of degenerate coding is focused on the problem of system identification and approximation in control theory, although reference will be made to other areas. After a brief illustration of the problem, a probability model is used to explain Hancock's findings, and is further elucidated with some experimental work involving a system identification problem.

2 Degeneracy: An Example

Consider the problem of modelling a 6th order system specified by the $z-$transform equation

$$H(z) = \frac{\theta_0 + \theta_1 z^{-1} + ... + \theta_6 z^{-6}}{\phi_0 + \phi_1 z^{-1} + ... + \phi_6 z^{-6}}$$

---

*Ping Dai is now at the Software Technology Research Laboratory, De Montfort University, Leicester, UK using a combination of 2nd order components. In any system model, there is always the problem of ensuring that the 'best' set of coefficients used actually corresponds to a stable system; checking this can involve a lot of effort for high-order models. By using 2nd order components, it is relatively easy to check that a given set of coefficients does represent a stable system. It may also be somewhat easier to realize the model in hardware using lower-order components.

The problems associated with degeneracy arise regardless of whether the system is a parallel and or a cascade form. In this case, we assume the case of a parallel design. Figure 1 illustrates the problem: two identical solutions to the problem can be specified by two different encodings.

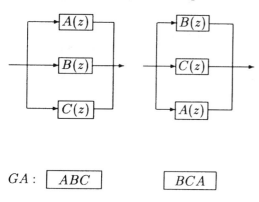

Fig. 1. Two parallel 3-component models of a 6th order system

Here $A(z), B(z), C(z)$ represent the z−transforms of the respective 2nd order components ($H(z) = A(z) + B(z) + C(z)$), specified by a set of coefficients. It is clear that for a given set of coefficients, there are $3! = 6$ alternative ways of specifying the same model. However, using the 'standard' concatenated binary coding for a GA, it would appear that each of these representations is different—a non-injective mapping. In general, with n components there is an $n!$-fold redundancy involved in adopting the concatenated binary coding. In Radcliffe's terminology [1], the two strings ABC and BCA of Figure 1 are members of the same forma—both chromosomes share the characteristics represented by A, B and C.

When these two chromosomes are recombined by crossover, the results would be something like ACA or BBC, in which ACA lost element B and BBC lost element A. This type of situation was characterized by Radcliffe [1] as showing a lack of 'respect' for the common traits of the parents, and he has argued

that this problem is likely to lead to poor performance of a GA which fails to take this problem into account.

In order to avoid this effect, an obvious solution would be first to search the two parent strings before crossover in order to identify common elements and then to re-order the string. For example, if each coefficient in the 2nd-order filter is specified by 3 bits, the following situation might arise:

	A	B	C
chrom1	101101	******	******
chrom2	******	******	101101

(Here the * symbols are simply arbitrary characters.) It can be seen that part A of chrom1 is same as the part C of chrom2 (101101), i.e., the parents have common traits which happen to be in different positions. They are members of the same 'forma', in Radcliffe's terminology [1]. However, if crossover were to be performed directly, the common traits of the parents (101101) might be lost in at least one offspring. This could lead the GA in the wrong direction. Therefore, having found a partial matching in the two chromosomes, the string order in one chromosome is re-arranged so that it matches the other one. In this case, the contents of part A could be exchanged with part C in chrom2, in order to match part A of chrom1. The result is as follows:

	A	B	C
chrom1	101101	******	******
chrom2	101101	******	******

Ordinary GA operators (crossover, mutation, etc.) can then be applied to the chromosomes chrom1 and chrom2, but common traits of the parents (101101) will be inherited by both of their offspring. That is, the modification helps the GA to meet Radcliffe's design principle of 'respect', since crossing two instances of any forma should produce another instance of that forma; i.e. any properties that parents share, and that are capable of expression in terms of the formae, should be passed on to all their children. This is effectively the modification described by Hancock [3].

3 A Probability Model

Hancock's experience was that the modification improved the GA performance in matching target neural net designs slightly, but hardly significantly—a result which was not altogether expected. We now develop a simple probability model that will help to explain this result.

It is obvious that the chance that the re-ordering will be found to be necessary will depend on the number of units making up the string (the higher, the more likely it is to occur), and on the number of bits used for each unit (the more bits, the less likely it is to occur). We will use the 3-filter example above as an illustration.

Suppose each chromosome in the population has three 2nd order filter components (A, B, C). Each unit (set of filter coefficients) is encoded using l-bit binary strings. The randomly chosen $Chrom1$ and $Chrom2$ are expressed as A_1, B_1, C_1 and A_2, B_2, C_2. The probability is $P\{E\}$ that any element A_1, B_1, C_1 in $Chrom1$ matches a different position of any element A_2, B_2, C_2 in $Chrom2$, as described in Table 1:

	A_1	B_1	C_1	probability
E_1	B_2			$P_1[A_1 = B_2] = 2^{-l}$
E_2	C_2			$P_2[A_1 = C_2] = 2^{-l}$
E_3		A_2		$P_3[B_1 = A_2] = 2^{-l}$
E_4		C_2		$P_4[B_1 = C_2] = 2^{-l}$
E_5			A_2	$P_5[C_1 = A_2] = 2^{-l}$
E_6			B_2	$P_6[C_1 = B_2] = 2^{-l}$

Table 1. Cases where re-ordering is necessary

Now each of these events has the same probability $P[E] = \beta = 2^{-l}$, and the probability that *at least one* match is found is thus

$$P_{match} = 1 - (1 - \beta)^6$$

if we assume that the events are all independent. In the general case, where there are n units (filters, neurons or whatever), the probability is obviously

$$P_{match} = 1 - (1 - \beta)^{n(n-1)}.$$

This probability can easily be calculated, and some examples are shown in Table 2, from which it can be seen that re-ordering is only likely to be necessary if very few bits are used to encode each unit, unless there are many units.

Thus, we would expect that re-ordering prior to applying recombination operators is likely to be effective in only a few cases. Of course, this calculation considers only the chance of finding matching units in 2 randomly chosen strings; the actual situation depends on other properties such as string

	No. of units (n)		
No. of bits (l)	3	5	10
6	0.090	0.270	0.758
8	0.023	0.075	0.297
12	0.001	0.005	0.022
16	0.0001	0.0003	0.0014

Table 2. Probability (P_{match}) that randomly-chosen strings have matching units in different parts of the string.

fitnesses and the population size. However, fitness will be problem-specific, and population size effects are unlikely to be significant unless the population is very large. String length and the number of units 8 generally be the the most important factors affecting the probability of re-ordering. Finally, while we realize that this argument strictly applies only in an initial random population, we would expect it to apply *a fortiori* for subsequent populations, when dominant strings have had a chance to exert their preference for a particular 'labelling convention', thus effectively reducing the value of β.

4 Experimental Results

As an example, we applied a GA to solve the problem presented originally by [6], in which a 6th order IIR filter was modelled by a bank of three 2nd order filters. The transfer function of their 6th order IIR filter used the following coefficients:

$\theta_0 = 3.0$	$\phi_0 = 1.0$
$\theta_1 = -7.5822$	$\phi_1 = -3.7911$
$\theta_2 = 7.9203$	$\phi_2 = 6.3959$
$\theta_3 = -3.9101$	$\phi_3 = -6.0223$
$\theta_4 = 0.7625$	$\phi_4 = 3.315$
$\theta_5 = 0$	$\phi_5 = -0.997$
$\theta_6 = 0$	$\phi_6 = 0.1248$

We used a parallel design, as in Figure 1, with the aim of approximating the behaviour of the full 6th order filter as closely as possible; i.e. the model was

$$H_{model}(z) = H_1(z) + H_2(z) + H_3(z)$$

where

$$H_i(z) = \frac{1}{1 - a_{1i}z^{-1} - a_{2i}z^{-2}} \quad \forall i.$$

In using a genetic algorithm, each string is thus composed of 3 sets of 2 coefficients represented in binary form. In our experiments each coefficient was coded in either 3, 4, 5, 6, 8 or 10 bits. The GA used the incremental (or steady-state) replacement strategy where each generation creates one new string, which is inserted in place of one of the worst $\alpha\%$ of the current population. Simple one-point crossover was used, and the various numerical parameters used were as shown in Table 3.

max. evaluations	500
population. size	50
mutation probability	0.005
deletion percentile (α)	10%
crossover probability	1.0

Table 3. The parameters of a GA

The fitness of each string was evaluated in this case by means of a simulation. Each system was simulated for T time-steps using a pseudo-random binary input signal. The first T_0 time-steps were ignored, in order to allow the effect of the initial conditions to 'wash out' of the system. For the remainder, the fitness was calculated by the mean of the squared errors between the output of the system (the fixed 6th order IIR filter y_S) and the output of the model (the three parallel 2nd order IIR filter components y_M):

$$MSE = \frac{\sum_{t=T_0+1}^{T}(y_S - y_M)^2}{T - T_0}.$$

Initially, 3 bits were chosen for each coefficient and 150 experiments were run, with the results shown in Table 4. Here OGA is the original GA, with no re-ordering; MGA is the same GA, but with re-ordering of units when degeneracy is discovered. The table also records the number of times a match was found during the course of the run.

	Matches	Better result	Average MSE
MGA	215	77%	0.393
OGA	N/A	23%	0.609

Table 4. Comparison between MGA and OGA using 3 bits (i.e. units of 6 bits). The results represent the averages of 150 runs, each limited to 500 function evaluations.

On average 215 matches were found during each experiment (out of a total of 500 evaluations). We expect most of this to occur early in the search—eventually one or more strings will start to dominate, and impose a particular 'labelling convention' on the rest of the population. In this case, MGA

found a better MSE on average than OGA, and 77% of the final MSE values were better.

In further simulations 4, 5, 6, 8, and 10 bits were used for each coefficient, respectively. Each was run for 30 experiments with the same initial random seeds. The results were as shown in Table 5. On the basis of the probability model developed above, we would expect that as more bits are used, the influence of re-ordering will be lessened. This indeed is what happens: fewer matches are found, and once more than 8 bits are used, MGA and OGA become indistinguishable. As it is also clear (again as expected) that using more bits gives scope for a better approximation, in practice the effort of checking for matching units and re-ordering is likely to be wasted.

4.1 System stability

When using a GA in system identification, the question of the stability of the model also needs to be considered. It is well-known that for stability, the coefficients of each 2nd order IIR filter must lie inside in a stability triangle bounded by

$$0 \leq |a_2| < 1; \qquad |a_1| \leq 1 - a_2.$$

Unfortunately, crossover and mutation may generate coefficients which do not satisfy these conditions. Thus, at each iteration, each component was checked, and if the conditions were not satisfied, the coefficients were replaced by 'shrinking' them in a suitable manner. Clearly there is more than one way to accomplish this: in this case a coefficient was randomly perturbed until the conditions were satisfied. This is not difficult to realize in the GA. A further alternative is to encode the poles of the filter directly in (r, θ) form, which eliminates the need to check for stability. Further experiments were carried out—not reported here in order to comply with space restrictions. However, if few bits are used, the degree of resolution in terms of the search is too coarse for accuracy, and encoding the numerical values of the coefficients turns out to be better. If more bits are used, it certainly seems better to use poles directly, but this has no relevance to the degeneracy issue. Detailed results and a full discussion can be found in [7].

5 Other Applications

It could be argued that asking for complete agreement between the units of two parent strings is too

Bits/coeff	3	4	5	6	8	10
Bits/unit	6	8	10	12	16	20
MGA best	22	17	11	11	0	1
OGA best	4	12	11	4	1	0
Identical	4	1	8	15	29	29
Av. no. matches	226	83	26	13	0.7	0.033
Av. MSE_MGA	0.614	0.279	0.490	0.230	0.196	0.141
Av. MSE_OGA	0.862	0.348	0.507	0.273	0.196	0.141

Table 5. Comparison between MGA and OGA using different numbers of bits. Averages are based on 30 runs.

stringent a requirement. In practice, we might accept the k most significant bits being the same. In such cases the phenotypic distance (on a particular dimension) between the chromosomes is fairly small, and we could reasonably argue that to a first approximation, the units represent the same values. (There is also a further assumption about the smoothness of the phenotype-fitness mapping here; but if this is notably unsmooth, then all bets are off). The above analysis stands, except that now $\beta = 2^{-k}$. Table 6 shows the effect of requiring the most significant $(l-2)$ bits to be the same. Although the effect is large for small l, the effect of using more bits is still such that we would not expect re-ordering to have a significant influence.

	No. of units (n)		
No. of bits (l)	3	5	10
6	0.321	0.725	0.997
8	0.090	0.270	0.758
12	0.006	0.019	0.084
16	0.0004	0.0012	0.0055

Table 6. Probability (P_{match}) that re-ordering occurs when only the k most significant bits are identical, for $k = l - 2$. The high probability for $l = 6$ is not surprising, since there are then only two most significant bits for each coefficient.

Although the effect is large for small l, the effect of using a larger resolution is still such that we would not expect re-ordering to have a significant influence.

5.1 Neural net connectivity

Another area where the degeneracy effect arises is in neural net applications [2, 3]. For instance, a common idea is to use a string to specify the connectivity of the input-to-hidden layer part of a neural net (a 1

meaning the link is present, a 0 that it is absent). An analysis of this type of encoding shows that degeneracy could cause rather more trouble than it did in system identification. Again, if complete agreement is required between different units of two parents, the likelihood of re-ordering having to be applied will be small unless the net has very few inputs. But if we ask for *at least* k connections to be the same, the chance of re-ordering being needed rises quite considerably, since now we have

$$\beta = \sum_{j=k}^{l} \binom{l}{j} 2^{-l}$$

which is typically rather larger than in the cases discussed above. Table 7 shows the effect of asking for at least k bits to be the same in each unit, for $k = l - 2$.

	No. of units (n)		
No. of bits (l)	3	5	10
6	0.920	1.000	1.000
8	0.608	0.956	1.000
12	0.110	0.323	0.827
16	0.012	0.041	0.172

Table 7. Probability (P_{match}) that re-ordering occurs when at least k bits are identical in the case of neural net connectivity, for $k = l - 2$. The result for $l = 6$ is to be expected, since the condition supposed is then highly likely to arise in any unit.

Again, we are making some assumptions about the importance of phenotypic distances in the overall picture. If we leave those aside, on the evidence of this table the chance of getting differently-labelled neurons with *similar* (but not identical) connectivity in an initial random population can be quite high if there are relatively few inputs. However, in such cases it would often be possible to enumerate the set of possible inputs anyway, so that it is unlikely a GA would be necessary. There may be some intermediate cases where the number of inputs is not small enough for enumeration to be efficient, and not large enough to rule out the occurrence of degeneracy, but overall, such problems are not likely to be a major area of GA application.

6   Conclusions

We have examined the problem of degeneracy in the coding of neural nets by means of some simple probability calculations, and shown that although it is theoretically a constraint on efficient implementation of genetic search, in practice it is unlikely to have a significant impact. We have confirmed this experimentally in the case of a system identification problem, and extended the calculation to a more relaxed condition on the nature of the chromosome units. As a further example, we have also considered the case of neural net connectivity, which potentially represents a greater problem than the case of system identification. Further theoretical and experimental work is being carried out to ascertain whether, and in what circumstances (e.g. population sizes) the latter case may cause difficulties.

References

[1] N.J.Radcliffe (1991) Equivalence class analysis of genetic algorithms. *Complex Systems*, **5**, 183-205.

[2] N.J.Radcliffe (1993) Genetic set recombination and its application to neural network topology optimisation. *Neural Computing and Applications*, **1**, 67-90.

[3] P.J.B.Hancock (1992) Genetic algorithms and permutation problems: A comparison of recombination operators for neural net structure specification. In L.D.Whitley and J.D.Schaffer (Eds.) (1992) *Proceedings of · COGANN-92: International Workshop on Combinations of Genetic Algorithms and Neural Networks*, L.D.Whitley and J.D.Schaffer (eds.), IEEE Computer Society Press, Los Alamitos, CA, 108-122.

[4] J.D.Schaffer, D.Whitley and L.J.Eshelman (1992) Combinations of genetic algorithms and neural networks: A survey of the state of the art. In L.D.Whitley and J.D.Schaffer (Eds.) (1992) *Proceedings of COGANN-92: International Workshop on Combinations of Genetic Algorithms and Neural Networks*, IEEE Computer Society Press, Los Alamitos, CA, 1-37.

[5] N.J.Radcliffe and P.Surry (1995) Formae and the variance of fitness. In D.Whitley and M.Vose (1995) *Foundations of Genetic Algorithms 3*, Morgan Kaufmann, San Mateo, CA, 51-72.

[6] R.Nambiar and P.Mars (1992) Genetic algorithms for adaptive digital filtering. *Proc. IEE Colloquium on Genetic Algorithms for Control and Systems Engineering.* Digest No.1992/106, IEE, London.

[7] Ping Dai (1997) *Hybrid Genetic Algorithms with Application to System Identification and Control.* MPhil dissertation, Control Theory and Applications Centre, Coventry University, UK.

Genetic algorithms for the identification of the generalised Erlang laws parameters used in systems dependability studies

Lamine Ngom[1], Claude Baron[1], André Cabarbaye[2],
Jean-Claude Geffroy[1], Linda Tomasini[2]

[1]INSA Toulouse - DGEI/LESIA/SFS. Complexe Scientifique Rangueil. 31400 Toulouse - France
{ngom, geffroy, baron}@dge.insa-tlse.fr
[2]CNES Toulouse - CT/AQ/SE/SF. 18, Avenue Edouard Bélin. 31401 Toulouse - France
{Andre.Cabarbaye, Linda.Tomasini}@cst.cnes.fr

Abstract

In systems dependability modelling, the absence of a fine knowledge on the failure dynamics for certain systems and on the multiple interactions which exist between the various subsystems, and also the difficulty to validly use some simplifying assumptions require to resort with the exploitation of experience feedback. In addition, one has approximate models and, the problem is then to find the parameters of these models which satisfy "as well as possible" the observed feedback data, according to the principle of maximum of probability or minimum of least squares (it depends on the nature of the obtained data). Certain identification heuristics were hitherto used, but they showed their limits when, for instance, the relief of the function to be optimised presents many local valleys. These difficulties led us to consider an approach totally different where the transition rules can allow to avoid local cavities. For that, we studied a certain number of operational research techniques and finally chose a resolution by genetic algorithms. Their major advantage is that they operate the search of an optimum starting from a population and not from only one single point, allowing thus a parallel search, effective on the whole solutions space. After a thorough presentation of the considered applicability and the obtained results in this study, we underline in this communication the observed advantages, difficulties and limits compared to more traditional techniques for the parametric identification.

1. Introduction

This communication proposes to show the interest of the use of genetic algorithms in dependability studies, particularly for the parametric identification of systems reliability curves.

Regarded as the science of failures and breakdowns, the reliability has an importance of very first plan like quality insurance and condition of systems acceptance. So it is a variable which must be supervised, analysed and controlled in all the stages of a system development. The preliminary analysis which is consequently necessary allows, by the mean of more or less well-known methods today (Failure Mode and Effects Analysis, Hazard analysis, Fault Tree Method, Stochastic Petri Nets Method, Markovian Analysis, Block Diagram Method...), to estimate a future reliability starting from considerations on the systems design and its components reliability. However, the reliability modelling is not always directly possible and designers have thus to work with approximated models. Among the various methods of systems reliability estimation analysis, one distinguishes the use of Markovian processes, well-known today, which make it possible to model the behaviour of many systems. The powerful computational tool that they constitute provides, with the help of some restrictions which will be clarified in section 1, precise results dependability (contrary to the Monte-Carlo simulation). In this first section, we will also give the definition of a Markovian process, and will point out the motivations that caused the development of the fictitious states method.

The following part tackle to be strictly accurate the problem of parametric identification by genetic algorithms. And finally the obtained results are presented in the last section. We underline here the observed advantages, difficulties and limits in comparison with more traditional techniques of parametric identification.

2. Markovian processes and fictitious states method

2.1. Markovian processes

A fundamental property for the quantitative evaluation of stochastic models is the markovian assumption, which allows, if checked, to considerably simplify the calculations leading to the probability vectors. A markovian process represents a stochastic process at continuous time and discrete states space for which the evolution of the random variables $X(t)$ for $t > t_n$ depends only on $X(t_n)$, what still returns to saying that the current state summarizes all the past of the system. The description of a markovian process passes then by the characterization of transition probability of a state towards another, which himself expresses by the transition rate defined in the following way :

$$\lambda_{ij}(t) = \lim_{\substack{dt \to 0}} \frac{1}{dt} P\left[X(t+dt) = j \, / \, X(t) = i\right]$$

(1)

Another simplifying assumption lies in the concept of homogeneity of a markovian process. For such a process, the transition probability from a state I to the state J between t1 and t2 depends only on the time interval between t1 and t2 and not on the individual values t1 and t2. The previously definite transition rate λ_{ij} is then constant.

These two probabilistic assumptions (homogeneous and markovian stochastic process) facilitate largely the obtaining of vectors of probability because of the simplicity of the Chapman-Kolmogorov equation which results from it:

$$P'(t) = P(t) * \Lambda \qquad (2)$$

where Λ is the infinitesimal generator (constant transition rates matrix)

Well-known techniques allow to solve this equation: spectral analysis which proceeds by searching for the eigen values and then the exponentiation of the transition matrix, Laplace transformation, and so on ...

These simplifying assumptions are acceptable to calculate the reliability of certain equipment including electronic components in particular. However, such assumptions cannot be considered anymore for other types of equipment such as servomechanisms, whose components, subjected to wear, have increasing failure rates. Moreover, if for certain components, the failure rate can be constant, it is never the same for the repair rates (the repair duration are generally distributed according to Normal laws).

2.2.   The fictitious states method

Because of the difficulties we mentioned above, the fictitious states method has been elaborated in order to be reduced to the homogeneous markovian case, at the price of certain modifications made to the model. This method consists in replacing in a markovian model an unspecified transition (at non constant transition rate) between two states by a combination of constant rates transitions between these two states and a certain number of fictitious states.

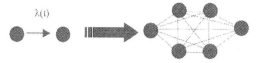

Figure 1. Fictitious graph

Thus, a complex transition such as that relating to a wearing law can be modelled at the price of some additional fictitious states.

It has be shown that any type of transition can be modelled by the use of fictitious states, and that the combination of transitions between fictitious states can be limited to the hereafter presented Cox model :

Figure 2. Cox model

The statistical law which results from it is known as of Cox. If the transition rates from the direct chain are identical, the law is known as generalised Erlang law. If, moreover, the set of transitions is limited to the direct chain, the resulting law is a simple Erlang law of parameter k (k = number of fictitious states between the states 1 and 2) corresponding to the convolution of k exponential laws. Near to a log-normal law, the simple Erlang law is often used to model repair rates. Its characteristics are the following ones:

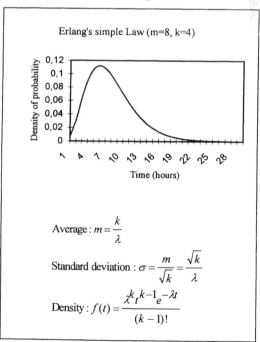

Figure 3. Characteristics of the simple Erlang law

216

3. Parametric identification of a generalised Erlang law and arising problems

One wishes to model the behaviour of a system starting from its reliability curve. For that, one directly seeks a matrix of Markov giving an approximation "as well as possible" of the systems observed dynamics, this matrix being related to several variable parameters and of given size. The markovian model which we considered here is a model of generalised Erlang law type, because this law makes it possible to model many distributions.

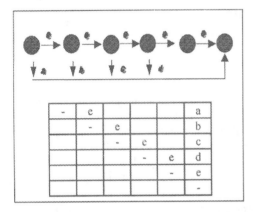

Figure 4. Schematic diagram of a generalised Erlang law.

The parametric optimisation which is carried out consists of a minimisation (by the difference of least squares) of the quadratic error between the observed reliability (by experience feedback) and that obtained for a given configuration of the several parameters (a, b, c...).

The hitherto used operational search algorithm for optimising a function of several variables consisted in a descent of gradient reiterated until the error is considered to be sufficiently weak. This algorithm can be compared to a walker in the fog who seeks the lowest point of the valley. Indeed, on the basis of an initial given point on the surface, one calculates the gradient at this point and one takes the direction of the greatest slope with a step which evolves according to a "well" established strategy [3].

In order to accelerate the convergence of the search process, the evolution principle of the step is the following: one evaluates initially the slopes according to each direction (i.e. according to the various parameters), then one moves in the direction of greatest slope with a step which is multiplied by 2 if the new obtained point is better than the precedent, and divided by 2 otherwise (to avoid a possible divergence). The used technique can be described by the following algorithm:

Initialisation

```
Choose a tuple M0 = (a0, b0, c0...);
nbiter := 0;   /* current iteration */
Evaluate the theoretical reliability for M0;
e* := quadratic error;
step := initial_step;
```

Iterative process

```
While (e* > epsilon ) and (nbiter > nbmax) do
    nbiter := nbiter + 1;
    For each direction i do
        Evaluate Gradient_i by finite differences;
    end for;
    Modify the parameter j of M0 corresponding
    to the highest gradient:
    j := j - step;
    Evaluate the new reliability of the tuple;
    e := quadratic error;
    if e < e* then
        step := step * 2;
        e* := e;
    else
        step := step / 2;
    end if;
end while;
```

The shelf, which one meets with this algorithm, is the blocking in valleys: the search cannot progress anymore as soon as a local cavity is reached. Indeed, the characterisation of the generalised law of Erlang lets appear a prevalent effect of the first two parameters, therefore a stronger slope in these two directions. These first two parameters (a and b) play a preponderant role in the failure rate curve from. The algorithm converges rather quickly towards the bottom of the valley (for these parameters) but it does not work practically anymore in the other directions, its step having become too weak.

More refined heuristics was then to divide the step by 2 if the maximum gradient changes sign, which actually corresponds to a jump between the 2 slopes of the steepsided valley. A more effective mean could still consist in having a specific step for each direction. However, a problem of slowness of the convergence process would insofar arise in any case as the progressions are done each time according to a single direction corresponding at the highest gradient.

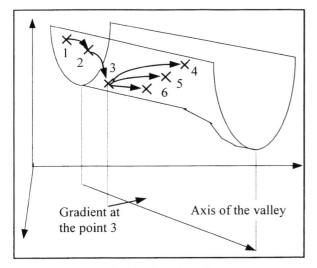

Figure 5. Blocking in valleys phenomenon

Initially, the algorithm converges very quickly in the valley (see figure 5), and then tries to move towards a better point. However, the gradient at the point 3 being slightly tilted compared to the axis of the valley, items 4, 5 and 6 are thus always worse than item 3, and the algorithm thus divides always its step by 2. And finally the process remains blocked at the point 3.

The difficulties mentioned above then led us to consider an approach totally different where the transition rules can allow avoiding the local cavities. For that, we studied a certain number of techniques of which the Taboo search method [7] and the technique of Simulated Annealing [8], to choose finally a resolution by the genetic algorithms. The major advantage of genetic algorithms is that they operate the search of a optimum starting from a population and not from only one single point, allowing thus a parallel search (either local but total and distributed on all the space of search) effective on the whole solution space (but sometimes at the price of an often non considerable resolution time according to the members coding).

Among the multiple other interests which one can find with this optimisation method, one can quote for example the fact that it does not require many assumptions, and that it directly handles the information brought by the objective function and not by its derivative or its gradient. In fact, what make the attraction of genetic algorithms, it is their capacity to accumulate some information on an unknown search space and to exploit this information to guide the future search in some promising subspaces [2,6].

The first results obtained with the genetic algorithms are already very satisfactory, and the continuation of committed works will allow us to draw up comparative analyses between the performances obtained and those before provided by the method of descent of gradient.

4. Genetic algorithm approach

A first genetic algorithm version has been used to solve the problem of identifying the parameters of generalised Erlang laws corresponding to an observed probability distribution. We discuss here the fundamental ingredients of this version which ensure a convergence of the genetic algorithm under the best conditions of speed (execution time machine) and of reliability (ability of the genetic algorithm to find the good solutions).

To accelerate the convergence of our genetic algorithm, we were interested in the principal ingredients of the genetic algorithms and on the possibility of choosing them in order to achieve our goal of performance. Also, we thought to make comparisons between real-coded genetic algorithms according to their selection, crossover and mutation used procedures.

For the choice of the selection procedure, we were satisfied with the result mentioned in [1] which stipulates that apart from the selection by lottery whose bad results quality has been proven, there is no particular recommendation relating to one or the other among the various selection techniques. As all these other sampling procedures approximately equivalent results, the influence on the good convergence of the genetic algorithm is not considerable. Consequently we chose that which seemed to us the more sympathetic and one of the easiest to program, namely the technique of stochastic sampling of the remaining part with replacement [1].

With regard to the crossover and mutation procedures, we compare them initially by measuring the impact of the choice of one or the other among them on the genetic algorithm convergence. In fact, what we tried to highlight, it is the capacity of local and total browsing of these recombination techniques whose principal goal is to introduce diversity into the population. If this capacity of browsing is obvious in the case of a binary coding of the individuals of the population, it is not the same when one places oneself within the framework of a real coding of the parameters.

For instance, in the crossover case, it is not difficult to prove that, in the real field, the used techniques do not allow to perform an effective browsing of the possible solutions space. Indeed, for the sharing of properties between individuals, all the real parameters crossover techniques under consideration give, on the basis of a pair of individuals belonging to a given interval, a pair of new individuals always belonging to the same interval. The role of the crossover is thus reduced to a simple local browsing of the individuals combination possibilities. There would be no more whereas the mutation phenomenon to allow a complete random browsing of the configurations space. Under these conditions, the empirical mutation rates generally used (about some thousandth) in genetic algorithms are some too ridiculous compared to the essential function of the mutation operator.

These first observations tend to support the thesis [9] that recommends a real genetic algorithm where the crossover is completely removed, and where the recombinations are only carried out using a mutation procedure, which uses the philosophy of, simulated annealing. One thus authorises significant jumps at the beginning of the genetic algorithm (completely blind browsing) and, at the opposite side, one controls by tightening the induced noise (targeted local browsing) when the genetic algorithm is supposed to have acquired a rather good knowledge on the ecological niches where are the required optima (i.e. when the number of generations carried out is sufficiently significant compared to the type of problem). Same manner, for the binary problems, the role of the crossover operator is dominating in the convergence process. Even in situations of great combinative complexity, the binary crossover allows a broad uniform browsing of the potential solutions space throughout generations [4].

As the comparison works between the various mutation and crossover procedures require significant execution times (one needs a hundred execution at least to have a representative percentage of the effectiveness of a procedure compared to others), it is considered a parallelisation of the genetic algorithms implemented for this comparison which would be carried out then simultaneously on several processors. This study being still at the embryonic stage of sharing the independent processes on various machines by using pvm (parallel virtual machine), more detailed results will appear, we hope for it, in a future communication. For the moment, we draw up below the elements, which we chose to constitute the nerve of our parametric optimisation genetic algorithm, with which we could obtain the results presented in the following section.

✓ Population size = 100 individuals representing each one a configuration (a, b, c, d, e) of the parameters to optimise,

✓ Crossover rate = 75 %,

✓ Mutation rate = 15% (this high rate is explained by the inefficiency of the total browsing that one obtains with the crossover in real numbers case),

✓ Maximum number of generations = 100,

✓ Maximum number of generations without improvement of the current solution = 10,

✓ Minimum value of the parameters = 0.0,

✓ Maximum value of the parameters = 0.01.

5. Experimental results

The first results we obtained with the genetic algorithm we have developed are very satisfactory. One notes that the problem of blocking does not take place anymore and the algorithm converges well towards a largely better solution within the meaning of the optimisation criterion, namely the difference of least squares. One arrives thus at a quadratic error value about 1e-02 instead of the strong values previously obtained by the method of gradient descent at single step. The example hereafter shows the very clear improvement that has been generated by the using genetic algorithms.

Time	Reliability
t=0	r=1
t=100	r=0,83
t=200	r=0,69
t=300	r=0,55
t=400	r=0,43
t=500	r=0,35

Table 1. Experience feedback example

Table 1 gives an example of experience feedback results observed for equipment. The goal is then to find the generalised Erlang model, which this distribution follows. For that, the optimisation criterion is to minimise the quadratic error defined as follow:

$$\varepsilon = \sqrt{\sum_{t=t_0}^{t=t_N} (f_t(x) - y_t)^2} \qquad (3)$$

where $f_t(x)$ is the reliability calculated at the time t for the parameters configuration x, and y_t the real observed reliability.

The fitness valuation of the individuals is performed by assigning to each of them the corresponding quadratic error. The value $f_t(x)$ is calculated by a matrix exponentiation:

$$f_t(x) = \begin{bmatrix} 1 & 0 & \cdots & 0 \end{bmatrix} \begin{bmatrix} Id + \Lambda t + \dfrac{\Lambda^2 t^2}{2!} + \dfrac{\Lambda^3 t^3}{3!} + \cdots + \dfrac{\Lambda^k t^k}{k!} \end{bmatrix} \begin{bmatrix} 1 \\ 0 \\ \vdots \\ 0 \end{bmatrix}$$

(4)

where Λ is the markovian model (generalised Erlang model) matrix given in figure 4, and the limited development order k is selected so as to obtain a good relative precision, i.e. $\left| \dfrac{f^k - f^{k-1}}{f^k} \right|$ is sufficiently weak.

Time (hours)	Reliability	Classic Approximation	G.A. Approach
t=0	r=1	1	1
t=100	r=0,83	0,746602182	0,832908
t=200	r=0,69	0,575494214	0,683025
t=300	r=0,55	0,458434552	0,551238
t=400	r=0,43	0,377100345	0,438779
t=500	r=0,35	0,31952047	0,345361
error:		*0,179383*	*0,01253*

Table 2: Comparative results

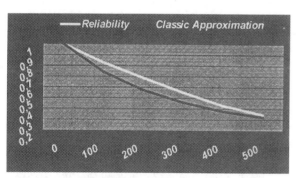

Figure 6: Classic approximation agreement curve

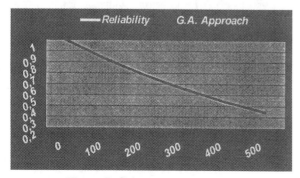

Figure 7: G.A. agreement curve

It is seen very clearly that the parameters found by genetic programming allow to obtain an almost perfect correspondence between the data of experience feedback and the values computed by the markovian model (generalised Erlang law).

6. Conclusion

We presented a genetic approach for the parametric optimisation of the models of reliability laws used in dynamic systems reliability estimation. This approach made it possible to obtain very satisfactory results compared to the step we previously used, which were based on a traditional approximation with descent of gradient. In own way of illustration of the suggested approach, we were based on a simple example, showing the advantages offered by the genetic method. This first work encouraged us then to explore the real filed of the potential use of genetic algorithms in reliability studies [5]. The applications are varied and enriching as well for quantitative evaluation as for forward-looking modelling.

References

[1]. D. E. Golberg, *Genetic Algorithms in Search, Optimisation, and Machine Learning*, Addison-Wesley, Reading (MA), 1989.

[2]. J. M. Renders, *Genetic Algorithms and Neural Networks*, Hermès Editions, Paris, 1995.

[3]. E. Lourme, *System Modelling by Markov Matrix*, Engineer degree report, ENSAE, Toulouse, 1995.

[4]. L. Ngom, C. Baron, J-C. Geffroy, *Genetic Simulation for Finite-State Machine Identification*, to appear in 32nd Annual Simulation Symposium, San Diego, USA, April 1999.

[5]. L. Tomasini, A. Cabarbaye, L. Ngom, S. Allibe, *Genetic Algorithms Supply to Safety and Systems Optimisation*, to appear in 3rd Pluridisciplinary Conference on Quality and Safety, Paris, France, March 1999.

[6]. K. De Jong, *Learning with Genetic Algorithms: An overview*, Machine Learning, vol. 3, pp. 121-138; 1988.

[7]. F. Glover, E. Taillard, D. De Werra, *A user's guide to tabu search*, Annals of Operational Research, volume 41, 1993, pp 3-28.

[8]. P. Siarry, *Simulated Annealing Method: application to electronic circuit design*, Ph. D. Thesis, University of Paris VI, 1986.

[9]. R. Cerf, *An asymptotic theory of genetic algorithms*, Ph. D. Thesis, University of Montpellier, France, 1993.

Multiple Traffic Signal Control Using A Genetic Algorithm

T. Kalganova, G.Russell, A. Cumming

Department of Computing, Napier University, Edinburgh, EH14 1DJ, UK
Email: {t.kalganova, g.russell, andrew}@dcs.napier.ac.uk

Abstract

Optimising traffic signal timings for a multiple-junction road network is a difficult but important problem. The essential difficulty of this problem is that the traffic signals need to co-ordinate their behaviours to achieve the common goal of optimising overall network delay. This paper discusses a novel approach towards the generation of optimal signalling strategies, based on the use of a genetic algorithm (GA). This GA optimises the set of signal timings for all junctions in network. The different efficient red and green times for all the signals are determined by genetic algorithm as well as the offset time for each junction. Previous attempts to do this rely on a fixed cycle time, whereas the algorithm described here attempts to optimise cycle time for each junction as well as proportion of green times. The fitness function is a measure of the overall delay of the network. The resulting optimised signalling strategies were compared against a well-known civil engineering technique, and conclusions drawn.

1 Introduction

One important and still difficult problem in traffic planning is the optimisation of traffic signals for a network of interconnected junctions, such that traffic moves through the network in an optimal way.

The main idea of traffic signal planning is that traffic signals in a road network can be co-ordinated so that the delay experienced by drivers is minimised over the duration of their journeys. By taking into account the available road-space at junctions and balancing the travel time between successive traffic signals, it is possible to derive widespread advantages in terms of free-flowing traffic and reduced overall journey time. Co-ordination between adjacent traffic signals involves designing a plan based on the occurrence and duration of individual signals and the time offsets (journey time between junctions). Sensor or survey information at junctions is also used to improve the overall timing plan of the network [1]. The research documented in this paper is concerned with the creation of optimum signal timing plans based on traffic survey information, with the fitness of the plan being calculated from the total delay incurred by the vehicles using the network. Delay is the additional time required to make a journey over and above that required to make the journey at normal driving speed without ever having to slow down.

There has been many studies carries out on the control of traffic signals [2, 3]. Some of the approaches are based on the existence of a precise traffic model and research methods are applied to get the optimal control parameters. In such approaches a set of control parameters is created for each possible scenario envisaged in the network (e.g. a lane closed due to roadworks). The main disadvantage of this approach is that there are only a finite number of scenarios for which a traffic plan can exist, and if a scenario occurs for which no plan exists it is down to the operator of the traffic network signals to change the plan dynamically by hand to suit.

Other approaches are based on designing multi-agent systems in which the evaluation of each agent is carried out in simulation [2]. The first attempt to apply genetic programming and genetic algorithms (GAs) to traffic control is discussed in [3]. A fuzzy logic approach to network optimisation have also been examined [4, 3], using heuristic rules as a way of adjusting signal timings.

Optimising a traffic network by optimising each junction in the network in turn does not usually result in a good solution, as each signal also should be optimised such that traffic flowing through a number of junctions in turn does not get stopped at each junction. This is important for arterial roadways through networks, where the objective is to allow vehicles to leave the network as quickly as possible. In addition, where sensor information is present at a junction, this may be utilized by other junctions in the surrounding area.

This paper presents a traffic control method based on GAs. The goal of the author's investigations was to generate near optimal traffic signal settings for each traffic controller in terms of minimal delay of an entire road network. The idea of this approach can be briefly described as follows:

1. Traffic signal timings are generated by a GA for the entire network.
2. A high-speed traffic simulator simulated the network using the traffic signal timings generated from the GA.
3. The simulator produced a fitness measure for the signal settings, and this was fed back into the GA.

This list was iterated over until the GA was stopped, which occurred either when a execution-duration maximum or an apparent fitness maximum was reached (depending on the experiment being performed). The best timing diagrams generated by the GA for a particular 3-junction network are given. These results are then compared with ones generated from theoretical analysis of the network.

2 A Genetic Algorithm For Traffic Signal Control

The objective in adjusting the timing of traffic control signals is to minimise the overall delay to traffic over an entire traffic network [5]. The GA described in this paper tackles the problem of obtaining good signal timings by evolving the signal proportions, offsets and the cycle time.

Each junction has a number of roads approaching it, and each approach gets some green time to allow cars into the junction. Consequentially, each approach also has red time during which cars on that approach must not enter the junction. The proportion of red time to green time for an approach is referred to as the signal proportion for that approach.

Each approach to a junction has a green period, and a red period. The total of the green and the red period is referred to as the cycle time.

Consider two junctions, one after another, along a particular route. Optimally, ignoring any other vehicles, a vehicle which was stopped at one junction and which then receives a green light should almost reach the second junction before receiving a green light at that junction, thus ensuring a delay-free journey. Thus it is useful to be able to describe when the next junction in a series goes green in comparison to the first junction. To perform this task, we use the idea of an offset time, which is basically the time between the whole network being switched on and approach 1 of the junction in question going green.

In the author's approach, signal timings for all the traffic controllers in the network are represented by a single chromosome. The fitness of the chromosome corresponds to the total delay for vehicles using the network over a fixed time period. The fitness is obtained from simulation, which uses random vehicle arrival rates into the edges of the network, constrained to meet a preset average vehicle flow. This allows us to develop optimal signal timing information which are robust for a variety of different traffic patterns.

2.1 Traffic Signal Parameters

Consider a road network containing n junctions. Each junction controls its own signals using the following parameters:

- the *inter-green period (IGP)* - in a junction, when one light changes from green to red there must be a short delay before another signal in the junction can change from red to green. This is to give vehicles already in the junction a chance to leave. This is especially true for vehicles which are trying to turn across oncoming traffic, which may be trapped in the centre of a junction until the lights turn red. In Britain, IGP = 3 seconds.
- the *green* and *red time* for a particular approach, t_g and t_r respectively;
- the *cycle time* for a particular junction, $t_c = t_r + t_g$;
- the *offset time*, t_o;

Signal timings are cyclic, in that they repeat themselves every t_c. It is useful to know how the separate junctions are synchronised with each other. The time t_o gives an indication of this. Effectively, this is the time between activation of the junction signal controller and the time that the controller begins to follow the signal configuration. Specifically, the time t_o is the difference between an absolute zero clock and the clock that turns the first of the signals in a junction to green. Each junction has an offset time. The traffic model is evaluated in the simulator for t_m simulated seconds of traffic flow. For the experiments performed in this paper $t_m = 960s$. The graphical representation of some of these parameters are shown in Figure 1.

2.2 Chromosome Representation

Chromosome C describes the behaviour of n traffic controllers which is represented by C, where $C = < C_1, C_2, ..., C_n >$. Each $C_i = < b^i \, c_1^i \, c_2^i \, c_3^i >$, $i=1, ..., n$ defines all the parameters needed by the i-th controller. In our experiments we assume that people drive on the left-hand-side. A junction diagram can be found in figure 1.

Signals are assumed to have only two colours: red and green. If b^i is 0, then when the junction is activated the light on the Main Street approach starts as red, and if it is 1 then the light starts at green. This is referred to as the *first colour* of the signal, and has a duration $(c_1^i + c_3^i)$. c_2^i is the duration of the *second colour*, which is the opposite colour to that of the first colour. Based on this type of chromosome representation we can compute the cycle time for the i-th junction and offset time as follows:

$$t_o = c_1^i \text{ and } t_c = c_1^i + c_2^i + c_3^i + 2 * \text{IGP}$$

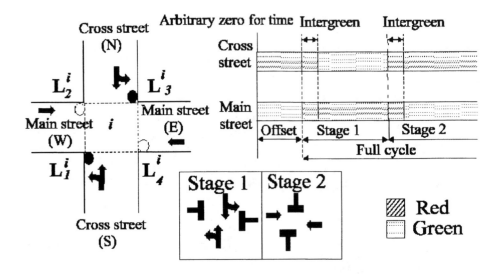

Fig. 1. Cycle time

If gene b^i indicates that the first colour is green, then Main Street is:

$$t_g = c_1^i + c_3^i \text{ and } t_r = c_2^i + 2 * \text{IGP}$$

otherwise:

$$t_r = c_1^i + c_3^i + 2 * \text{IGP} \text{ and } t_g = c_2^i$$

Cross Street t_g is simply Main Street $t_r - 2 * (IGP)$.

2.3 Single-Junction Network Representation

Consider the chromosome representation for the one-junction network shown in Figure 1. In this case chromosome $C = < C_1 >$. By way of an example, assume $C_1 = < 0\ 15\ 25\ 10 >$. First consider the traffic lights on Main Street. The traffic lights $L_1(i)$ and $L_3(i)$ will have the same signal. The start signal is red ($b = 0$), thus during the first *15s* ($c_1 = 15$) the traffic lights located on Main Street with be red. The IGP is *3s*, so for the next *3s* the red signal is maintained. Then the signal is changed to green for *25s* ($c_2 = 25$). As the actual red time in this case is represented by the sum of the first and third genes, and take into account the double IGP (changing signals from red to green and then from green to red), the next red signal last for *31s* ($c_1 + c_3 + 2 * \text{IGP}$). The Cross Street signals are green IGP seconds after Main Street goes red, and remain green till IGP seconds before Main Street return to green. The cycle time is *56s*. The timing diagram is shown in Figure 1.

2.4 Multiple-Junction Network Representation

Consider the three-junction road network shown in Figure 2. In this case the chromosome contains $< C_1, C_2, C_3 >$, as the network has three junctions. An example of the chromosome representation for this network is given in Figure 2. The behaviour of each junction is described by a single C_i. Based on this the signal timings for each traffic light in every junction can be computed, and is shown in Figure 3. Note that in this representation each of the junctions can have a different cycle time and offset time.

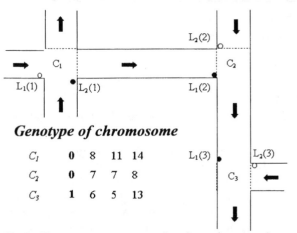

Genotype of chromosome

C_1	0	8	11	14
C_2	0	7	7	8
C_3	1	6	5	13

Fig. 2. Chromosome representation for a three-junction network

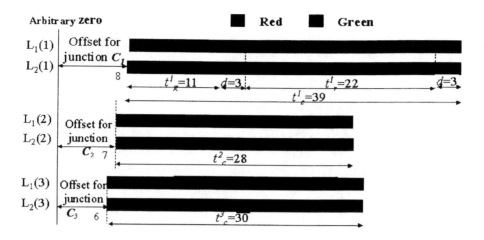

Fig. 3. Timing diagram for a three-junction network

2.5 Preparation Of String Encoded Parameters

A set of randomly initialised parameters $b^{ij}, c_1^{ij}, c_2^{ij}$ and $c_3^{ij}, i=1, ..., n; j=1,..., m$ is prepared such that $c_1^{ij} + c_2^{ij} + c_3^{ij} + 2IGP \leq 130$, where m is the number of chromosomes in population. In Britain all cycle times must lie between 20 and 130 seconds. The gene b^{ij} defining the start signal for $L_1(i)$ traffic light is randomly initialised and takes value 0 or 1. c_k^{ij} is randomly initialised in the range [0, 130], where $k = 1, 2, 3$.

2.6 Calculating Optimum Signal Settings

The j-th chromosome $C(j)$ specifically the parameter set $\forall_i \forall_k c_k^{ij}$ is applied to the junction controllers $B_i, i=1, ...,n$ and the parameters used in a software simulation of duration t_m. The simulator returns the total delay incurred by vehicles travelling in the simulated network.

2.7 The Genetic Search Operators

Evolution of the population in the GA involves two basic steps: (1) a *selection mechanism* that implements a survival of the fittest strategy and (2) a *genetic recombination* of the selected chromosomes to produce offspring for the new population. Recombination is affected through the two genetic operators of crossover and mutation. A *uniform mutation* operator is used, where the mutated gene takes a randomly generated integer value from a specific range. In order to obtain the highest convergence within the GA different types of selection and crossover were investigated, as well as different sizes of population.

Two types of selection operators were used: *roulette wheel* and *probabilistic binary tournament*. With roulette wheel, probability of selection is proportional to fitness. With tournament selection, the probability of the tournament victor being selected is the *tournament discriminator*. Experimentation showed that the optimal tournament discriminator was 0.8. This experimentation also indicated that tournament selection significantly outperformed roulette wheel.

Three different types of crossover were implemented: (1) *uniform*, (2) *one-point* and (3) *two-point*. Cut points were chosen so that individual junctions were not perturbed. The best solution was obtained using uniform crossover, but the behaviour of mean fitness is better for two-point crossover. The convergence of the GA is faster for one-point crossover, and in the experiments discussed in the next section the one-point crossover operator is used. A number of experiments were carried out to discover the optimum mutation rate. The results of this indicate an optimum mutation rate of 0.03. Using an adaptation technique for mutation rate, where the rate is increased when the fitness of the population reaches a local maxima, appears to improve the optimal solutions obtained during our experiments.

3 Experimental Results

In order to evaluate the usefulness of the GA in optimizing traffic signals, an experiment was constructed where the GA controlled the settings of the three simulated signalised junctions shown in Figure 2. Each of these junctions use a two phase signal configuration.

The experiment was constructed so that the GA selected a sequence of traffic signals, and then passed these to the simulation engine. Traffic was then simulated in the traffic network using the specified signal timings, and a measure of how well the traffic moved through the network was obtained. This measure was then returned to the GA as a fitness value, and the cycle repeated.

3.1 The Simulated Road Network

Ideally the GA would have been best connected to a real-world set of traffic signals, but the expense of this, not to mention the inconvenience to real road users as the GA tried to find an optimum signal setting means that getting permission to using this approach would have been difficult. Instead, we have used a urban traffic simulator, JUDGE [6], which was developed by the authors during research performed a few years ago at Strathclyde University. It makes use of probabilistic traffic arrival times, and simulates traffic using a microscopic discrete modelling technique [7, 8]. With this system we can simulate a hour of traffic flow in a few seconds.

For the purposes of the evaluation, the JUDGE traffic simulator was programmed with the three-junction traffic network shown in Figure 2. JUDGE allows networks to be built up using a "train-set" approach, linking up roads with other constructs (such as junctions) by placing components next to each other on a two-dimensional surface.

The simple network used in the simulation is designed to be hard enough so that the solution is not easily solved without simulation, but is easy enough to understand what the GA has tried to achieve. The authors intend to look at more complicated networks with a higher traffic loading as a next stage in developing this research.

The GA produced traffic signalling information, which is then simulated by JUDGE for 16 minutes of traffic time (this takes about 1 second of CPU time). JUDGE also analyses the performance of the network, calculated vehicle delay, flows, average speeds and queue lengths. This information is then collated and formed into a fitness measure for the network. For simplicity, the fitness function is based solely on vehicle delay measures, although the use of the other measurements is also being investigated.

3.2 Flows In The Network

There is a number of possible settings for the signals in this network that produce optimal fitness measurements. No fitness should ever reach zero, as even on a straight roadway cars will interfere with each other (e.g. a fast vehicle may be slowed down by a slower- moving vehicle, causing delays).

The input flows for the network is shown in table 1. In traffic modelling, flows are measured in pcu/hr. A pcu, or *Passenger Car Unit* is a number which describes how big a vehicle is. For instance, a bicycle is 0.3 pcu, whereas the standard household vehicle is 1.0 pcu.

There are also turning probabilities for each approach road to junction 1, indicating how much of each traffic flow goes down each of the two roads which leave junction 1. The result is that 700 pcu/hr travel from junction 1 to junction 2. The flow from junction 2 to junction 3 should be around 1000 pcu/hr, and the output from junction 3 is around 1300 pcu/hr. All roadways have a capacity of 1800 pcu/hr, and a speed limit of 46 km/hr.

3.3 Expected Results

In civil engineering, the design of traffic signaling strategies can often be thought of as an art rather than a science. There are a large number of theoretical models which describe ways to calculate how good a particular signal strategy actually is. None of the theoretical models are recognised as being an accurate model of real-world traffic. Even the best software simulations of traffic flow must be, by definition, be a simulation of some of the important factors of real-world traffic networks. Thus, it is an almost impossible task to confirm that the results generated by the GA are in fact good, near-optimum results.

In order to give some confidence to the results produced by the GA, two separate analyses of the data was performed. Firstly and most simply, a study was made as to the proportion of traffic flows which arrive at a particular junction. This should give an indication as the the proportion of the green times which each approach should expect. A table depicting this information for the network used in analysing the GA can be found in table 1.

Secondly, a respected theoretical model for steady-state single junction networks, Webster [9], was used to get a better understanding of the cycle times and green times produced by the GA. This theoretical model is based on queuing theory, and takes no account of the varying distribution of vehicles as they approach the junction (which happens in real-life traffic). Optimum signal settings for the network, as calculated from the Webster equation is also shown in table 1:

Note that neither of these two data analysis techniques take into account the fact that the junctions are combined together in a single network, but instead treat them all as separate entities. This is unfortunate, as related junctions will have to make compromises in their configuration in order to achieve a global optimal setting for the network. Nevertheless, it should give some sort of indication as to how well the GA has performed.

the traffic network using the specified signal timings, and a measure of how well the traffic moved through the network was obtained. This measure was then returned to the GA as a fitness value, and the cycle repeated.

3.1 The Simulated Road Network

Ideally the GA would have been best connected to a real-world set of traffic signals, but the expense of this, not to mention the inconvenience to real road users as the GA tried to find an optimum signal setting means that getting permission to using this approach would have been difficult. Instead, we have used a urban traffic simulator, JUDGE [6], which was developed by the authors during research performed a few years ago at Strathclyde University. It makes use of probabilistic traffic arrival times, and simulates traffic using a microscopic discrete modelling technique [7, 8]. With this system we can simulate a hour of traffic flow in a few seconds.

For the purposes of the evaluation, the JUDGE traffic simulator was programmed with the three-junction traffic network shown in Figure 2. JUDGE allows networks to be built up using a "train-set" approach, linking up roads with other constructs (such as junctions) by placing components next to each other on a two-dimensional surface.

The simple network used in the simulation is designed to be hard enough so that the solution is not easily solved without simulation, but is easy enough to understand what the GA has tried to achieve. The authors intend to look at more complicated networks with a higher traffic loading as a next stage in developing this research.

The GA produced traffic signalling information, which is then simulated by JUDGE for 16 minutes of traffic time (this takes about 1 second of CPU time). JUDGE also analyses the performance of the network, calculated vehicle delay, flows, average speeds and queue lengths. This information is then collated and formed into a fitness measure for the network. For simplicity, the fitness function is based solely on vehicle delay measures, although the use of the other measurements is also being investigated.

3.2 Flows In The Network

There is a number of possible settings for the signals in this network that produce optimal fitness measurements. No fitness should ever reach zero, as even on a straight roadway cars will interfere with each other (e.g. a fast vehicle may be slowed down by a slower- moving vehicle, causing delays).

The input flows for the network is shown in table 1. In traffic modelling, flows are measured in pcu/hr. A

pcu, or *Passenger Car Unit* is a number which describes how big a vehicle is. For instance, a bicycle is 0.3 pcu, whereas the standard household vehicle is 1.0 pcu.

There are also turning probabilities for each approach road to junction 1, indicating how much of each traffic flow goes down each of the two roads which leave junction 1. The result is that 700 pcu/hr travel from junction 1 to junction 2. The flow from junction 2 to junction 3 should be around 1000 pcu/hr, and the output from junction 3 is around 1300 pcu/hr. All roadways have a capacity of 1800 pcu/hr, and a speed limit of 46 km/hr.

3.3 Expected Results

In civil engineering, the design of traffic signaling strategies can often be thought of as an art rather than a science. There are a large number of theoretical models which describe ways to calculate how good a particular signal strategy actually is. None of the theoretical models are recognised as being an accurate model of real-world traffic. Even the best software simulations of traffic flow must be, by definition, be a simulation of some of the important factors of real-world traffic networks. Thus, it is an almost impossible task to confirm that the results generated by the GA are in fact good, near-optimum results.

In order to give some confidence to the results produced by the GA, two separate analyses of the data was performed. Firstly and most simply, a study was made as to the proportion of traffic flows which arrive at a particular junction. This should give an indication as the the proportion of the green times which each approach should expect. A table depicting this information for the network used in analysing the GA can be found in table 1.

Secondly, a respected theoretical model for steady-state single junction networks, Webster [9], was used to get a better understanding of the cycle times and green times produced by the GA. This theoretical model is based on queuing theory, and takes no account of the varying distribution of vehicles as they approach the junction (which happens in real-life traffic). Optimum signal settings for the network, as calculated from the Webster equation is also shown in table 1:

Note that neither of these two data analysis techniques take into account the fact that the junctions are combined together in a single network, but instead treat them all as separate entities. This is unfortunate, as related junctions will have to make compromises in their configuration in order to achieve a global optimal setting for the network. Nevertheless, it should give some sort of indication as to how well the GA has performed.

Junction	Approach Direction	Flow	Total	Green Proportions $\sigma_g^{predicted}$	Webster's Predicted Green Time	Cycle Time
Junction 1	West	500	1400	35%	21	62
Junction 1	South	900	1400	65%	35	62
Junction 2	North	300	1000	30%	9	31
Junction 2	West	700	1000	70%	16	31
Junction 3	North	1000	1300	76%	31	49
Junction 3	East	300	1300	24%	12	49

Table 1. Green time proportions and Webster Predictions

3.4 Obtained Results

Table 2 contain the data analysis of several best signal timings defined by the genetic algorithm described above. The proportion of green time is defined as σ_g and computed as follows:

$$\sigma_g = \frac{t_g}{t_c - 2 * d} * 100\%.$$

The proportion of green time for each traffic light in the three-junction network is given in Table 2 for different signal timing diagrams. The difference between the expected proportion of green time and that obtained by experimentation, Δ_p, is calculated as follows:

$$\Delta_p = \sigma_g^{predicted} - \sigma_g^{obtained}.$$

When Δ_p is minimal over all traffic lights of network then the best fitness is obtained. It is interesting to note that the biggest differences are mostly found in the second junction, and that in the majority of cases the cycle time for this junction is also smaller than for the others. The best fitness (54.24) had been obtained with a population of 80 chromosomes, a genetic algorithm with tournament selection, and using a one-point crossover and uniform mutation. The mutation rate was 0.3 and crossover rate was 0.25. Elitism was used to improve the GA convergence.

The table also shows Δ_w, which it the difference between the $\sigma_g^{obtained}$ and that predicted by the Webster equation.

The analysis of these results indicates that the GA is excellent in finding near-optimum σ_g values, as the corresponding Δ_p and Δ_w values equate to only about +/- 1 second in terms of the cycle times involved. This is to be expected, as neither the proportional study nor the Webster equation is as accurate as the software simulation.

What is interesting is that the optimum cycle times produced by the GA are for the most part all much lower than that predicted by Webster equation. Post analysis has given three main reasons for this; the fitness of a junction is governed more by green proportions than cycle time, the equations used treat each junction separately but yet in real-life junction settings must be changed due to the effect of surrounding junctions, and that the genome encoding used cannot modify cycle times without changing the green time proportions for a junction.

3.5 Fitness Variations

Two graphs have been created in order to demonstrate why the fitness measure was dominated by green time proportions rather than cycle times. They have been generated from the Webster equation.

In Figure 4, graph b shows the effect of varying the green time for a junction (in this case Junction 2), while keeping the cycle time fixed, results in an obvious "saddle" area where the delays are quite similar in value. In this area, a green time of between 6 and 10 would produce similar delays. However, moving outside this area in either direction and the delay increases dramatically.

Figure 4, graph a shows the effect of varying the cycle time for a junction (this time Junction 3), while keeping the green time proportions fixed. There is a obvious "knee" area in the graph. In this area, the minimum delay is to be found, but it is significantly effected by noise. This is due to rounding effects causes by a simulation constraint that signal timings must be rounded to the nearest second. To the left of this area, the delay increases significantly, and to the right the delay increases more gradually.

The genome in the GA was such that the cycle time was calculated from the red and green time components of the genome. In this way the cycle time was inseparable from the green proportions. Increasing or decreasing the cycle time could only be achieved by changing the green proportions in the genome. As there is only a narrow area where green proportions had similar delay measures, and outside this area delay increased dramatically,

Junction	σ_g %	Δ_p %	Δ_w %	t_c s	t_o s	σ_g %	Δ_p %	Δ_w %	t_c s	t_o s	σ_g %	Δ_p %	Δ_w %	t_c s	t_o s
C_1, West	33.3	1.7	4.2	39	8	34.5	-0.5	3.0	35	8	33.33	1.7	4.2	33	8
C_1, South	66.6	-1.7	-4.1			65.5	0.5	-3.0			66.64	-1.7	-4.1		
C_2, North	31.8	-1.8	4.2	28	7	41.2	-11.2	-5.2	40	8	31.6	-1.6	4.4	25	9
C_2, West	68.2	1.8	-4.2			58.8	11.2	5.2			68.4	1.6	-4.4		
C_3, North	79.2	-3.2	-8.1	30	6	73.5	2.5	-2.4	40	12	75	1.0	-3.9	30	5
C_3, East	20.8	3.2	8.1			26.5	-2.5	2.4			25	-1.0	3.9		
Fitness	56.8417					60.3319					54.24				

Table 2. The best traffic signal control results obtained by GA

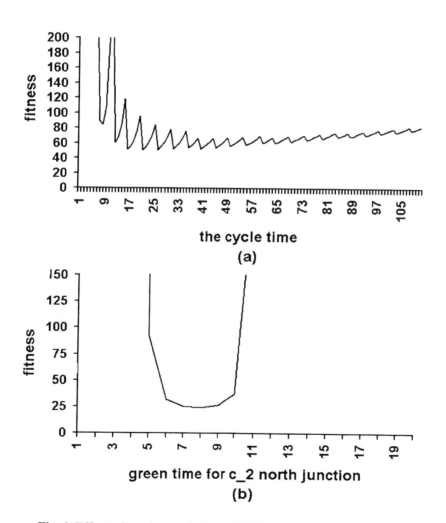

Fig. 4. Effect of varying cycle time while keeping green time proportions fixed

attempts to increase the cycle time (and thus change the green proportions) would initially be given low fitness values, and would thus be considered as highly undesirable members of the population. In addition, as the change in delay caused by changes in cycle time are small in comparison to the effect of changes in green proportions, even a genome designed so that cycle time was independent of green proportions may mean that a GA may not have spent enough effort in producing an optimum cycle time.

From this analysis it is clear that, as a next step, cycle time should be isolated from the green proportions in some way, perhaps encoding it as a percentage green, percentage red, and a cycle time. However, some way to promote the breeding of better cycle times rather than concentrating on green proportions must also be developed.

4 Conclusion

Our novel approach to signal optimisation allow us to configure traffic controllers for a variety of geometric configurations and traffic conditions, by adapting the signal parameters of each traffic junction controller in terms of overall network delay. These parameters include cycle times, offsets, and green proportions. Our approach can be easily extended for different types of junctions and roadways. This is one of the future directions for our investigations.

The actual signal configurations produced by the GA were similar to that expected by the Webster equation; a traditional theoretical model for junction optimisation. Where the GA configured signals differed from Webster the cause can be traced to the way the chromosome was encoded, and to the fact that the Webster equation is based on a number of assumptions which are not present is either real traffic flows or the JUDGE simulator-based traffic model.

The GA investigation showed that the choice of genetic operators and their parameters has a marked effect upon GA performance, and that the choice is dependent on the GA components used. For the given chromosome representation the tournament selection operator performed better in comparison to the uniform operator. It was discovered that the best GA convergence was achieved when the tournament discriminator equalled 100%. However, the best fitness result can be achieved for a tournament discriminator equalling 80%.

Many key questions remain unanswered. These include: (1) how using multi-agent techniques instead of traditional genetic algorithm will influence the obtained result, (2) how our approach handles a larger number of junctions, (3) how our approach compares in performance terms with other adaptive control approaches.

References

1. C.A. O'Flaherty, editor. *Transport Planning and Traffic Engineering.* Arnold, 1997. 544 p.
2. Sadoyoshi Mikami and Yukinori Kakazu. Self-organised control of traffic signals through genetic reinforcement learning. In *Proceedings of IEEE Intelligent Vehicles,* pages 113–118, 1993.
3. David J. Montana and Steven Czerwinski. Evolving control laws for a network of traffic signals. In John R. Koza, David E. Goldberg, David B. Fogel, and Rick L. Riolo, editors, *Proceedings of the first annual conference on Genetic Programming, Stanford University,* pages 333–338. A Bradford Book, July 1996.
4. Suh-Wen Chion. Optimisation of area traffic control subject to user equilibrium traffic assignment. In *Proceedings of the 25th European Transport Forum. Transportation Planning Methods,* volume 2, pages 53–64. Springer, 1995.
5. Carol Ashley. *Traffic and Highway Engineering for Developments.* Blackwell Scientific publications, 1994. 174 p.
6. Gordon Russell, Paul Shaw, and Neil Ferguson. Accurate rapid simulation of urban traffic using discrete modelling. Technical Report RR-96-1, Napier University, January 1996.
7. Gordon Russell, Paul Shaw, John McInnes, Neil Ferguson, and George Milne. The rapid simulation of urban traffic using field-programmable gate arrays. In *Proceedings of the International Conference on the Application of New Technologies to Transport Systems.* Australasian Road Research Board Ltd, May 1995.
8. Martin Bate, Alex Cowie, George Milne, and Gordon Russell. Process algebras and the rapid simulation of highly concurrent systems. In *Proceedings of the 18th Australasian Computer Science Conference,* February 1995.
9. R. J. Salter. *Highway Traffic Analysis and Design.* MacMillan Education, second edition, 1990.

Evolving Musical Harmonisation

Somnuk Phon-Amnuaisuk, Andrew Tuson[1], and Geraint Wiggins

Music Informatics Research Group, Division of Informatics, University of Edinburgh

80 South Bridge, Edinburgh EH1 1HN, Scotland, UK.

Email: {somnukpa,andrewt,geraint}@dai.ed.ac.uk

Abstract

We describe a series of experiments in generating traditional musical harmony using Genetic Algorithms. We discuss some problems which are specific to the musical domain, and conclude that a GA with no notion of meta-level control of the reasoning process is unlikely to solve the harmonisation problem well.

1 Introduction

In recent years, the use of evolutionary techniques such as Genetic Algorithms (GAs) has generated significant interest in the artificial intelligence and computer science communities. This has been reflected in a number of publications in the computer music world, some of which will be discussed later.

In this paper, we explore two aspects of the application of GAs to music:

1. the use of knowledge-rich structures and procedures within the algorithm itself, as opposed to the more traditional use of GA components which are not problem-specific;

2. the strict use of objective methods, in the sense that any reasoning encoded in the GA should be stated explicitly, rather than being implicit in the expressed opinion of a human user.

These criteria are important because we are working in the wider context of simulating and understanding aspects of human behaviour, so we are not interested just in achieving a musical result: we wish to be able to examine the internal behaviour of our methods, and attempt to form some notion of *why* the answer we achieve is produced. In particular, we wish to compare the behaviour of our harmonisation system with human behaviour, and attempt to explain any discrepancies.

This paper is structured as follows. We present a brief statement on the issues interaction *vs.* non-interaction in GAs from the point of view of this study. We then outline existing applications of GAs in computer music. We present a case study of a knowledge-rich musical GA, including a discussion of some significant problems, and then draw conclusions about the implications of the work for musical GAs in general.

2 Interactive GAs

The *Interactive GA* (IGA) approach, sometimes taken by GA applications in the musical domain (see section 3, below), uses a human listener as a means to evaluate the fitness of chromosomes. This approach is inappropriate to our ends in this research, for the following reasons.

1. It is subjective, because it relies on individual preferences – we want an objective measure of what is going on in our system, so we can properly judge its performance;

2. Human listeners tend to become more open to a given piece of music on repeated hearings, and are prone to other inconsistencies based on mood, attention span, and so on;

3. An IGA does not allow us to study the fitness function itself, to determine how faithful it is to our chosen task – so using an IGA would be removing a major object of interest in this study.

A further tenable position on this issue, lying part-way between the two poles of interactive and non-interactive, is the idea of using a corpus of existing works and some form of machine learning system to infer a fitness function. This approach has been applied by, for example, Spector and Alpern [14], Burton and Valdimirova [4] and Johanson and Poli [12]. Again, for our purposes, this approach is inappropriate, as we are primarily interested in the nature of the search space and how to control search, rather than the nature of the result (given that it be

---

[1]Andrew Tuson may now be contacted care of the Department of Computing, City University, London, UK.

acceptable music). In fact, use of neural networks makes the fitness judgements even more inscrutable, since at least one can ask a human judge for the motivation behind his or her judgements.

In the experiments described here, music-theoretical knowledge is used to construct a fitness function in objective and consistent logical terms, which allows us to examine the behaviour of the system more scientifically than would an IGA. The GA can then be used to compare theories (psychological or otherwise) of musical behaviour by observing the search patterns and results produced. In particular, it is to be emphasised that the encoded musical knowledge does not directly constrain the search *path* – it merely constrains the solution. So our interest focusses on what our GA can tell us about the search paths which arise from this unconstrained setup.

3 Existing Work on GAs in Music

GAs have been used in music generation elsewhere. Examples include Horner and Goldberg [10] who used a GA for thematic bridging; Biles [1], who used an IGA for Jazz improvisation; Jacob [11], who devised a composing system using an IGA; and McIntyre [13] and Horner and Ayers [9]. McIntyre used a GA to generate a four part harmonisation of an input melody, focusing on Baroque harmony, while Horner and Ayers focussed on the harmonisation of chord progressions using GAs.

A main aim of our harmonisation project is to investigate the potential of a knowledge-rich GA and its performance in the musical domain. So our solution space is not artificially constrained as in McIntyre's system (which only used a C major scale); nor is there problem abstraction as in Horner and Ayres' system, (which uses the GA to generate parts, given a chord progression, which is a significantly simpler task). Our work aims to harmonise input melodies with no explicit cues as to the required harmony, and does not limit itself to a specific key or scale; and it works at the level of individual voices, with all the extra constraints this entails.

Finally, for a more complete summary of GA work in music, see Burton and Vladimirova [3].

4 Harmonising Chorale Melodies by GA

In this section, we present the results of a study on the use of GAs in generating four-part homophonic tonal harmony for user-specified melodies. The domain-specific (*i.e.*, musical) knowledge in this system is implemented in three parts of the GA. These are described in turn, and followed by an overview of the GA configuration used.

4.1 Chromosome Representation

Generally speaking, keys and chords are the main concepts in harmonisation of western tonal music. Harmonisation rules are expressed in terms of relationships between triads, and between degrees of scale within a key signature (*e.g.*, tonic-dominant, etc) but not the absolute pitch. Therefore, in this implementation, musical information (*e.g.*, pitch, interval, time, duration) is represented after *normalisation* with respect to key – that is, absolute pitch information is abstracted out. Then, pitch is expressed in terms of scale degree. To express all twelve semitones, the standard five accidentals are used. Different octaves are distinguished by an associated integer. Finally, time intervals are represented as integers. The representation conforms to the CHARM specification of Wiggins *et al.* [16].

As noted earlier, a knowledge-rich and directly meaningful representation is used in our chromosome representation. This representation may be thought of as a matrix, which consists of five strings of equal, fixed length. The top four strings contain soprano (fixed), alto, tenor and bass parts, with the fifth describing the durations of the chords. The user inputs the soprano information (assumed to be the melody); the GA will then harmonise the input soprano, homophonically, with a further three voices in conjunction with the musical domain knowledge encoded in its operators and fitness function.

This approach is illustrated in Figure 1, using the completed harmonisation for the first two bars of "O Come, All Ye Faithful"; see Figure 4 for the score rendition of the corresponding output.

				chromosome length		
Soprano	[0,0,3]	[0,0,3]	[4,0,2]	[0,0,3]	[1,0,3]	[4,0,2]
Alto	[2,0,2]	[2,0,2]	[2,0,2]	[2,0,2]	[4,0,2]	[1,0,2]
Tenor	[4,0,1]	[4,0,1]	[2,0,1]	[0,0,2]	[7,0,1]	[1,0,1]
Bass	[2,0,1]	[0,0,1]	[0,0,1]	[4,0,1]	[4,0,1]	[7,0,0]
Duration	1	2	1	1	2	2

Fig. 1. Schematic Diagram of a Four-Voice Harmony Chromosome

In the figure, we can see the matrix arrangement produced by the direct representation of time (left to right along the structure) and vocal part and note duration (top to bottom down the structure). For the purposes of reproduction operators, we view the five horizontal layers as inseparable.

4.2 Reproduction Operators

The following crossover and mutation operators are used in this implementation, described here in musical terms. The reader less familiar with musical jargon may find solace in Taylor [15].

Splice: One point crossover between two chromosomes – selects a crossover point between successive notes of the melody and corresponding chords.

Perturb: Mutate by allowing alto, tenor and bass to move up or down by one semitone or tone. The selection of the various possible mutations is random.

Swap: Mutate by swapping two randomly picked voices between alto, tenor or bass. This gives the effect of changing the chord between different open and closed positions, and of changing inversions.

Rechord: Mutate to a different chord type. This mutation generates a new chord from the melody data. A chord is built with the soprano note as root, 3rd or 5th. Doubling (necessary for a four note chord) can be in any position.

PhraseStart: Mutate the beginning of each phrase to start with tonic root position on a down beat.

PhraseEnd: Mutate the end of each phrase to end with a chord in root position.

4.3 Fitness Function

The fitness function judges the fitness of each chromosome according to the following criteria derived directly from music theory. Within individual voices (as opposed to between voices), we prefer stepwise progression over large leaps, and we keep the voice within its proper range. We penalise progression to dissonant chords, and we avoid leaps of major and minor 7ths, of augmented and diminished intervals, and of intervals larger than one octave.

Between voices, we apply the following criteria: we avoid parallel unison, parallel perfect 5ths, and parallel octaves; we forbid progression from diminished 5th to perfect 5th (though the converse is allowed); we avoid hidden unison; we forbid crossing voices; and we forbid hidden 5th and octave in the outer voices, when soprano is not progressing stepwise.

Solutions are penalised for note doubling and omission, in the primary major and minor triads: doubling of the root (tonic) is preferred, while doubling of the 3rd is penalised; doubling of the 3rd is forbidden in a dominant chord; if it is necessary to omit a voice,

omit the fifth only, except in 1st inversion; in inverted chords, doubling of the bass is preferred; and we penalise doubling of tones which give a strong harmonic tendency, such as leading notes.

In this implementation, the system does not have enough knowledge to plan for large scale harmonic progression. The fitness function determines only the plausible harmonic movement between two adjacent chords. The fitness function prefers (in decreasing order of preference): descending 5th movement; progression towards the *tonic*; retrogression; and repetition.

4.4 GA Configuration

In most of these experiments, exceptions being mentioned explicitly here, a generational GA, in the style of Davis [5], with a panmitic (unstructured) population model was used. Strings were initialised by randomly picking chords containing the relevant soprano pitch. Finally, a population size of at least 50 was used, with binary tournament selection.

5 Results and Analysis

All the output of the system was assessed by Dr. John Kitchen, a senior lecturer in the Faculty of Music at the University of Edinburgh, according to the criteria he uses for 1st year undergraduate students' harmony. This example scored 5 out of 10 – a clear pass. While other examples were less successful (most earning around the 30% mark), according to the assesser, this was mostly due to the lack of coherent large-scale musical progression – which was *not evaluated in the fitness function*, so this is not a surprise. The system was judged by the assesser to be better than student harmonisers at getting the basic rules right.

Figure 2 shows a harmonisation by our system of the first eight notes of "Joy to the world". The output is not perfect, but it is surprisingly good given the limited, local nature of the rules built into the system.

Fig. 2. Harmonisation of the First Line of *Joy to the World*

Some experimentation was carried out with various GA parameter settings, as shown in Figure 1; the penalties given here are all-or-nothing penalties, except where two numbers are given, in which case either value may be applied as appropriate. As expected, the weights of the various penalties applied in the fitness function have a significant effect on the solution. Other parameters, such as crossover rate, mutation rate, and different selection schemes appear to affect the time taken for the population to converge, and do little for the solution quality. This is due to the fact that it is the fitness function which primarily defines the knowledge in the system pertaining to what does or does not constitute a good piece of music, while the other parameters define the search strategy.

Table 1. Probabilities and penalties used in the experiments

Population:	30-200
Migration interval:	20 generations
Operator Probabilities	
P(Splice):	0.3
P(Perturb):	0.1
P(Swap):	0.2
P(Rechord):	0.3
P(PhraseStart):	0.05
P(PhraseEnd):	0.05
Fitness Penalties	
Invalid Pitch:	1
Invalid Chord:	10,1
Invalid Range:	10
Invalid Interval:	10
Invalid Doubling:	10
Voice Crossing:	10
Hidden Unison:	10
Single voice progression:	10
Dual voice progression:	10
Harmonic progression:	10,1
Harmonic Analysis:	100,20

What is most significant – and problematic – is that, with the current evaluation functions and reproduction operators, the GA still cannot satisfy all the constraints within 300 generations. Figure 3 illustrates a typical fitness profile of the best solution in each generation. The data is from the first phrase of the hymn "O come, all ye faithful", which is twenty four notes long. The figure shows the distribution of penalties along the chromosome – in other words, the score of each harmonic movement or chord in the piece – as the generations proceed. The higher the contour, the less acceptable the chord at that point.

In the figure, we can see that the distribution of

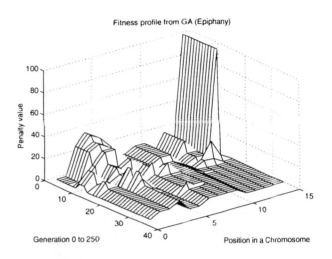

Fig. 3. A Typical Fitness Profile Landscape

penalties at the beginning is quite random. After a few generations, it starts to shape to a certain pattern. However, the GA could not reduce all the penalties in the fitness function even with enlarged populations and when run for large amounts of time. In an attempt to solve this problem, an experiment with an *island model* [7] with four population groups was carried out, to determine whether different groups might be able to preserve their own salient cultures, and so bring the GA to a more globally acceptable solution. However, though the experiment showed an improvement in search efficiency, the GA still could not reduce all of the penalties at once, and the general shape of the unacceptable contour above was maintained. The musical output associated with that final contour is shown in Figure 4.

Why, then, do these problems arise? They arise because the structure of a harmonisation of this kind is very specific, in that individual variations in chord are very strongly context dependent. Therefore, it is often the case that one cannot change any given chord without changing the chords around it. In terms of the GA, this means that reduction of a fitness penalty in one position is likely to increase penalties in other positions, because the movement from one chord to the next is not considered with respect to overall movement in the phrase. Human composers solve this problem by structuring the construction process itself in an explicit way and designing an overall harmonic framework, and then filling in the most crucial parts first – in short, a least commitment strategy is taken. This kind of structured reasoning is not a feature of a simple GA of the kind used here, and so the results are sub-optimal.

To restate this argument in terms of the search space: we believe that the problem is due to a mul-

Fig. 4. A GA Harmonisation of "O Come, All Ye Faithful"

timodal fitness landscape, characterised by many local optima in basins of attraction which separated by very high barriers, due to the interactions described above. Before the GA can move from one basin of attraction to another, multiple factors leading to a fitness penalty need to be changed. Such a simultaneous change is very unlikely to occur.

There are various solutions which might be applied to solve this problem in a GA context. Niching [6] might be expected to help with the problems of multimodality, and we expect to look at this in future. Linkage learning [8] is unlikely to help as the spatial arrangement of the genes in our chromosome is already optimised, and so changing it is unlikely to help the search. Further, the nature of the epistatic interactions is well understood, and therefore this knowledge can be better utilised directly, rather than by having the GA learn it.

6 Conclusions

It is quite clear from the experiments here and elsewhere that GAs can be applied successfully in the musical domain – up to a point. Looking at the output of our systems from an aesthetic viewpoint, the results are still far from ideal: the harmonisation produced by the GA has neither clear plan nor intention. This is not a surprise as the discussion above suggested that we cannot expect large scale structure to arise from the kind of programming inherent in a GA containing (even in this experiment) relatively little domain knowledge.

However, we would claim that they are surprisingly successful musically, within this limitation. The

inclusion of well-established musical laws and constraints within the search seems to yield a fairly lightweight but effective level of search control.

In summary, therefore, we conclude that while GAs can be surprisingly good at small, constrained musical tasks, their performance, at least in this context of simulating human musical behaviour, is currently limited by two issues.

1. GAs are a stochastic, heuristic search method, so one cannot be sure that an optimal solution will be reached, even if there is one. In particular, in a problem of this kind, they tend to get stuck in local optima in the search space.

2. GAs of the form used here lack structure in their reasoning. On the other hand, composers have developed complex and subtle methods over several centuries involving different techniques for solving the problems highlighted here. No musician would seriously suggest that an author of hymn tunes works in the same way as our GA. Therefore, while we may be able to produce (near) acceptable results with a GA, doing so says little about the working of the compositional mind.

Other approaches to using GAs for this kind of task may be more successful. For example, one might take the approach of optimising a set of instructions to plan a harmonisation, rather than actually operating on the musical score directly; this approach is similar to the *indirect* representational approach used by Burke [2] and others for timetabling and scheduling problems.

Genetic programming approaches may also be fruitful, but we suggest that in order to produce music which is coherent within any accepted musical system (*e.g.*, the tonal system used in the vast majority of rock and pop music), there will need to be some encoding of musical practice in the GP operators. An example of a GP music system which suffers from lack of such knowledge is the *GP-music* program of Johanson and Poli [12].

An alternative solution in the context of our GA would be to introduce even more knowledge-rich mutations, which would possess knowledge about the entire harmonic structure of each candidate solution, and thus would be able to leap directly across the barriers in the fitness landscape described earlier in a single bound. However, it would be hard indeed to make a clear differentiation between this approach and a conventional KBS, if indeed one could apply such a rule without being unacceptably *ad hoc*.

We conclude, therefore, that neighbourhood search methods such as GAs are fundamentally limited in the musical harmonisation domain by the non-local nature of the harmonisation problem itself. This said, at least intuitively, GAs seem to offer an interesting approach to the study of creativity. It would appear that, if GAs are to both improve in musical performance, and allow us to gain insights into the compositional mind, dealing with the issues raised here is an urgent task. We end with a suggestion that it is likely that much can be gained in this particular problem by somehow combining a GA with a conventional rule-based system.

Acknowledgements

Thanks to Dr. John Kitchen for his help in assessing the harmonisation system. Andrew Tuson is supported by EPSRC studentship, reference number 95306458.

References

[1] J. A. Biles. GenJam: A genetic algorithm for generating jazz solos. In *ICMC Proceedings 1994*. The Computer Music Association, 1994.

[2] E. Burke, D. Elliman, and R. Weare. The Automated Timetabling of University Exams using a Hybrid Genetic Algorithm. In *The AISB Workshop on Evolutionary Computing*, 1995.

[3] A. R. Burton and T. Vladimirova. Applications of genetic techniques to musical composition, 1997. Available by WWW at http://www.ee.surrey.ac.uk/Personal/ A.Burton/work.html.

[4] A. R. Burton and T. Vladimirova. A genetic algorithm for utilising neural network fitness evaluation for musical composition. In *Proceedings of the 1997 International Conference on Artificial Neural Networks and Genetic Algorithms*, pages 220–224, 1997.

[5] L. Davis, editor. *Handbook of Genetic Algorithms*. New York: Van Nostrand Reinhold, 1991.

[6] David E. Goldberg and Jon Richardson. Genetic algorithms with sharing for multimodal function optimization. In *Proceedings of the Second International Conference on Genetic Algorithms and Their Applications*, pages 41–49. San Mateo: Morgan Kaufmann, 1987.

[7] V.S. Gordon, D. Whitley, and A. Böhn. Dataflow parallelism in genetic algorithms. In R. Männer and B Manderick, editors, *Parallel Problem Solving from Nature 2*, pages 553–42, Amsterdam, 1992. Elsevier Science.

[8] G. R. Harik and D. E. Goldberg. Learning linkage. In *Foundations of Genetic Algorithms IV*, pages 270–85. San Mateo: Morgan Kaufmann, 1996.

[9] A. Horner and L. Ayers. Harmonisation of musical progression with genetic algorithms. In *ICMC Proceedings 1995*, pages 483–484. The Computer Music Association, 1995.

[10] A. Horner and D. E. Goldberg. Genetic algorithms and computer-assisted music composition. Technical report, University of Illinois, December 1991.

[11] B. L. Jacob. Composing with genetic algorithms. Technical report, University of Michigan, September 1995.

[12] B. Johanson and R. Poli. Gp-music: An interactive genetic programming system for music generation with automated fitness raters. In *Proceedings of the 3rd International Conference on Genetic Programming, GP'98*. MIT Press, 1998.

[13] R. A. McIntyre. Bach in a box: The evolution of four-part baroque harmony using a genetic algorithm. In *First IEEE Conference on Evolutionary Computation*, pages 852–857. 1994.

[14] L. Spector and A. Alpern. Criticism, culture, and the automatic generation of artworks. In *Proceedings of the 12th National Conference on Artificial Intelligence*, 1994.

[15] E. Taylor. *The **AB** Guide to Music Theory*. The Associated Board of the Royal Schools of Music, London, 1996.

[16] G. A. Wiggins, M. Harris, and A. Smaill. Representing music for analysis and composition. In M. Balaban, K. Ebcioglu, O. Laske, C. Lischka, and L. Sorisio, editors, *Proceedings of the 2nd IJCAI AI/Music Workshop*, pages 63–71, Detroit, Michigan, 1989. Also from Edinburgh as DAI Research Paper No. 504.

A Niched-Penalty Approach for Constraint Handling in Genetic Algorithms

Kalyanmoy Deb*and Samir Agrawal

Kanpur Genetic Algorithms Laboratory (KanGAL), Department of Mechanical Engineering,
Indian Institute of Technology Kanpur, PIN 208 016, India
E-mail: {deb,samira}@iitk.ac.in

Abstract

Most applications of genetic algorithms (GAs) in handling constraints use a straightforward penalty function method. Such techniques involve penalty parameters which must be set right in order for GAs to work. Although many researchers use adaptive variation of penalty parameters and penalty functions, the general conclusion is that these variations are specific to a problem and cannot be generalized. In this paper, we propose a niched-penalty approach which does not require any penalty parameter. The penalty function creates a selective pressure towards the feasible region and a niching maintains diversity among feasible solutions for the genetic recombination operator to find new feasible solutions. The approach is only applicable to population-based approaches, thereby giving GAs (or other evolutionary algorithms) a niche in exploiting this penalty-parameter-less penalty approach. Simulation results on a number of constrained optimization problems suggest the efficacy of the proposed method.

1 Introduction

Real-world search and optimization problems are written as nonlinear programming (NLP) problem of the following kind [2, 16]:

$$
\begin{aligned}
\text{Minimize} \quad & f(\vec{x}) \\
\text{Subject to} \quad & g_j(\vec{x}) \geq 0, \qquad j = 1, \ldots, J, \\
& h_k(\vec{x}) = 0, \qquad k = 1, \ldots, K, \\
& x_i^l \leq x_i \leq x_i^u, \quad i = 1, \ldots, n.
\end{aligned}
\tag{1}
$$

In the above NLP problem, there are n variables (that is, $\vec{x}$ is a vector of size n), J greater-than-equal-to type inequality constraints, and K equality constraints. The function $f(\vec{x})$ is the objective function, $g_j(\vec{x})$ is the j-th inequality constraints, and $h_k(\vec{x})$ is the k-th equality constraints. The i-th variable varies in the range $[x_i^l, x_i^u]$. A solution $\vec{x}$ that satisfies all the above equality and inequality constraints and above variable bounds is called a *feasible* solution. Other solutions are called *infeasible* solutions.

Most classical search and optimization methods handle above NLP problems by using a penalty function, where infeasible solutions are penalized depending on

the amount of constraint violation. Although there exist a number of other specialized constraint handling techniques which are applicable to only a particular type of constraints [2, 16], the following simple procedure is generic and most popular [2]:

$$
P(\vec{x}, \vec{R}, \vec{r}) = f(\vec{x}) + \sum_{j=1}^{J} R_j \langle g_j(\vec{x}) \rangle^2 + r_k \sum_k [h_k(\vec{x})]^2,
\tag{2}
$$

where $\langle \ \rangle$ denotes the absolute value of the operand, if the operand is negative and returns a value zero, otherwise. The parameter R_j is the penalty parameter of the j-th inequality constraint and r_k is the penalty parameter for k-th equality constraint.

The above penalized function $P(\vec{x}, \vec{R}, \vec{r})$ makes the constrained NLP problem of equation 1 into a unconstrained minimization problem. Optimization of $P(\vec{x}, \vec{R}, \vec{r})$ largely depends on the penalty parameter $\vec{R}$ and $\vec{r}$. The optimal solution of $P(\vec{x}, \vec{R}, \vec{r})$ is close to the true constrained optimal solution of equation 1 for very large values of $\vec{R}$ and $\vec{r}$. But using large values of $\vec{R}$ and $\vec{r}$ does not emphasize the objective function $f(\vec{x})$ and thus most effort is spent in finding feasible solutions. Classical gradient-based methods suffer from computation of meaningless gradients for large $\vec{R}$ and $\vec{r}$ values [2]. Although GAs do not use gradients, a different scenario happens. Since too much emphasis is given to feasible solutions, GAs usually prematurely converge around the feasible solutions found early on in the simulation. In most GA simulations using the above penalty function approach, researchers usually experiment to find a correct combination of penalty parameters $\vec{R}$ and $\vec{r}$. However, in order to reduce the number of penalty parameters, often the constraints are normalized and only one penalty parameter R is used [2]:

$$
P(\vec{x}, R) = f(\vec{x}) + R \left[\sum_j \langle \bar{g}_j(\vec{x}) \rangle^2 + \sum_k [\bar{h}_k(\vec{x})]^2 \right],
\tag{3}
$$

Even in this case, fixing a correct penalty parameter R is an essential element of a successful simulation.

The importance of the penalty parameter in obtaining any reasonable solution, leave alone an optimal solution, is evident from a plethora of research in designing

---

*An Alexander von Humboldt Fellow.

penalty parameters for a problem. Homaifar et al. [10] designed a multi-level penalty function, depending on the level of constraint violations. Joines and Hauck [11] used a dynamic penalty parameter, which is varied with generation. Michalewicz and Attia [13] updated penalty parameters based on a temperature-based evolution used in simulated annealing techniques. Michalewicz and Schoenauer [14] suggested specific recombination operators which preserve feasibility of solutions for some specific type of constraints. All these research suggests that constraint handling techniques used in GAs are still problem-specific and one technique may work in few problems, but may not work as well in other problems.

In this paper, we develop a constraint handling method based on penalty function approach which does not require any penalty parameter. The pair-wise comparison used in tournament selection is exploited to make sure that (i) when two feasible solutions are compared the one with better objective function value is chosen, (ii) when one feasible and one infeasible solutions are compared, the feasible solution is chosen, and (iii) when two infeasible solutions are compared, the one with smaller constraint violation is chosen. This approach is only applicable to population-based search methods such as GAs or other evolutionary computation methods.

In the remainder of the paper, we present the performance of GAs with the proposed constraint handling method on five test problems, including one engineering design problem. The results are also compared with the best-known solutions obtained using earlier GA implementations or using classical methods.

2 Proposed Constraint Handling Method

Constraints are handled by using a suitable fitness function which depends on the current population. Solutions in a population are assigned a fitness so that feasible solutions are emphasized more than infeasible solutions. The proposed method uses tournament selection operator where two solutions are chosen from the population and one is chosen. The following three criteria are satisfied during the selection operator:

1. Any feasible solution wins over any infeasible solution,

2. Two feasible solutions are compared only based on their objective function values.

3. Two infeasible solutions are compared based on the amount of constraint violations.

Although there exist a number of implementations [12, 15, 17] where these criteria are imposed in their constraint handling approaches, all of these implementations used different measures of constraint violations which still needed a penalty parameter for each constraint.

In the proposed method, we choose a constraint violation measure from a practical standpoint. In order to evaluate a solution, any designer will first check if the solution is feasible. If the solution is infeasible (that is, at least one constraint is violated), the designer will never bother to compute its objective function value (such as cost of the design). It does not make sense to compute the objective function value of an infeasible solution, because the solution simply cannot be implemented in practice. Motivated by this argument, we device the following penalty term where infeasible solutions are compared based on only their constraint violation values:

$$F(\vec{x}) = \begin{cases} f(\vec{x}), & \text{if } g_j(\vec{x}) \geq 0, \ \forall j \in J, \\ f_{\max} + \sum_{j=1}^{J} \langle g_j(\vec{x}) \rangle, & \text{otherwise.} \end{cases}$$
(4)

The parameter $f_{\max}$ is the maximum function value of all feasible solutions in the population. Let us illustrate this constraint handling technique on a single-variable constrained minimization problem, shown in Figure 1. In the figure, the unconstrained minimum solution is not

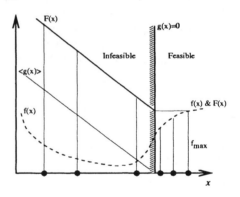

Fig. 1: The proposed constraint handling scheme is illustrated. Six solid circles are solutions in a GA population.

feasible. The objective function $f(\vec{x})$, constraint violation $\langle g(\vec{x}) \rangle$, and the fitness function $F(\vec{x})$ are shown in the figure. It is important to note that $F(\vec{x}) = f(\vec{x})$ in the feasible region. When tournament selection operator is applied to a such a fitness function $F(\vec{x})$, all three criteria mentioned above will be satisfied and there will be selective pressure towards the feasible region. The figure also shows how the fitness value of six arbitrary solutions will be calculated. Thus, under this constraint handling scheme, the fitness value of an infeasible solution may change from one generation to another, but the fitness value of a feasible solution will always be the same. Since the above constraint handling method assigns a hierarchy to infeasible solutions and tournament selection does not depend on the exact fitness values, instead their relative difference is important, any arbitrary penalty parameter will work the same. In fact, there is no need of any explicit penalty parameter. This is a major advantage of the proposed method over earlier penalty function implementations. However, to avoid any bias from any particular constraint, all constraints are normalized (a usual practice in constrained optimization [2]) and equation 4 is used. It is important to note that such a constraint handling scheme without the need of a penalty parameter is possible because GAs use a population of solutions and pair-wise comparison of solutions is possible using the tournament selection. For the same reason, such schemes cannot be used with classical point-by-point search and optimization methods.

The above constraint handling method seems interesting and eliminate the need of any penalty parameter. But, there are two other aspects that must also be used to qualify this method as an efficient constraint handling methods.

2.1 Use real-coded GAs

It is intuitive that the feasible region in constrained optimization problems may be of any shape (convex or concave and connected or disjointed). In real-parameter constrained optimization using GAs, schemata specifying contiguous regions in the search space (such as (110*...*) may be considered to be generally more important than schemata specifying discrete regions in the search space (such as (*1*10*...*). In a binary GA under a single-point crossover operator, all common schemata corresponding to both parent strings are preserved in both children strings. Since, any arbitrary contiguous region in the search space cannot be represented by a single Holland's schema and since the feasible search space can usually be of any arbitrary shape, it is expected that the crossover operator used in binary GAs may not always be able to create feasible children solutions from two feasible parent solutions.

However, the floating-point representation of variables in a GA and a search operator that respects contiguous regions in the search space may be able to perform better than binary GAs in constrained optimization problems with continuous search space. In this paper, we use real-coded GAs with simulated binary crossover (SBX) operator and a parameter-based mutation operator [3], for this purpose. We present procedures of these operators in the Appendix. SBX operator is particularly suitable here, because the spread of children solutions around parent solutions can be controlled using a distribution index η_c. A large value of η_c allows only near-parent solutions to be created, whereas a small value of η_c allows distant solutions to be created. Another aspect of this crossover operator is that it is adaptive, allowing any solution to be created in the beginning and allowing

to have a more focused search when the population is converging.

2.2 Use a Niching Technique

With real-coded GAs having SBX operator any arbitrary contiguous region can be searched, provided there are enough diversity maintained among the feasible parent solutions. There are a number of ways diversity can be maintained in a population. Among them, sharing methods [4] and use of mutation are popular ones. In this paper, we use either or both of the above methods of maintaining diversity among the feasible solutions. A simple sharing strategy is implemented in the tournament selection operator. When comparing two feasible solutions (i and j), a normalized Euclidean distance d_{ij} is measured between them. If this distance is smaller than a critical distance $\bar{d}$, the solutions are compared with their objective function values. Otherwise, they are not compared and another solution j is checked. The normalized Euclidean distance is calculated as follows:

$$d_{ij} = \sqrt{\frac{1}{n} \sum_{k=1}^{n} \left(\frac{x_k^{(i)} - x_k^{(j)}}{x_k^u - x_k^l} \right)^2}. \tag{5}$$

This way, the solutions that are far away from each other are not compared to each other and diversity among feasible solutions is maintained.

3 Results

In this section, we apply GAs with the proposed constraint handling method to five different constrained optimization problems that have been studied in the literature.

In all problems, we run GAs 50 times from different initial populations. Fixing a correct population size in a GA is as important as anything else. Taking the cue from previous studies on population sizing [6, 7], we use a population size which linearly varies with the number of variables (we use $N = 10n$, where n is the number of variables in a problem). In all problems, we use binary tournament selection operator without replacement. We use a crossover probability of 0.9. SBX operator with a distribution index of $\eta_c = 1$. In the real-coded mutation, the amount of perturbation in a variable can be controlled by fixing a parameter η_m [5]. The probability of mutation is varied with generation number so that initially only one variable is expected to get mutated for each solution with larger perturbance and at the specified maximum generation, all variables are mutated with a smaller perturbance. For sharing operator, we use $\bar{d} = 0.1$.

3.1 Test Problem 1

To investigate the efficacy of the proposed constraint handling method, we first choose a two-dimensional constrained minimization problem:

$$\text{Minimize} \quad f_1(\vec{x}) = (x_1^2 + x_2 - 11)^2 + (x_1 + x_2^2 - 7)^2,$$
$$\text{Subject to} \quad g_1(\vec{x}) \equiv 4.84 - x_1^2 - (x_2 - 2.5)^2 \geq 0,$$
$$g_2(\vec{x}) \equiv (x_1 - 0.05)^2 + (x_2 - 2.5)^2 - 4.84 \geq 0,$$
$$0 \leq x_1 \leq 6, 0 \leq x_2 \leq 6. \tag{6}$$

The unconstrained objective function $f_1(\vec{x})$ has a minimum solution at $\vec{x} = (3, 2)$ with a function value equal to zero. However, due to the presence of constraints, this solution is no more feasible and the constrained optimum solution is $\vec{x}^* = (2.246826, 2.381865)$ with a function value equal to $f_1^* = 13.59085$. The feasible region is a narrow crescent-shaped region (approximately 0.7% of the total search space) with the optimum solution lying on the second constraint, as shown in Figure 3.

Sharing and mutation operators are not used here. We have run GAs till 50 generations. Table 1 shows the consequence of 50 runs of three different GAs. First, a binary-coded GA is used with single-point crossover operator. Variables are coded in 20 bits each. The proposed constraint handling method is used. The table presents the number of times (out of 50 runs) a GA finds a solution within ϵ% of the best-known optimal function value. Different values of ϵ is used, as shown in the table.

Since the feasible space is non-convex, binary-coded GAs with single-point crossover cannot find a near-optimal solution many times. Whereas, real-coded GAs with the proposed constraint handling method find a solution within 1% of true optimal solution in 29 of 50 GA runs. When a similar constraint handling technique [15] (termed as 'PS' in the table) is used with real-coded GAs, the performance is not as good as the proposed method. In 11 of 50 GA runs, no infeasible solution was found.

To show the working of the proposed constrained handling technique (compared to PS technique), we show the population histories in generations 0, 10, and 50 in Figures 2 and 3. The initial population of 50 random solutions show that initially solutions exist all over the search space (no solution is feasible in the initial population). After 10 generations, GAs with Powell and Skolnick's constraint handling strategy (with $R = 1$) could not drive the solutions towards the narrow feasible region. Instead, the solutions get stuck at a solution $\vec{x} = (2.891103, 2.11839)$ with a function value equal to 0.41708, which is closer to the unconstrained minimum (albeit infeasible) solution at $\vec{x} = (3, 2)$. When a real-coded GA with the proposed constraint handling strategy (TS-R) is applied on the identical initial populations of 50 solutions (rest all parameter settings are also the same as that used in the Powell and Skolnick's algo-

rithm), the GA distributes well its population around and inside the feasible region (Figure 2) after 10 generations. Finally, GAs converge near the true optimum solution at (2.243636, 2.342702) with a function value equal to 13.66464 (within 0.54% of the true optimum solution).

3.2 Test Problem 2

This problem has eight variables and six inequality constraints [12]:

$$\text{Minimize} \quad f_2(\vec{x}) = x_1 + x_2 + x_3$$
$$\text{Subject to} \quad g_1(\vec{x}) \equiv 1 - 0.0025(x_4 + x_6) \geq 0,$$
$$g_2(\vec{x}) \equiv 1 - 0.0025(x_5 + x_7 - x_4) \geq 0,$$
$$g_3(\vec{x}) \equiv 1 - 0.01(x_8 - x_5) \geq 0,$$
$$g_4(\vec{x}) \equiv x_1 x_6 - 833.33252 x_4 - 100 x_1$$
$$+ 83333.333 \geq 0,$$
$$g_5(\vec{x}) \equiv x_2 x_7 - 1250 x_5 - x_2 x_4 + 1250 x_4 \geq 0,$$
$$g_6(\vec{x}) \equiv x_3 x_8 - x_3 x_5 + 2500 x_5 - 1250000 \geq 0$$
$$100 \leq x_1 \leq 10000,$$
$$1000 \leq x_2, x_3 \leq 10000,$$
$$10 \leq x_i \leq 1000, \quad i = 4, \ldots, 8. \tag{7}$$

The optimum solution is

$$\vec{x}^* = (579.3167, 1359.943, 5110.071, 182.0174,$$
$$295.5985, 217.9799, 286.4162, 395.5979),$$
$$f_2^* = 7049.330923.$$

All six constraints are active at this solution.

Table 2 shows the performance of GAs with different constraint handling methods. Michalewicz [12] experienced that this problem is difficult to solve. The best solution obtained by any method used in that study (with approximately 350,000 function evaluations) had an objective function value equal to 7377.976, which is about 4.66% worse than the true optimal objective function value. The proposed approach consistently finds solutions very close to the true optimum with only 80,080 function evaluations (population size 80, maximum generations 1,000). However, the best solution obtained by GAs with sharing and mutation and with a maximum of 320,080 function evaluations (population size 80, maximum generations 4,000) has a function value equal to 7060.221, which is only about 0.15% more than the true optimal objective function value. This shows the efficacy of the proposed approach on this rather complex constrained optimization problem.

Table 1: Number of runs (out of 50 runs) converged within $\epsilon\%$ of the optimum solution for GAs with two constraint handling techniques—proposed method with binary GAs (TS-B) and with real-coded GAs (TS-R) and Powell and Skolnick's method [15] (PS) with $R = 1$ on test problem 1.

Meth- od	ϵ							Infea- sible	Best	Median	Worst
	$\leq 1\%$	$\leq 2\%$	$\leq 5\%$	$\leq 10\%$	$\leq 20\%$	$\leq 50\%$	$> 50\%$				
TS-B	2	2	5	6	9	13	37	0	13.59658	37.90495	244.11616
TS-R	29	31	31	32	33	39	11	0	13.59085	13.61673	117.02971
PS ($R = 1$)	17	19	20	20	24	32	7	11	13.59108	16.35284	172.81369

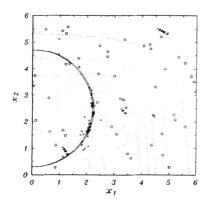

Fig. 2: Population history at initial generation (marked with open circles), at generation 10 (marked with 'x') and at generation 50 (marked with open boxes) using the proposed scheme. The population converges to a solution very close to the true constrained optimum solution on a constraint boundary.

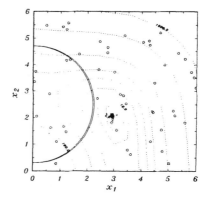

Fig. 3: Population history at initial generation (marked with open circles), at generation 10 (marked with 'x') and at generation 50 (marked with open boxes) using Powell and Skolnick's [15] method ($R = 1$). The population converges to a wrong, infeasible solution.

3.3 Test Problem 3

This problem has five variables and six inequality constraints [8, 14]:

Minimize $f_3(\vec{x}) = 5.3578547x_3^2 + 0.8356891x_1x_5$
$+37.293239x_1 - 40792.141$,

Subject to $g_1(\vec{x}) \equiv 85.334407 + 0.0056858x_2x_5$
$+0.0006262x_1x_4 - 0.0022053x_3x_5 \geq 0$,
$g_2(\vec{x}) \equiv 85.334407 + 0.0056858x_2x_5$
$+0.0006262x_1x_4 - 0.0022053x_3x_5 \leq 92$,
$g_3(\vec{x}) \equiv 80.51249 + 0.0071317x_2x_5$
$+0.0029955x_1x_2 + 0.0021813x_3^2 \geq 90$,
$g_4(\vec{x}) \equiv 80.51249 + 0.0071317x_2x_5$
$+0.0029955x_1x_2 + 0.0021813x_3^2 \leq 110$,
$g_5(\vec{x}) \equiv 9.300961 + 0.0047026x_3x_5$
$+0.0012547x_1x_3 + 0.0019085x_3x_4 \geq 20$,
$g_6(\vec{x}) \equiv 9.300961 + 0.0047026x_3x_5$
$+0.0012547x_1x_3 + 0.0019085x_3x_4 \leq 25$,
$78 \leq x_1 \leq 102$,
$33 \leq x_2 \leq 45$,
$27 \leq x_i \leq 45, \quad i = 3, 4, 5.$

(8)

The best-known optimum solution [8] is

$$\vec{x}^* = (78.0, 33.0, 29.995, 45.0, 36.776),$$
$$f_3^* = -30,665.5.$$

At this solution, constraints g_2 and g_5 are active. The best-known GA solution to this problem obtained elsewhere [10] using a multi-level penalty function method is

$$\vec{x}^{GA} = (80.49, 35.07, 32.05, 40.33, 33.34),$$
$$f_3^{GA} = -30,005.7,$$

which is about 2.15% worse than the best-known optimum solution.

Table 3 presents the performance of GAs with the proposed constraint handling method with a population size 10×5 or 50. The presence of sharing and mutation produces the best performance. The best solution is as follows:

$$\vec{x} = (78, 33, 29.995, 45, 36.776).$$

As many as 47 or 50 GA runs have found solutions within 1% of the best-known solution. It is also interesting to note that all GAs used here have found solutions

Table 2: Number of runs (out of 50 runs) converged within ϵ% of the best-known solution using real-coded GAs with the proposed constraint handling scheme on test problem 2.

Muta-tion	Shar-ing	ϵ							Infea-sible	Best	Median	Worst
		$\leq$ 1%	$\leq$ 2%	$\leq$ 5%	$\leq$ 10%	$\leq$ 20%	$\leq$ 50%	$>$ 50%				
Maximum generation = 1,000												
No	No	2	3	8	14	29	47	3	0	7063.377	8319.211	13738.276
No	Yes	3	5	7	14	29	47	2	1	7065.742	8274.830	10925.165
Maximum generation = 4,000												
Yes	Yes	17	23	33	36	42	50	0	0	7060.221	7220.026	10230.834

Table 3: Number of runs (out of 50 runs) converged within ϵ% of the best-known solution using real-coded GAs with the proposed constraint handling scheme on test problem 3.

Muta-tion	Shar-ing	ϵ						Best	Median	Worst
		$\leq$ 1%	$\leq$ 2%	$\leq$ 5%	$\leq$ 10%	$\leq$ 20%	$\leq$ 50%			
No	No	18	34	50	50	50	50	−30614.814	−30196.404	−29606.596
No	Yes	28	44	50	50	50	50	−30651.865	−30376.906	−29913.635
Yes	Yes	47	48	50	50	50	50	−30665.537	−30665.535	−29846.654

better than that reported earlier [10], solved using binary GAs with a multi-level penalty function method.

3.4 Test Problem 4

This problem has 10 variables and eight constraints [12]:

$$\text{Minimize} \quad f_4(\vec{x}) = x_1^2 + x_2^2 + x_1 x_2 - 14x_1 - 16x_2$$
$$+(x_3 - 10)^2 + 4(x_4 - 5)^2 + (x_5 - 3)^2$$
$$+2(x_6 - 1)^2 + 5x_7^2 + 7(x_8 - 11)^2$$
$$+2(x_9 - 10)^2 + (x_{10} - 7)^2 + 45,$$

$$\text{Subject to} \quad g_1(\vec{x}) \equiv 105 - 4x_1 - 5x_2 + 3x_7 - 9x_8 \geq 0,$$
$$g_2(\vec{x}) \equiv -10x_1 + 8x_2 + 17x_7 - 2x_8 \geq 0,$$
$$g_3(\vec{x}) \equiv 8x_1 - 2x_2 - 5x_9 + 2x_{10} + 12 \geq 0,$$
$$g_4(\vec{x}) \equiv -3(x_1 - 2)^2 - 4(x_2 - 3)^2 - 2x_3^2$$
$$+7x_4 + 120 \geq 0,$$
$$g_5(\vec{x}) \equiv -5x_1^2 - 8x_2 - (x_3 - 6)^2$$
$$+2x_4 + 40 \geq 0,$$
$$g_6(\vec{x}) \equiv -x_1^2 - 2(x_2 - 2)^2 + 2x_1 x_2$$
$$-14x_5 + 6x_6 \geq 0,$$
$$g_7(\vec{x}) \equiv -0.5(x_1 - 8)^2 - 2(x_2 - 4)^2 - 3x_5^2$$
$$+x_6 + 30 \geq 0,$$
$$g_8(\vec{x}) \equiv 3x_1 - 6x_2 - 12(x_9 - 8)^2 + 7x_{10} \geq 0,$$
$$-10 \leq x_1 \leq 10, \quad i = 1, \ldots, 10.$$
$$(9)$$

The optimum solution to this problem is as follows:

$$\vec{x}^* = (2.171996, 2.363683, 8.773926, 5.095984,$$
$$0.9906548, 1.430574, 1.321644, 9.828726,$$
$$8.280092, 8.375927), \quad f_4^* = 24.3062091.$$

The first six constraints are active at this solution.

Table 4 shows the performance of GAs with the proposed constraint handling scheme with a population size 10 × 10 or 100. In this problem, GAs with and without sharing performed equally well. However, GA's performance improves drastically with mutation, which provided the necessary diversity among the feasible solutions. This problem was also solved by Michalewicz [12] by different constraint handling techniques. The best reported method had its best, median, and worst objective function values as 24.690, 29.258, and 36.060, respectively, in 350,070 function evaluations. This was achieved with a multi-level penalty function approach. With a similar maximum number of function evaluations, GAs with the proposed constraint handling method have found better solutions (best: 24.372, median: 24.409, and worst: 25.075). The best solution is within 0.27% of the optimal objective function value. Most interestingly, 41 out of 50 runs have found a solution having objective function value within 1% (or $f(\vec{x})$ smaller than 24.549) of the optimal objective function value.

3.5 Welded beam design

Next, we apply the proposed method to solve a welded beam design problem, which has been attempted to solve by using a number of classical optimization methods [16] and by using GAs [1].

The objective of the design is to minimize the fabrication cost of a beam which is required to be welded in a column. There are four design variables $\vec{x} = (h, \ell, t, b)$ and five constraints involving stress, deflection and bucking restrictions. The resulting NLP problem is shown

Table 4: Number of runs (out of 50 runs) converged within ϵ% of the best-known solution using real-coded GAs with the proposed constraint handling scheme on test problem 4.

Muta-tion	Shar-ing	ϵ							Infea-sible	Best	Median	Worst
		$\leq 1\%$	$\leq 2\%$	$\leq 5\%$	$\leq 10\%$	$\leq 20\%$	$\leq 50\%$	$> 50\%$				
Maximum generation = 1,000												
No	No	0	0	8	16	28	47	3	0	24.81711	27.85520	42.47685
No	Yes	0	0	9	25	36	45	5	0	24.87747	26.73401	50.40042
Maximum generation = 3,500												
Yes	Yes	41	41	50	50	50	50	0	0	24.37248	24.40940	25.07530

below:

$$
\begin{aligned}
\text{Minimize} \quad & f_w(\vec{x}) = 1.10471 h^2 \ell + 0.04811 tb(14.0 + \ell), \\
\text{Subject to} \quad & g_1(\vec{x}) \equiv 13,600 - \tau(\vec{x}) \geq 0, \\
& g_2(\vec{x}) \equiv 30,000 - \sigma(\vec{x}) \geq 0, \\
& g_3(\vec{x}) \equiv b - h \geq 0, \\
& g_4(\vec{x}) \equiv P_c(\vec{x}) - 6,000 \geq 0, \\
& g_5(\vec{x}) \equiv 0.25 - \delta(\vec{x}) \geq 0, \\
& 0.125 \leq h \leq 10, \\
& 0.1 \leq \ell, t, b \leq 10.
\end{aligned}
$$

(10)

The terms $\tau(\vec{x})$, $\sigma(\vec{x})$, $P_c(\vec{x})$, and $\delta(\vec{x})$ are given below:

$$
\begin{aligned}
\tau(\vec{x}) &= \sqrt{\tau'^2 + \tau''^2 + \ell\tau'\tau''} / \sqrt{0.25(\ell^2 + (h+t)^2)}, \\
\sigma(\vec{x}) &= \frac{504,000}{t^2 b}, \\
P_c(\vec{x}) &= 64,746.022(1 - 0.0282346 t) tb^3, \\
\delta(\vec{x}) &= \frac{2.1952}{t^3 b},
\end{aligned}
$$

where

$$
\begin{aligned}
\tau' &= \frac{6,000}{\sqrt{2} h \ell}, \\
\tau'' &= \frac{6,000(14 + 0.5\ell)\sqrt{0.25(\ell^2 + (h+t)^2)}}{2\{0.707 h \ell (\ell^2/12 + 0.25(h+t)^2)\}}.
\end{aligned}
$$

The optimized solution reported in the literature [16] is $h^* = 0.2444$, $\ell^* = 6.2187$, $t^* = 8.2915$, and $b^* = 0.2444$ with a function value equal to $f_w^* = 2.38116$. Binary GAs are applied on this problem in an earlier study [1] and the solution $\vec{x} = (0.2489, 6.1730, 8.1789, 0.2533)$ with $f_w = 2.43$ (within 2% of the above best solution) was obtained with a population size of 100. However, it was observed that the performance of GAs largely dependent on the chosen penalty parameter values.

We use the proposed constraint handling technique here. Table 5 presents the performance of GAs with a population size 80. Real-parameter GAs without sharing is good enough to find a solution within 2.6% of the optimal objective function value. However, with the introduction of sharing, 28 runs out of 50 runs have found

a solution within 1% of the optimal objective function value and this has been achieved with only a maximum of 40,080 function evaluations. When more number of function evaluations are allowed, real GAs with the proposed constraint handling technique and mutation operator perform much better—all 50 runs have found a solution within 0.1% of the true optimal objective function value. This means that with the proposed GAs, one run is enough to find a satisfactory solution close to the true optimal solution. In handling such complex constrained optimization problems, any designer would like to use such an efficient yet robust optimization algorithm.

3.6 Summary of Results

Here, we summarize the best GA results obtained in this paper and compare that with the best reported results in earlier studies. For test problems 2 and 4, earlier methods recorded the best, median, and worst values for 10 GA runs only. However, the corresponding values for GAs with the proposed method have been presented for 50 runs.

It is clear that in most cases the proposed constraint handling strategy has performed with more *efficiency* (in terms of getting closer to the best-known solution) and with more *robustness* (in terms of more number of successful GA runs finding solutions close to the best-known solution) than previous methods. Since solutions are compared either with objective function values or with the amount of constraint violation, penalty coefficients are not needed in the proposed approach.

4 Conclusions

The major difficulty in handling constraints using penalty function methods in GAs and in classical optimization methods has been to set an appropriate value for penalty parameters. This often requires users to experiment with different values of penalty parameters. In this paper, we have developed a constraint handling method for GAs which does not require any penalty pa-

Table 5: Number of runs (out of 50 runs) converged within ε% of the best-known solution using real-coded GAs (TS-R) with the proposed constraint handling scheme.

Method	Muta-tion	Shar-ing	≤ 1%	≤ 2%	≤ 5%	≤ 10%	≤ 20%	≤ 50%	> 50%	Best	Median	Worst
Maximum generations = 500												
TS-R	No	No	0	0	1	4	8	16	34	2.44271	3.83412	7.44425
TS-R	No	Yes	28	36	44	48	50	50	0	2.38119	2.39289	2.64583
Maximum generations = 4,000												
TS-R	No	Yes	28	37	44	48	50	50	0	2.38119	2.39203	2.64583
TS-R	Yes	Yes	50	50	50	50	50	50	0	2.38145	2.38263	2.38355

Table 6 Summary of results of this study. A '−' indicates that information is not available.

Prob No.	True optimum	Best-known GA			Results of this study		
		Best	Median	Worst	Best	Median	Worst
1	13.59085	−	−	−	13.59085	13.65413	117.02971
2	7049.331	7485.667	8271.292	8752.412	7060.221	7220.026	10230.834
3	−30665.5	−30005.7	−	−	−30665.537	−30665.535	−29846.654
4	24.306	24.690	29.258	36.060	24.372	24.409	25.075
Weld	2.381	2.430	−	−	2.381	2.383	2.384

rameter. Infeasible solutions are penalized in a way so as to provide a search direction towards the feasible region. This has been possible mainly because of the population approach of GAs and ability to have pair-wise comparison of solutions using the tournament selection operator. On a number of test problems including an engineering design problem, GAs with the proposed constraint handling method have repeatedly found solutions closer to the true optimal solutions than earlier GAs. The results of this study are interesting and show promise for a reliable and efficient constrained optimization task through GAs.

References

[1] Deb, K.: Optimal design of a welded beam structure via genetic algorithms, AIAA Journal, vol. 29, no. 11, pp. 2013–2015, 1991.

[2] Deb, K.: Optimization for engineering design: Algorithms and examples. New Delhi: Prentice-Hall 1995.

[3] Deb, K., Agrawal, R. B.: Simulated binary crossover for continuous search space. Complex Systems, vol. 9, pp. 115–148, 1995.

[4] Deb, K., Goldberg, D. E.: An investigation of niche and species formation in genetic function optimization, Proc. Third International Conference on Genetic Algorithms, Washington D. C., pp. 42–50, 1989.

[5] Deb, K., Goyal, M.: A combined genetic adaptive search (GeneAS) for engineering design, Computer Science and Informatics, vol. 26, no. 4, pp. 30–45, 1996.

[6] Goldberg, D. E., Deb, K., Clark, J. H.: Genetic algorithms, noise, and the sizing of populations, Complex Systems, vol. 6, pp. 333–362, 1992.

[7] Harik, G., Cantu-Paz, E., Goldberg, D. E., Miller, B. L.: The gambler's ruin problem, genetic algorithms, and the sizing of populations, Proc. 1997 IEEE International Conference on Evolutionary Computation, Orlando, FL, pp. 7–12, 1997.

[8] Himmelblau, D. M.: Applied nonlinear programming, New York: McGraw-Hill, 1972.

[9] Hock, W., Schittkowski, K.: Test examples for nonlinear programming code, Lecture Notes on Economics and Mathematical Systems, 187, Berlin: Springer-Verlag, 1980.

[10] Homaifar, A., Lai, S. H.-Y., Qi., X.: Constrained optimization via genetic algorithms, Simulation, vol. 62, no. 4, 242–254, 1994.

[11] Joines, J. A., Houck, C. R.: On the use of nonstationary penalty functions to solve nonlinear constrained optimization problems with GAs, Proc International Conference on Evolutionary Computation, Orlando, FL, pp. 579–584, 1994.

[12] Michalewicz, Z.: Genetic algorithms, numerical optimization, and constraints, Proc. Sixth International Conference on Genetic Algorithms, Carnegie-Mellon, PA, pp. 151–158, 1995.

[13] Michalewicz, Z., Attia, N.: Evolutionary optimization of constrained problems, Proc. Third Annual Conference on Evolutionary Programming, pp. 98–108, 1994.

[14] Michalewicz, Z., Schoenauer, M.: Evolutionary algorithms for constrained parameter optimization problems, Evolutionary Computation, vol. 4, no. 1, pp. 1–32, 1996.

[15] Powell, D., Skolnick, M. M.: Using genetic algorithms in engineering design optimization with nonlinear constraints, Proc. Fifth International Conference on Genetic Algorithms, Urbana, IL, pp. 424–430, 1993.

[16] Reklaitis, G. V., Ravindran, A., Ragsdell, K. M.: Engineering optimization methods and applications, New York: Wiley, 1983.

[17] Richardson, J. T., Palmer, M. R., Liepins, G., Hilliard, M.: Some guidelines for genetic algorithms with penalty functions, Proc. Third International Conference on Genetic Algorithms, Washington D. C., pp. 191–197, 1989.

Appendix

A Simulated Binary Crossover (SBX) and Parameter-based Mutation

The procedure of computing children solutions c_1 and c_2 from two parent solutions y_1 and y_2 under SBX operator is as follows:

1. Create a random number u between 0 and 1.

2. Find a parameter β_q using a polynomial probability distribution, developed in [3] from a schema processing point of view, as follows:

$$\beta_q = \begin{cases} (u\alpha)^{\frac{1}{\eta_c+1}}, & \text{if } u \le \frac{1}{\alpha}, \\ \left(\frac{1}{2-u\alpha}\right)^{\frac{1}{\eta_c+1}}, & \text{otherwise,} \end{cases} \quad (11)$$

where $\alpha = 2 - \beta^{-(\eta_c+1)}$ and β is calculated as follows:

$$\beta = 1 + \frac{2}{y_2 - y_1} \min[(y_1 - y_l), (y_u - y_2)].$$

Here, the parameter y is assumed to vary in $[y_l, y_u]$. The parameter η_c is the distribution index for SBX and can take any non-negative value. A small value of η_c allows solutions far away from parents to be created as children solutions and a large value restricts only near-parent solutions to be created as children solutions.

3. The children solutions are then calculated as follows:

$$\begin{aligned} c_1 &= 0.5\left[(y_1 + y_2) - \beta_q|y_2 - y_1|\right], \\ c_2 &= 0.5\left[(y_1 + y_2) + \beta_q|y_2 - y_1|\right]. \end{aligned}$$

It is assumed here that $y_1 < y_2$. A simple modification to the above equation can be made for $y_1 > y_2$. For handling multiple variables, each variable is chosen with a probability 0.5 and the above SBX operator is applied variable-by-variable. In all simulation results here, we have used $\eta_c = 1$.

A polynomial probability distribution is used to create a solution c in the vicinity of a parent solution y under the mutation operator [5]. The following procedure is used for a parameter $y \in [y_l, y_u]$:

1. Create a random number u between 0 and 1.

2. Calculate the parameter δ_q as follows:

$$\delta_q = \begin{cases} \left[2u + (1-2u)(1-\delta)^{\eta_m+1}\right]^{\frac{1}{\eta_m+1}} - 1, \\ \qquad \text{if } u \le 0.5, \\ 1 - \left[2(1-u) + 2(u-0.5)(1-\delta)^{\eta_m+1}\right]^{\frac{1}{\eta_m+1}}, \\ \qquad \text{otherwise,} \end{cases} \quad (12)$$

where $\delta = \min[(y - y_l), (y_u - y)]/(y_u - y_l)$. The parameter η_m is the distribution index for mutation and takes any non-negative value.

3. Calculate the mutated child as follows:

$$c = y + \delta_q(y_u - y_l),$$

Using above equations, we can calculate the expected normalized perturbance $((c - y)/(y_u - y_l))$ of the mutated solutions in both positive and negative sides separately. We observe that this value is $O(1/\eta_m)$. In order to get a mutation effect of 1% perturbation in solutions, we set $\eta_m = 100 + t$ and the probability of mutation is changed as follows:

$$p_m = \frac{1}{n} + \frac{t}{t_{\max}}\left(1 - \frac{1}{n}\right),$$

where t and $t_{\max}$ are current generation number and the maximum number of generations allowed, respectively. Thus, in the initial generation, we mutate on an average one variable ($p_m = 1/n$) with an expected 1% perturbance and as generations proceed, we mutate more variables with lesser expected perturbance.

A Genetic Algorithm with Dynamic Population Size

Márton-Ernő Balázs
Department of Computer Science
Williams College
Williamstown, MA, USA
mebalazs@cs.williams.edu

Abstract

This paper introduces a modified version of a simple genetic algorithm (SGA) in which the size of the population changes according to a model inspired from mathematical biology. The primary purpose of this research described is to use the growth of the population to vary the selective pressure through the scaling of the fitness function. The experimental results presented in the paper are used to demonstrate the behavior of the proposed algorithms in comparison with a SGA using fixed size population.

1 Introduction

It has been proven that in several applications (see for instance [Filipic 1992]) fitness scaling can significantly improve the performance of genetic algorithms (GA's). Those results were however mostly experimental, without any theoretical explanation and without well-founded suggestions on how fitness scaling has to be applied. In [Balázs 1995] it was shown how the selective pressure in a simple genetic algorithm (SGA) using proportional selection depends on the convexity of the fitness function. Based on those results it was suggested that during the genetic search the convexity of the fitness function has to be increased (from concave to convex). This would gradually increase the selective pressure from very low to very high. While this result is interesting and gives an explanation of why fitness scaling works for a particular class of GA's it still lacks any suggestion of how the change of the convexity of the fitness function (and through this of the selective pressure) can be controlled.

The current paper proposes an approach to controlling the selective pressure by gradually changing the convexity of the fitness function in a SGA using proportional selection. The approach is based on an assumption inspired form a mathematical model of the dynamics of a population composed of a single species. According to this assumption, in an environment with limited resources (e.g. food) inhabited by a single species, the selective pressure during evolution increases with the increase of the size of the population. This mathematical model suggests a modified version of the SGA in which the size of the population varies with time. Other GA's with dynamic population sizes have been suggested earlier [Smith 1993], however in those approaches the variation of the population size was rather a goal than a means. The paper first presents the mathematical model used for varying the population size explaining why that specific a model was chosen. Then a modified SGA with dynamic population size is described. The performance of the presented algorithm is compared to that of a generic SGA using a set of examples. Finally it is suggested how the variation of the population size can be used to vary the convexity of the fitness function and by this the selective pressure during different stages of the genetic search. The paper concludes with suggesting further benefits from using a SGA with a dynamic population size.

2 A Population Model for Single Species

Part of the classical results of mathematical biology [Murray 1989] refers to various models for the growth of natural populations under certain constraints. For many of these cases the constraints stabilize in some sense the otherwise exponential growth of the population size.

Probably the simplest, never the less interesting, models is the self-limiting process proposed by Verhulst [Murray 1989] which can be described by the following differential equation:

$$\frac{dN}{dt} = rN \left(1 - \frac{N}{K} \right)$$

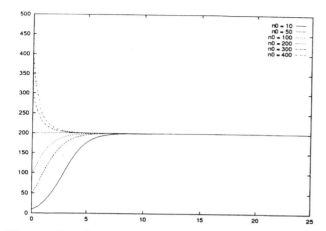

Fig. 1. *Population dynamics for $K = 200, r = 1$ and different sizes for the initial population*

where r and K are positive constants, r being the growth rate of the population and K the carrying capacity of the environment which is determined by the available sustaining resources. In this model the per capita birth rate depends at any point in time on the size of the population and it is $rN \left(1 - \frac{N}{K}\right)$.

There are two steady states for this model $N = 0$ and $N = K$, of which only the second one is of interest (since the other one refers to a trivial - empty - population):

$$X(t) = \frac{K \cdot N0}{N0 + (K - N0) \cdot e^{-r(t-t_0)}}$$

This results in a very interesting behavior of the size of the population: no matter what the size of the initial population is, it will eventually stabilize at about K individuals. If the size of the initial population is greater than K then it will decrease to K and if it is smaller than K, it will increase to K. Figure 1. illustrates the shapes of the graphs for different initial population sizes.

In this paper we will only consider in the case when the size of the initial population is smaller than K. If we analyze the graphs in Figure 1. for this case we can see that the population size seems to grow almost exponentially for some time (the convex portion of the graph) after which the growth decreases gradually (the convex portion of the graph) till it stabilizes at approximately 0 rate. This suggests that in the first stage of the evolution of the population its growth is almost "free" (quite likely with a low selective pressure), while later this growth slows down till it completely stops (as a consequence of a high selective pressure).

The above observations suggested us to propose a genetic algorithm in which the size of the population

would be controlled by the presented model and used further on to control the selective pressure through the convexity of the fitness function used.

3 A SGA with Dynamic Population Size

Modifying the SGA such that the population size is dynamically changed is straightforward and is given below:

Algorithm
```
  read(NO, r, K);
  t <- 0;
  N <- NO;
  Initialize population P(t, N);
  Evaluate members of P(t, N);
  while Stopping Condition False do
    N' <- F(N, r, K);
    Generate P'(N') from P(t, N)
    t <- t + 1;
    N <- N';
    P(t, N) = P'(N');
    Evaluate members of P(t, N);
  end (while);
  Select result;
end (algorithm)
```

where $F(N, r, K)$ computes the size of the next generation $(N + 1)$.

It remains however the issue of determining the parameters $N0$, r and K as well as of specifying how population `P'(N')` is generated from population `P(t, N)`.

The first note that we need to make is that no matter what the size of the initial population and the growth rate of the population are the population size will eventually stabilize at K. Due to this fact we suggest that K be chosen first, and that it be chosen as being the target size of the population. Choosing K may be as hard as choosing the population size in any GA ([Goldberg 1989][Mahfoud 1995]) however we suspect (based on some preliminary experimental results) that due to the way our algorithm works it will be less critical.

Choosing the growth rate (r) will influence the speed with which the size of the population reaches its stable size and is most likely to be chosen on an experimental basis just like the crossover rate in most of the GA's.

Finally $N0$ has to be chosen such that the chromosomes in the initial populatio contain all the

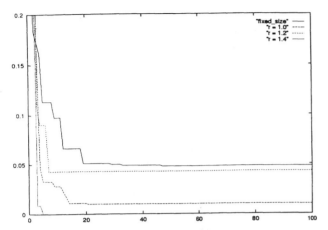

Fig. 2. *Evolution of the best individula for De Jong's function F2, measured for a fixed size population GA and a dynamic population size GA with different birth rate values.*

"genetic material" needed for producing the solution(s). For instance in a binary string representation the initial population can consist of just two individuals with represented by a string of 1's and a string of 0's respectively.

4 Experimental Results

We ran a set of experiments with standard function optimization benchmark problems (e.g. De Jong functions) to compare the results and performances produced by out GA with dynamic population size and a SGA.

For the experiments we used the following parameter settings:

1. both of the algorithms used only crossover to eliminate the random effect of mutation, and the crossover rate was set to 1;

2. we used proportional selection because our theoretical results on fitness scaling ([Balázs 1995]) refer to this type of selection;

3. the population size in the fixed size population GA was set to the same value as the carrying capacity (K) in the dynamic population GA;

4. in both algorithms at any point in time the "old" generation and the newly generated individuals were merged to produce the new generation (the best N' of the two survived).

5. we measured for both of the algorithms the best individual per generation, the average fitness and the standard deviation of the fitness.

6. for each problem and each population size we ran the dynamic population GA for different values of the growth rate (r).

In Figure 2. We illustrate the evolution of the best individual for the two GA's applied to De Jong's function F2. For the dynamic population GA it shows the graph for different values of r.

In our experiments the dynamic population GA gave results comparable to the fixed size one. While we do not claim that this is true for any problem, we believe that it is sufficient support for the feasibility of this direction.

5 Fitness Scaling based on Population Dynamics

The primary purpose of this paper is to demonstrate that the genetic algorithm with dynamic population size proposed performs just as well as the one with fixed size population. The final goal of our research is however to find a means to control selective pressure in a genetic algorithm through fitness scaling. In the followings we make some suggestions how the population dynamics presented earlier can be used for this purpose.

Based on the interpretation of the mathematical model used in our algorithm, we suggest that the size of the population determines the selective pressure at any point in time. Since we know how to control the selective pressure using the convexity of the fitness function ([Balázs 1995]), our purpose is to build a model for the dependency between population size and convexity of the fitness function. Such a dependency should be defined as a function of the form ,

$$f_t(x) = f(t, x)$$

that is at any point in time a different member of a family of fitness functions will be used.

The family of fitness functions has to be determined such that its members are concave (low selective pressure) till a certain point in time and concave (after that), with the convexity decreasing continuously with time. The point where the switch from concave to convex has to happen is naturally given by the inflection point on the graph of the population growth function.

There are several families of functions that could be used. For instance, assuming that the objective function is to be maximized and its range is the interval [0, 1], the family

$$f(t, x) = x^t (t > 0)$$

could be used.

6 Conclusions and Future Work

This paper proposes a variation of a SGA that uses dynamic population size controlled by a model inspired from mathematical biology. It is demonstrated through experimental results that such an algorithm performs comparably with GA's using fixed size populations.

The paper also suggests a way to control the selective pressure in a GA by dynamically changing the size of the population, which in turn varies the convexity of the fitness function. This approach is essentially a suggestion of the way theoretical results on fitness scaling presented in [Balázs 1995] can be used to improve the performance of GA's.

Part of our future work will involve experimenting with further families of fitness functions that can be used in our algorithm. Another direction of research is the experimenting with populations of different initial sizes as well as with different ways of generating the initial population. Finally will test our algorithm on further examples and applications

References

1. Balázs, M. E. (1995) An Approach to the Study of Sensitivity for a Class of Genetic Algorithms, in Foundations of Genetic Algorithms 3, (L.D. Whitley and M.D. Vose eds.), Morgan Kaufmann.

2. Filipic, B. (1992) Enhancing Genetic Search to Schedule a Production Unit, Proceedings of ECAI'92, (Neumann, B. ed.), John Wiley & So.

3. Goldberg, D. E. (1989) Sizing Populations for Serial and Parallel Genetic Algorithms, Proceedings of the Third International Conference on Genetic Algorithms, ICGA'89.

4. Goldberg, D. E., Deb, K. & Clark, J. H. (1991) Genetic Algorithms, noise and sizing of Populations, IlliGAL Technical Report No. 91010, University of Illinois at Urbana-Champaign.

5. Holland, J.H., (1992) Adaptation in Natural and Artificial Systems, Ann Arbor, MIT Press.

6. Mahfoud, S.W. (1995) Population Size and Genetic Drift in Fitness Sharing, in Foundations of Genetic Algorithms 3, (L.D. Whitley and M.D. Vose eds.), Morgan Kaufmann, 1995.

7. Michalewicz, Z.(1996) Genetic Algorithms + Data Structures = Evolution Programs (Third, Revised and Extended Edition), Springer Verlag, 1996.

8. Murray, J.D. (1989). Mathematical Biology, Springer Verlag, 1989.

9. Smith, R. E. (1993). Adaptively Resizing Populations: An Algorithm and Analysis, TCGA Report No. 93001, February 9, 1993

A Genetic Algorithm with Dynamic Niche Clustering for Multimodal Function Optimisation

Justin Gan and Kevin Warwick
Department of Cybernetics,
University of Reading,
Whiteknights, Reading,
Berkshire (England), RG6 6AY
Email : J.G.R.Gan@reading.ac.uk

Abstract

Genetic algorithm's (GA's) have become a powerful search tool pertaining to the identification of global optima within multimodal domains. Many different methodologies and techniques have been developed to aid in this search, and facilitate the efficient location of these optima. What has become known as Goldberg's standard fitness sharing methodology is inefficient and does not explicitly identify or provide any information about the peaks (niches) of a fitness function. In this paper, a mechanism is formulated that will identify the peaks of a multimodal fitness function in a one-dimensional parameter space, using a hybrid form of clustering in the framework of a genetic algorithm. It is shown that the proposed Dynamic Niche Clustering scheme not only performs as well as standard nicheing, but works in $O(nq)$ time, rather than $O(n^2)$ time. In addition to this, it explicitly provides statistical information about the peaks themselves. The Dynamic Niche Clustering scheme is also shown to have favourable qualities in revealing multimodal function optima when there is little or no knowledge of the fitness function itself a priori.

1 Introduction

Determining the location of global optima within a fitness landscape has been the subject of much research. Many different algorithms and techniques have been developed which, to a greater or lesser degree, achieve this aim. Genetic Algorithms offer not only an efficient way of locating these global optima, but have the added ability of being able to explore much more of the problem space, and therefore run less of a risk of becoming trapped at a local optimum. With the introduction of the concept of niches and speciation, the field of multimodal function optimisation has been opened up. Now, not only can a GA provide a user with the global optimum, it can also effectively identify all the additional local maxima within the fitness landscape as well. This too, has lead to the development of a plenitude of multimodal function optimisation algorithms and techniques in GA's, each with their own merits and shortcomings.

In this paper, a new multimodal function optimisation technique employing clustering and two co-evolving populations, is described. It attempts to address some of the deficiencies of other techniques, and truly overcome the requirement for a priori knowledge of the fitness landscape in order to function. The algorithm is shown to work on a standard test-bed function, and is compared to standard fitness sharing. In the first instance, however, some of the existing sharing techniques are reviewed.

2 Niche Methods

Goldberg and Richardson [4] defined a *fitness sharing* mechanism, where highly similar individuals in the population are penalised by a reduction in fitness. This causes population diversity pressure that allows the population to maintain individuals at local optima. The sharing function, $sh(d_{ij})$, is used to derate an individual's fitness, where d_{ij} is the distance (usually phenotype distance) between two individuals, i and j, and is defined as:

$$sh(d_{ij}) = \begin{cases} 1 - \left(\dfrac{d_{ij}}{\sigma_{share}} \right)^{\alpha_{share}} & \text{if } d_{ij} < \sigma_{share} \\ 0 & \text{otherwise} \end{cases} \quad (1)$$

Here, σ_{share} is defined as the maximum distance between strings required to form a predicted number of niches in the parameter space. The niche count, m_i, for each individual i, is calculated by summing the sharing function values over all the individuals in the population:

$$m_i = \sum_{j=1}^{N} sh(d_{ij}) \quad (2)$$

The shared fitness of an individual i, is given by dividing the individual's raw fitness by the niche count, m_i, for that individual:

$$f_{sh,i} = \frac{f_i}{m_i} \quad (3)$$

This scheme allows stable subpopulations to form around the different optima in the parameter space. However, the scheme is computationally expensive; of the order $O(n^2)$, and is restricted by a fixed choice of the value of σ_{share}. The value of σ_{share} determines the maximum number of peaks that the GA can populate, and also assumes that each peak is of equal height and width. In order to select this value effectively, prior knowledge of the fitness landscape is required, and this is not always available.

Yin and Germay [10] described a nicheing algorithm that used a form of adaptive clustering in order to remove some of the need for *a priori* knowledge of the fitness function. This was achieved by using MacQueen's adaptive KMEAN clustering algorithm to divide the population up into k clusters of individuals, corresponding to k niches. Here, the value k is determined by the algorithm. The shared fitness of an individual, i, is determined using equation (3), as in Goldberg and Richardson's *sharing* scheme, except that m_i is calculated as the approximated number of individuals in the cluster to which the individual i belongs:

$$m_i = n_c - n_c * \left(\frac{d_{ic}}{2d_{max}} \right)^\alpha \qquad x_i \in C_c \qquad (4)$$

Where α is a constant, d_{ic} is the distance between the individual i and the cluster's centroid $G(C_c)$, and n_c is the number of individuals in cluster c. Two further parameters are required, d_{min} and d_{max}, which define the minimum and maximum radii of a cluster. In order to choose these values effectively, the algorithm needs prior knowledge about the fitness landscape. After determining the positions of the clusters, the clusters are then compared. Two clusters are merged if the distance between their centroids is smaller than the threshold value, d_{min}. If an individual is further than d_{max} from all existing clusters, a new cluster is formed with that individual as a member. This method allows the formation of stable subpopulations, but it does require initial values of k, d_{max} and d_{min}. In addition to this, it does not provide any explicit information about the individuals in each of the clusters.

Pictet et al [9] took Yin and Germay's sharing scheme using cluster analysis a step further by modifying the sharing function, $f_{sh,i}$. Here, the sharing function penalises clusters with a large variance of the individual fitness values, and it also penalises clusters with too many solutions concentrated inside too small a region:

$$f_{sh,i} = \overline{f_c} - \left(\frac{N_c}{N_{av}} + \frac{1 - r_d}{r_d} \right) * \sigma(f_c) \qquad \forall x_i \in C_c \qquad (5)$$

Where N_c is the number of individuals in cluster c, $\overline{f_c}$ is the average fitness value of the individuals in cluster c, and $\sigma(f_c)$ is the standard deviation of the fitness values of the individuals in cluster c. The term N_c/N_{av} is used to control the number of individuals inside each cluster. The second term containing r_d is used to penalise

clusters with a concentration of individuals around their centroid that is too high. The value r_d is defined as:

$$r_d = \sqrt{\frac{1}{N_c} * \sum_{i=1}^{N_c} \frac{d_{ic}}{d_{max}}} \qquad (6)$$

This sharing scheme shifts the selection pressure from the individual to the cluster. Thus, the GA will attempt to find subpopulations of solutions with an average high fitness, instead of the best individual solution. Unfortunately, this scheme suffers from the same requirements as Yin and Germay's clustering scheme, with the addition of the choice of the N_{av} parameter.

Hanagandi and Nikolaou [7] proposed a similar scheme using clustering, except that here, there is no sharing function. Törn's clustering is applied to the population to identify the clusters, and the best member of each of these clusters is retained. The remainder of the population is then reinitialised, i.e. their chromosomes are randomised. This technique suffers from the stochastic sampling errors and genetic drift inherent in GA's, so the local elitism that the scheme employs may not be enough to maintain a cluster from generation to generation. However, this scheme by its very nature will, potentially, explore more of the parameter space in later generations than any of the other sharing mechanisms described above. Thus this scheme will never suffer from genetic stasis, but it does require a suitably large enough population size so that any two adjacent, randomly generated individuals will not be farther than the maximum radius of a cluster's boundary from each other. Otherwise, this means that a generation will consist of a number of isolated individuals and no clusters will be identified.

All of these techniques suffer, to a greater or lesser degree, from the requirement of at least some prior knowledge of the fitness landscape in order to set the appropriate values of variables that dictate restrictions, such as the maximum and minimum sizes of a cluster, within the schemes.

3 The Proposed Dynamic Niche Method

The proposed Dynamic Niche Clustering algorithm explicitly maintains a separate population of existing niches, in addition to the normal population. However, unlike Goldberg and Wang's *Adaptive Nicheing via Coevolutionary Sharing* [6], the niches in this scheme are not subject to modification by genetic operators. The existence of the niches is based entirely on the position and spread of the individuals in parameter space at a given point in time. Also, unlike Yin and Germay's *Clustering* [10], niches are retained from generation to generation, and not completely recalculated from the current population.

In this scheme, a niche consists of a number of variables. Each niche has a σ_{share} value, which is the niche radius in parameter space. In addition to this, each niche has a *midpoint* value, which indicates where the centre of the niche is in parameter space.

$$mid_j - \sigma_{share_j} \leq v_i \leq mid_j + \sigma_{share_j} \qquad (7)$$

An individual is considered to be a member of a niche if it falls within the radius defined by the *midpoint* and σ_{share} values for a particular niche. So an individual i, is a member of niche j, if equation (7) is true. Here, v_i is the phenotype value of individual i, and mid_j and σ_{share_j} are the midpoint and niche radii of niche j, respectively. A further value β_{share}, defines the inner niche radius that will be used to determine whether or not two niches will be merged together (see Sections 3.2 and 3.3). A niche also stores information defining the generation at which the niche was spawned, its original midpoint value, and a reference to each of the individuals that constitute its current members. This methodology allows an individual to be a member of more than one niche.

In this scheme, the values of σ_{share} and β_{share} are allowed to change, thus allowing the formation of non-equal hyper-volume niches. In order to prevent the unlimited growth or reduction of a niche's radius, four values; σ_{max}, σ_{min}, β_{max}, and β_{min}, were introduced. The parameters σ_{share} and β_{share} may not exceed these boundaries. A value of 0.15 was chosen for σ_{max}, and a value of 0.05 chosen for β_{max}. The values of σ_{min} and β_{min}, however, are dependent on the initial choice of σ_{share}, which itself is dependent on the population size and size of the parameter space.

In this methodology, the initial choice of value for σ_{share} is defined by the population size. The larger the population size, the smaller the initial value of σ_{share}, and hence the more niches may be found. A number of different schemes for the initial choice of σ_{share} were tried and are reported in [3], but for the purposes of this paper's conciseness, only three of these schemes will be described (See Table 1).

Table 1. Initial Choices of Niche Radii

Scheme	σ_{share}	β_{share}
Fixed Radius	0.1	0.05
Inverse Power Law	$1/2pop^{0.4}$	$1/4pop^{0.4}$
Inverse Law	$4/pop$	$2/pop$

In all schemes, $\sigma_{min} = \beta_{min}$ and $\beta_{min} = \beta_{share}/2$. Figure 1 shows how the initial choice of σ_{share} varies as population size increases for each of the three schemes.

Using a fixed niche radius scheme or a nicheing scheme with a minimum niche radius in a GA with fixed phenotype space, will only allow the formation of

a finite number of niches. This is especially the case in methodologies where niches (or clusters) can be merged. Figure 2 shows the maximum number of niches that may be found by Yin and Germay's *clustering* scheme [10], and the proposed schemes in Table 1.

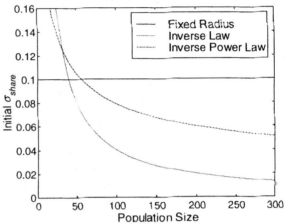

Fig. 1. Initial choice of σ_{share}

Figure 2 assumes that the midpoints of each niche will be the minimum distance from one another, and that the parameter space is normalised and fixed.

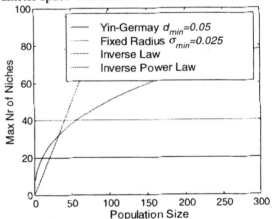

Fig. 2. Maximum Number of Niches

In the first generation, a niche is added to the niche set for each individual within the population, with its midpoint centred on that individual. In the initial and each subsequent generation, the following process is performed replacing any fitness scaling procedures:
1) Recalculate the midpoints of each of the niches in the current niche set. Four schemes have been proposed, see [3], but only *Fitness Distribution from Midpoint* will be described here. The midpoint of niche j will be modified according to:

$$mid_j = mid_j + \frac{\sum_{i=1}^{n_j}(v_i - mid_j)*f_i}{\sum_{i=1}^{n_j}f_i} \qquad (8)$$

Where mid_j is the midpoint of niche j, v_i is the phenotype value of individual i, f_i is the fitness of individual i, and n_j is the number of individuals within niche j. So here, the midpoint of each niche will be moved to the highest density of most fit individuals within the niche. If a niche has no members, then it is dropped from the current niche set.

2) The niche members are recalculated. If an individual is not a member of any niche, a new niche is formed, centred on that individual.

3) Each niche in the existing niche set is compared to every other niche in the niche set, twice. On the first pass:

a) If the midpoint of a niche is within β_{share} of the midpoint of another niche, then the two niches are *merged* together (See section 3.2).

On the second pass:

b) If the midpoint of a niche is further than β_{share} from the midpoint of another niche, but the two niche's radii overlap, then the two niches are *separated* (See section 3.3).

The *merge* and *separate* rules are applied until there are no further changes within the niche set.

4) The niche members are recalculated, and then the sharing function described in Section 3.1 is applied to each of the individuals within each of the niches.

3.1 The Sharing Function

A simple sharing function was used for this methodology, with the option of adapting it at a later date. For this investigation, the niche count is simply the number of individuals within the niche. The sharing function, $m_{csh,i}$, is defined as:

$$m_{csh,i} = \begin{cases} n_j & \text{if individual } i \in \text{niche } j \\ 0 & \text{otherwise} \end{cases} \quad (9)$$

Where n_j is the number of individuals within niche j. This is similar to the sharing function used by Miller and Shaw [8]. So now, the shared fitness of each of the individuals in niche j can be defined as:

$$f_{csh,i} = \frac{f_i}{m_{csh,i}} \quad (10)$$

3.2 Merging Two Niches

Two niches, i and j, will be merged together into a new niche if their midpoints are within β_{share} of one another, i.e.

$$mid_i + \beta_{share_i} \geq mid_j \geq mid_i - \beta_{share_i} \quad (11)$$

The new niche midpoint is given by using one of five schemes. Only the *Fitness Distribution from Naïve Midpoint* scheme is described in this paper, the rest may be found in [3]. In this scheme the midpoint is moved to the average weighted distance of each of the

individuals within both the niches, from the naïve midpoint of both the niches. The naïve midpoint is defined as:

$$mid_{naive} = mid_i + \frac{mid_j - mid_i}{2} \quad (12)$$

Where $mid_j > mid_i$. The new niche midpoint is defined as:

$$w_i = \sum_{a=1}^{n_i} (v_a - mid_{naive}) * f_a \quad \forall a \in i$$

$$w_j = \sum_{a=1}^{n_j} (v_a - mid_{naive}) * f_a \quad \forall a \in j \quad (13)$$

$$mid_{new} = mid_{naive} + \frac{w_i + w_j}{\sum_{a=1}^{n_i} f_a + \sum_{b=1}^{n_j} f_b}$$

Where n_i is the number of individuals in niche i, f_a is the fitness of individual a, and v_a is the phenotype value of individual a. The niche radius of the new niche is determined by comparing the new niche midpoint to the midpoints of the two original niches. There are three possible cases that may occur:

1) $mid_{new} < mid_{naive}$, i.e. the new midpoint is nearer to niche i. In this case, $\sigma_{sharenew}$ is calculated as the distance to the left-most extent of niche i or j. The value of $\beta_{sharenew}$ is calculated as the old value of β_{share}, plus half the change in $\sigma_{sharenew}$ and the old value of σ_{share}. If $mid_i - \sigma_{sharei} < mid_j - \sigma_{sharej}$, then:

$$\sigma_{share_{new}} = mid_{new} - (mid_i - \sigma_{share_i})$$

$$\beta_{share_{new}} = \beta_{share_i} + \frac{\sigma_{share_{new}} - \sigma_{share_i}}{2} \quad (14)$$

Otherwise:

$$\sigma_{share_{new}} = mid_{new} - (mid_j - \sigma_{share_j})$$

$$\beta_{share_{new}} = \beta_{share_j} + \frac{\sigma_{share_{new}} - \sigma_{share_j}}{2} \quad (15)$$

2) $mid_{new} > mid_{naive}$, i.e. the new midpoint is nearer to niche j. In this case, $\sigma_{sharenew}$ is calculated as the distance to the right-most extent of niche i or j. The value of $\beta_{sharenew}$ is calculated as the old value of β_{share}, plus half the change in $\sigma_{sharenew}$ and the old value of σ_{share}. If $mid_i + \sigma_{sharei} > mid_j + \sigma_{sharej}$, then:

$$\sigma_{share_{new}} = (mid_i + \sigma_{share_i}) - mid_{new}$$

$$\beta_{share_{new}} = \beta_{share_i} + \frac{\sigma_{share_{new}} - \sigma_{share_i}}{2} \quad (16)$$

Otherwise:

$$\sigma_{share_{new}} = (mid_j + \sigma_{share_j}) - mid_{new}$$

$$\beta_{share_{new}} = \beta_{share_j} + \frac{\sigma_{share_{new}} - \sigma_{share_j}}{2} \quad (17)$$

3) The new midpoint is equal to the naïve midpoint, i.e. $mid_{new} = mid_{naive}$. In this case, $\sigma_{sharenew}$ is calculated as half the distance between the outermost extents of niche's i and j.

The value of $\beta_{sharenew}$ is calculated as the largest value of β_{share} of niche i or j, plus half the change in $\sigma_{sharenew}$ and the largest value of σ_{share} of niche i or j.

$$\sigma_{sh_{new}} = \frac{(mid_{right} + \sigma_{sh_{right}}) - (mid_{left} - \sigma_{sh_{left}})}{2} \quad (18)$$

In all three cases above, if $\sigma_{sharenew} > \sigma_{max}$ then $\sigma_{sharenew} = \sigma_{max}$. Similarly for $\beta_{sharenew}$, if $\beta_{sharenew} > \beta_{max}$ then $\beta_{sharenew} = \beta_{max}$.

$$\beta_{share_{new}} = \beta_{sh_{l\,arg\,est}} + \frac{\sigma_{share_{new}} - \sigma_{sh_{l\,arg\,est}}}{2} \quad (19)$$

3.3 Separating Two Niches

Two niches, i and j, will be *separated* if their niche radii overlap, but the niches themselves do not *merge* (See Section 3.2), i.e.

$$mid_i - \sigma_{sh_i} < mid_j - \sigma_{sh_j} < mid_i + \sigma_{sh_i} \quad (20)$$

Each niche's σ_{share} value will be reduced by an amount so as to stop the niche radii overlapping. Three schemes have been proposed, but only the *Niche Fitness Separate* scheme will be described here. A complete description of the other schemes may be found in [3]. In this scheme, the niche radii will be reduced proportionally in favour of the niche with highest average fitness.

$$crossover = (mid_i + \sigma_{sh_i}) - (mid_j - \sigma_{sh_j})$$

$$\sigma_{sh_i} = \sigma_{sh_i} - crossover * \frac{\overline{f_j}}{\overline{f_i} + \overline{f_j}} \quad (21)$$

$$\sigma_{sh_j} = \sigma_{sh_j} - crossover * \frac{\overline{f_i}}{\overline{f_i} + \overline{f_j}}$$

Here, $\overline{f_i}$ is the average fitness of all the individuals within niche i. Thus, the niche with higher average fitness will not have its niche radius reduced by as much as the less fit niche. The values of β_{share} for both niches are reduced by half the change in the niche's radius.

$$\beta_{share_i} = \beta_{share_i} - \frac{\sigma_{share_i^{new}} - \sigma_{share_i^{old}}}{2} \quad (22)$$

Where $\sigma_{share_i}{}^{old}$ is the old niche radius of niche i, and $\sigma_{share_i}{}^{new}$ is the new niche radius of niche i. If $\sigma_{share} < \sigma_{min}$ then $\sigma_{share} = \sigma_{min}$. Similarly for β_{share}, if $\beta_{share} < \beta_{min}$ then $\beta_{share} = \beta_{min}$.

3.4 Complexity Issues

In the initial generations, there will be an equivalent number of niches and individuals. The number of comparisons between niches and individuals equates to approximately $O(n^2)$. In later generations, when both populations of individuals and niches have stabilised, the number of comparisons is much less; of the order $O(nq)$, where q is the number of peaks in the fitness landscape. The proposed scheme is greatly sped up by sorting the population in phenotype space using *quicksort*, prior to any niche clustering. Thus, the number of comparisons made between niches and individuals can be minimised. So now, the overall complexity of the proposed scheme is of the order $O(ngq)$, where g is the number of generations for which the scheme is run.

4 Test Functions

The Dynamic Niche Clustering algorithm has been tested on many standard test-bed functions, see [3]. In this paper, the results from tests performed on function $F1$, as defined in [1] and [5], are described. Function $F1$ has five peaks of decreasing height in the range $0 \leq x \leq 1$, and is defined as:

$$F1(x) = e^{-2\log(2)*\left(\frac{x-0.1}{0.8}\right)^2} * \sin^6(5\pi x) \quad (23)$$

The exact positions and values of the maxima are given in Table 2.

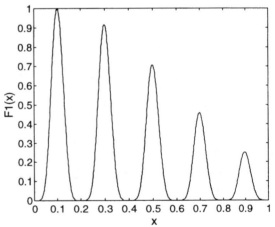

Fig. 3. Decreasing Maxima Function $F1$

5 Results

In order to provide a performance criterion, the chi-square like error distribution measure developed in [2] is used to compare the actual distribution of individuals in the parameter space after 50 generations, to the ideal distribution of individuals for function $F1$. The chi-square like measure is defined as:

$$Chi-square = \sqrt{\sum_{i=1}^{q+1}\left(\frac{x_i - \mu_i}{\sigma_i}\right)^2} \quad (24)$$

Where x_i is the actual distribution of niche i, μ_i is the ideal distribution, and σ_i is the standard deviation for niche i. An individual is considered to be a member of a peak if it has a fitness value higher than $\in=80\%$ of the maximum fitness of the peak.

Table 2. *F1* peaks

Peak	X	Fitness
1	0.1	1
2	0.29942	0.917236
3	0.49883	0.707822
4	0.69825	0.459546
5	0.89767	0.251013

In order to present a fair comparison of the proposed schemes and standard nicheing, the same genetic operators were used. Each individual consisted of one 30-bit chromosome which was mapped into the interval $0 \leq x \leq 1$. The mutation rate, $p_m=0.0$, the crossover rate $p_c=1.0$ with 1 point crossover. There was no elitist selection, and remainder stochastic sampling was employed. Population sizes of 50, 100, 150, 200 and 300 were tested, and the overall performance of each of the schemes is the average chi-squared performance over 10 runs for each population size. In each case, the generation gap, $G_{gap}=1.0$.

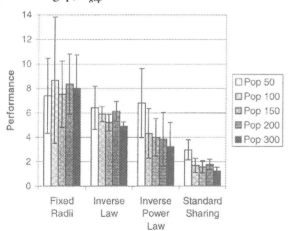

Fig. 4. Chi-Squared Performance

Figure 4 shows the average chi-squared performance of each of the four schemes (three proposed schemes and classic sharing) over the 10 runs made. As can be seen, standard sharing performs better than the proposed scheme, however, computational time is not included in this performance criterion. Figure 5 shows the average time taken for a run. It is obvious that the time taken for the standard sharing scheme is increasing exponentially as population size increases, whereas the time taken for the proposed scheme is increasing linearly as population size increases.

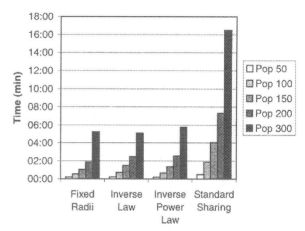

Fig. 5. Average Time per Run

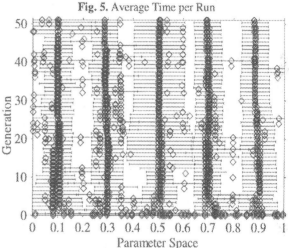

Fig. 6. Population and Niche Spread, Inverse Power Law, with population size of 50

Figure 6 shows the spread of individuals and niches throughout the parameter space, as it varies from generation to generation. The individuals are shown by the ◊ symbol, and the midpoint of a niche is given by the X symbol. Each niche has its niche radius shown by an error bar to either side of the midpoint. Niches are shown slightly above the generation to which they belong. It can be seen that there are concentrations of individuals at each of the peaks in the parameter space of the fitness function throughout the lifetime of the population. In addition to this, the midpoints of each of the niches can also be found at the positions of the peaks in parameter space. Figures 7 and 8 show close-ups of the beginning and end generations.

The individuals in the initial generation 0, are randomly spread about the parameter space. For every individual in the population a niche is spawned, centred on that individual. This can be seen by looking at the line of niches just above the initial generation 0, in Figure 7.

The final generation for this run can be seen to

consist of five distinctly separate niches, each with a midpoint at a peak in the parameter space (See Figure 8).

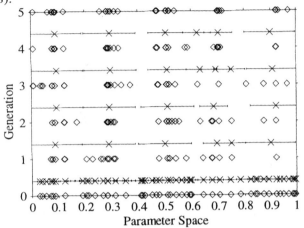

Fig. 7. Initial Population and Niche Spread, Inverse Power Law, with population size of 50

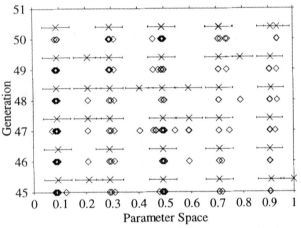

Fig. 8. Final Population and Niche Spread, Inverse Power Law, with population size of 50

Figure 9 shows the final population spread against fitness for the *Inverse Power Law* scheme with a population size of 50.

If an inappropriate choice of initial σ_{share} is made, i.e. the population size is too large, then the proposed scheme will attempt to find too many niches. This can be seen in Figure 10 with the formation of a number of *striations* at each of the peaks. These *striations* are formed when niches overlap at the peaks. Individuals are doubly penalised by the simple sharing function, $m_{chs,i}$, see equation (9), for being members of more than one niche, so individuals that fall under two niches will tend not to be selected for the next generation. Thus, where two or more niches overlap, the individuals will tend to be clustered around the midpoints of each of the niches and not under the overlapping area of the niches. These *striations* are more evident in Figure 11.

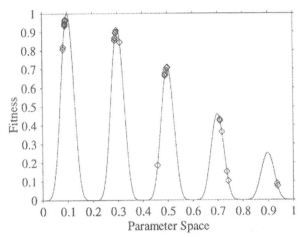

Fig. 9. Final Population Spread, superimposed on fitness function

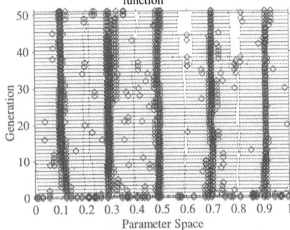

Fig. 10. Population and Niche Spread, Fixed Radius Scheme, with population size of 50

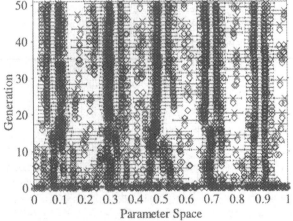

Fig. 11. Population and Niche Spread, Inverse Law, with population size of 150

The proposed Dynamic Niche Clustering scheme will also output the fitness statistics of the niches that it has found. Table 3 is the output of the *Inverse Power Law* scheme with a population size of 50.

Table 3. Discovered Niches

C	mid_c	σ_{share_c}	n_c	$\bar{f_c}$	$\sigma(f_c)$
1	0.089587	0.052281	17	0.909253	0.06531
2	0.293458	0.052281	13	0.876215	0.01998
3	0.493237	0.052281	10	0.637813	0.1518
4	0.712071	0.052281	8	0.343388	0.12759
5	0.935814	0.052281	2	0.080303	0.00618

6 Conclusion and Future Research

A Dynamic Niche Clustering scheme utilising a hybrid form of clustering and two co-evolving populations has been presented in this paper. It has not only been shown that the proposed scheme performs almost as well as standard sharing, but it is also more efficient than Goldberg's standard sharing scheme. This is due to the fact that the complexity of the proposed scheme is of the order $O(nq)$. It has also been shown that the scheme is less restrictive in its requirements for *a priori* knowledge of the fitness landscape. Because the initial values of niche radii are dependent on the population size, and this determines the maximum number of niches that may be discovered, the only decision that needs to be made is on choice of population size. This is less restrictive than defining an explicit minimum niche radius value that is fixed, regardless of population size.

However, the scheme is far from complete, and presently suffers from a number of restrictions. First, the proposed scheme requires that the phenotype space is one-dimensional, normalised and fixed. So for n-dimensional fitness functions, the n-dimensions must be mapped down into one dimension. Second, if an inappropriate population size is chosen, the scheme may potentially find more niches than there are peaks. This leads to the formation of the *striations* seen in Section 5. Third, the scheme assumes that each of the peaks is symmetrical about its maxima, and that the radii extend the same distance either side of the niche midpoint. This could potentially be a problem if the fitness landscape consists of irregular, non-symmetrical peaks. Proposals for further work to solve some of these restrictions are described below.

The *striations* that were seen in the population spreads in Section 5 can be eliminated by the use of a more sophisticated sharing function, $m_{csh,i}$, equation (9). Yin and Germay's approximated number of individuals in the niche (see equation (4) in [10]) will be used. But here, instead of the d_{max} parameter, the proposed scheme will use the current niche's radius value, σ_{share}.

Elitist selection in multimodal landscapes has already been implemented in Dynamic Niche Clustering. The niche set has two additional variables; the minimum niche count, c_{min}, and the elitist selection, *elite*. Here, if a niche has at least c_{min} members, then the top *elite* members of the niche are automatically retained in the next generation. The values of c_{min} and *elite* are fixed for all niches. The preliminary results are favourable and niche losses are dramatically reduced.

The proposed scheme will merge and separate two niches naïvely, without analysing the fitness statistics of the members of either niche. Work on a *selective merge* scheme is currently underway, where the choice to merge two niches is based on the average fitness of the members to either side of each niche's midpoint. This can be seen to act as a form of fitness gradient analysis.

References

[1] Beasley, D. & Bull, R.R. & Martin, R.R.: A Sequential Niche Technique for Multimodal Function Optimization, Evolutionary Computation 1(2), pp101-125, MIT Press 1993.

[2] Deb, K. & Goldberg, D.E.: An Investigation of Niche and Species Formation in Genetic Optimization, Proc 3rd Inter. Conf. Genetic Algorithms, pp42-50, 1989.

[3] Gan, J.: A Genetic Algorithm with Dynamic Niche Clustering for Multimodal Function Optimisation, (Internal Report No. 98-001) Cybernetics Dept, Reading University, 1998.

[4] Goldberg, D.E. & Richardson, J.: Genetic Algorithms with Sharing for Multimodal Function Optimization, Proc. 2nd Inter. Conf. Genetic Algorithms, pp41-49, 1987.

[5] Goldberg, D.E.: Genetic Algorithms in Search Optimization & Machine Learning, Addison-Wesley 1989.

[6] Goldberg, D.E. & Wang, L.: Adaptive Nicheing Via Co-evolutionary Sharing, Quaglianell et al (Eds) Genetic Algorithms in Engineering and Computer Science, pp21-38, John Wiley and Sons, Ltd. 1997.

[7] Hanagandi, V. & Nikolaou, M.: A Hybrid Approach to Global Optimization using a Clustering Algorithm in a Genetic Search Framework, Computers and Chemical Engineering 1995.

[8] Miler, B.L. & Shaw, M.J.: Genetic Algorithms with Dynamic Niche Sharing for Multimodal Function Optimization, IEEE International Conference on Evolutionary Computation, pp786-791, Piscataway, NJ: IEEE Press 1995.

[9] Pictet, O.V. & Dacarogna, M.M. & Davé, R.D. & Chopard, B. & Schirru, R. & Tomassini, M.: Genetic Algorithms with Collective Sharing for Robust Optimization in Financial Applications, Olsen & Associates Working Paper (OVP.1995-02-06) 1995.

[10] Yin, X. & Germay, N.: A Fast Genetic Algorithm with Sharing Scheme Using Cluster Analysis Methods in Multimodal Function Optimization, Proc. Inter. Conf. Artificial Neural Nets and Genetic Algorithms, pp450-457, Innsbruck, Austria 1993.

Time and Size Limited Harvesting Models of Genetic Algorithm

Subbiah Baskaran[1,2,3] and David Noever[3]
[1]Institut fuer Theoretische Chemie, Waehringerstrasse 17, A-1090, Wien, Austria
[2]Raytheon ITSS [3]Biophysics Branch ES76
National Aeronautics and Space Administration
George C. Marshall Space Flight Center Huntsville, AL-35812 USA
Email: {subbiah.baskaran,david.noever}@msfc.nasa.gov

Abstract

In this paper we formulate and investigate a novel model of a Genetic Algorithm (GA) in which the genetic population is allowed to grow with a delay in selection. And during selection, the excess growth over a preset constant size is harvested. Two possible delay modes result in two harvesting schemes called time and size limited harvesting. The two schemes generalize the standard genetic algorithm in the direction of treating population size as a stochastic parameter. If the delay threshold is one, then both schemes reduce to the standard genetic algorithm. The retention of low fitness members for extended period in the evolving population promotes preservation of schema pathways which enable escape from local optima and also help alleviate premature convergence. The extended model is successfully applied to a difficult two-dimensional non-stationary problem for tracking time-varying optima in real time.

1 Standard Genetic Algorithm

In the standard generational binary Genetic algorithm (GA) the following dynamical steps are done: 1) a fixed size initial population is generated. The individuals of the population are initialized as fixed-length strings of 0's and 1's in a random way, the length being determined by the number of objective variables and their precision in the underlying optimization problem. Thus each string essentially encodes a trial solution to the problem under optimization. 2) Through operators of crossover and mutation, new genetic strings called offsprings are generated. In crossover genomic segments are exchanged between two selected strings, and in mutation one or more bits of the same string are modified. During this process only a finite number of new members are allowed to be generated. 3) the population is evaluated by applying a fitness function to every member of the population (fitness represents the quality of the coded solution of an individual). 4) Reproductive trials are given to a member according to its fitness relative to the fitness of the remaining population. During this selection phase some of the old (poor fitness) members are deleted from further evolution. 5) Steps (2 to 4) are repeated until a certain prescribed condition is met. This stopping condition will mostly be either a fixed number of generations or a certain alloted quantum of processing time. Usually the best member will be decoded and its solution used as the outcome of the genetic algorithm.

2 The Harvesting Models

Holland's original formalism [1] as well as other extensions of the genetic algorithm evolve with fixed size populations. In them the size of evolving population is kept fixed for the entire course of the algorithm and for that at the end of each generation some of the members of the old population must be deleted to accommodate the new members created by genetic operators. This corresponds to viewing evolution as a self organization process for fitness improvement under constant population size with birth and death of individuals.

However recent discoveries in molecular biology reveal that the genetic information in the cell machinery is stored in large quantities and only a small portion of that is actively used in evolution, and the rest remains as a dormant or junk reserve[2]. At present we do not have a very clear hypothesis on their existence in the cell machinery. It appears that during evolution functionally connected clusters of genomes may become active and inactive at large intervals in time. They mostly accumulate without deterministic elimination based on the total carrying capacity of the genetic reservoir. The present model is based on the above concept of "pseudo genes" [2]. Here we use harvesting as a means of dynamically marking the active members from the inactive rest. If we imagine that a constant value is preset for the size of the active region and all the genomes are linearly arranged according to their fitness starting with the most fit first, then growth corresponds to moving the inactive members into the active region, and harvesting corresponds to sorting the genomes up to the growth and resetting the size of the active region to the already preset value.

In the language of the standard Genetic Algorithm this corresponds to allowing the genetic population to grow stochastically during evolution. The algorithm starts with a fixed population size, P_0 and a maximum size, P_M for the genome bank as a constant multiple of P_0 such that $P_M = \epsilon P_0$. Here P_M is analogous to the carrying capacity of an organism, ϵ the explosion factor. During each generation a small number of new members are generated stochastically by operators like crossover and mutation and added to the population without deleting any old ones. And the selection phase is not applied immediately at the end of each generation. Hence the population accumulates monotonically retaining members of differing fitnesses. Such a growth is regulated or harvested in two ways. This delay in selection is allowed until either a given number of generations are gone, or the excess growth of population exceeds a certain threshold.

During selection the population is down sized or harvested to the preset constant value. At the end of a growth prior to harvesting, the population is sorted by fitness and the best members filling up to the population size are marked as active and the rest as dormant or inactive genomes. This constitutes one harvesting cycle. Many such harvesting cycles are executed before halting the genetic algorithm. In the early cycles, harvesting will correspond to the dynamic relabeling of active and inactive genomes and the growth will be allowed to accumulate up to the carrying capacity without physical deletion of genetic members. In general with each harvesting cycle the total genomes in the gene pool will increase. However after several cycles the carrying capacity of the pool will be exceeded and the members at the poor end of the sorted spectrum in the dormant reserve will be overflown into extinction.

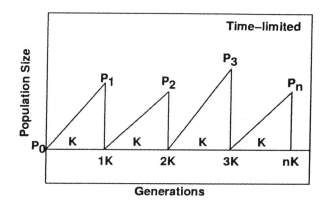

Fig. 1. Time Limited Harvesting Model. The saw tooth curve has constant widths and different heights.

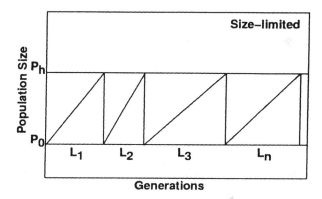

Fig. 2. Size Limited Harvesting Model. The saw tooth curve has different widths and constant heights.

The concept of harvesting has been successfully applied by the authors to explain the algorithm behavior of an another genetic algorithm model called, "Excursion Set Mediated Genetic Algorithm" [3].

The dormant genomes are kept silent or inactive in the genetic reservoir for several generations until with a very low probability every now and then a member from the dormant pool is mutated and selected to enter the recombination operation with those of the active members. This enables occasional testing of a new trial solution (strategy) and its consequent incorporation into normal evolution. The present algorithm will not in principle suffer from premature convergence as it allows the coexistence of poor as well as fitter members in the population during evolution.

2.1 Scheme-A Time Limited Harvesting

Figure 1 shows schematically the time limited harvesting procedure. This may also be called the generation limited harvesting. In this scheme, we start with a finite population size. During each generation crossover and mutation are done producing a small number of additional members which are just added to the population with no deletion of old members. This continues until we pass a fixed number of generations K, called the harvesting threshold which means that there are K generations in a harvesting cycle. Let g_1, g_2, ..., g_K be the individual growths at generations 1, 2, ..., K within a harvesting cycle and let P_1, P_2, ..., P_n be the growth accumulations at the end of cycles 1, 2, ..., n, then $P_j = \sum_{i=1}^{K} g_i$ is the accumulated growth for the jth cycle. Since K is a constant for the time limited scheme, harvesting is done at generations:

$$K, 2K, 3K, \ldots, nK$$

where n is the number of cycles performed. If Δt refers to the elapsed generations after a previous harvest then $\Delta t = K$ which is the length of the harvesting cycle.

During accumulation phase, the population grows steadily, but not to infinite size, as reproduction is done only probabilistically on selected pairs. And the population growth will be a saw tooth curve with unequal tooth height and equal tooth widths (figure 1). At the end of harvesting cycle, selection is applied and population is harvested to the original constant value. This process is repeated until the algorithm obtains steady state convergence or a prescribed number of cycles are done.

2.2 Scheme-B Size Limited Harvesting

Figure 2 shows size limited harvesting schematics. All the algorithmic steps are same as the previous scheme except that the growth is continued until the accumulated growth sum exceeds the prescribed threshold size, P_h. Here L_1, L_2, ..., L_n are the number of generations contained in the harvesting cycles 1, 2, ..., n respectively and are not equal to each other as in the previous scheme. Let g_1, g_2, ..., g_L be the individual growths at generations 1, 2, ..., L, and let $P_L = \sum_{i=1}^{L} g_i$ be the accumulated growth up to generation L. If P_h refers to the constant threshold then size limited harvesting takes place whenever the growth sum P_L exceeds the constant threshold P_h. Thus the harvesting condition reads:

$$P_L \geq P_h$$

Since genetic operators reproduce new members stochastically in small random amounts there will be always an excess stochastic production $\Delta P_j = P_j - P_h$ in any harvesting cycle, j. The population growth curve will be a saw tooth with unequal tooth widths and constant heights when this excess productions are subtracted out (figure 2).

Schemes A and B represent two stochastic harvesting schemes which are complementary to each other: in scheme A the excess growth determines the harvesting condition whereas in scheme B the condition was the fixed elapsed generations between two successive over growths. For both schemes the quantities $g_i's$ always refer to the growth excess over P_0 which is the population size preset in the beginning of the algorithm as a constant for the entire course of the algorithm. If the delay period is one, then both reduce to the standard genetic algorithm. In both schemes the delay in selection permits the co existence of poor as well as fitter members in the population during growth prior to harvesting. In addition at the end of any harvesting cycle the less fit genomes are not totally removed from the population reservoir. Instead these members are marked dormant and allowed to remain at the poor end of the sorted population spectrum.

3 Non-stationary Optimization

Several earlier studies developed genetic algorithm variants for non-stationary optimization including models based on diploid/dominance with homologous allele hypothesis, immune systems, thermodynamic models, and structured genetic algorithms with differing degrees of success [1, 5, 4, 6, 7, 8]. Mostly tests were done for one dimensional problems only.

The retaining of a definite portion of genetic material constantly in the computational apparatus is expected to show properties like long term memory useful in tracking time varying optimal solutions [10, 11]. In addition the co-existence of poor as well as fitter members in the population during evolution may also promote pathways to lost schemas that will enable global exploration through occasional testing of different solutions in addition to exploiting local information for appropriate sustaining of optima hitherto discovered.

In this section we test our model on a two dimensional non-stationary optimization problem defined by the following fitness landscape:

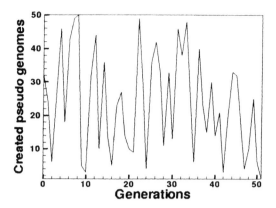

Fig. 3. The plot of pseudo-genomes added at each generation in size limited harvesting which accumulate until harvesting.

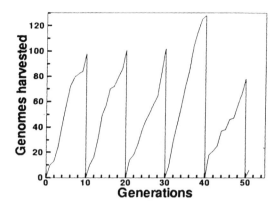

Fig. 4. A typical time limited harvesting profile for the test problem.

$$f_t(x,y) = 0.2((x-\alpha_t)^2 + (y-\beta_t)^2) \\ + 4sin(10(x-\alpha_t+y-\beta_t)) + 4$$

where

$$\alpha_t = 5 + 2cos(2\pi t/100.0); \quad \beta_t = 5 + 2sin(2\pi t/100.0)$$

and the variables (x,y) are defined in the square $[0.0, 10.0] \times [0.0, 10.0]$.

For all time, t, there exists a unique global optimum of f_t and is close to (α_t, β_t). The trajectory of this time-varying minimum is a circle of radius 2.0 centered at a point close to $(5.0, 5.0)$. The algorithm is supposed to locate them. This is a very hard test problem for non-stationary optima tracking algorithms. Although there is no hope of effective adaptation in the worst case for search, some forms of fitness variability and initial conditions do provide opportunities for successful adaptation and discovery of optima. The results for the locus of the moving optima found by our algorithm is shown in Figure 7.

For the model simulation reported here the population size was kept at 100, the population explosion factor ϵ was kept at 3 which allowed a maximum carrying capacity of $P_M = \epsilon P_0 = 300$. For the size limited harvesting the threshold was kept at 100. The upper limit for stochastic growth was kept at

50 in any generation. The time limited harvesting threshold was kept at 10 generations. The gray coding was used for the two objective variables. The mutation rate was kept at 0.125, and the uniform crossover rate was kept at 0.800. For selection two tournament with a tournament probability of 0.800 was applied. The starting population was initialized near origin $(0.0, 0.0)$ where the landscape offers maximum resistance for optimization. We obtained similar results when the initial population was generated uniformly randomly from the entire square. However experiments for very low mutation rates (0.001 and 0.01) showed less robust results by getting trapped in few or several of the possible optima along the ring instead of distributively discovering most of them as for the case of appreciably high mutation rates reported here. This indicates that the current algorithm may show better robust performance with a form of adaptive mutation operator that permits high rates at the beginning of the search and slowly decays to lower rates as the search proceeds to the end and this needs to be tested.

4 Results and Discussion

Figure 3 shows the number of the pseudo-genomes, $g_i's$ created in each generation by mutation and crossover during a typical size limited harvesting run for the problem. This stochastic growth is the essential source of fitness diversity and randomness in the algorithm.

Figure 4 shows a typical time limited harvest-

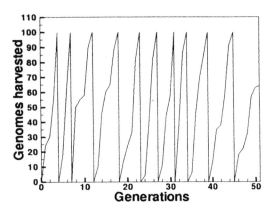

Fig. 5. A typical size limited harvesting profile for the test problem.

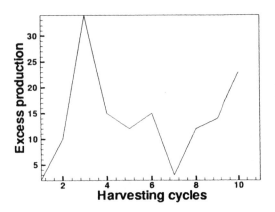

Fig. 6. Excess genome production for size limited harvesting.

ing profile for the problem studied. As predicted in the schematics the population growth curve is a saw tooth curve with unequal tooth heights and constant tooth widths (10 generations). Here the growth between two harvesting cycles is a stochastic variable. The graph shows only the profile up to 50 generations. However the algorithm was run for several harvesting cycles before halting it.

Figure 5 on the other hand shows the size limited harvesting profile for the same problem. As expected from the theoretical schematics this profile exhibits unequal saw tooth width corresponding to the stochastic time intervals between successive harvests. In this scheme the harvesting condition is met almost always as an inequality. And hence there will always be an excess growth of genomes over the threshold and these are subtracted in the figure. In other words the actual experimental saw tooths will in general be unequal in heights signaling this stochastic excess production.

Figure 6 in fact shows exactly the corresponding excess production over the harvesting threshold as a function of harvesting cycles. We note that there are 10 harvesting cycles contained within the first 50 generations. This excess stochastic production introduces an additional source of genomic variability and is characteristic of the algorithm. In both schemes harvesting ensures maintaining the growing population to a near constant value so that genetic algorithm dynamics proceeds normally except the additional delay in selection.

Figure 7 shows the results of the optima tracking

by the proposed algorithm for the problem studied. The 12 insets in the figure depict the evolution of time dependent solutions at different generations as marked in the respective insets. From them one can easily observe that a robust self-organization emerges around generation 500 and afterwards the solutions steadily organize themselves on the solution ring. If we imagine each one of the square as one of the possible strategies, and those on the ring as corresponding to all the non stationary solutions then the current algorithm has succeeded in locating them. The proposed model is clearly robust in efficiently tracking not only a single optima (corresponding to a source) but in fact successfully located a set of moving optima (multiple moving sources). One note worthy observation is that the genetic population is not uniformly distributed around the annular ring as seen in the inset for generation, $g = 1250$ and a higher density population cluster might be indicative of the landscape region where the time dependent evolution is currently in progress. However a complete characterization of the algorithm for all possible parameter combinations remains to be done.

From the properties of this model we are able to conjecture a hypothesis on pseudo genes themselves. Their accumulation and presence in the biological organism might be to combat sudden and drastic changes in the environment that might otherwise wipe away the entire organism (or species) from existence. This is the first time a concept modeled from biology when studied in its abstract setting has

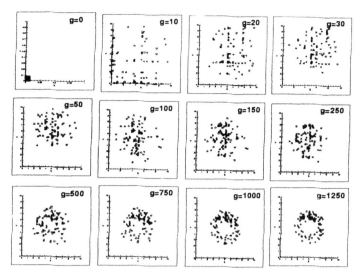

Fig. 7. Evolution of Optima tracking for a time-dependent fitness landscape in real time. Results from size limited harvesting. The g's in the insets refer to generations.

yielded light on its biological significance itself. This encouraging result will be further investigated using appropriate mathematics. Although the latent optimization behavior of these harvesting schemes has similarities with steady state algorithms (in having a large population readily available for genetic manipulations) their operational dynamics are evidently different [9, 10, 11].

5 Conclusions

In the present paper a novel model of genetic algorithm with time and size limited harvesting features based on delayed selection is forwarded. The performance of the new model is found to be robust when applied to a model non-stationary landscape for tracking moving optima in real time. Generic slowing of convergence is observed in both models during evolution. A mathematical analysis of these models is beyond the scope of the current paper and results of such an analysis will be reported elsewhere.

References

[1] Holland, J.H: Adaptation in Natural and Artificial Systems, Ann Arbor, University of Michi-gan Press, 1975.

[2] Watson, J.D et.al.,: Recombinant DNA, Second Edition, W.H. Freeman and Company, NY, 1992.

[3] Noever, D., Baskaran, S., Schuster, P.: Understanding Genetic Algorithm Dynamics using harvesting Strategies, Physica D, Vol. 72, pp. 456-435, 1995.

[4] Smith, R.E., Goldberg, D.E.: Diploidy and dominance in artificial genetic search, Complex Systems, Vol. 6, pp. 21-28, 1992.

[5] Gobb, H.G., Grefenstette, J.J.: Genetic Algorithms for Tracking Changing Environments, Proc. of 5th ICGA, Vol. 5, pp. 523-529, 1993.

[6] Mori, K., Tsukiyama, M., Fukuda, T.: Immune Algorithm with Searching Diversity and its Application to Resource Allocation Problem, Trans. IEE of Japan, Vol. 113-C, No. 10, pp. 872-878, 1993.

[7] Mori, N., Kita, H., Nishikawa, Y.: Application to a changing environment by means of the Thermodynamical Genetic Algorithm, Proc. of. 4th PPSN (PPSN'96), Vol. 4, pp. 513-522, 1996.

[8] Dasgupda, D., Mcgregor, D.P.: Non-stationary Function Optimization using the Structured Genetic Algorithm, Proc. of. 2nd PPSN (PPSN'92), Vol. 2, pp. 145-154, 1992.

[9] Whitley, D.: The genitor algorithm and selection pressure: why rank based allocation of reproductive trials is best, In Proceedings of the Third Int'l. Conference on Genetic Algorithms and their Applications, 116-121, J.D.Schaffer, (ed.), Morgan Kaufmann, June, 1989.

[10] Cerf, R.: The dynamics of mutation selection algorithms with large population sizes, Ann. Inst. H. Poincaré Probab. Statist., Vol. 32, pp. 455-508, 1996.

[11] Leung, Y., Gao, Y., Xu, Z.B.: Degree of population diversity - A perspective on premature convergence in genetic algorithms and its Markov chain analysis, IEEE Trans. Neural Networks, Vol. 8, pp. 1165-1175, 1997.

A General Model of Co-evolution for Genetic Algorithms

Jason Morrison and Franz Oppacher
{morrison,oppacher}@scs.carleton.ca

Intelligent Systems Lab
School of Computer Science
Carleton University
Ottawa, ON K1S 5B6

Abstract. Compared with natural systems, Genetic Algorithms have a limited adaptive capacity, i.e. they get quite frequently trapped at local optima and they are poor at tracking moving optima in dynamic environments. This paper describes a general, formal model of co-evolution, the Linear Model of Symbiosis, that allows for the concise, unified expression of all types of coevolutionary relations studied in ecology. Experiments on several difficult problems support our assumption that the addition of the Linear Model of Symbiosis to a canonical Genetic Algorithm can remedy the above shortcomings.

1 Introduction

Although Genetic Algorithms (GAs) have demonstrated their robustness and efficiency as search and learning techniques in many application domains, they often suffer from the problem of premature convergence: as the individuals in an evolving population approach a local optimum, the resulting loss of genetic diversity may prevent the GA from finding the global optimum[1].

A related problem arises in dynamic environments even if the landscape is unimodal. The loss of diversity in the population resulting from successfully converging to an optimum renders the GA incapable of tracking the optimum if it should drift to another point in gene space. Currently there are a limited number of mechanisms available, short of a restart, that allow the GA to find a new optimum (see [2, 4]).

It is the hypothesis of this paper that the limited adaptive capacity of the GA - which has no counterpart in natural systems and which is acknowledged by the widespread interest in the problem of premature convergence - can be overcome by enhancing the canonical GA with a 'plug-in', tunable co-evolution module.

Section 2 describes related work in coevolutionary computing, Section 3 gives the definitions necessary to express the coevolutionary relations studied in ecology, Section 4 presents our Linear Model of Symbiosis, Section 5 describes the results of our experiments with two classical GA test functions, and Section 6 concludes with a discussion of the properties and capabilities of our model.

2 Previous Work in Co-evolutionary Computing

In a classical GA each individual in the (single) population is considered a potential solution to the problem the GA attempts to solve, and an individual's fitness, which determines its probabilities of surviving and/or reproducing, measures how well it solves the problem. In co-evolutionary systems, by contrast, the fitness of an individual depends not only on how well it solves a problem but also on other individuals.

Since the introduction of co-evolution into evolutionary computing in [7, 8] many variants of co-evolution have been implemented. These variants fall into two broad classes.

The first and more popular class comprises systems that rely on a tournament approach to calculating 'Competitive Fitness Functions' (CFFs) (see e.g. [1, 6–8, 13, 17, 19]). To calculate a CFF requires two (or more) individuals to compete in a game and their fitness values are then based on relative performance in the game. The key idea is that given different pairs (or groups) of individuals the fitness value of a specific individual varies. Hence an individual's fitness is affected by other individuals in its population and therefore GAs that use CFFs are co-evolutionary.

The second class contains systems using multiple populations (see e.g. [15, 9]). Typically a problem broken into components and each component is assigned a different GA. The fitness value of an individual is calculated by randomly choosing the necessary individual(s) from the other population(s) and then evaluating how well this group of individuals

---

[1] e.g., [16, 12].

solves the problem. The individual being evaluated is then assigned its fitness according to how well it assisted in the solution or according to how well the entire group solved the problem. Since the fitness of an individual depends on the other individuals chosen, this type of system is also co-evolutionary.

Several authors, e.g. [7, 8, 17, 13, 14], claim significant successes for their co-evolutionary systems - they can find solutions that regular GAs missed or they can achieve more accurate solutions after fewer fitness evaluations.

Unfortunately, it is difficult to generalize previous work on co-evolutionary computation because it is plagued by inconsistent terminology and conflicting classification schemes[2] and because the various architectures are designed to deal with specific problems. We attempt to remedy these drawbacks by presenting a simple set of definitions that are in agreement with an influential portion of the ecological literature and by proposing a single model of co-evolution that can express all types of co-evolutionary relations studied in the ecological literature and that can be easily added to a regular GA.

3 A Definitional Framework for Co-evolutionary Relations

We follow [11] and [18] in adopting a wide notion of symbiosis which encompasses, e.g., parasitism and contrasts with interactions with the abiotic environment[3].

Definition 1. Symbiosis *is a relationship between two (or more) individuals such that the fitness of one individual directly affects the fitness of the other individual(s).*

While it is in principle possible to study symbiotic relationships between arbitrary and varying numbers of individuals, we confine ourselves here to look only at pairs of individuals.

Definition 2. *A symbiotic connection,* $A \rightarrow B$ *(A affects B) between two individuals A and B exists if and only if the fitness of A has a direct effect on the fitness of B.*

Most forms of symbiosis can be expressed in terms of two specializations of this broad notion of a symbiotic connection.

Definition 3. $A \stackrel{\pm}{\rightharpoonup} B$ *(A protagonizes B) if and only if there exists a connection* $A \rightarrow B$ *such that as the fitness of A increases, the fitness of B increases, and as the fitness of A decreases, the fitness of B decreases.*

Definition 4. $A \stackrel{\mp}{\rightharpoonup} B$ *(A antagonizes B) if and only if there exists a connection* $A \rightarrow B$ *such that as the fitness of A increases, the fitness of B decreases, and as the fitness of A decreases, the fitness of B increases.*

Since we wish to take the effects of individuals on the fitness of other individuals into account, we need to distinguish two notions of fitness. Absolute fitness coincides with the usual notion of fitness, i.e., it measures how well an individual solves the problem the GA is working on, and expressed fitness includes the effects of other individuals.

Definition 5. *The* absolute fitness *of individual* x_i *at time t,* $f_i^a(t)$, *depends only on the genotype and phenotype of* x_i *at t, and excludes the effects of other individuals.*

Definition 6. *The* expressed fitness *of individual* x_i *at time t,* $f_i^e(t)$, *depends on the genotype and phenotype of* x_i *at t, and on all individuals* x_j *such that* $x_j \rightarrow x_i$.

With these definitions in place it is possible to describe the various forms of symbiosis that exist in nature[4], including complex varieties of indirect mutualism[5] and indirect competition between more than two species or individuals. However, the following definitions are confined to the symbiotic relationships between pairs of individuals.

Definition 7. Amensalism *occurs between two individuals, Host and Amensal, if and only if Host* $\stackrel{\mp}{\rightharpoonup}$ *Amensal and* $\neg(Amensal \rightarrow Host)$.

Definition 8. Commensalism *occurs between two individuals, Host and Commensal, if and only if Host* $\stackrel{\pm}{\rightharpoonup}$ *Commensal and* $\neg(Commensal \rightarrow Host)$.

---

[2] e.g., Hillis' use of 'parasitism' to describe a model involving competition.

[3] Note also that we interpret co-evolutionary relations to obtain primarily between individuals, not species.

[4] We use [3, 11, 18] as base classification schemes and include all cases of pairwise relationships.

[5] See [10], who analyzes symbiotic relationships among more than two species. As an example of such a relation that we could handle in our framework, consider the following case: although species A competes with B, and B competes with C, the net interaction between A and C is mutualistic.

Definition 9. Competition *occurs between two individuals, CompetitorA and CompetitorB, if and only if CompetitorA $\overset{-}{\rightharpoonup}$ CompetitorB and CompetitorB $\overset{-}{\rightharpoonup}$ CompetitorA.*

Definition 10. Predation *occurs between two individuals, Predator and Prey, if and only if Predator $\overset{-}{\rightharpoonup}$ Prey and Prey $\overset{+}{\rightharpoonup}$ Predator.*

Definition 11. Mutualism *occurs between two individuals, SymbiontA and SymbiontB, if and only if SymbiontA $\overset{+}{\rightharpoonup}$ SymbiontB and SymbiontB $\overset{+}{\rightharpoonup}$ SymbiontA.*

The next relation is not classified in the ecological literature but can be usefully incorporated into evolutionary algorithms. It characterizes the case where an individual's current expressed fitness is affected by previous changes in its expressed fitness (as, e.g., in an individual's immune system).

Definition 12. Adaptism *occurs if and only if IndividualA $\overset{+}{\rightharpoonup}$ IndividualA or IndividualA $\overset{-}{\rightharpoonup}$ IndividualA.*

4 The Linear Model of Symbiosis

We begin the description of our model with Equation 1 which calculates expressed fitness at time t or generation t. In words, the expressed fitness of individual x_i at time t, $f_i^e(t)$, equals the expressed fitness of individual x_i at time $t-1$, $f_i^e(t-1)$, plus the total change, $\mathcal{C}_i(t)$, in the fitness of x_i at time t.

$$f_i^e(t) = f_i^e(t-1) + \mathcal{C}_i(t) \qquad (1)$$

This enigmatic "total change", $\mathcal{C}_i(t)$, can be broken down into two components: i) changes due to connections to individuals and ii) a change in absolute fitness of the individual.

Connections, are assumed to be independent of one another so that the first component of change, i.e., the total effect of all connections, is the sum of all effects due to individual connections. We represent the effect of connection $x_j \rightharpoonup x_i$ at time t by $c_{ij}(t)$. The second component of total change can effectively be ignored for most GAs because absolute fitness (i.e. how well an individual solves the problem) is static.

To determine initial conditions, assume that a new individual x_i is born at time t. Since x_i has not been affected by other individuals, its initial expressed fitness is its absolute fitness, $f_i^e(t) = f_i^a(t)$.

Similarly, since there was no previous fitness the total change is equal to the initial fitness $\mathcal{C}_i(t) = f_i^a(t)$. Using this idea of birth, and assuming that all individuals are born at time $t = 0$, the initial conditions of expressed fitness can be stated in Equation 2.

$$\begin{aligned} f_i^e(0) &= f_i^a(0) \\ \mathcal{C}_i(0) &= f_i^a(0) \end{aligned} \qquad (2)$$

Equation 3 represents the full formulation for $\mathcal{C}_i(t)$.

$$\mathcal{C}_i(t) = \sum_{j=1}^{S} c_{ij}(t) + (f_i^a(t) - f_i^a(t-1)) \qquad (3)$$

This leads to Equation 4 which is the defining equation for our model of symbiosis.

$$f_i^e(t) = f_i^e(t-1) + \sum_{j=1}^{S} c_{ij}(t) + (f_i^a(t) - f_i^a(t-1)) \qquad (4)$$

Using the simplest function for a connection leads to the linear connection. Equation 5 shows that the linear connection c_{ij} is a protagonist connection if $\alpha_{ij} > 0$ and an antagonist connection if $\alpha_{ij} < 0$. Throughout the remainder of this paper the values α_{ij} will be referred to as connection strengths or weights.

$$c_{ij}(t) = \alpha_{ij}\mathcal{C}_j(t-1) \qquad (5)$$

The Linear Symbiosis Model of Co-evolution models co-evolution using only linear connections. Given the form of the linear connections it is possible to restate the basic equation of the symbiosis model (Equation 4). Thus let $F^a(t)$ be the vector of absolute fitnesses for all S individuals in all populations $[f_1^a(t), \ldots, f_i^a(t), \ldots, f_S^a(t)]$. Similarly let $F^e(t)$ be the vector $[f_1^e(t), \ldots, f_S^e(t)]$ and $C(t)$ be the vector $[\mathcal{C}_1(t), \ldots, \mathcal{C}_S(t)]$. Finally let $\mathcal{A}$ be the matrix with components α_{ij}. Using these definitions with Equations 5 and 4 produces the defining equation of the linear model given in Equation 6.

$$\begin{aligned} F^e(t) = F^e(t-1) &+ AC(t-1) \\ &+ F^a(t) - F^a(t-1) \qquad (6) \\ \text{where} \quad C(t) &= F^e(t) - F^e(t-1) \end{aligned}$$

Thus a GA with our linear model of co-evolution uses the vector $F^e(t)$ as the fitnesses of all individuals instead of $F^a(t)$.

Table 1. The values in the column time $t-1$ are assumed to have been taken from midpoint in the individuals' lives. $\alpha_{ij} = 0.4$ and $\alpha_{ji} = 0.5$. It is important to note the change in $C_j(t)$ to 0 once the death has occurred. The bracketed value is the value that would have been appropriate had the death not occurred.

Quantity	$t-1$	t	$t+1$	$t+2$	$t+3$
$f_i^a(t)$	10	10	10	10	10
$f_i^e(t) = f_i^e(t-1) + c_{ij}(t) + f_i^a(t) - f_i^a(t-1)$	19.6	20.2	20.2	21.4	21.4
$c_{ij}(t) = \alpha_{ij}C_j(t-1)$	1.6	0.6	0	1.2	0
$C_i(t) = f_i^e(t) - f_i^e(t-1)$	1.6	0.6	0	1.2	0
$f_j^a(t)$	5	5	11	11	11
$f_j^e(t) = f_j^e(t-1) + c_{ji}(t) + f_j^a(t) - f_j^a(t-1)$	13.5	14.3	11.3	11.3	11.9
$c_{ji}(t) = \alpha_{ji}C_i(t-1)$	2.5	0.8	0.3	0	0.6
$C_j(t) = f_j^e(t) - f_j^e(t-1)$	2.5	(0.8) 0	3	0	0.6
Event		Death of x_j	Birth of new x_j		

5 Implementation and Experiments

In order to implement the expressed fitness of an individual, several questions concerning births and deaths must be answered. These questions are: How does the death of an individual in symbiosis affect its partners? How does a newborn affect expressed fitnesses? How does the system change as a whole because of births and deaths?

Suppose a child x_j is born at time $t+1$. At the time of birth the expressed fitness of x_j should be equal to the absolute fitness of x_j plus any effects felt at time t. However, since the child was not alive in the previous time then $C_j(t) = 0$. The remainder of the equations apply. In the model explored here each child is assigned its connections at birth and they remain constant throughout its life. When an individual dies the new child that is created to take its place inherits the dead individual's connections. This is possible because the population size is kept constant in the implemented GA.

Death and its effects are also critical to the implementation of the model. Consider again Equation 4. Suppose at time t an individual x_j is alive and affects individual x_i, and at time $t+1$ x_j is dead. This means that the change $C_j(t)$ in x_j between times $t-1$ and t will not affect individual x_i at time $t+1$ (i.e., $C_j(t) = 0$, regardless of $f_j^e(t)$ and $f_j^e(t-1)$).

Suppose that an individual x_i is in symbiosis with x_j and that x_j lives at time t and is dead at time $t+1$. Further suppose that at time $t+1$ a new individual is born and takes x_j's place. Since the death and birth both indicate that $C_j(t) = 0$ there is nothing further to model. If x_i and x_j are connected with the αs given in the caption, then their fitnesses would be calculated as shown in Table 1.

In this paper we present tests of a few basic types of symbiosis patterns that include feedback between individuals, i.e., Mutualism, Predation, Competition and Adaptism. In the case of Adaptism each individual is simply given a connection to itself and no other connections are made. The other cases involve pairing individuals x_i and x_{i+1} were $i = 0 \pmod 2$. Each pair is then given connections appropriate to the type of symbiosis being tested. Thus in a population of S individuals there are $\frac{S}{2}$ symbioses of the same type. Furthermore all connection weights are assigned a single magnitude (i.e. $|\alpha_{ij}| = \alpha$). This allows a direct comparison of the effect of varying connection strengths between otherwise equivalent experiments.

Our GA implementation uses standard parameter settings to avoid unfairly favouring the coevolutionary GA[6].

Due to space limitations, we describe only two problems, i.e., Rosenbrock's and Griewangk's Functions (Equations 7 and 8 respectively). Both of these classical Operations Research functions are difficult to solve with normal line search techniques due to non-linearity and non-separability. For both of these problems an individual x_i is represented by a binary string of 32 bits which is mapped to two numbers y_{i1} and y_{i2}.

For Rosenbrock's function, Max(y_{i1}, y_{i2}) is (2.048, 2.048), Min(y_{i1}, y_{i2}) is (-2.048, -2.048), maximum fitness occurs are (1, 1) and equals 0.

---

[6] Our GA uses rank selection and is elitist with a 90% replacement rate in a population of 100 individuals. Mutation and uniform crossover rates are 0.001 and 0.60 respectively.

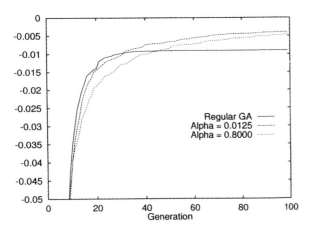

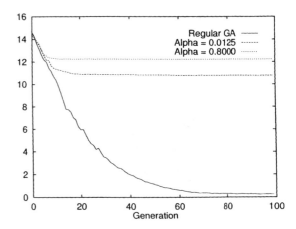

(a) Best absolute fitness acheived

(b) Average Hamming distance of survivors to the current best survivor

Fig. 1. Adaptism tested on Rosenbrock's Function.

For Griewank's function, $\text{Max}(y_{i1}, y_{i2})$ is $(511, 511)$, $\text{Min}(y_{i1}, y_{i2})$ is $(-512, -512)$, maximum fitness occurs are $(0, 0)$ and equals 0.

$$f_i^a(t) = -100(y_{i1}^2 - y_{i2})^2 - (1 - y_{i1})^2 \qquad (7)$$

$$f_i^a(t) = -1 - \sum_{j=1}^{2} \frac{y_{ij}^2}{4000} + \Pi_{j=1}^2 \left(\cos \left(\frac{y_{ij}}{\sqrt{i}} \right) \right) \qquad (8)$$

Several measures of performance are used to describe the experiments: the best fitness (in the sense of "absolute fitness") of every individual (best ever fitness) and the best fitness of current individuals (current best fitness) express the system's potential for optimization, and the average Hamming distance of the survivors to the best current survivor measures the genetic diversity.

Before describing the specific results for the different problems a description of behaviour common to all problems is warranted. Since the effects on expressed fitness are cumulative, an individual's expressed fitness may become higher than the global optimum of the absolute fitness. This leads to two possible stages in a co-evolutionary GA.

In the first stage, during the early generations, the co-evolutionary GA behaves similarly to a regular GA. The exact behaviour seems dependent on the form of symbiosis used as well as the problem itself. However, the interesting stage occurs in the later generations.

After the co-evolutionary GA has converged for a while, the amount of change introduced by the new individuals is lower, i.e., the average $C_i(t) = f_i^e(t) - f_i^e(t-1)$ is smaller. This implies that new individuals survive less frequently into the next generation. At some point the value of $C_i(t)$ is too small to allow new individuals to survive. It is at this point that the co-evolutionary GA changes to its second stage of behaviour. From this point onward the surviving population freezes, i.e., changes become more and more infrequent and eventually stop. This means that after this point the genetic diversity remains constant. Improvement in the best ever fitness, as compared to a standard GA, is achieved because of the constant random recombination of the existing genetic diversity into new individuals. Thus the second stage of behaviour is equivalent to a random search.

The beginning of the change between the two stages and the length of time during which the change occurs depend on the form of co-evolution, the strength of the connections and the problem structure. The results show that co-evolutionary GAs can outperform the regular GA. That is, the best individual found (in absolute fitness) by the co-evolutionary GA is on average better than that found by the regular GA.

Adaptism, Mutualism and Predation show very similar behaviour to one another. Each of these symbiosis patterns preserves genetic diversity and can outperform the regular GA. Since in Competition the effects of connections balance one another, Com

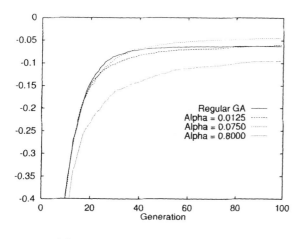 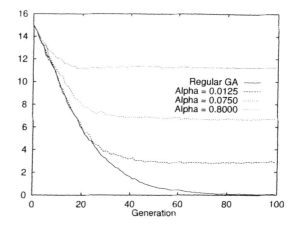

(a) Best absolute fitness achieved

(b) Average Hamming distance from survivors to the current best survivor

Fig. 2. Predation tested on Griewangk's Function.

petition is not significantly different from the regular GA. The results given here have been selected because they display typical behaviour. A more detailed presentation of results and analysis will be available in a forthcoming paper. For each combination of connection strength (α), symbiosis type and problem, the results show the average of 200 experiments.

Figure 1a) shows the best absolute fitness achieved by Adaptism while optimizing Rosenbrock's Function. Clearly both extremes of α perform better than the regular GA. It is important to note that the co-evolutionary GAs continue to improve while the regular GA stagnates at a sub-optimal point.

Figure 1b) shows the average Hamming distance from the survivors to the current best survivor achieved by Adaptism while optimizing Rosenbrock's Function. For this problem even the weakest connection weight causes a dramatic increase in genetic diversity. Also important is the fact that genetic diversity plateaus very early (7th to 20th generation) and never decreases further.

Figure 2a) shows the best absolute fitness achieved by Predation while optimizing Griewangk's Function. The curve for $\alpha = 0.8000$ shows that the survivors freeze too early. Despite a lot of diversity the co-evolutionary GA is unable to outperform the regular GA. The curve for $\alpha = 0.0125$ shows that the population freezes too late. Despite converging to a high fitness level the lack of genetic diversity once the survivors do freeze restricts the overall performance of the co-evolutionary GA. Finally $\alpha =$

0.0750 is the compromise that outperforms all others tested.

Figure 2b) shows the average Hamming distance from survivors to the current best survivor achieved by Predation while optimizing Griewangk's Function. For this problem there is a wide range of increase in genetic diversity with varying α. Also important is the fact that once genetic diversity plateaus in the co-evolutionary GA it never decreases further.

6 Conclusions

We have described a co-evolutionary model that successfully maintains genetic diversity in a population. This is not the first attempt to use co-evolution (see Section 2); however, our model seems to be the first systematic approach to co-evolution. Using the Linear Model of Symbiosis, it is possible to add co-evolution to any GA.

More importantly, this co-evolutionary "add-on" is a tunable system, such that an increase in the strengths of connections results in an increase in the sustained genetic diversity. This relationship holds across a wide variety of possible absolute fitness functions.

Fortunately the same solution that solves the local optima problem also greatly improves the ability of the GA to gain back diversity quickly to allow it to find a new global optimum, and should allow for quick response time to enable even a converged GA to track a moving optimum.

References

1. P. J. Angeline and J. B. Pollack. Competitive Environments Evolve Better Solutions for Complex Tasks. In *Fifth International Conference on Genetic Algorithms*, pages 264–270, 1993.

2. T. Bäck. Self-adaptation. In T. Bäck, D. Fogel, and Z. Michalewicz, editors, *Handbook of Evolutionary Computation*. Oxford University Press, 1997.

3. D. Boucher, editor. *The Biology of Mutualism: Ecology and Evolution*. Croom Helm, 1985.

4. H. Cobb and J. Grefenstette. Genetic Algorithms for Tracking Changing Environments. In S. Forrest, editor, *International Conference on Genetic Algorithms: ICGA'93*, pages 523–530, 1993.

5. L. Eshelman, editor. *Proceedings of the Sixth International Conference on Genetic Algorithms*, San Mateo, California, 1995. Morgan Kaufmann.

6. G. Hartvigsen and W. Starmer. Plant-herbivore coevolution in a spatially and genetically explicit mode. *Artificial Life*, 2(2):129–156, 1995.

7. W. Hillis. Co-evolving parasites improve simulated evolution as an optimization procedure. In C. Langton, C. Taylor, J. Farmer, and S. Rasmussen, editors, *Artificial Life II, SFI Studies in the Sciences of Complexity*, volume 10, pages 313–323. Addison-Wesley Publishing Co., 1991.

8. W. Hillis. Co-evolving parasites improve simulated evolution as an optimization procedure. In C. Langton, C.Taylor, J. Farmer, and S. Rasmussen, editors, *Artificial Life II*, pages 264–270. Massachusetts: Addison-Wesley, 1992.

9. P. Husbands and F. Mill. Simulated Co-Evolution as The Mechanism for Emergent Planning and Scheduling. In *Fourth International Conference on Genetic Algorithms*, pages 264–270, 1991.

10. L. Lawlor. Direct and indirect effects of n-species competition. *Oecologica*, 43:355–364, 1979.

11. D. Lewis. Symbiosis and mutualism. In Boucher [3], pages 29–39.

12. J. Maresky, Y. Davidor, D. Gitler, and G. Aharoni. Selectively destructive re-start. In Eshelman [5], pages 144–150.

13. J. Paredis. Co-evolutionary Constraint Satisfaction. In H. S. Y. Davidor and R. Männer, editors, *Parallel Problem Solving from Nature - PPSN III*, pages 46–55, Berlin, Oct 1994. Springer-Verlag.

14. J. Paredis. The symbiotic evolution of solutions and their representations. In Eshelman [5], pages 359–365.

15. M. A. Potter and K. A. DeJong. A Cooperative Co-evolutionary Approach to Function Optimization. In H. S. Y. Davidor and R. Männer, editors, *Parallel Problem Solving from Nature - PPSN III*, pages 249–257, Berlin, Oct 1994. Springer-Verlag.

16. H. Sakanashi, H. Suzuki, and Y. Kakazu. Controlling dynamics of ga through filtered evaluation function. In H. S. Y. Davidor and R. Männer, editors, *Parallel Problem Solving from Nature - PPSN III*, pages 239–248, Berlin, Oct 1994. Springer-Verlag.

17. K. Sims. Evolving 3D Morphology and Behavior by Competition. In R. Brooks and P. Maes, editors, *Artificial Life IV*, pages 28–39, 1994.

18. M. Starr. A generalized scheme for classifying organismic associations. *Symposia of the Society for Experimental Biology*, 29:1–20, 1975.

19. X. Yao and P. J. Darwen. Evolving Robust Strategies for Iterated Prisoner's Dilemma. In X. Yao, editor, *Progress in Evolutionary Computation*, pages 276–292, 1994.

Selection of Informative Inputs Using Genetic Algorithms

Primož Potočnik, Igor Grabec

University of Ljubljana, Faculty of Mechanical Engineering,
Aškerčeva 6, POB 394, SI–1000 Ljubljana, Slovenia
E-mail: primoz.potocnik@fs.uni-lj.si, igor.grabec@fs.uni-lj.si

Abstract

Modeling of processes with many input variables requires selection of informative inputs in order to construct less complex models with good generalization abilities. In this paper two feature selection methods are compared: mutual information (MI) based feature selection and genetic algorithm (GA) based feature selection. As a modeling structure a hybrid linear-neural model is used. The methods are applied to a case study: modeling of an industrial antibiotic fermentation process. It is shown that both feature selection methods can lead to similar results. 8s based feature selection can be applied to problems where only few data exist and MI can not be calculated. In GA based feature selection it is possibile to adjust the objective function in order to control the propperties of the method.

1 Introduction

Mathematical modeling of complex industrial processes can be helpful for solving optimization problems or designing control strategies. Particularly for the processes where adequate analytical knowledge doesn't exist, empirical modeling approach can be applied. Construction of a model involves a selection of input variables, a selection of a model structure, a learning procedure to calculate model parameters and a validation procedure to test the applicability of the model.

Input variables which provide little or no information about the predicted variable should be eliminated in order to reduce the variance of model parameters and the model complexity. By using only informative input variables we can expect to build more reliable and accurate model. This paper is focused on methods for selection of informative inputs combined with a hybrid modeling approach. We are interested in predictive importance of input variables which is concerned with the increase in generalization error when the input is omitted from the model [1]. We compare the feature selection method based on mutual information [2] and the method based on evolutionary computation [3]. We show that both methods can lead to similar results and investigate the conditions for the use of the one or the other. As a modeling structure we propose a hybrid model [4] which is a combination of a linear regression with a nonlinear error correction. For the nonlinear part of the modeling procedure, a radial basis function network [5] is used.

The proposed approach is illustrated with an industrial case study where the feature selection methods are applied to modeling of a secondary metabolic fermentation process, designed for the production of antibiotics.

2 Hybrid Linear-Neural model

A hybrid parallel linear-neural model is schematically shown in Fig. 1 and is also referred to as a linear model with a non-linear error correction. The motivation to combine linear and neural models comes from different properties of the two models. While neural networks have good approximation and interpolation properties, they show limited extrapolation capacity. On the other hand, linear models are limited to deal with nonlinearities but often demonstrate more robust extrapolation behavior. Therefore, the combination of a linear model with nonlinear neural-network-based error correction is used in our study.

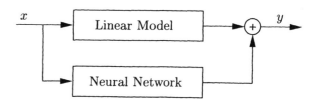

Fig. 1. Hybrid parallel linear-neural model

The model output is composed of a linear part and a non-linear part which is modeled by a radial basis function network. A hybrid model is described by the following equation

$$\widehat{y}(x) = \theta^T x + \sum_{k=1}^{K} w_k \, \phi_k(x) + C \qquad (1)$$

where $\widehat{y}$ denotes an output, x an input vector, θ^T a transposed parameter vector of a linear model, ϕ_k radial basis function, w_k weight, K a number of radial basis functions and C a constant. The two step learning procedure involves formation of the linear model at the first step and training of the radial basis function network at the second step. The target values for the neural network learning are the residual non-linearities of the modeled process.

3 Selection of Inputs Using Mutual Information

Mutual information (MI) is a natural measure of input variable relevance because it has the capability to measure general dependencies between the variables. It can be used to reduce the number of input variables either by rejecting inputs with low information content or high redundancy with respect to other inputs. Application of MI for selection of informative input variables in neural-net-based modeling was proposed in [6]. We are following the more elaborated approach [2] which consists of:

- MI as a relevance measure;
- nonparametric kernel density estimation for calculation of MI;
- forward selection as an input variable search method.

Mutual information measure is based on Shannon's entropy

$$H(Y) = - \sum_{y \in \mathcal{Y}} p(y) \log p(y) \qquad (2)$$

which can be viewed as a measure of uncertainty related to observation of variable Y. Mutual information $I(X;Y)$ is expressed by the entropy $H(Y)$ and the conditional entropy $H(X|Y)$

$$\begin{aligned} I(X;Y) &= H(Y) - H(Y|X) \qquad (3) \\ &= - \sum_{y \in \mathcal{Y}} p(y) \log p(y) + \\ &\quad \sum_{x \in \mathcal{X}} \sum_{y \in \mathcal{Y}} p(x,y) \log p(y|x) \qquad (4) \end{aligned}$$

and quantifies information provided by variable X about the variable Y. An unbiased estimator for

calculation of mutual information from the set of samples $\{x_i, y_i; \ i = 1, \ldots, N\}$ is given by

$$\hat{I}(X;Y) = \frac{1}{N} \sum_{i=1}^{N} \log \frac{p(x_i, y_i)}{p(x_i)p(y_i)}. \qquad (5)$$

Multi-dimensional Epaneschnikov product kernels

$$\mathcal{K}(x) = \begin{cases} \prod_{i=1}^{d_x} \frac{3}{4}(1 - x_i^2) & \text{for } ||x||_\infty < 1 \\ 0 & \text{otherwise} \end{cases} \qquad (6)$$

are used for the nonparametric probability density estimation

$$P(x) = \frac{1}{N} \frac{1}{\sigma^{d_x}} \sum_{n=1}^{N} \mathcal{K}\left(\frac{1}{\sigma}(x - x_n)\right), \qquad (7)$$

where the kernel width σ is determined by likelihood cross-validation. A forward selection procedure is used to search for the subset of the most relevant input variables. The iterative procedure starts with one input variable and subsequent variables are appended as long as mutual information is increasing.

4 Selection of Inputs by Genetic Algorithms

The method proposed is model based and searches through the space of possible combinations of input variables by means of evolutionary computation [3]. A binary selector S_x is introduced for the representation of selected inputs. It is a binary string with single bits determining inclusion ("1") or exclusion ("0") of the corresponding input variable. An example of a binary selector operating on a set of 10 input variables is shown in Fig. 2.

The set of inputs: $X = \{x_1, x_2, \ldots, x_{10}\}$

Binary selector: $S_x = \boxed{0\,|\,1\,|\,0\,|\,0\,|\,1\,|\,1\,|\,0\,|\,0\,|\,1\,|\,0}$

$\qquad\qquad\qquad \downarrow \qquad \downarrow\,\downarrow \qquad\quad \downarrow$

$\qquad\qquad\qquad x_2 \qquad x_5\, x_6 \qquad x_9$

Selected inputs: $X_S = \{x_2, x_5, x_6, x_9\}$

Fig. 2. Operation of a binary selector S_x: The initial set of input variables X is mapped to a subset X_s which contains only selected variables.

The subset of input variables X_S is evaluated by an objective function $f(X_S)$ which returns a scalar value indicating the relevance of the evaluated subset. The goal is to find the small subset of informative input variables which renders possible construction of a reliable model $\mathcal{M}$ with good generalization properties. The objective function is defined by

$$f(X_S) = g(\mathcal{M}|X_S)\ h(L_1, p) \qquad (8)$$

where the functions $g(\cdot)$ and $h(\cdot)$ have the following meaning:

g – generalization error of the model $\mathcal{M}$ which is build with the subset of input variables X_S. Generalization error is estimated with a cross-validation procedure.

h – penalty function, limiting the number of included variables L_1. The p-value in interval $[0, 1]$ determines the percentage, for which the generalization error g is increased if all inputs are selected. Penalty function $h(L_1, p)$ is described by equation

$$h(L_1, p) = 1 + \frac{p}{2}\left(1 + \tanh\left(\frac{L_1 - L/2}{L/5}\right)\right) \quad (9)$$

where L denotes the number of all input variables and L_1 the number of selected input variables. Fig. 3 shows the penalty function for the p-values $p = 0.1$ and $p = 0.5$ which correspond to 10% and 50% increase of the objective function when all inputs are selected.

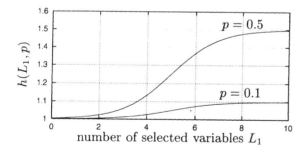

Fig. 3. Penalty function $h(L_1, p)$ is used to increase the return value of the objective function (Eq. 8), if many input variables are selected. The penalty curves for the p-values $p = 0.1$ and $p = 0.5$ with respect to the number of selected variables L_1 are shown.

The selection of inputs by genetic algorithm consists of the following steps:

1. An initial population of random binary selectors $\{S_x(n), n = 1, 2, \ldots N_p\}$ is generated. N_p denotes the size of the population.

2. Initial selectors are evaluated by the objective function (Eq. 8) where the model is built and validated by cross-validation method.

3. Numeric optimization of the initial binary selectors is performed by genetic algorithm. A binary selector which achieves the minimum of the objective function determines the most informative subset of input variables.

5   A Case Study: Production of Antibiotics

The proposed methods are applied to an industrial secondary metabolic fermentation process which is designed for the production of antibiotics. The process runs in a fed-batch mode which refers to continuous, variable rate of feeding while the product is removed at the end of the process. The antibiotic is produced as a secondary metabolite by the microorganisms. The average batch lasts about 140 hours and, after the product recovery, the process is restarted.

Several references reported successful applications of neural networks to modeling of various fermentation processes. Applications include biomass estimation in the penicillin industrial fermentation [7, 8], on-line prediction of fermentation variables [9] and neuro-identification of an industrial secondary metabolic fermentation [10, 11]. In our previous work [3], modeling of the antibiotic production in the fermentation sample space was investigated where the fermentation batch was regarded as one sample and the suitability of the hybrid linear-neural model structure was demonstrated.

The present study is focused on the current state of the fermentation process with the aim to predict 24-hours ahead the product concentration C_p. Ten process variables are measured in 3-hours time intervals to represent the current state of the process: age of the batch t, pH, percentage of mass volume pmV, viscosity η, nitrogen C_{N_2}, phosphor C_{phos}, glucose C_{glu}, glycerol C_{gly}, soya oil C_{soya} and the product concentration C_p. An available data base consists of 64 fermentation batches with 1861 input-output samples. The set of possible input variables for the modeling of the process is defined by

$$
\begin{aligned}
X(t) = \{&t, pH(t), pmV(t), \eta(t), \\
&C_{N_2}(t), C_{phos}(t), C_{glu}(t), \\
&C_{gly}(t), C_{soya}(t), C_p(t)\} \qquad (10)
\end{aligned}
$$

and the predicted output variable is

$$y(t) = C_{\mathrm{p}}(t + 24). \tag{11}$$

Mutual information between single input variables $\{X_i(t); \ i = 1, \ldots, 10\}$ and the output variable $y(t)$ is presented in Fig. 4. We can observe the high value of MI for the variables C_{p}, t and η, followed by pmV, C_{soya} and C_{gly}. MI for the variables pH and C_{phos} is negligible.

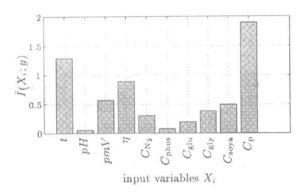

Fig. 4. Mutual information (MI) analysis for the fermentation process. MI values between the predicted output variable $y(t) = C_{\mathrm{p}}(t + 24)$ and the individual input variables $X_i(t)$ are shown.

A hybrid linear-neural model is used as a model structure for 24-hours ahead prediction of the product concentration and informative inputs are selected by mutual information and by genetic algorithm. Results of modeling with MI feature selection and GA feature selection are presented in Table 1.

Selection methods include MI based feature selection, GA based feature selection with penalty value $p = 0.5$ and GA based feature selection with penalty value $p = 0.1$. Selected process variables for the three methods are indicated and the cross-validation root mean square errors of prediction (RMS) are given. Results obtained by MI selection and GA selection with $p = 0.5$ are identical, resulting in RMS $= 0.215$. GA selection with $p = 0.1$, imposing softer constraint on the number of selected inputs, gives only slight improvement of 5% but requires inclusion of two additional input variables.

Based on the results in table 1 we conclude that for 24-hours ahead prediction of the fermentation product concentration, three informative variables exist: t, η and C_{p}. Using additional input variables increases the complexity of a model and contributes to only slight improvement in prediction accuracy. Therefore we conclude that only three se-

lected input variables are sufficient for our modeling purpose.

Results of 24-hours ahead prediction of the fermentation product concentration $y(t) = C_{\mathrm{p}}(t + 24)$, based on the selected inputs $X_S = \{t, \eta, C_{\mathrm{p}}\}$, are presented in Fig. 5. A scatter plot of predictions is shown where the values of $\hat{y}(t)$ and $y(t)$ refer to zero-mean variance-one normalized data.

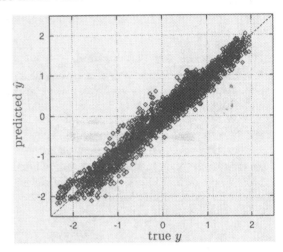

Fig. 5. Results of 24-hours ahead prediction of the fermentation product concentration $y(t) = C_{\mathrm{p}}(t + 24)$. A scatter plot of predictions $\hat{y}(t)$ versus observed values $y(t)$ is shown.

6 Conclusions

Modeling approach with selection of informative inputs and application of a hybrid linear-neural model is presented. Two feature selection methods are compared: mutual information (MI) and selection by genetic algorithm (GA). Both methods measure arbitrary relationships between variables and are more general than methods based on linear relations. The methods are applied to modeling of an industrial secondary metabolic fermentation process, designed for the production of antibiotics. It is shown that both feature selection methods can lead to similar results. Regarding the comparison between MI based feature selection and GA based feature selection we draw the following conclusions:

- mutual information is model independent but needs large amount of data for accurate estimations, therefore the method is suitable for problems where enough data exist;

- GA based feature selection is suitable for data sets with many potential inputs and few sam-

Table 1. Selection of input variables for 24-hours ahead prediction of the fermentation product concentration $C_p(t+24)$. Feature selection methods include mutual information and genetic algorithm (GA) with two levels of penalizing the number of inputs, determined by penalty value $p = 0.5$ and $p = 0.1$. Selected process variables and the cross-validation root mean square errors of prediction (RMS) are indicated.

Selection method	Selected process variables										RMS error
	t	pH	pmV	η	C_{N_2}	C_{phos}	C_{glu}	C_{gly}	C_{soya}	C_p	
Mutual Information	X			X						X	0.215
GA with penalty $p = 0.5$	X			X						X	0.215
GA with penalty $p = 0.1$	X			X				X	X	X	0.205

ples, thus being applicable in the domain where it is not possible to apply MI;

- GA method is model based and requires high computational costs. Consequently it is suitable for models with fast learning procedures.

- The selection of inputs by GA based method can be controlled by adjusting the penalty function, given by Eq. (9). This property renders possible suiting the objective function to particular problems.

References

[1] W. S. Sarle, "How to measure importance of inputs?," 1997. URI = `ftp://ftp.sas.com/pub/neural/importance.html`

[2] B. Bonnlander, "Nonparametric selection of input variables for connectionist learning", PhD thesis, University of Colorado, 1996.

[3] P. Potočnik and I. Grabec, "Neural–genetic system for modeling of antibiotic fermentation process," in Proceedings of the International ICSC Symposium on Engineering of Intelligent Systems EIS'98, Volume 2, (Tenerife, Spain), pp. 307–313, 1998.

[4] P. Potočnik, "Nonparametric modeling of a fermentation process," Master's thesis, University of Ljubljana, Faculty of Mechanical Engineering, Ljubljana, 1997. (in Slovenian).

[5] J. Moody and C. J. Darken, "Fast learning in networks of locally-tuned processing units," Neural Computation, vol. 1, pp. 281–294, 1989.

[6] R. Batitti, "Using mutual information for selecting features in supervised neural net learning," IEEE Transactions on Neural Networks, vol. 5, no. 4, pp. 537–550, 1994.

[7] M. J. Willis, C. D. Massimo, G. A. Montague, M. T. Tham, and A. J. Morris, "Artificial neural networks in process engineering," IEE Proceedings–D, vol. 138, no. 3, pp. 256–266, 1990.

[8] C. D. Massimo, G. A. Montague, M. J. Willis, M. H. Tham, and A. J. Morris, "Towards improved penicillin fermentation via artificial neural networks," Computers & Chemical Engineering, vol. 16, no. 4, pp. 283–291, 1992.

[9] J. Thibault, V. V. Breusegem, and A. Chéruy, "On-line prediction of fermentation variables using neural networks," Biotechnology and Bioengineering, vol. 36, pp. 1041–1048, 1990.

[10] D. Tsaptsinos and J. R. Leigh, "Modelling of a fermentation process using multi-layer perceptrons: Epochs vs pattern learning, sigmoid vs linear transfer function," Journal of Microcomputer Applications, vol. 16, pp. 125–136, 1993.

[11] D. Tsaptsinos, R. Tang, and J. R. Leigh, "Neuroidentification of a biotechnological process: Issues and applications," Neurocomputing, vol. 9, pp. 63–79, 1995.

Parallel Evolutionary Algorithms with SOM-Like Migration and their Application to Real World Data Sets

Th. Villmann°, R. Haupt•, K. Hering• and H. Schulze•

Universität Leipzig, Germany

°Klinik für Psychotherapie, 04107 Leipzig, K.-Tauchnitz-Str.25
villmann@informatik.uni-leipzig.de

•Institut für Informatik, 04109 Leipzig, Augustusplatz 10/11
{haupt,hering,schulze}@informatik.uni-leipzig.de

Abstract

We introduce a multiple subpopulation approach for parallel evolutionary algorithms the migration scheme of which follows a SOM-like dynamics. We succesfully apply this approach to clustering in both VLSI-design and psychotherapy research. The advantages of the approach are shown which consist in a reduced communication overhead between the subpopulations preserving a non-vanishing information flow.

1 Introduction

Evolutionary Algorithms (EAs) are a biologically motivated stochastic iterative optimization method. In EAs the manipulation of objects, which are called individuals $s \in \Pi$, is separated from their evaluation by a fitness function F. The set Π is called population. There exist two basic manipulation operators: mutation as random change of parameters and the crossover as merging of two individuals. Usually the μ individuals of Π generate λ new ones with $\lambda > \mu$. After these manipulations F judges how proper the individuals fulfil the considered task.[1] If in one time step all individuals have gone under manipulation and evaluation, from these the *selection operator* extracts a new generation for the next iteration step. For a detailed overview we refer to [10].

Several approaches were developed to improve this basic EA-scheme. Especially, multiple subpopulation approaches are widely considered [4, 11]. Thereby the basic population Π is divided into subpopulations Π_i

---

[1]Thereby, the fitness function may contain explicit expert knowledge regarding to the optimization task which may be difficult to code otherwise.

which have more or less communication during the evolution which may be realized as migration [15].

Yet, high communication frequency causes large effort, especially if one uses a multi-processor system for the evaluation in such a way that each subpopulation is evaluated by one processor. However, the high information flow through the set Π of subpopulations increases the genetic diversity which, in general, accelerates the development of the fitness of the individuals during the first generations. In contradiction, in the convergence phase the problem of homogeneity between the subpopulations may occur. This problem is more relevant for discrete optimization tasks, however it is also addressed elsewhere [4, 10, 11]. On the other hand, if we allow only a small amount of communication the performance of a multi-processor system will be improved. Moreover, because of the relative independence between the subpopulations each of them searches in a different region of the solution space. However, the improvement rate of the fitness of the individuals during the first generation is reduced which leads, in general, to a lower fitness.

Therefore, in the present contribution we focus on a merging of both strategies. For this purpose we apply the concept of collective learning with a progressing separation during the time development as it is known from the concept of *neural maps* considered by KOHONEN [8]. In addition, we prefer a mixture of both the (μ, λ)– and the $(\mu + \lambda)$–strategy (in the notation of SCHWEFEL, [13]) for the selection operator. We show the success of this approach for two real world applications (in the area of VLSI-design and psychotherapy research).

2 EAs with SOM-Like Migration for Clustering

2.1 Clustering as a Partitioning Task

Mapping a set $\mathcal{S}$ onto clusters $C \in \mathcal{C}$ in the context of EAs can be taken as a *partitioning problem* under constraints which are specified by the fitness function F. Thereby, a *partitioning* of a nonempty set $\mathcal{S}$ related to a nonempty set $\mathcal{C}$ is an unique and surjective mapping

$$\Phi : \mathcal{S} \to \mathcal{C} \quad . \qquad (2.1)$$

Then a *partition* Ψ_Φ of $\mathcal{S}$ related to the partitioning Φ is given by

$$\Psi_\Phi = \left\{ \Phi^{-1}(C) \mid C \in \mathrm{cod}(\Phi) \right\} \qquad (2.2)$$

whereby $\mathrm{cod}(\Phi)$ is the range of Φ [7].

For solving the partitioning task an individual in a generation of an EA describes a certain partition. In the present paper we assume that the number of clusters to be built is predefined as $c_{\max}$. Furthermore, we consider $\mathcal{S}$ to be discrete containing $s_{\max}$ elements. Then we can take each individual as a string of length $s_{\max}$ the components of which contain the cluster index onto which the respective component has to be mapped. Mutation of an individual is defined as a random change of the mapping for a randomly selected individual component and the crossover is a cut of two individuals at the same point followed by crossed sticking together.

2.2 EAs with SOM-Like Migration Scheme

We consider a *multiple subpopulation approach* whereby a set $\mathbf{\Pi}$ of subpopulations Π_i is arranged on a topological order which is often chosen to be a regular lattice, for instance a ring or a quadratic lattice[2]. Between these subpopulations a *migration scheme* was introduced, originally developed by TOTH&LÖRINCZ [15] in a basic variant, and applied here in an extended approach. In this extended approach a visit (migration) from individuals between neighboring subpopulations is allowed for a short time *regarding to the topological order* Ω in $\mathbf{\Pi}$ with respect to a time-dependent neighborhood function

$$h_{i^*}(t, k) = (1.0 - \epsilon_h) \cdot \exp\left(-\frac{r_{i^*,k}}{2\left(\sigma_h(t)\right)^2} \right) + \epsilon_h \quad (2.3)$$

with a small positive number ϵ_h. During the evalua-

---

[2]In general, other arrangements are also admissible. Then the lattice can be defined by a connection matrix which describes the neighborhood relations.

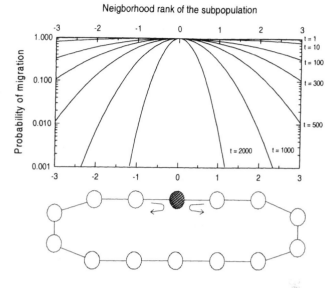

Fig. 1: Plot of the neighborhood function $h_{i^*}(t, k)$ from eq.(2.3) determining the migration scheme.

tion of a certain subpopulation Π_{i^*} the neighborhood function h is applied to determine the number of visiting individuals from each other subpopulation. Because of $h_{i^*}(t, k) \in (0, 1]$ we can interpret $h_{i^*}(t, k)$ as a *probability for migration* of an individual of the subpopulation Π_k into the actual evaluated Π_{i^*}. The value $r_{i^*,k}$ is defined as the rank of neighborhood

$$r_{i^*,k} = \mathrm{rank}\left(\Pi_{i^*}, \Pi_k, \Omega\right) \qquad (2.4)$$

between the actually evaluated subpopulation Π_{i^*} and another Π_k according to the topologial order Ω in $\mathbf{\Pi}$. At the beginning t_0 of the evolution process the range h of the neighborhood nearly comprises the complete set of subpopulations and decreases exponentially during the time determined by

$$\sigma_h(t) \underset{t \to \infty}{\to} 0 \qquad (2.5)$$

in (2.3) (see Fig. 1).

Following this approach one can take the individuals of all subpopulations at the beginning as an uniform population which performs a first rough adaptation process. During the further development the various subpopulations become more and more separated from each other and, hence, search in *different regions* of the solution space. This approach of first rough adaptation of neighboring subpopulations together with a more fine tuning in the further process by more but not completely independent subpopulations arranged on a topological structure shows an analogy to the concept of *self-organizing maps* (SOM) in the field of neural maps [8]. Thereby, the neural units are also placed on a topological structure (lattice) and the collective dynamics during the learning procedure follows the same idea as the subpopulations in the present article: a first rough adaptation of the neuron weights takes place changing to a precise adjustment simultaneously with loosing the strong neighborhood conditions in the lattice.

On the other hand, ϵ_h in (2.3) preserves a remaining probability for migration. In this way one has a non–vanishing information flow through the topological ordered set Π of subpopulations which *accelerates* the adaptation process, especially, if the search space possesses many local minima [10] (see sect. 4.2). Thereby, neighboring subpopulations contain similar in fitness but genetically different individuals whereas usual EAs tend to homogeneity. Hence, if the fitness function F only incompletely describes the expert knowledge as in sect. 4.1, the expert can choose one of these similar individuals as the final solution. In the present applications the topological order in Π for the SOM-like migration scheme is a ring.

3 The $(\mu * \lambda)$–Approach for the Selection Scheme in EAs

For selection of the offspring generation we have used a mixture of the (μ, λ)– and the $(\mu + \lambda)$–strategy (in the notation of SCHWEFEL, [13]). While in the (μ, λ)-strategy where μ individuals produce λ children with $\mu < \lambda$ only the μ best of the λ children form the new population, in the $(\mu + \lambda)$–strategy all $\mu + \lambda$ individuals are allowed for the selection process. Whereas in the second strategy the best solution is preserved but the evolution tends to stagnate into a local minimum, in the first one the convergence is decelerated to allow reaching deeper minima (near the global minimum) but good solutions may be lost during the evolution. Balancing the advantages of both strategies [10] the $(\mu * \lambda)$–approach was introduced to solve hard partitioning problems in VLSI–design [7] with many widely distributed local minima.

In the $(\mu * \lambda)$–approach again μ individuals produce the λ preliminary offsprings. However, in the selection step the μ_t best individuals of the old generation and the λ new ones are allowed for comparison with respect to their fitness to generate the final offspring generation of μ individuals. Thereby μ_t depends on time t of evolution appearing as the number of generations performed:

$$\mu_t = \text{int} \left[(\mu - \mu_\tau) \cdot \gamma(t) \right] + \mu_\tau \quad , \qquad (3.1)$$

with $\text{int}[x]$ stands for the integer value of x. The function $\gamma(t)$ is of decreasing sigmoid type with $0 \leq \gamma(t) \leq 1$ here chosen as the Fermi function

$$\gamma(t) = 1/(1 + \exp((t - t_a)/t_b)) \qquad (3.2)$$

to switch near the t_a-*th* generation from the $(\mu + \lambda)$–strategy to the (μ, λ)–strategy in a definite range of generation steps ($\approx 4t_b$). We have $\mu_0 = \mu$ for the initial value and

$$\lim_{t \to \infty} \mu_t = \mu_\tau \text{ with } (\mu_\tau \ll \mu) \qquad (3.3)$$

coding a minimal survival probability for the parent individuals. In this way we get a smoothed switch from the $(\mu + \lambda)$– to the (μ, λ)–strategy, what we call $(\mu * \lambda)$–strategy, combining the advantages of both strategies and, additionally, always preserving the best μ_τ individuals (slightly different from the original (μ, λ)-strategy).

4 Applications and Results

4.1 Clustering of Psychological Categories

One of the mostly used methods for acquisition of structures of interpersonal relationships in the area of psycho–dynamic psychotherapy research is the method of the 'Core Conflictual Relationship Theme' (CCRT) developed by LUBORSKY [9]. The method investigates so–called *relationship–episodes*, which are often reported by the patients in their therapeutical sessions. For each of these episodes the components *wish of the subject* (W), *response of the object* (RO) and *response of the subject* (RS) are encoded which then are used to perform the CCRT. BARBER ET AL. [2] determined a system S^w of $s^w_{\max} = 34$ so–called *standard categories* S^w_j to classify the wishes which are collected in a set $\mathcal{C}^w$ of $c^w_{\max} = 8$ *clusters* C^w_k [2]. The number and the interpretation of the clusters as well as the assignment of the standard categories are obtained from the experience of several psychotherapists using conventional statistical methods. Analogously they explained $s^{ro}_{\max} = 30$ categories $S^{ro}_j \in S^{ro}$ for encoding the RO and $s^{rs}_{\max} = 30$ categories $S^{rs}_j \in S^{rs}$ for encoding the RS which are collected in $c^{ro}_{\max} = 8$ clusters $C^{ro}_k \in \mathcal{C}^{ro}$ and $c^{rs}_{\max} = 8$ clusters $C^{rs}_k \in \mathcal{C}^{rs}$, respectively.

However, as mentioned in [1] the clusters are still correlated again what *leads to low reliability rates*.

Hence, the task is to reform the clusters of standard categories to improve the reliability. For this purpose several raters judge a large number of episodes and determine the most relevant standard category $S_{j*}^w(i)$ of each wish and, *in addition*, a second one (denoted as $S_{j+}^w(i)$), which has to be different from the first one but is also well describing. All pairs $p_i^w = \left(S_{j*}^w(i), S_{j+}^w(i)\right)$, $i = 1 \ldots N_w$, $N_w = 7922$, form a database P^w which implicitly contains the correlation information between the standard categories. Analogously we obtained a database P^{ro} of $N_{ro} = 9651$ pairs p_i^{ro} and a database P^{rs} of $N_{rs} = 11629$ pairs p_i^{rs}.

If the respective clusters are determined in a reliable way according to an arbitrary clustering algorithm, both the most and the second relevant standard category should belong to the same cluster after this procedure, otherwise we denote this fact as misclassification. Let us denote

$$\phi_{j*}^i = \Phi\left(S_{j*}(i)\right) \quad, \quad \phi_{j+}^i = \Phi\left(S_{j+}(i)\right) \qquad (4.1)$$

for a given cluster solution according to (2.1). Instead of applying the misclassifications $\mathcal{M}$, we involved the conditional probabilities

$$\rho_{j*,j+}^i = p\left(S_{j*}(i) \mid S_{j+}(i)\right) \quad, \qquad (4.2)$$

estimated from the database, into the fitness function:

$$\hat{F}(s) = \sum_{\substack{i=1 \\ \phi_{j*}^i \neq \phi_{j+}^i}}^{N} \max\left(\rho_{j*,j+}^i, \rho_{j+,j*}^i\right) \qquad (4.3)$$

However, we have to pay attention to the balancing of the clusters which is measured by the variance of the number of the standard categories belonging to the clusters as suggested in [16]. Without this constraint the optimal solution would be to collect all standard categories into only one cluster. Furthermore, some explicitly known therapeutical knowledge about the parallel appearance of standard categories is coded. Both aspects are included in an additional term $\tilde{F}$. Thus the final fitness function is obtained by

$$F = \hat{F} + \tilde{F} \quad . \qquad (4.4)$$

The cluster solution found by our approach shows more coherence than the original one from a psychological point of view. The clusters itself are more consistent and, on the other hand, now there is a greater differentiation between the clusters in psychological meaning. For a more detailed psychological consideration we refer to [1].

Table 1: Different values for the weighted concordance coefficient κ and the respective meaning for the agreement of the appearance of the considered observables

κ–coefficient	meaning
$\kappa < 0.1$	no agreement
$0.1 \leq \kappa < 0.4$	weak agreement
$0.4 \leq \kappa < 0.6$	clear agreement
$0.6 \leq \kappa < 0.8$	strong agreement
$0.8 \leq \kappa$	nearly complete agreement.

In addition, the new clusters are compared with the original ones via the *weighted* concordance coefficient κ as a mathematical measure [3] which scores the simultaneous appearance of the respective clusters ϕ_{j*}^i and ϕ_{j+}^i of a pair $p_i \in P$ obtained according to (4.1). At first we computed κ for the original clusters $\mathcal{C}^w$, $\mathcal{C}^{ro}$ and $\mathcal{C}^{rs}$ of the respective standard categories as defined in [2]. We found the values $\kappa_w = 0.334$, $\kappa_{ro} = 0.330$ and $\kappa_{rs} = 0.492$ whereby the standard deviations were $\nu_w = 0.00626$, $\nu_{ro} = 0.00600$ and $\nu_{rs} = 0.00601$, respectively. Hence, the κ–values for W and RO are related to an only weak agreement whereas the κ_{rs} refers to a clear agreement according to Tab.1 [12]. The values κ_w, κ_{ro} and κ_s correspond to $\mathcal{M}_w = 4662$ (58.8%), $\mathcal{M}_{ro} = 5410$ (56.1%) and $\mathcal{M}_{rs} = 4910$ (46%) misclassifications in the respective databases P^w, P^{ro} and P^{rs}. Our parallel SOM-like EA–approach together with the $(\mu * \lambda)$-strategy yields the following results:

	$\mathcal{M}$	κ	ν	
P^w	3764 (47.5%)	0.429	0.00675	
P^{ro}	4917 (50.9%)	0.421	0.00578	(4.5)
P^{rs}	4352 (37.4%)	0.562	0.00525	

In fact, now all κ–values refer to a clear agreement according to Tab.1 which is a strong improvement. Here we mention again that we did not use κ as fitness function itself because of the more adequate description (in psychotherapeutic sense) by the conditional probabilities. However, κ is the mostly used mathematical measure in psychotherapy research. Hence we have to present it for comparison.

4.2 VLSI Model Partitioning for Logic Simulation

For the logic design of whole microprocessor structures time-extensive cycle-based simulation processes are necessary [14]. Time spent for simulation can be drastically reduced using *parallel simulation* based on *model partitioning*. The model partitioning problem can be formulated as a combinational optimization problem. In this context partitions are characterized by a complex cost function [5] which estimates the run-time of one parallel simulation cycle of the corresponding hardware model parts. This cost function has to be minimized to reduce the expected parallel simulation time. Hierarchical model partitioning for parallel system simulation allows a successful application of EAs at the second hierarchy level [7].

The components of a suited prepartition (super-cones) obtained from the first hierarchy level are used as basic set for the partitioning at the second level yielding a final set of clusters (blocks). To apply EAs at the second level, a set of initial partitions is required as start population. They are built by special algorithms, for instance by *MOCC* [7], using expert knowledge for qualified but genetically different individuals (i.e. partitions) to start with a good population of high diversity. Each component of an individual (partition) represents a super-cone coded by an integer as block index. The *fitness F* of an individual is identified by the *estimated run-time* for parallel simulation with respect to the corresponding partition and is calculated using two kinds of hypergraphs – *Overlap and Communication Hypergraph* as explained in [6].

To achieve better partitions in shorter time we have parallelized EAs using the multiple subpopulation approach with the SOM-like migration scheme mentioned above. For each subpopulation the $(\mu * \lambda)$-strategy is applied. The migration is implemented by nonblocking Point-to-Point communication in the frame of the *Message Passing Interface*.

Experimental results are given for an *IBM S/390* processor model which is partitioned into 15 blocks via the *STEP* prepartitioning algorithm resulting in 250 super-cones [5]. At the second hierarchy level three different EA strategies are realized with the parameters $\mu = 294$, $\lambda = 700$ in each run. A sequential one-population EA is opposed to two parallel multiple subpopulation EAs each consisting of 7 subpopulations (i.e. $\tilde{\mu} = 42$, $\tilde{\lambda} = 100$ for each subpopulation). We compare the SOM-like migration scheme with an *all-to-all communication scheme* where the communication effort is much higher than for the SOM-like case. Considering the partitioning effort for the multiple subpopulation approach, in Fig.2 the fitness of the best individual is plotted over the time t_{part} spent for the EA partitioning.

Parallel partitioning drastically reduces the parti-

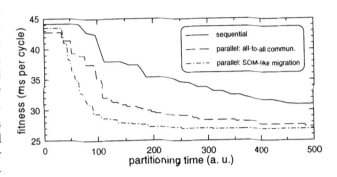

Fig. 2: Fitness of the best individual of all subpopulations comparing a sequential run with two parallel ones in dependence on the *partitioning time* t_{part} measured in arbitrary units (a. u.).

tioning time because the partitioning effort is distributed to 7 processors. The fitness of the best individual significantly faster decreases than in the sequential case. But the all-to-all communication scheme is accompanied by a high communication overhead. Using our SOM-like migration approach this communication effort can be reduced in such a way that better individuals (partitions) are obtained in shorter partitioning time t_{part}. So, in the example discussed here an estimated run-time of $27\,ms$ per cycle (fitness) is reached in the half of the partitioning time comparing to the all-to-all communication scheme. Although the individuals of our initial population are already equipped with expert knowledge and not randomly produced, EAs yield a reduction of estimated run-time from $\approx 44\,ms$ down to $27\,ms$ (see Fig.2).

5   Concluding Remarks

We have developed a SOM-like migration scheme for multiple subpopulation systems for EAs which is inspired by neuron dynamics of self–organizing maps in the area of neural computation. Thereby the subpopulations are arranged on a topological structure and the possibility for migration depends on the neighborhood rank and decreases during time with a remaining rest probability. Additionally, we have introduced a balance between the classical $(\mu + \lambda)$- and (μ, λ)-strategy. We successfully applied this approach to two real world data sets from *VLSI-design* and psychotherapy research and demonstrated its advantages.

Acknowledgments. The authors would like to thank D. Pokorny (Universität Ulm) for giving the approach of the weighted concordance coefficient and C. Albani / A. Körner (Universität Leipzig) for psychological data and discussions. Moreover, R. Haupt and H. Schulze are grateful for the support by DEUTSCHE FORSCHUNGSGEMEINSCHAFT (DFG) under grant SP 487/1-2.

References

[1] C. Albani, B. Villmann, T. Villmann, A. Körner, M. Geyer, D. Pokorny, G. Blaser, and H. Kächele. Kritik und erste Reformulierung der kategorialen Strukturen der Methode des Zentralen Beziehungs–Konflikt–Themas (ZBKT). *Psychotherapie, Psychosomathik und medizinische Psychologie - G.-Thieme-Verlag Stuttgart, New York*, page to appear, 1998.

[2] J. Barber, P. Crits-Christoph, and L. Luborsky. A guide to the CCRT Standard Categories and their classification. In L. Luborsky and P. Crits-Chrostoph, editors, *Understanding Transference*, pages 37–50. Basic Books New York, 1990.

[3] J. Cohen. Weighted kappa. *Psychological Bulletin*, 70:213–220, 1968.

[4] D. B. Fogel. *Evolutionary Computation: Towards a New Philosophy of Machine Intelligence.* IEEE Press, Piscataway, NJ, 1995.

[5] R. Haupt, K. Hering, U. Petri, and T. Villmann. Hierarchical model partitioning for parallel VLSI–simulation using evolutionary algorithms improved by superpositions of partitions. In K. Lieven, editor, *Proceedings of European Congress on Intelligent Techniques and Soft Computing (EUFIT'97)*, pages 804–808, Aachen, Germany, 1997.

[6] R. Haupt, K. Hering, and T. Siedschlag. Integration of a Local Search Operator into Evolutionary Algorithms for VLSI-Model Partitioning. In K. Lieven, editor, *Proceedings of European Congress on Intelligent Techniques and Soft Computing (EUFIT'98)*, pages 377 – 381, Aachen, Germany, 1998. ELITE Foundation.

[7] K. Hering, R. Haupt, and T. Villmann. Hierarchical Strategy of Model Partitioning for VLSI–Design Using an Improved Mixture of Experts Approach. In *Proc. Of the Conference on Parallel and Distributed Simulation (PADS'96)*, pages 106–113. IEEE Computer Society Press, Los Alamitos, 1996.

[8] T. Kohonen. *Self-Organizing Maps.* Springer, Berlin, Heidelberg, 1995. (Second Extended Edition 1997).

[9] L. Luborsky. The core conflictual relationship scheme. In N. Freedman and S. Grand, editors, *Communicative Structure and Psychic Structures.* Plenum Press New York, 1977.

[10] Z. Michalewicz. *Genetic Algorithms + Data Structures = Evolution Programs.* Springer–Verlag Berlin Heidelberg New York, third, revised and extended edition, 1996.

[11] H. Mühlenbein, M. Gorges-Schleuter, and O. Krämer. Evolution Algorithm in Combinatorial Optimization. *Parallel Computing*, (7):65–88, 1988.

[12] L. Sachs. *Angewandte Statistik.* Springer Verlag, 7-th edition, 1992.

[13] H.-P. Schwefel. *Numerical Optimization of Computer Models.* Wiley and Sons, 1981.

[14] W. G. Spruth. *The Design of a Microprocessor.* Springer Berlin, Heidelberg, 1989.

[15] G. J. Tóth and A. Lőrincz. Genetic algorithm with migration on topology conserving maps. In S. Gielen and B. Kappen, editors, *Proc. ICANN'93, Int. Conf. on Artificial Neural Networks*, pages 605–608, London, UK, 1993. Springer.

[16] T. Villmann, B. Villmann, and C. Albani. Application of evolutionary algorithms to the problem of new clustering of psychological categories using real clinical data sets. In B. Reusch, editor, *Computational Intelligence - Theory and Applications - Proc. Of the International Conference on Computational Intelligence - 5th. Dortmunder Fuzzy–Days*, pages 311–320, Berlin, New York, Heidelberg, 1997. Lecture Notes in Computer Science 1226, Springer–Verlag.

Skewed Crossover and the Dynamic Distributed Database Problem

M. Oates[1], D. Corne[2], R. Loader[2]

[1]British Telecommunications Laboratories, Martlesham Heath, Suffolk, England, IP5 3RE
[2]Department of Computer Science, University of Reading, Reading, RG6 6AY, UK
Email: moates@srd.bt.co.uk, {D.W.Corne, Roger.Loader}@reading.ac.uk

Abstract

The automatic self-management of large, distributed databases is a significant problem area for providers of global management information systems and services. Finding a way of dynamically balancing changing load over a number of globally distributed servers can be an arduous task, particularly when communications costs and overheads are also considered. Previous work has shown that this problem can prove a difficult search space to negotiate for Genetic Algorithms. This paper introduces a skewed form of 2-point crossover which appears to give exceedingly encouraging results, particularly on scenarios which have previously been categorised as 'problematic'. Whilst the advantages of this form of crossover may prove to be problem (or even solution) specific, it is likely to be of use in problems where sub-sequence information is an important feature of schemata (such as scheduling problems), or where optimal solutions contain repetition of either individual or short sequences of alleles across numerous gene positions. The technique effectively provides an additional, orthogonal source of genetic diversity, apparently reducing the need for either excessive initial population size or high levels of mutation. It is also shown to be effective as part of the mutation operator in Simulated Annealing.

1   Introduction

Global Management Information Systems, providing access to data such as customer details, or facilities such as 'service provision', are increasingly being provided via corporate intranets. These applications require users at many globally distributed sites to access and update large quantities of shared information. Performance and resilience requirements usually force this information to be replicated on a number of servers, which can then be used to distribute the user load. However as the usage rates from different groups of users around the world changes over a 24 hour cycle, the 'ideal' distribution (or configuration of users to servers) will change. Indeed, over the 24 hour workday cycle, there may be considerable variance in inter-node communications costs or performance, and this too must be considered when trying to determine configurations which will give optimum performance under current conditions.

Literature search shows that Evolutionary Computing approaches have already been considered on similar problems to this [4,10,2] but these have either not considered communications costs or have considered only static, 'once off' distributions applied at the database design stage. More recent results in [6] and in [1] are beginning to show that EC techniques can be applied to distributed, multi-fragment databases, dynamically optimising data distribution and user access configurations in near-real time. For simple database performance criteria (such as minimising access latency to the poorest performing server), a simple Hillclimber has been shown to give impressive results. However once the fitness function is enhanced to give a more balanced measure of user perception of the system (such as by adding to the figure to be minimised, a measure of 'average' access time as seen by other users) the search space becomes considerably more complex and rugged, leaving Hillclimber stuck at local optima. A comparison of various techniques including Hillclimber, Simulated Annealing, Breeder [5] and Tournament style [3] Genetic Algorithms is given in [8], together with results from a range of performance measures / fitness functions. The effects of some GA parameter choices are also demonstrated, in particular focusing on population size and mutation rate. This is also further explored in [7].

This comparison categorised combinations of 'usage scenario' and 'performance measure' based on how many times, out of 1000 independent runs, a Simulated Annealer (SA) given 5000 evaluations, could find industrially acceptable solutions. Industrially acceptable was defined as having a fitness value within 5% of the known globally optimal solution. Values more than 30% off the known globally optimal solution were deemed wholly unacceptable, and thus it was deemed critical to not only consistently give 'moderate' solutions, but also to demonstrate exceedingly low likelihood of delivering poor solutions. Combinations of 'scenario' and 'performance measure' for which SA nearly always scored industrially acceptable solutions, without any lapses into bad performance, were deemed to present an 'easily searchable' fitness landscape, whilst combinations which failed on either or both of these criteria were categorised as increasingly 'hard to search' landscapes. A reduced summary of these results is reproduced below in Table 1, and an explanation of the models and scenarios is given in the following 'Model and Method' section of this paper.

Table 1. - Summary of search space difficulty

Model	Scenario A	Scenario B
basic	Very Easy	Moderate
plus avg	Easy	Very Easy
plus used	Very Easy	Very Hard
plus all	Easy	Very Hard

In an effort to improve the performance of the optimisers against the 'very hard' combinations, a new crossover operator termed 'skewed' was devised and very preliminary results with this were presented in [9]. This operator is explained in detail herein and the original results are significantly expanded upon in this paper, exploring the effects of different population sizes and mutation rates on the performance of a Tournament style GA over two different (but deemed 'hard') examples of the distributed database problem. Results from an initial trial of a Simulated Annealer, using this 'skewed, 2-point crossover' operator as part of its mutation operator, are also given together with a brief discussion on possible applicability of the operator in other problem domains.

2 Model and Method

By the use of a model of the distributed information system covering performance of both database servers and communications networks, the performance of the system as seen by individual client applications, can be estimated for current load conditions over a range of different access and server configurations. The choice of configuration is determined by an optimisation algorithm which produces 'solution vectors' stating, for each client, which server that client should currently connect to for 'read' access ('update' access must be made to all servers maintaining a copy of the database). This is then used as a direct representation (i.e. a k-ary representation, not binary) where the cardinality of each gene position is equal to the number of servers in the system. Our experiments have so far looked at 10 node scenarios where each node can act as both a client and a server. This therefore creates an n^n search space (where n=10). The system model is based on work at BT Labs and is based on assumptions that the performance of both servers and communications

links degrade under load according to Little's Law and the principles of MM1 queues. Two key formulae are utilised :

$$\text{Degraded Response Time} = 1 / ((1 / BTT) - TAR)$$

Where :   BTT = server Base Transaction Time
              TAR = Transaction Arrival Rate

$$CTR = (((\Sigma\, TRn) - Max(TRn)) * CV) + Max (TRn)$$

Where :   CTR = Combined Transaction Rate
              TRn = Transaction Rates over a range of sources
              CV = Contention Value between sources

Details of the model are on the ECTELNET website (http://www.dcs.napier.ac.uk/evonet/). ECTELNET is the Telecommunications subgroup of the European Network of Excellence in Evolutionary Computation.

For each node, a retrieval rate and update rate are defined, together with a measure of the amount of concurrency between these two methods of access. This defines the client workloads for the system. For each node, a base transaction time is defined indicating the undegraded performance of that node when dealing with a client transaction. Also for each node, a concurrency factor is included to ease modelling of multi-processor systems. Finally, for each potential client / server communications link, a base communications time is specified giving the undegraded comms performance for traffic between that client / server pairing. Together with the 'solution vector' given by the optimiser, these values are used to calculate the degraded performance of each server and comms link and hence the transaction performance as seen by each client.

In the 'basic' model, the fitness of a solution vector is the worst response time seen by any client from any server over whichever comms link is utilised. 'Write Access' to all nodes is considered in the 'basic model' calculation, since even if a node is currently not being accessed by a client for 'Read' operations, it is still receiving regular updates, which affect its quality as a 'standby' server if another node fails. A variant of this model however, adds 10% of the access time for all nodes, to the 'least worst' server time returned. This strikes a balance between minimising worst performance and aggregate server performance and is referred to as 'plus avg'. Another variant model considers applying updates only to those nodes that are currently being accessed as servers, and adds 10% of the comms access time seen by all clients on the worst server, divided by the number of servers used. This adds a bias based on aggregate user perception but with weighting in favour of over-duplication of data. This was argued for on the basis of enhanced resilience and is referred to as 'plus used'. Finally, the 'plus all' variant, considering only used servers, adds the average of all client accesses weighted by their usage rate to all used servers. This last variant is seen as probably the most realistic in terms of representing user perception of Quality of Service.

These results are interpreted as an inverse measure of the 'Quality of Service' for the underlying information system. The optimisation algorithm seeks to minimise this value by finding a suitable configuration of client accesses. This is a classic 'minimax' scenario. These and other models are described in more detail in [8].

A simple 'Tournament' GA was used here, employing 3 way tournament selection, where 3 members of the population were chosen at random, ranked, and the best and second best used to create a 'child' which automatically replaces the third member chosen in the tournament. This GA used uniform crossover [12] followed by uniformly distributed allele replacement mutation and is later referred to as 'TNT'.

The performance of a similar 3-way Tournament GA using a variant of 2-point crossover employing a skewing mechanism was then compared to the simple uniform crossover 'TNT' GA. This crossover operator starts with the child being an exact copy of the first parent. A random start position in the second parent is then chosen, together with a random length (with wrap-around) of genes, and these are overlaid into the child starting at yet another randomly chosen position. This is then followed by uniformly distributed allele replacement mutation. This gives a skewing effect and is later referred to as 'SKT'. It is demonstrated below.

Gene Position :	1	2	3	4	5	6	7	8	9	10
First Parent :	A	B	C	D	E	F	G	H	I	J
Second Parent :	a	b	c	d	e	f	g	h	i	j

Random start position in second parent   : 8
Random length chosen from second parent : 5
Random start position in child   : 4

Resulting Child	A	B	C	h	i	j	a	b	I	J

It can be argued that this operator is similar to the 'shift' or 'shunt' operator suggested in [11] or 'partial schedule exchange' in [13], often used in 'permutation' problems. These can be particularly effective when operating with single parents, where a sub-section of the chromosome ('shift' using only a single gene) is extracted and removed from the 'exact copy' child starting at a random start position, with random length, and re-inserted back into the child at a different random start position. In single parent operations, this operator should not create fundamentally illegal permutations, whilst corrective procedures are often devised to allow 2 parent use. The principle difference between 'shunt' and 'skewed' crossover is that with 'shunt', the sub-sequence to be inserted does not directly overwrite allele values in the child, it is genuinely 'inserted', with the allele values in the original gene positions being 'pushed along' the chromosome. Skewed crossover is of course only directly applicable to allelic representations in which the allowed allele-range is the same for each gene. Where this is not the case, simple amendments to the mechanism can of course be applied.

The two Tournament style GA optimisers, were tested against 2 scenarios. The first, where the performance of all nodes and communications links in the model were of similar performance (varying only by a factor of 2 from least to worst) is referred to as Scenario A. This problem (with the basic evaluation function) has many 'optimum' solutions each of the same fitness value.

The second scenario, referred to as Scenario B, is where the nodes are effectively split into 2 geographic regions, having low communications costs between nodes within the same region, but high costs between regions. Within each region is a 'supernode' whose basic performance is 10 times that of the other nodes in the region. This problem has only one optimal solution, with each node accessing its region's supernode.

The performance of the GAs was examined over population sizes ranging from 2 chromosomes to 100 chromosomes, in steps of 2, and over mutation rates ranging from 0% chance of mutation per gene to 50% chance in steps of 2%. For each combination of population size and mutation rate, 50 trials were run, each starting with different, randomly generated initial populations, with the GA running for 5000 evaluations, noting the fitness of the best solution found, and the evaluation number at which this was first found. For industrial applicability, the GA must be able to consistently find good solutions in the shortest time possible. Therefore the number of evaluations taken to find the best solution that the GA found, was then weighted by an offset proportional to the square of the difference between the best fitness value found in that trial versus the globally optimal fitness value. The results for the 50 trials was then averaged, and this 'average weighted number of evaluations' then plotted against population size and mutation rate giving an inverse measure of the 'quality' of the optimiser. The lowest values on these 3D plots therefore denote 'tuned' GAs which consistently find good solutions in the minimum number of evaluations.

Finally, the 'skewed' operator was used in a 'single parent' form in a Simulated Annealer, which utilised this operator prior to single gene, uniformly distributed allele replacement mutation.

3   Results

Figures 1 and 2 show the results on the uniform crossover tournament GA on scenarios 'A' and 'B' respectively using the 'plus used' model. On the 'A' scenario, the 'very easy' search space category is demonstrated by the wide range of mutation rates and population sizes which are able to return good solutions in a minimum number of evaluations (in region of 1000 to 2000 evaluations). There is a slight 'diagonal trough' effect where, as population size increases, less mutation is needed and indeed too much mutation causes degraded performance, and this has previously been explored and explained for Breeder GAs in [7]. By contrast, the scenario 'B' plot has a 'weighted evaluations' scale of 3 orders of magnitude greater, showing that the error function is totally dominant in the 'quality' measure. This implies that of the 50 runs, many produce 'best found solutions' that are wholly unacceptable. Here, starting at the highest levels of mutation plotted (50%), performance steadily 'improves' as mutation is reduced towards 30% chance of mutation per gene. Below 30% mutation, we enter an unstable region where performance can improve, but increasingly can be catastrophically bad. Below 12% mutation, performance is reliably bad. The 'diagonal trough' is again apparent, allowing lower mutation rates to give better results at greater population size, indeed there is the suggestion that considerably higher population sizes could give better results and this will be investigated in due course. The

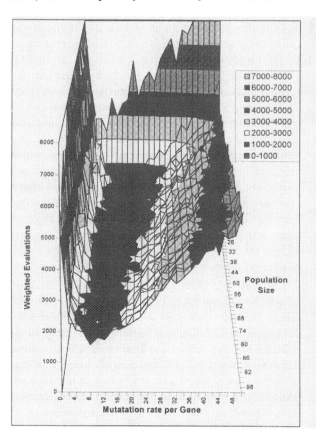

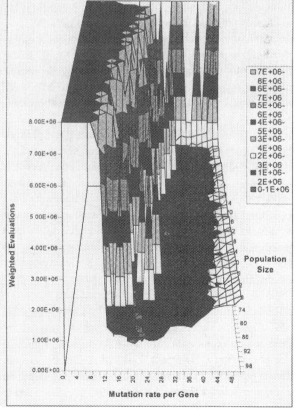

Fig. 1. Uniform Crossover, '+ used', Scenario 'A'   **Fig. 2**. Uniform Crossover, '+ used', Scenario

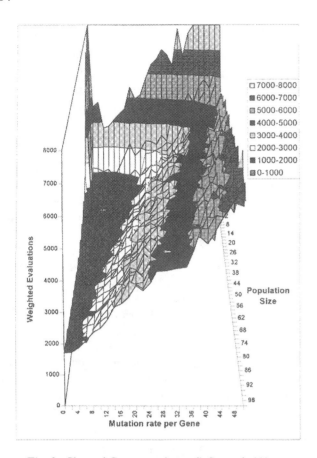

Fig. 3. Skewed Crossover, '+ used', Scenario 'A'

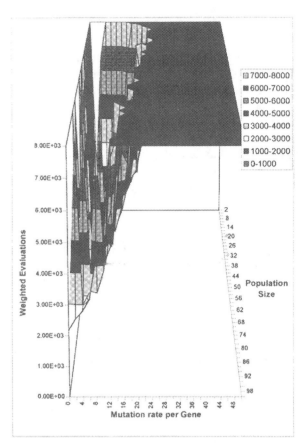

Fig. 4. Skewed Crossover, '+ used', Scenario 'B'

important feature however in both these plots is that a high level of genetic diversity is required by the search algorithm, provided by significantly high levels of mutation rate, and little dependence on population size.

Figures 3 and 4 show the results from using the skewed crossover operator with the tournament selection GA, on the 'plus used' model against scenarios 'A' and 'B' respectively. Figure 3 can be seen to compare favourably with Figure 1, both being Scenario 'A', except that the bottom of the trough is now lower, and at values of lower levels of mutation. Indeed the lowest point in the valley of Figure 3 is at mutation rates of around 6% and population sizes of 18 chromosomes. Contrasting Figures 2 and 4, the latter now shows regions of fully acceptable performance on Scenario 'B'. The 'Weighted Evaluations' scale is now the same as for the 'A' scenario (not 3 orders of magnitude greater as was the case for Figure 2). Figure 5 gives a 'birds eye' view of Figure 4, showing clearly a large stable region of performance for population sizes in the range 6 to 70 chromosomes, with mutation rates in the range 2% to 12%.

The bottom of the trough is in a region of low population size (6 to 18 chromosomes), with mutation in the range 2% to 6%. This is in stark contrast to the performance of the uniform crossover operator. Figures 6 and 8 show performance of the skewed tournament GA against the 'plus all' model variant for Scenarios 'A' and 'B'. From Fig 6 we can see that little or no mutation actually gives best performance, but Fig 7 (the 'bird's eye' equivalent) shows that there is also another good region at a population of 4 or 6 chromosomes with a wide range of useable mutation rates (2% to 30%). Interestingly, there is no simple trade-off here between pop size and mutation rate, with combinations of say 12 chromosomes at 12% mutation performing significantly worse. This is in contrast to Figs 1 and 3, and indeed earlier published results and will be investigated in due course. Fig 8 again shows that the skewed operator is able to deliver acceptable results on what has been previously categorised as a 'hard problem'. Fig 9 (the 'bird's eye' of Fig 8) again shows that the optimum GA tuning parameters are both low population size and low mutation rate, 4 to 12 chromosomes with 2% to 8% mutation. In all cases, it

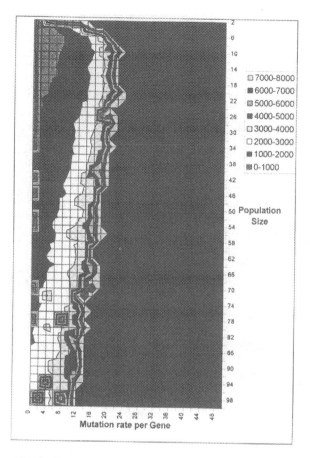

Fig. 5. Skewed Crossover, '+ used', Scenario 'B'

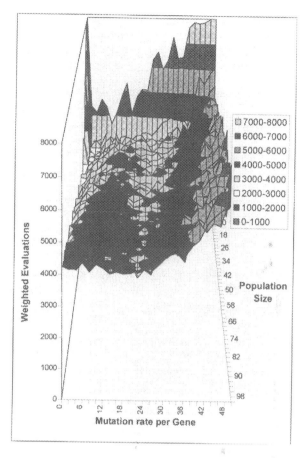

Fig. 6. Skewed Crossover, 'plus all', Scenario 'A'

is clear that if skewed crossover is used, the best results are achieved at lower mutation rates than uniform crossover, see also [7,8]. This is perhaps not surprising because, temporarily ignoring mutation, uniform crossover can only exploit the diversity of allele values in equivalent gene positions available in the initial population. Even this will rapidly be reduced as dominant sub-schema in other gene positions cause premature extinction of alleles in less dominant gene positions. However skewed crossover is able to re-inject alleles from other gene positions into the available pool for all gene positions, thus acting as a potential source of additional diversity, albeit biased.

Given the apparent success of low population sizes with skewed crossover, this operator was trailed in a Simulated Annealer, followed by single gene, uniformly distributed allele replacement mutation. Out of 1000 trials each of 5000 evaluations against Scenario 'B', 'plus used', the skewed SA found the globally optimal solution 99.8% of the time, with the remaining 0.2% all having 'best found' fitness values within 30% of the global optimum. A tournament GA

with the skewed operator achieved 99.7% and by contrast, a mutation-only SA achieved only 22.4% global optimum hits, the remaining 77.6% all being worse than 30% off the optimum. A uniform crossover tournament GA was down to 14.9% global optimum hits, all others being more than 30% off.

The success of skewed crossover on this problem has led us to look into its general applicability to different optimisation problems. Clearly, skewed crossover attempts to exploit a particular feature of the chosen representation: namely, that good contiguous 'chunks' of allele values in one part of the chromosome may also work well as a building block elsewhere in the chromosome. Skewed crossover may therefore work well as a first-choice operator if we have an a priori intuition that the latter may be the case for the problem in hand. Also, of course, in problems for which we have no such intuition, or no intuition at all, this may or may not be true; so skewed crossover may be a useful extra tool to join the standard set of operators which can be tried on such problems.

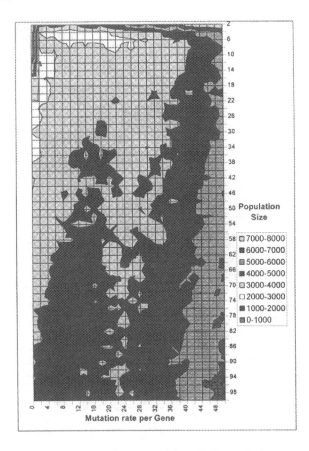

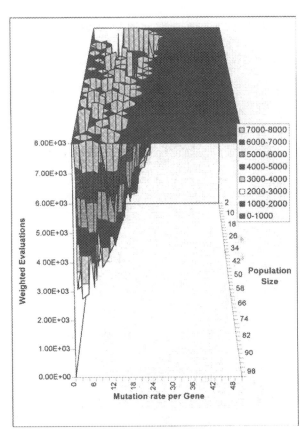

Fig. 7. Skewed Crossover, 'plus all', Scenario A

Fig. 8. Skewed Crossover, 'plus all', Scenario 'B'

We have begun to test these ideas by looking at an optimisation problem in which we had an a priori expectation that the `skew' mechanism would more successfully exploit building blocks than standard operators. The results are described very briefly here. The problem is a hard benchmark problem in structural chemistry [14], and involves finding the 2-dimensional structure which minimises the energy (as estimated by Lennard-Jones potentials) of a bonded string of atoms. Here, the chromosome represents a list of adjacent bond angles, and the true optimum in each case is a close-to spiral structure, in which adjacent angles are similar to each other (but not exactly the same), with repeated contiguous patterns of angles along the string.

Table 2. - 10 atoms case:

1 pt:	20	19	19	19	19	19	19	19	19	19
2 pt:	19	18	19	18	20	19	18	19	19	19
unif:	19	20	19	17	19	18	18	19	19	20
skew:	20	20	20	20	20	20	20	20	20	20

Coded as a maximisation problem, tables 2 and 3 summarise the results over 10 trials, in both the 10

atoms and 20 atoms cases, for each of one-point, two-point, uniform, and skewed crossover. Skewed crossover seems clearly best on this problem, as we expected. We have yet to experiment with skew on a wider range of problems.

Table 3. - 20 atoms case:

1 pt:	37	40	38	37	41	41	40	42	37	39
2 pt:	41	39	40	39	40	41	40	40	41	42
unif:	43	43	37	39	41	41	39	42	41	42
skew:	47	47	45	47	47	46	46	45	47	47

4   Conclusions

Whilst the results from using skewed crossover are apparently very encouraging, it must be borne in mind that for the database problem, there are only 2 supernodes in Scenario B. A solution vector containing only these two server choices, duplicated for each client in the same region, gives the globally optimal fitness value. The 'skewed' crossover operator is particularly effective at 'copying' allele values into different gene positions across the length of the chromosome. In its defence however, three points are

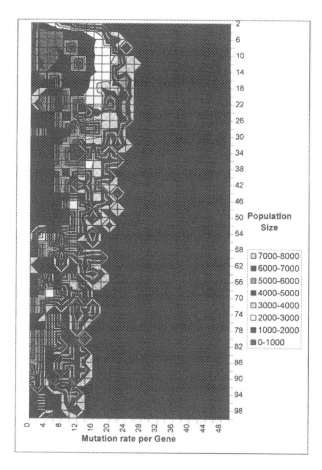

Fig. 9. Skewed Crossover, 'plus all', Scenario 'B'

worthy of note. Firstly, that SKT does give acceptable performance on Scenario 'A' where alleles across the chromosome have high diversity in 'good solutions'. Secondly, that Scenario B probably represents a more 'typical' scenario when considering real world applications. Finally, good results have been demonstrated on a completely unrelated problem, however once again the problem is likely to have 'good' solutions containing similar or repeated sequences of alleles. Further work is needed to get a clearer understanding of the general and specific applicability of the skewed crossover operator.

5   Acknowledgements

The authors wish to thank British Telecommunications Ltd for their ongoing support for this research.

6   References

[1] G Bilchev and S Olafsson (1998), *Comparing Evolutionary Algorithms and Greedy Heuristics for Adaption Problems*, in Proceedings of the 1998 IEEE Int. Conf. on Evolutionary Computation, pp. 458-463.

[2] W Cedeno and V.R. Vemuri (1997), Database Design with Genetic Algorithms, in D. Dasgupta and Z. Michalewicz (eds.), *Evolutionary Algorithms in Engineering Applications*, Springer-Verlag, pp. 189-206.

[3] D. E, Goldberg (1989), *Genetic Algorithms in Search Optimisation and Machine Learning*, Addison Wesley.

[4] S.T. March and S Rho (1995), *Allocating Data and Operations to Nodes in Distributed Database Design*. IEEE Transactions on Knowledge and Data Engineering 7(2), April 1995, pp. 305-317.

[5] H. Mühlenbein and D. Schlierkamp-Voosen (1994), *The Science of Breeding & its application to the Breeder Genetic Algorithm*, Evolutionary Computation 1, pp. 335-360.

[6] M Oates, D Corne and R Loader (1998), *Investigating Evolutionary Approaches for Self-Adaption in Large Distributed Databases*, in Proceedings of the 1998 IEEE ICEC, pp. 452-457.

[7] M Oates and D Corne (1998-ECAI), *QoS based GA Parameter Selection for Autonomously Managed Distributed Information Systems*, in Procs of ECAI 98, the 1998 European Conference on Artificial Intelligence, pp. 670-674.

[8] M Oates and D Corne (1998-PPSN), *Investigating Evolutionary Approaches to Adaptive Database Management against various Quality of Service Metrics*, LNCS, Procs of PPSN-V, pp. 775-784.

[9] M Oates (1998), *Autonomous Management of Distributed Information Systems using Evolutionary Computing Techniques*, invited paper to CASYS'98, 2nd Int. Conf on Computing Anticipatory Systems, 1998

[10] S Rho and S.T. March (1994), *A Nested Genetic Algorithm for Database Design*, in Proceedings of the 27th Hawaii Int. Conf. on System Sciences, pp. 33-42.

[11] P Ross and A Tuson (1997), *Directing the search of evolutionary and neighbourhood search optimisers for the flowshop sequencing problem with an idle-time heuristic*, in Corne and Shapiro (eds), Evolutionary Computing: Selected papers from the 1997 AISB International Workshop, Springer LNCS 1305, pp. 213-225.

[12] G Syswerda (1989), *Uniform Crossover in Genetic Algorithms*, in Schaffer J. (ed), Procs of the Third Int. Conf. on Genetic Algorithms, . Morgan Kaufmann, pp. 2 – 9

[13] M Gen, Y Tsujimura and E Kubota (1994), *Solving job-shop scheduling problems using genetic algorithms*, in Gen and Kobayashi (eds), Proc of the 16th Int'l Conf on Computers and Industrial Engineering, Japan, pp. 576-579.

[14] Pullan, W.J. "Genetic Operators for a two-Dimensional Bonded Molecular Model" Computers and Chemistry vol 22 (1998), pp. 331-338.

Improved Pseudo-Relaxation Learning Algorithm for Robust Bidirectional Associative Memory

K. Hasegawa[1] and M. Hattori[2]

[1]NEC Shizuoka Ltd., 4-2 Shimomata, Kakegawa, Shizuoka, 436-8501 Japan

[2]Yamanashi University, 4-3-11 Takeda, Kofu, Yamanashi, 400-8511 Japan

E-mail: hasegawa@pc.snec.nec.co.jp, hattori@pine.ese.yamanashi.ac.jp

Abstract

In this paper, we propose Improved Pseudo-Relaxation Learning Algorithm for Bidirectional Associative Memory (IPRLAB). Since the proposed IPRLAB is based on the conventional PRLAB, it can guarantee the recall of all training pairs and has high storage capacity. Furthermore, the proposed IPRLAB can much improve the noise reduction effect of the BAM and contribute to construct a robust memory. A number of computer simulation results show the effectiveness of the proposed learning algorithm.

1 Introduction

Recently, in order to mimic human memory, several memory models have been proposed [1]-[5], [12]-[14]. Among them, a Bidirectional Associative Memory (BAM) [4] has attracted much attention because of its simple structure and its similarity to brain; for example, there exists reverberation in the BAM. The BAM consists of two layers of neurons and in order to store a set of training vector pairs, it is trained by Hebbian learning. However, Hebbian learning does not guarantee the recall of all training pairs unless the training vectors are orthogonal. Namely, the BAM trained by Hebbian learning suffers from low memory capacity. In [7], a multiple training concept which improves the memory capacity has been proposed, while it does not guarantee the recall of all training pairs. Recently, Wang *et al.* have formulated the learning problem into a system of linear inequalities [8]. In order to solve the system of linear inequalities, several learning algorithms using relaxation methods have been proposed [8]-[11]. Owing to these learning algorithms, both the storage capacity and the learning speed of the BAM have greatly improved. In these learning algorithms, however, the capability of association from fuzzy input, which we have wanted to realize in artificial associative memories, has not been investigated sufficiently.

In this paper, we propose Improved Pseudo Relaxation Learning Algorithm for BAM (IPRLAB). The proposed algorithm has the same characteristics as the PRLAB's, that is, it can guarantee the recall of all training pairs and it can much improve the storage capacity of the BAM in comparison with the Hebbian Learning. In addition, the proposed IPRLAB can greatly improve the capability of association from noisy or incomplete inputs.

2 Bidirectional Associative Memory

2.1 Learning of BAM

As shown in Figure 1, a Bidirectional Associative Memory (BAM) [4] consists of two layers.

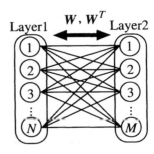

Fig. 1. Structure of BAM.

Consider an N-M BAM with N neurons in the first layer and M neurons in the second layer. Let W_{ij} be the connection weights between the i-th neuron in the fist layer and the j-th neuron in the second layer. Let $\boldsymbol{V} = \{\boldsymbol{X}^{(k)}, \boldsymbol{Y}^{(k)}\}_{k=1,\cdots,p}$ be a set of training vector pairs, where $\boldsymbol{X}^{(k)} \in \{-1, 1\}^N$ and $\boldsymbol{Y}^{(k)} \in \{-1, 1\}^M$. In the learning of the original BAM [4], the connection weights are trained by a correlation matrix, or Hebbian learning as follows:

$$\boldsymbol{W} = \sum_{k=1}^{p} \boldsymbol{X}^{(k)T} \boldsymbol{Y}^{(k)}. \tag{1}$$

In the BAM learned by Hebbian learning, however, the recall of all training pairs can not be guaranteed unless the training pairs are orthogonal. Namely, the memory capacity of the BAM becomes extremely low.

2.2 Pseudo-Relaxation Learning Algorithm

The vectors in V are guaranteed to be recalled if the following system of linear inequalities is satisfied for all $k = 1, \cdots, p$ [8]-[11].

$$(\sum_{i=1}^{N} W_{ij} X_i^{(k)} - \theta_{Y_j}) Y_j^{(k)} > 0 \quad \text{for } j = 1, \cdots, M \quad (2)$$

$$(\sum_{j=1}^{M} W_{ij} Y_j^{(k)} - \theta_{X_i}) X_i^{(k)} > 0 \quad \text{for } i = 1, \cdots, N \quad (3)$$

where we assume that each neuron has its own threshold: θ_{X_i} shows the threshold for the i-th neuron in the first layer and θ_{Y_j} shows the threshold for the j-th neuron in the second layer.

As seen in Ineqs.(2) and (3), if we can find a feasible solution to the system of linear inequalities, the recall of all training pairs can be guaranteed. Since it is very difficult to solve the inequalities systematically, several learning algorithms using relaxation methods have been proposed [8]-[11]. Among them, Pseudo-Relaxation Learning Algorithm for BAM (PRLAB) [9] is effective because the algorithm is very simple and it can converge fast.

The PRLAB is summarized as follows:

> **Stage1.** Choose the initial values of weights and thresholds arbitrary.
> **Stage2.** For the neurons in the first layer, if $S_{X_i}^{(k)} X_i^{(k)} \leq 0$

$$\Delta W_{ij} = -\frac{\lambda}{1+M}(S_{X_i}^{(k)} - \xi X_i^{(k)}) Y_j^{(k)} \quad (4)$$

$$\Delta \theta_{X_i} = +\frac{\lambda}{1+M}(S_{X_i}^{(k)} - \xi X_i^{(k)}) \quad (5)$$

and for the neurons in the second layer, if $S_{Y_j}^{(k)} Y_j^{(k)} \leq 0$

$$\Delta W_{ij} = -\frac{\lambda}{1+N}(S_{Y_j}^{(k)} - \xi Y_j^{(k)}) X_i^{(k)} \quad (6)$$

$$\Delta \theta_{Y_j} = +\frac{\lambda}{1+N}(S_{Y_j}^{(k)} - \xi Y_j^{(k)}) \quad (7)$$

where

$$S_{X_i}^{(k)} = \sum_{j=1}^{M} W_{ij} Y_j^{(k)} - \theta_{X_i},$$

$$S_{Y_j}^{(k)} = \sum_{i=1}^{N} W_{ij} X_i^{(k)} - \theta_{Y_j},$$

the relaxation factor λ is a constant between 0 and 2 and the normalizing factor ξ must be positive. The above procedure is iterated for all $k = 1, \cdots, p$ until it converges.

3 Improved Pseudo-Relaxation Learning Algorithm

3.1 Robustness of BAM

Figure 2 shows the geometrical image of the PRLAB in a two dimensional case. In general, each inequality of Ineqs.(2) and (3) defines a halfspace and the solution space is determined by each hyperplane which represents the border of the halfspace. In the PRLAB, the weights are changed toward the pseudo solution space determined by the parameter ξ [9]. As described in Sect.2, however, the PRLAB terminates when the weights reach the real solution space. Therefore, the weights may converge very close to one of the hyperplanes in the solution space (See Figure 2). In such a case, the robustness, or the noise reduction effect of the BAM declines because it is considered that the distance between the weights and the nearest border indicates the robustness of the BAM.

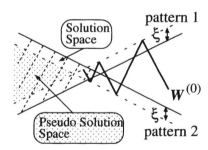

Fig. 2. Geometrical image of weights renewals of PRLAB in a two dimensional case($\lambda = 1.0$).

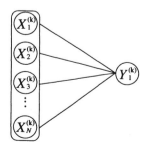

Fig. 3. Structure of perceptron.

In order to confirm this assumption, we trained the perceptron shown in Figure 3 by the PRLAB with $\lambda = 1.9$ and $\xi = 10.0$ because a BAM is composed of several perceptrons.

Fig. 4. Training data. The desired outputs for C, G and O were set to 1, -1 and 1, respectively.

Figure 4 shows the training data used in this experiment. Table 1 shows the result when the noisy patterns corrupted by randomly reversing each bit with a probability of 10% were applied as inputs. The Euclid distance between the weights and the hyperplane of the k-th pattern was determined by

$$D^{(k)} = \frac{|S_{Y_1}^{(k)} Y_1^{(k)}|}{\sqrt{N+1}}. \tag{8}$$

Table 1. Relation between recall rate and Euclid distance.

Pattern	Recall Rate (%)	Euclid Distance
C	60.5	0.0432
G	86.2	0.1847
O	99.6	0.5066

As we can see in the table, the farther the distance from the border is, the better the noise reduction is. Therefore, we can consider that the distance between the weights and the border of the solution space indicates the noise margin.

3.2 IPRLAB

It is considered that the robustness of the BAM can be improved by putting the weights as far as possible from each border. Here, we propose two relaxation learning algorithms based on the PRLAB which can much improve the robustness of the BAM.

IPRLAB I. The idea of the improved PRLAB I is very simple: the weights are changed until they reach the pseudo-solution space. In order to realize this, the relaxation factor λ is set to a constant between 1 and 2 $(1 < \lambda < 2)$ and the convergence criterion is modified in the IPRLAB I.

In the IPRLAB I, the weights are learned as follows:

For the neurons in the first layer, if $S_{X_i}^{(k)} X_i^{(k)} - \xi \leq 0$

$$\Delta W_{ij} = -\frac{\lambda}{1+M}(S_{X_i}^{(k)} - \xi X_i^{(k)}) Y_j^{(k)} \tag{9}$$

$$\Delta \theta_{X_i} = +\frac{\lambda}{1+M}(S_{X_i}^{(k)} - \xi X_i^{(k)}) \tag{10}$$

and for the neurons in the second layer, if $S_{Y_j}^{(k)} Y_j^{(k)} - \xi \leq 0$

$$\Delta W_{ij} = -\frac{\lambda}{1+N}(S_{Y_j}^{(k)} - \xi Y_j^{(k)}) X_i^{(k)} \tag{11}$$

$$\Delta \theta_{Y_j} = +\frac{\lambda}{1+N}(S_{Y_j}^{(k)} - \xi Y_j^{(k)}). \tag{12}$$

Figure 5 shows the geometrical image of the IPRLAB I in a two dimensional case. As seen in the figure, the weights can be prevented to converge close to borders in the IPRLAB I.

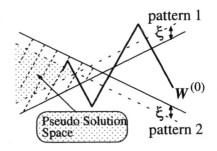

Fig. 5. IPRLAB I.

IPRLAB II. In the IPRLAB II, the weights are changed toward the point which has the similar distance from each border of the solution space. Namely, the weights are desired to converge at the apex of the convex polyhedron determined by the pseudo solution space. In this case, it is considered that the robustness of the BAM can be improved with respect to all training data.

The learning procedure of the IPRLAB II is similar to the PRLAB's described in Sect.2 except $0 < \lambda < 1$. Since the IPRLAB II requires infinite weights renewals to reach the apex of the convex polyhedron, we define the convergence criterion as follows:

The IPRLAB II terminates if the following condition is satisfied:

$$|D' - D| < D \times \delta \tag{13}$$

where

$$D = \sum_{k=1}^{p}(D_1^{(k)} + D_2^{(k)}) \tag{14}$$

$$D_1^{(k)} = \sum_{j=1}^{M} \frac{|S_{Y_j}^{(k)} Y_j^{(k)}|}{\sqrt{N+1}} \tag{15}$$

$$D_2^{(k)} = \sum_{i=1}^{N} \frac{|S_{X_i}^{(k)} X_i^{(k)}|}{\sqrt{M+1}} \tag{16}$$

and D' is the newly obtained distance between weights and borders. As the parameter δ is chosen smaller, the weights reach closer the apex of the convex polyhedron.

Figure 6 shows the geometrical image of the IPRLAB II.

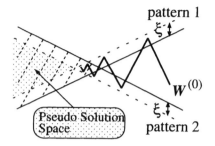

Fig. 6. IPRLAB II.

4 Computer simulation results

4.1 Noise reduction effect

It is known that the robustness of the BAM learned by the PRLAB can be improved by choosing large ξ [10]. In this experiment, we first examined the noise reduction effect of the PRLAB on the parameter ξ.

Fig. 7. Training pairs.

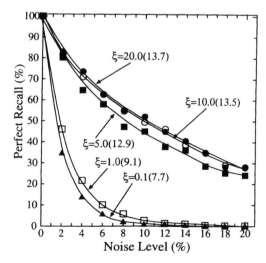

Fig. 8. Relation between ξ and noise reduction effect. Numbers in parentheses show the averaged learning epochs.

Figure 7 shows 10 pairs of the training data learned by the PRLAB. The relaxation factor λ was

set to 1.9 and the initial weights were chosen randomly between -1 and 1 as shown in [9], [10]. Figure 8 shows the result based on 100 trials.

We can see that the noise reduction effect becomes better as the parameter ξ becomes larger. However, when ξ became more than 10.0, we could not observe the improvement of the noise reduction effect. Therefore, we used $\xi = 10.0$ in the experiments described below.

Figure 9 shows the noise reduction effect of the BAM when the training data shown in Figure 7 were learned by the PRLAB, IPRLAB I and IPRLAB II. In this experiment, we used $\lambda = 1.9$ for the PRLAB and the IPRLAB I, $\lambda = 0.99$ and $\delta = 0.001$ for the IPRLAB II. As shown in the figure, the proposed IPRLAB can greatly improve the noise reduction effect in comparison with the PRLAB. Table 2 shows the required learning epochs of each algorithm. We can see that the proposed IPRLAB I and the PRLAB required almost the same learning epochs.

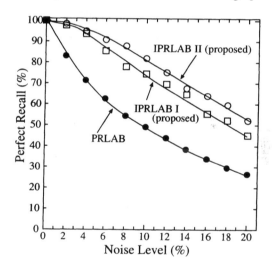

Fig. 9. Sensitivity to noise: 10 pairs of letters were stored in a 49-49 BAM ($\xi = 10.0$).

Table 2. Required learning epochs.

	PRLAB	IPRLAB I	IPRLAB II
Learning epochs	11	12	22

Figure 10 shows the perfect recall rate of each training pair when input patterns were corrupted by randomly reversing each bit with a probability of 10%. By using the proposed IPRLAB II, the noise reduction effect are much improved for each training pair.

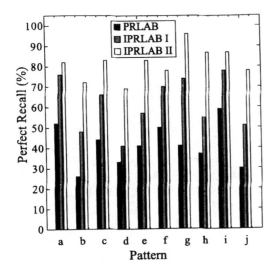

Fig. 10. Sensitivity to noise for each training pair: Capital letters were corrupted by randomly reversing each bit with a probability of 10% and then applied as inputs.

4.2 Improvement of basin of attraction

In this experiment, we applied a noisy input to the BAM and examined the direction cosine between the actual output and the desired one during the recall process.

The direction cosine, which is one of indices for the basin of attraction [6] is defined by

$$a = \frac{1}{n} \boldsymbol{x} \cdot \boldsymbol{t} \tag{17}$$

where $\boldsymbol{x}$ and $\boldsymbol{t}$ show the actual output of the BAM and the desired output, respectively. n indicates the dimension of the pattern.

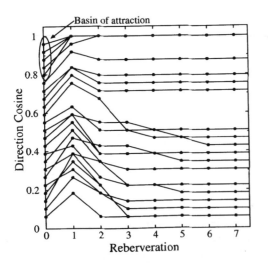

Fig. 11. Basin of attraction (PRLAB).

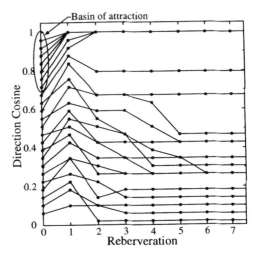

Fig. 12. Basin of attraction (IPRLAB I).

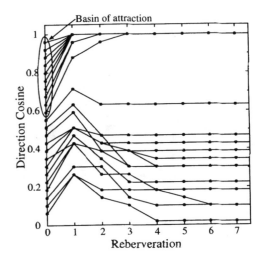

Fig. 13. Basin of attraction (IPRLAB II).

Figures 11-13 show the direction cosine for pattern "A" when the training data shown in Figure 7 were stored by the PRLAB, the proposed IPRLAB I and IPRLAB II. As seen in the figures, the BAM learned by the IPRLAB has the larger basin of attraction than that learned by the PRLAB.

5 Conclusions

In this paper, we have proposed two Improved Pseudo-Relaxation Learning Algorithms for BAM, IPRLAB I and IPRLAB II. The proposed IPRLAB has the following features:

(1) It can guarantee the recall of all training data because it is based on the PRLAB.
(2) It has high storage capacity.

(3) It can greatly improve the robustness of the BAM for noisy inputs.

(4) It can expand the basin of attraction for each training data.

(5) The BAM learned by the IPRLAB II becomes more robust than that learned by the IPRLAB I.

The proposed IPRLAB as well as the conventional PRLAB can be applied to various associative memories such as Hopfield memory [2] and Multidirectional Associative Memory [5]. In the future research, we will examine the characteristics of the proposed learning algorithms theoretically.

Acknowledgment

The authors would like to thank Professor Hiroshi Ito and Masayuki Morisawa of Yamanashi University. A part of this research was supported by the Telecommunications Advancement Organization of Japan.

References

[1] Nakano, K.: Associatron - a model of associative memory. IEEE Trans. Syst. Man. Cybern., *SMC-2*, no.1, pp.380-388 (1972).

[2] Hopfield, J.J.: Neural Networks and Physical Systems with Emergent Collective Computational Abilities. Proc. of National Academy Sciences USA, *79*, pp.2554-2558 (1982).

[3] Hirai, Y.: A model of human associative processor (HASP). IEEE Trans. Syst. Man. Cybern., *SMC-13*, no.5, pp.851-857 (1983).

[4] Kosko, B.: Bidirectional Associative Memories. IEEE Trans. Syst. Man. Cybern., *18*, no.1, pp.49-60 (1988).

[5] Hagiwara, M.: Multidirectional Associative Memory. Proc. of IJCNN'90, Washington, D.C., *1*, pp.3-6 (1990).

[6] Morita, M., Yoshizawa, S. and Nakano, K.: Analysis and Improvement of the Dynamics of Autocorrelation Associative Memory. IEICE Japan, *J73-D-II*, no.2, pp.232-242 (1990).

[7] Wang, Y.F., Cruz, Jr., J.B and Mulligan, Jr., J.H.: Two Coding Strategies for Bidirectional Associative Memory. IEEE Trans. Neural Networks, *1*, no.1, pp.81-91 (1990).

[8] Wang, Y.F., Cruz, Jr., J.B and Mulligan, Jr., J.H.: Guaranteed Recall of All Training Pairs for Bidirectional Associative Memory. IEEE Trans. Neural Networks, *2*, no.6, pp.559-567 (1991).

[9] Oh, H. and Kothari, S.C.: Adaptation of the Relaxation Method for Learning in Bidirectional Associative Memory. IEEE Trans. Neural Networks, *5*, no.4, pp.576-583 (1994).

[10] Hattori, M., Hagiwara, M. and Nakagawa, M.: Quick Learning for Bidirectional Associative Memory. IEICE Japan, *E77-D*, no.4, pp.385-392 (1994).

[11] Hattori, M. and Hagiwara, M.: Intersection Learning for Bidirectional Associative Memory. Proc. of ICNN'96, *1*, pp.555-560 (1996).

[12] Hattori, M. and Hagiwara, M.: Episodic Associative Memory. Neurocomputing, *12*, pp.1-18 (1996).

[13] Hattori, M. and Hagiwara, M.: Neural Associative Memory for Intelligent Information Processing. Proc. of KES'98, Adelaide, *2*, pp.377-386 (1998).

[14] Hattori, M. and Hagiwara, M.: Multimodule Associative Memory for Many-to-Many Associations. Neurocomputing, *19*, pp.99-119 (1998).

Self-Adaptation of Evolutionary Constructed Decision Trees by Information Spreading

Vili Podgorelec, Peter Kokol

Faculty of Electrical Engineering and Computer Science, University of Maribor, Smetanova 17, 2000 Maribor, Slovenia

Email: {Vili.Podgorelec, Kokol}@uni-mb.si

Abstract

Decision support systems that help physicians are becoming very important part of medical decision making. They are based on different models and the best of them are providing an explanation together with an accurate, reliable and quick response. One of the most viable among models are decision trees, already successfully used for many medical decision making purposes. Although effective and reliable, the traditional decision tree construction approach still contains several deficiencies. Therefore we decided to develop a self-adapting evolutionary decision support model, that uses evolutionary principles for the induction of decision trees.

We constructed a multi-population decision model with information spreading and inter-population competition as the self-adaptive method with the aim to improve the quality of the obtained solution. Several solutions were evolved for the classification of mitral valve prolapse syndrome. A comparison has been made with the traditional induction of decision trees. Our approach can be considered as a good choice for different kinds of real-world medical decision making, with respect to the advantages of our model and the quality of the results that we obtain, especially in various medical applications.

1 Introduction

As in many other areas, decisions play an important role also in medicine, especially in medical diagnostic processes. Decision support systems (DSS) helping physicians are becoming a very important part in medical decision making, particularly in those where decision must be made effectively and reliably [1,2]. Since conceptual simple decision making models with the possibility of automatic learning should be considered for performing such tasks [3], according to recent reviews [1,4,5] decision trees are a very suitable candidate. They have been already successfully used in DSS for many decision making purposes, but some problems still persist in their traditional way of induction, which has not changed much since their introduction. Considering the advantages of the evolutionary methods in performing complex tasks, we decided to overcome these problems by inducing decision trees with genetic algorithms.

1.1 Objective and Scope of the Paper

In this paper we would like to present a new approach to building decision trees for DSS and discuss the results. We developed a multi-population decision support model that uses information spreading and inter-population competition as a self-adapting mechanism which improves the quality of the obtained solutions. The developed evolutionary model has among others been used for the mitral valve prolapse prediction in children and a comparison has been made with the traditionally induced decision trees. Decision trees have been already used in DSS for the prediction of mitral valve prolapse [1], as well as for many other medical decision making purposes, which gives us a sound base for the direct comparison of the quality of our model with the existing ones. Besides, the selected problem is very relevant for other reasons. It is a real-world problem from medicine, where decision making should be maximally reliable without any deficiencies or errors. When analyzing the obtained results not only statistical comparisons have been done, but the results have been objectively evaluated by experts from the field of pediatric cardiology, with no regard to the methods being used. Furthermore, the selected problem is very complex, consisting of many details that have to be considered. And last, but not least we had access to a large database of solved cases, which were used for learning and testing purposes of our new method.

In the paper the MVP problem will be introduced first, following by a short introduction to genetic algorithms and decision trees. Then the evolutionary approach to the induction of decision trees will be explained and some facts will be given that justify our approach. The

This work has been partially financed by the Ministry of Science and Technology of Slovenia, grant number L2-1640-0796-99.

whole evolutionary process will be explained in detail with emphasis on the self-adaptation mechanism. The results obtained with the described model for the MVP prediction will be presented. A discussion about important findings, problems and analysis of the results will follow after which the paper will conclude with a short summary of achievements and several directives for further research work.

1.2 Main Contributions of the Paper

There are two essential contributions of this paper. The first one is our new approach to the induction of decision trees with the use of genetic algorithms. We tried to unite the effectiveness and robustness of evolutionary methods with the simplicity and popularity of decision trees, and in this manner improve the quality of the obtained solutions. The second contribution is the introduction of the self-adapting mechanism in the evolutionary process which adapts the evaluation function. In this way the effort and time spent on the definition of a proper evaluation function can be minimized, and the quality of the solutions is increased.

2 Mitral Valve Prolapse

Prolapse is defined as the displacement of a bodily part from its normal position. The term mitral valve prolapse (MVP) [7-10], therefore, implies that the mitral leaflets are displaced relative to some structure, generally taken to be the mitral annulus. The silent prolapse is the prolapse which can not be heard with the auscultation diagnosis and is especially hard to diagnose. The implications of the MVP are the following: disturbed normal laminar blood flow, turbulence of the blood flow, injury of the chordae tendinae, the possibility of thrombus' composition, bacterial endocarditis and finally hemodynamic changes defined as mitral insufficiency and mitral regurgitation.

Mitral valve prolapse is one of the most prevalent cardiac conditions, which may affect up to five to ten percent of normal population and one of the most controversial one. The commonest cause is probably myxomatous change in the connective tissue of the valvar liflets that makes them excessively pliable and allows them to prolapse into the left atrium during ventricular systole. The clinical manifestations of the Syndrome are multiple. The great majority of patients are asymptomatic. Other patients, however may present atypical chest-pain or supraventricular

tachyarrhythmyas. Rarely, patients develop significant mitral regurgitation and, as with any valvar lesions, bacterial andocarditis is a risk.

Uncertainty persists about how it should be diagnosed and about its clinical importance. Historically, MVP was first recognised by auscultation of mid systolic "click" and late systolic murmur, and its presence is still usually suggested by auscultatory findings. However, the recognition of the variability of the auscultatory findings and of the high level of skill needed to perform such an examination has prompted a search for reliable laboratory methods of diagnosis. M-mod echocardiography and 2D echocardiography have played an important part in the diagnosis of mitral valve prolapse because of the comprehensive information they provide about the structure and function of the mitral valve.

Medical experts propose [7-9] that echocardiography enables properly trained experts armed with proper criteria to evaluate mitral valve prolapse (MVP) almost 100%. Unfortunately, however, there are some problems concerned with the use of echocardiography. The first problem is that current MVP evaluation criteria are not strict enough [private communication and 1]. The second problem is the incidence of the MVP in the general population and the unavailability of the expensive ECHO - machines to general practitioners. According to above problems we have decided to develop a decision support system enabling the general practitioner to evaluate the MVP using conventional methods and to identify potential patients from the general population.

3 Decision Trees

Inductive inference is the process of moving from concrete examples to general models, where the goal is to learn how to classify objects by analyzing a set of instances (already solved cases) whose classes are known. Instances are typically represented as attribute-value vectors. Learning input consists of a set of such vectors, each belonging to a known class, and the output consists of a mapping from attribute values to classes. This mapping should accurately classify both the given instances and other unseen instances.

A decision tree [11-13] is a formalism for expressing such mappings and consists of tests or attribute nodes linked to two or more subtrees and leafs or decision nodes labeled with a class which means the decision. A test node computes some outcome based on the attribute values of an instance, where each possible outcome is associated with one of the subtrees. An instance is classified by starting at the root node of the tree. If this node is a test, the outcome for the instance is determined and the process continues using the appropriate subtree. When a leaf is eventually encountered, its label gives the predicted class of the instance.

4   Genetic Algorithms

Genetic algorithms are adaptive heuristic search methods which may be used to solve all kinds of complex search and optimisation problems [14-16]. They are based on the evolutionary ideas of natural selection and genetic processes of biological organisms. As the natural populations evolve according to the principles of natural selection and "survival of the fittest", first laid down by Charles Darwin, so by simulating this process, genetic algorithms are able to evolve solutions to real-world problems, if they have been suitably encoded [17]. They are often capable of finding optimal solutions even in the most complex of search spaces or at least they offer significant benefits over other search and optimisation techniques.

5   Construction of Decision Trees

5.1   Reasons for Using the Evolutionary Method

Before getting into details, let us outline some of the reasons in favor of using genetic algorithms for the construction of decision trees. Genetic algorithms are generally used for very complex optimization tasks [16], for which no efficient heuristic method is developed. Construction of decision trees is a complex task, but an exact heuristic method exists, that usually works efficiently and reliably [12,13]. At a first glance there is no reason to use genetic algorithms. Nevertheless, there are some justifying our evolutionary approach. First, genetic algorithms provide a very general concept, that can be used in all kinds of decision making problems. Because of their robustness they can be used also on incomplete, noisy data (which often happens in medicine because of measurement errors, unavailability of proper instruments, risk to the patient, etc). and which are not very successfully solvable by traditional techniques of decision tree construction. Furthermore, genetic algorithms use evolutionary principles to evolve solutions, therefore solutions can be found that can be easily overlooked otherwise. Another important advantage of genetic algorithms is the possibility of

optimizing the decision tree's topology and the adaptation of class intervals for numeric attributes, simultaneously with the evolution process. One further advantage of the evolutionary approach should not be overlooked: not only one, but several equally qualitative solutions are obtained for the same problem (in most cases). In this way an expert can decide which of the given solutions will be used. In our case a physician can decide which decision tree to use for a specific patient, based on the examinations already performed or selecting the examinations that can be performed easier and/or sooner, than the others. In this way, by using an automatic scheduling system for patients [19], the whole diagnostic process can be optimized. And last but not least, by weighting different parameters, searching can be directed to the situation that best applies to current needs, particularly in multi-class decision making processes, where we have to decide for which decision the reliability should be maximized.

5.2   Evolutionary Process

Before the evolution process can begin, we have to define the internal representation of individuals within the population, together with the appropriate genetic operators that will work upon the population. It is important to assure the feasibility of all solutions during the whole evolution process, therefore we decided to present individuals directly as decision trees. This approach has some important features: all intermediate solutions are feasible, no information is lost because of conversion between internal representation and the decision tree, the evaluation function can be straightforward, etc. The problem with direct coding of solution may bring some problems in defining of genetic operators. As decision trees may be seen as a kind of simple computer programs (with attribute nodes being conditional clauses and decision nodes being assignments) we decided to define genetic operators similar to those used in genetic programming where individuals are computer program trees [18].

The evolution process starts with the seeding of the initial population. Since the evolution will work upon a constant size population, first we have to construct a certain number (population size) of random decision trees. This is performed by randomly choosing attributes and connecting them to a randomly selected branch in a growing decision tree. To prevent a tree becoming deteriorated, we defined certain probabilities of selecting individual branches when constructing a tree.

The first phase of the repeating evolution process is selection. We chose a slightly modified exponential ranking selection scheme with a predefined elitism value (number of preserved unchanged best individuals in each generation). Before the actual selection can take place, all the individuals have to be evaluated for their fitness score. As the evaluation function, a weighted sum of wrong decisions for testing objects was selected and also the complexity of evolved decision tree itself was considered (size and depth of a tree), but its participation in the function was very small. Wrong decisions occur when a learning object is not correctly classified (the correct classifications for all learning objects are known, but hidden for the developing method). Since not all wrong decisions are equally important, we sorted them as follows: the most important errors are made when patients with prolapse or silent prolapse are classified as "no prolapse"; when a decision for such a patient can not be made is the next on the scale of importance; but it is less serious when a healthy patient is classified as with prolapse or silent prolapse, because this only means that the patient will be sent to be further examined with more specialized equipment.

Second phase of the repeating evolution process is the application of genetic crossover. As two individuals are selected, a node is randomly chosen in the first tree and it is searched for the same attribute node in the second tree. Only if the same attribute node exists in both selected trees can the crossover be applied, preventing any two individuals from producing offspring. If the same attribute node is found, a branch at the resulting node in the first tree is cut and is then replaced by a branch at the same attribute node in the second tree. Next it is checked for possible existence of two attribute nodes of the same kind on the same branch in the resulting tree; in such a case the second one is extracted.

The third, and last phase of the repeating evolution process is mutation. The genetic mutation operator is divided into three parts with separate mutation probabilities. The first part works as the exchange of a randomly chosen attribute node in a decision tree with another, randomly selected one from the list of all attributes. The only limitation in attribute nodes exchange is that the both of them have to have the same number of successors in order to preserve all branches from the selected node in a mutated tree. The second part of the mutation works on the leaf nodes of a tree, namely decisions. When a decision node is randomly selected it is exchanged with one of the other possible decisions, also randomly selected. A problem may arise when applying the first part of mutation

operator. While it works if all attributes (or at least a few large groups of attributes) have the same number of classes, this approach may cause problems in the case of discrete attributes with different numbers of possible values. For example, if there are only two discrete attributes both having six values, then only these attributes can be exchanged. Because this limitation may seriously constrain the action of mutation operator, we included also the third part. It constructs the complete new subtree (in the similar way as the trees are constructed for the initial population) which replaces the old subtree from selected attribute.

As the evolution repeats, more qualitative solutions are obtained regarding the chosen evaluation function. The evolution stops when an optimal or at least an acceptable solution is found or if the fitness score of the best individual does not change for a predefined number of generations.

5.3 Self-Adaptation by Information Spreading in Multi-Population Model

One of the most difficult tasks in our evolutionary approach is the definition of an appropriate evaluation function. We have spent a lot of time and effort to experimentally define weights of the evaluation function. The quality and reliability of obtained results depend a great deal on the selected evaluation function which brought us to the conclusion that the definition of an appropriate evaluation function is vital for the successful solution of the given problem. Therefore it is reasonable to expect that automatic adaptation of an evaluation function would result in better results. We decided to develop a mechanism for automatic adaptation of the evaluation function that would lead to self-adaptation of evolved decision trees and in this manner provide better results overall.

One basic question has lead us to the development of a multi-population model for the construction of decision trees: how to provide a method for automatic adaptation of the evaluation function that would still assure the quality of evolved solutions? Namely, if the modification of the evaluation function is unsupervised, the evaluation function can easily become inappropriate which of course gives bad solutions. Our idea was that the evaluation function should evolve together with the decision trees, where solutions are regulated in that they are being compared with another set of decision trees, where the evaluation function is predefined and tested. This brings us to our multi-population model (Figure 1).

298

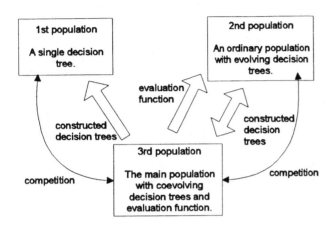

Fig. 1. The multi-population decision support model with information spreading.

The model consists of three independent populations, one main population and two competing ones. First population is actually a single decision tree, initially induced in traditional way with C4.5 algorithm [13]. It serves as a kind of milestone to indicate when the evolving solution in the main population has reached (or eventually outperformed) its quality level. To understand why it is important to have such competing populations, we have to see first how self-adaptation by information spreading actually works. The best individual from the main population, which serves as the general solution, is competing with the best individuals from other populations in solving a given problem upon a learning set. When it becomes dominant over the others (in the sense of the accuracy of the classification of learning objects) it spreads its "knowledge" to them. In this way the other solutions become equally successful and the main population has to improve further to beat them. One global evaluation function (different from those used in the evolutionary process to evaluate the fitness of individuals) is used to determine the dominance of one population over another. It is predetermined as a weighted sum of accuracy, sensitivity and specificity (look at the results in the next section) of the given solution. How information is exchanged between populations depends on which population spreads it to another and will be explained for each case separately.

Now, the single decision tree population does not change for as long as general solution does not outperform its quality level by a certain percentage, regarding the global evaluation function. When it happens, the single decision tree is simply replaced by the evolved general solution (which is the best decision tree from the main population) and general solution has to improve again to reclaim its dominance.

The second population evolves accordingly to the described method in the evolution process (5.2). As it has been shown this process alone would be enough to construct very good solutions, but in this model it serves as another competing population against the main one. Since it also evolves solutions, this population not only competes with the main one, but the knowledge about constructed decision trees can be passed to the main population. When the best individual in the main population does not improve for several generations and its fitness regarding the global evaluation function is lower than that of the second competing population, elite decision trees from the second competing population are copied into the main population, replacing its weakest individuals. In this manner the main population is improved, which also means the improvement of the general solution.

The third and last one, is the main population, whose best individual actually represents a general solution to the given problem. Beside evolving decision trees, also the evaluation function is being adapted simultaneously in order to find the best possible solution. One of the recent and very promising aspect of genetic algorithms, that is still unexploited, is using the natural phenomenon of coevolution. In nature it often happens that two or more species have to evolve with respect to each other to become more successful fighting for resources in the environment as a whole. We used the principle of coevolution to solve the simultaneous evolution of decision trees and the adaptation of evaluation function in the main population of our multi-population model. As already said before, the best constructed decision tree from the main population can be used to replace the single decision tree in the first population. This action does not improve the quality of the evolved decision trees in the main population directly, but rather serves as an indicator of how rapidly the quality of the general solution improves. Decision trees are spread to the second competing population in the same way as opposite, just that the evaluation function is passed together with decision trees and replaces the old evaluation function in the second population.

To conclude the description of the self-adapting model we still have to explain the adaptation of evaluation function in the main population. When seeding the initial population beside random decision trees, an initial set of evaluation functions should be constructed. Since the evaluation of quality of evaluation functions would be very difficult we decided to evolve only one evaluation function upon which only the directed mutation operator will be applied in the evolution process. The initial evaluation function is equal to the one used in the second competing population. The directed mutation operator depends on currently evolved decision trees and works as follows: if the number of wrong decisions for learning objects is very high for one of the possible categories (for example when silent prolapse is classified many times when the prolapse is the correct decision) then the weight for the parameter, describing this situation, is obviously set to low in the evaluation function and therefore it is increased for a small random percent of the weight value.

6   Application and Results

Using the Monte Carlo sampling method we have selected 900 children and adolescents representing the whole population under eighteen years of life. All of them were born in the Maribor region and all were white. Routinely they were called for an echocardiography with no prior findings. We examined 631 volunteers.

They all had an examination of their health state in the form of a carefully prepared protocol specially made for the Syndrome of MVP. The protocol consisted of general data, mothers' health, fathers' health, pregnancy, delivery, post-natal period, injuries of chest or any other kind, cronical diseases, sports, physical examination, subjective difficulties like headaches, chest-pain, palpitation, perspiring, dizziness etc., auscultation, phonocardiography, EKG and finally ECHO. In that manner, we gathered 103 parameters that can possibly indicate the presence of MVP.

All 631 patients were divided into a learning and a testing set, 500 patients were selected for the learning and the rest, 131, for the testing set. First we constructed decision tree traditionally. Then a few decision trees were evolved through our new evolutionary approach for the same learning and testing sets. A comparison of the best traditionally constructed decision tree (Table 1), with the simple evolutionary constructed solution (Table 2) and the new self-adapting evolutionary constructed solution from the final model (Table 3) has been made. A very significant improvement has been achieved with our new model, especially regarding the classification of prolapse (5 instead of 3 correct classifications mean improvement of 67 percent), which is also the most important one beside silent prolapse.

A comparison of the decision trees has shown that similar attributes were chosen as the most important

(nearer to the root of a decision tree), but the less important ones (deeper in tree, near to decision attributes) were quite different. This gives us a hint that the presence of MVP can be determined in several ways, with help of different examinations and data - the same opinion is shared by some medical experts who have tried to predict the MVP in different ways. Above results favor our evolutionary approach that is able to find those different solutions as opposed to the traditional approach.

Table 1. The results of MVP classification by traditionally constructed decision tree (C4.5 algorithm) with 0% tolerance and post-pruning upon testing set.

	Classified as prolapse	Classified as silent prolapse	Classified as no prolapse
prolapse	3	0	4
silent prolapse	0	4	0
no prolapse	8	4	108

Table 2. The results of MVP classification by simple evolutionary constructed decision tree.

	Classified as prolapse	Classified as silent prolapse	Classified as no prolapse
prolapse	3	0	4
silent prolapse	0	4	0
no prolapse	6	3	111

Table 3. The results of MVP classification by new self-adapting evolutionary constructed decision tree.

	Classified as prolapse	Classified as silent prolapse	Classified as no prolapse
prolapse	5	0	2
silent prolapse	0	4	0
no prolapse	3	1	116

The next important improvement, maybe even more important than the accuracy of suggested decisions, is the topology and size of the constructed decision trees (Table 4). The number of attribute nodes is about 20 percent lower in a simple evolutionary constructed decision tree compared with the traditional one, and even lower in a self-adapting evolutionary constructed decision tree, in spite of the post-pruning performed in the traditionally constructed one. This result supports our hypothesis about the optimization of tree topology and complexity and is very important because it enables a physician to predict the presence of MVP with two examinations less on average than before - a result that really made the medical experts enthusiastic.

Table 4. A comparison of the complexity between traditionally, evolutionary and self-adapting evolutionary constructed decision trees (TDT=traditionaly constructed decision tree; EDT=decision tree, constructed with simple evolutionary method; SAEDT=decision tree, constructed with self-adapting evolutionary method).

	TDT	EDT	SAEDT
Num. of attribute nodes	34	26	23
Num. of decision nodes	52	38	34
maximum depth	10	6	6
average depth	5,73	4,13	4,05

Table 5. A comparison of effectiveness for different decision tree induction methods (TDT=traditionaly constructed decision tree; EDT=decision tree, constructed with simple evolutionary method; SAEDT=decision tree, constructed with self-adapting evolutionary method).

	TDT	EDT	SAEDT
accuracy	78.79	90.08	93.89
sensitivity	63.64	63.64	81.82
specificity	90.00	92.50	95.00

A comparison of the effectiveness of different diagnostic methods is usually described in the field of machine learning by accuracy and in the field of medicine by sensitivity and specificity (Table 5). The accuracy of a diagnostic method is calculated as the relation between the correctly classified and all testing objects. Sensitivity is based on ratio of correctly classified positive patients (patients having prolapse or silent prolapse in our case) compared to all positive patients, and specificity is based on ratio of correctly classified negative patients (without prolapse or silent prolaspe) compared to all negative patients. All values are given as percentages. From table 5 it is obvious that the evolutionary approach greatly improves all measures of effectiveness, especially when the self-adapting model is considered.

It is interesting to see how the quality of constructed decision trees improves through generations (Figure 2). When considering the quality of the best individual, its fitness score improves rapidly at the start as expected. The few steep falls that occur later in the process are the consequence of a successful crossover where one bad branch is replaced by a much better one. The average fitness score for a simple evolutionary model and self-adapting evolutionary model are quite similar at the beginning and change after several hundreds of generations.

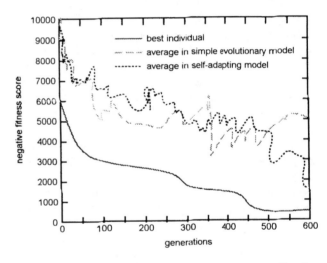

Fig. 2. Changing of fitness score over generations. The few steep falls of fitness score for the best individual mean sudden change of an important branch.

7 Discussion

Several interesting discoveries have been made while developing the evolutionary approach to the construction of decision trees, some of them expected and the others surprising. Population size is quite important: if it is too small solutions tend to overfitting, and if it is too big solutions improve very slowly through the generations. Also mutation is very important: if its probability is too low evolution converges prematurely, and if it is too high solutions tend to extreme overfitting. The use of a single evaluation function that coevolves with decision trees is in our opinion the weakest point of an otherwise very successful self-adapting evolutionary model. In this manner in the future we should find a way to effectively evaluate the quality of an evolving evaluation function (or even better a population of evolving evaluation functions) and in this way even increase the quality of obtained solutions and the stability of the method.

The analysis of obtained results shows that our new model is a bit better than the traditional regarding the reliability of suggested decisions and considerably better regarding the complexity of constructed decision trees what makes it especially useful for those purposes where a single step toward final decision is exacting and expensive. Regarding its generality it should be easily used for any decision making purpose where a problem can be described by a set of attributes, a list of possible decisions, and a set of already solved cases.

8 Conclusion

In the paper we present a new approach in building DSS based on decision trees. We applied it to the prediction of MVP in children, but it can be easily used for other kinds of medical decision making purposes. Some reasons in favor of using evolutionary principles for the construction of decision trees have been stated. A new multi-population decision support model has been developed which uses the inter-population competition and information spreading to achieve the maximum possible quality of the evolved solutions. The whole evolution process is described: seeding of initial population, selection, crossover, mutation, and evaluation function, as well as the idea behind the self-adaptation by information spreading. Results of MVP classification by evolutionary constructed decision trees are compared with those obtained by traditionally constructed ones. Because of the advantages of our model and the quality of the obtained results, in our opinion this approach can be successfully used for different kinds of medical diagnosis prediction.

There are two essential contributions of the paper. The first one is the evolutionary approach to the induction of decision trees that unites the effectiveness and the robustness of evolutionary methods with the simplicity and the popularity of decision trees, and, what is the most important, gives very good results. The second contribution is the way in which the adaptation of the evaluation function leading to the self-adaptation of the model is achieved. In our opinion the results show that it can be used to improve the quality of existing and more importantly future systems based on evolutionary principles, but there is still a lot to be done in this area.

References

[1] Kokol P, et al: Decision Trees and Automatic Learning and Their Use in Cardiology, Journal of Medical Systems 19(4), 1994.

[2] Kokol P, Podgorelec V, Malcic I: Diagnostic Process Optimisation with Evolutionary Programming, Proceedings of the 11th IEEE Symposium on Computer-based Medical Systems CBMS'98, pp. 62-67, Lubbock, Texas, USA, June 1998.

[3] Kokol P, Stiglic B, Zumer V: Metaparadigm: a soft and situation oriented MIS design approach, International Journal of Bio-Medical Computing, 39:243-256, 1995.

[4] Kokol P, et al: Spreadsheet Software and Decision making in Nursing, Nursing Informatics'91, (Honvenga E J S et al. eds.), Springer Verlag, 1991.

[5] Quinlan J R: Decision Trees and Decision making, IEEE Trans System, Man and Cybernetics 20(2)339-346, 1990.

[6] Podgorelec V, Kokol P: Evolutionary Construction of Medical Decision Trees, Proceedings of the 20th Annual International Conference of the IEEE Engineering in Medicine and Biology Society EMBS'98, Hong Kong, 1998.

[7] Anderson H R, et al: Clinicians Illustrated Dictionary of Cardiology. Science Press, London, 1991.

[8] Markiewicz W, et al: Mitral valve Prolaps in One Hundred Presumably Young Females, Circulation 53(3)464-473, 1976.

[9] Barlow J B, et al: The Significance of Late Systolic Murmurs, AHJ 66:443-452, 1963.

[10] Devereoux R: Diagnosis and Prognosis of Mitral Valve Prolaps, The New England Journal of Medicine 320(16)1077-1079, 1989.

[11] Quinlan J R: Induction of decision trees, Machine learning, No. 1, pp. 81-106, 1986.

[12] Quinlan J R: Simplifying decision trees, International journal of man-machine studies, No. 27, pp. 221-234, 1987.

[13] Quinlan J R: C4.5: Programs for Machine Learning, Morgan Kaufmann, 1993.

[14] Bäck T: Evolutionary Algorithms in Theory and Practice, Oxford University Press, Inc., 1996.

[15] Forrest S: Genetic Algorithms, ACM Computing Surveys, pp. 77-80, Vol. 28, No. 1, March 1996.

[16] Goldberg D E: Genetic Algorithms in Search, Optimization, and Machine Learning, Addison Wesley, Reading MA, 1989.

[17] Holland J H: Adaptation in natural and artificial systems, MIT Press, Cambridge MA, 1975.

[18] Koza J R: Genetic Programming: On the Programming of Computers by Natural Selection, MIT Press, 1992.

[19] Podgorelec V, Kokol P: Genetic Algorithm Based System for Patient Scheduling in Highly Constrained Situations, Journal of Medical Systems, Plenum Press, Volume 21, Num. 6, pp. 417-427, December 1997.

Evolutionary Programming of Near-Optimal Neural Networks

D. Lock[1] and C. Giraud-Carrier[2]

[1]RCMS Ltd, Windsor House, Milbrook Way, Colnbrook SL3 0HN, UK
[2]Department of Computer Science, University of Bristol, Bristol BS8 1UB, UK
Email: digby.lock@rcms.com, cgc@cs.bris.ac.uk

Abstract

A genetic algorithm (GA) method that evolves both the topology and training parameters of backpropagation-trained, fully-connected, feed-forward neural networks is presented. The GA uses a weak encoding scheme with real-valued alleles. One contribution of the proposed approach is to replace the needed but potentially slow evolution of final weights by the more efficient evolution of a single weight spread parameter used to set the initial weights only. In addition, the co-evolution of an input mask effects a form of automatic feature selection. Preliminary experiments suggest that the resulting system is able to produce networks that perform well under backpropagation.

1 Introduction

It is well recognised that the performance of backpropagation-trained neural networks depends highly on their topology and the values of a number of training parameters. Unfortunately, there are no comprehensive analytical methods of determining the optimal topology and training parameters for a particular task, which often hampers the wider application of neural networks to real-world applications. Traditionally, the design of a neural network is achieved by trial and error and requires the involvement of an expert. As systems become more complex this approach becomes more costly in computational and human terms, giving no guarantee that a near optimal design has been chosen. These factors detract heavily from the initial attraction of using neural networks to derive problem solutions.

In addition to these obstacles, standard backpropagation is beset by the problems of slow convergence and local minima. To alleviate these problems, a number of enhancements have been suggested, including, for example, the addition of a momentum term. However, although they do tend to result in improved performance, these extensions also produce a multiplication of the number of user-controlled parameters. The potential user of backpropagation is soon faced with nearly a dozen parameters whose values may have (unpredictable) effects on the performance of the network. If, for any given application, the set of possible combinations of parameter values is viewed as a search space, then it is evident that the search space is extremely large, highly dimensional and multimodal. Manual tuning is clearly not a practical option.

Given such characteristics for the search space, one suitable alternative consists in using a genetic algorithm (GA) [4,6]. This paper shows how a GA is used to automate the design of nearly optimal (with respect to predictive accuracy) backpropagation-trained, fully-connected, feed-forward neural networks [13]. The GA, which uses a weak encoding scheme with real-valued alleles, optimises parameters of both the topology and the learning algorithm. In particular, the system addresses efficiently the problem of weight setting. It is well known that the initial setting of weights often has a dramatic impact on a backpropagation-trained neural network. In many GA approaches, the weights are evolved along with the rest of the topology, thus seriously affecting the efficiency of the evolutionary process. In our approach, weights are not evolved. Instead, a single weight spread parameter is evolved and used for initial weights setting. These weights are then updated by backpropagation training.

The paper is organised as follows. Section 2 briefly motivates the evolutionary approach to neural network programming. Section 3 describes the actual genetic algorithm and shows how the weight spread parameter is used to improve efficiency, whilst retaining most of the effects of a guided search of the space of initial weights. Section 4 reports the results of empirical studies. Finally, section 5 concludes the paper.

2 Evolutionary NN Programming

A standard backpropagation-trained neural network is characterised by its topology and learning rate [13]. From these, it implements a simple gradient descent on the mean-squared error over the training set. Changes to the weights on connections to a node N are proportional to the error attributed to N, the output of N and the learning rate. This standard approach suffers from severe drawbacks, including slow convergence, local minima and poor generalisation ability. As a result, it is generally extended with a number of useful techniques, including weight decay and additive momentum terms (e.g., see [2,7,12,13]). In our chosen implementation, the neural network's training is characterised by five parameters:

- Learning rate,
- Momentum,
- Weight decay,
- Delta-Bar-Delta and
- Initial weight spread,

and the network's topology is defined in terms of four parameters:

- Number of inputs,
- Number of hidden layers
- Number of hidden units per hidden layer and
- Number of outputs.

Although neural networks are not programmed in the traditional sense, the adequate setting of both training and topology parameters can be regarded as the programming phase in the application of neural networks. However, whilst traditional programming is slowly emerging as a proper engineering discipline, with tried and tested, systematic techniques, neural network programming remains an art more than a science. It has indeed proven very difficult to develop comprehensive analytical methods for determining the optimal topology and training parameters of neural networks. Consequently, neural network programming remains mostly *ad hoc* and often based on trial and error. Indeed, in the absence of a suitable methodology, the user is left with what amounts to be a complex search problem: the discovery, in the space of all possible combinations of parameter values, of a near-optimal one.

Given the parameters listed above, it is clear that the search space is huge, highly dimensional and multimodal. In addition, the search space has no real exploitable underlying structure, which means that there does not exist simple, informed heuristics to guide the search for a solution. Hence, traditional search techniques (e.g., branch-and-bound, best-first, A*) are impractical and manual tuning, even by an expert, is inadequate in all but the simplest of applications. On the other hand, genetic algorithms (GAs) are search algorithms loosely based on evolutionary principles, which are well suited for large, irregular search spaces [4,6]. Through a combination of pseudo-random global search and guided local search, GAs are often able to produce near-optimal solutions where other methods have failed. It is our contention, therefore, that GAs offer a viable approach to the problem of neural network programming. One such approach is described in the following section.

3 GA-optimised NN Programming

In order to use GAs for neural network programming, it is necessary to design an adequate representation for networks. There are at least two possible encoding schemes generally considered: strong encoding and weak encoding [11]. Strong encoding allows the representation of explicit connectivity patterns, but does not scale well and tends to produce long chromosomes. In weak encoding, connectivity is implicit and developmental rules must be used to turn chromosomes into network topologies. Weak encoding supports large networks, but tends to be less flexible. Traditionally, whether strongly or weakly encoded, chromosomes are binary strings. This is in part due to the fact that theoretical results, such as the schema theorem [4], are only valid with binary strings. However, recent work with more structured GAs, such as marker-based encoding [3], real-valued encodings (e.g., [1]) and genetic programming [8], suggests that good results can be achieved with real-valued strings.

In this work, a real-valued, weak encoding is used. The advantages of scalability and ease of application of genetic operators outweigh the lack of control on connectivity. The gene sequence on each chromosome is as follows:

$$n \quad i_1 ... i_n \quad h_1 ... h_3 \quad l \quad m \quad d \quad a_1 \quad a_2 \quad w$$

where n is the number of potential inputs, $i_1 ... i_n$ is a binary mask for input selection, $h_1 ... h_3$ are the numbers of units in the first, second and third hidden layers, respectively, l is the learning rate, m is the momentum term, d is the weight decay term, a_1 and a_2 are adaptive learning rates (i.e., Delta-Bar-Delta terms), and w is the weight spread term.

The input mask effects a form of automatic feature selection as it identifies which inputs must actually be used in the final network's design. In traditional neural network design, the number of inputs is fixed *a priori* by the designer. As a result, in applications where the problem domain is ill understood and where the contribution of each input is unclear, there is a tendency to include all inputs that may have some relevance, the argument being that, if the network has too few inputs, it will be unable to learn the target input-output mapping. Ideally, however, a network should use as few inputs as are strictly necessary to solve the problem, since this produces smaller networks, reduces the computation time required for training and speeds up recall when the system is used online. Selecting inputs is part of the evolutionary process in our system. The first gene holds the maximum number of possible inputs, whilst the mask determines which of these inputs must be used. For example, the sequence $(5,1,0,1,0,0)$ represents a network using only the first and third of 5 possible inputs. Note that, for a given problem, only the mask is evolved; n is the same for all chromosomes and is used only for ease of implementation.

For each gene, the GA also has access to the range of allelic values that gene can take and the granularity of that range. This information is provided by the user. For binary-valued genes, the range is simply the set $\{0,1\}$ and the notion of granularity does not apply. For real-valued genes, however, the range takes the form $[lb, ub]$ and the granularity is captured by a step size value s, which indicates that the gene can take any value between lb and ub in increments of s, i.e., lb, $lb+s$, $lb+2s$, ..., ub. The smaller s, the larger the number of possible values. One advantage of using real values in this way is that, at run time, the user has a high degree of control over the boundaries within which the GA will operate whilst retaining an acceptable chromosome length.

The raw fitness is based on the network's predictive accuracy (i.e., number of correctly classified examples) on an independent validation set following training. In addition, penalty terms can be added to account for the size of the network and/or the number of elapsed training cycles. In this way the genetic algorithm can be tuned to evolve smaller networks with faster training times. The following detail the genetic operators used and the implicit evolution of weights through the single weight spread parameter.

3.1 Reproduction

The system uses a fitness-proportionate selection with elitism (see Section 3.2). However, in order to avoid the problem of premature convergence, the raw fitness is adjusted as follows. First, all the chromosomes in the population are ranked according to their raw fitness, from the least fit to the most fit. Then, the raw fitness of each individual is scaled and its new fitness is computed by:

$$F' = 2 - SP + 2(SP - 1)(Pos - 1)/(PopSize - 1)$$

where SP is a selection pressure parameter set to increase with each generation of the GA, Pos is the chromosome's ranking position in the raw fitness ordering and $PopSize$ is the size of the population. Finally, the probability of selection is computed based on F'. The use of F' reduces the influence of super-individuals in the selection process whilst still maintaining the needed bias in favour of fitter individuals. In addition, the selective pressure can be adjusted by the user.

3.2 Crossover

Parents selected by the reproduction operator are placed in an intermediary pool. The crossover operator then picks parents from that pool and, for each pair, produces offsprings. The offsprings replace their parents to form the next generation. In addition, the fittest chromosome in each generation is passed on unchanged to the next generation.

Here, single-point crossover is used. However, since the system allows the user to select arbitrary subsets of parameters to co-evolve (including all of them of course), a small change is necessary to avoid offsprings that are identical copies of their parents. In order to do so, the crossover operator ensures that the crossover point (or *locus*) falls between the first (i.e., leftmost) and last (i.e., rightmost) genes taking part in the evolution. Parameters that are not evolved keep the same fixed values in all chromosomes and are not affected by crossover.

Clearly, if unconstrained, crossover may produce non-viable offsprings (i.e., invalid neural network configurations). There are two solutions available: cure and prevention. The *cure* approach allows non-viable offsprings to be created but immediately penalises them with the lowest possible fitness, thus making it virtually impossible for them to be selected for reproduction. The *prevention* approach consists of preventing the creation of such offsprings in the first place. Here, we adopt the second approach. Given the fairly high probability of producing non-viable offsprings through unconstrained crossover in this context, the second approach is adopted. Impossible offsprings are prevented from creation by ensuring that the network structure has at least one active/selected input and at least one hidden neuron in one hidden layer. The ranges of acceptable values of all parameters are also defined so that they do not produce impossible settings. For example, the learning rate is not allowed to equal 0.

3.3 Mutation

Following crossover, offsprings may undergo random mutations. Like the crossover operator, the mutation operator only affects the genes that have been selected for evolution.

For binary-valued parameters, mutation simply toggles the current value. For real-valued parameters, mutation consists in replacing the current value by another valid value chosen at random within the parameter's allowed range. Thus, the smaller the step size, the larger the number of possible values to choose from in mutation.

3.4 Implicit Weights Evolution

Backpropagation-trained neural networks start with random initial weights. It is well known that this initial setting of weights often has a dramatic impact on performance and random setting is not guaranteed to produce "good" weights. In terms of optimisation, there are two basic alternatives: either all parameters are optimised to be resistant to initial weight setting or the weights themselves are optimised. Given the huge size of the weight space, it is not practical to evolve parameters that are robust to weight changes in initial weights. Indeed, experiments with random initial weights show that the system is unable to optimise the other parameters [9]. Hence, weights must be included in the evolutionary process.

The fundamental problem with searching the initial weight space is that the fitness value of the chromosome must be entirely attributable to the gene sequence on that chromosome. If one is trying to search for the best initial weight randomisation then the chromosome must encode for information about the weights it was initialised with. As a result, many GA approaches evolve the actual weights along with the rest of the topology. However, this seriously affects the efficiency of the evolutionary process.

In our approach, weights are not evolved explicitly. Instead, a single weight spread parameter is evolved and used for initial weights setting. The weights are then updated by backpropagation training. A similar approach is also adopted in [5], with the exception that here, fully-connected networks are assumed instead of allowing the pattern of network connectivity to evolve. One can argue, however, that network connectivity may be determined by the weights. Indeed, a weight value of 0 is identical to a missing connection, so that it is possible to start from a fully connected topology and change the connectivity by weight evolution only.

Initial weights are determined as follows. A single series of pseudo-random numbers, whose range is determined by a seeding number is generated. A given seeding number always produces the same series. It is chosen at the beginning of the GA process and remains unchanged throughout the whole evolutionary process. To initialise a network's weights, a sequence of length equal to the number of connections in the network is extracted from the series of weights created by the seeding number. Hence, a network with 10 connections will take the first 10 weights from the series, a network with 15 connections the first 15 weights, and so on. Networks that share the same underlying topology are thus given the same weights. However, prior to installing the weights into the network, each network modifies the weights it receives from the series by applying to them its particular weight spread range, as encoded in its corresponding chromosome.

One of the features of this mechanism is that, although each network has different actual weights (assuming they have different weight spread values), the values of these weights can be determined simply by looking at the weight spread gene. In other words, however large the network, all the information about its initial weights is contained in one gene. Most importantly, the fitness of a chromosome is due entirely to its genotype. When the weight spread gene of one parent is passed on to one of its offsprings, then that offspring will get exactly the same initial weights as its parent. In effect, the chromosomes now contain a gene that encodes for inheritance of initial weights. It follows that the space of initial weights can be searched effectively through the application of the normal GA operators on the weight spread gene.

On the other hand, it is important that the encoding may not make the system more efficient. The implicit encoding of the initial weights means that final weights must be computed through backpropagation. Hence, the complexity associated with evolving weights explicitly as part of the GA does not completely disappear. It is transferred in part to the training phase now needed by the evolved neural networks. The trade-off is between GA search and gradient-descent search of the weight space.

4 Preliminary Experiments

As a proof of concept, experiments were conducted with two simple problems, XOR and Lenses [10]. Figures 1 and 2 show the results of evolving all parameters, except weight decay. The crossover rate is 0.6 and the mutation rate is 0.1.

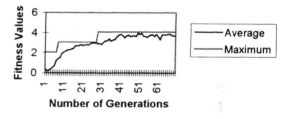

Fig. 1. XOR Neural Network Optimisation

The average fitness of the population rises sharply, which shows that the genes in the fitter chromosomes are being propagated throughout the population. Moreover, the constant rise of the plot indicates that the fitness of an individual is due entirely to its genotype and that the GA is not subject to environmental effects over which it has no control or information about (e.g., random weight initialisation at each generation).

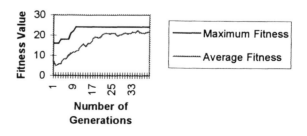

Fig. 2. Lenses Neural Network Evolution

306

For the XOR problem, the best evolved network has 2 (out of 2) selected inputs, 3 hidden layers (6-8-9), learning rate of 1.2, momentum term of 1, adaptive learning rates of 1.6 and 0.2, and a weight spread of 0.7 (i.e., weight values range over [-0.35, +0.35]). The network learns the problem (i.e., reaches 100% accuracy on the validation set) in 150 cycles. The large size of the network is due to the fact that there was no attempt at optimising either for size or training time.

For the Lenses problem, the best network has 4 (out of 4) selected inputs, 3 hidden layers (8-6-10), learning rate of 0.75, momentum term of 0.65, adaptive learning rates of 0.8 and 0.6, and a weight spread of 0.25 (i.e., weight values range over [-0.125, +0.125]). Here, a penalty term was added against networks with large number of training cycles. The best network trains in only 122 epochs.

5 Conclusion

This paper describes a method for genetic-based neural network programming, based on a weak encoding. In particular, a simple weight initialisation method is presented, that allows the GA to search the whole space of initial weight combinations by encoding a weight spread parameter on a single real-valued gene. In addition, the system evolves an input mask thus allowing a form of automatic feature selection.

Preliminary experimental results demonstrate promise. Further experiments with more complex problems and the full range of network parameters are needed to fully validate the proposed approach. It is also necessary to compare more formally the efficiency trade-off between the weak encoding proposed here and existing strong encoding approaches.

References

[1] Burdsall, B. and Giraud-Carrier, C. (1997). Evolving Fuzzy Prototypes for Efficient Data Clustering. In *Proceedings of the Second International ICSC Symposium on Fuzzy Logic and Applications (ISFL'97)*, 217-223.

[2] Fahlman, S.E. (1988). An Empirical Study of Learning Speed in Backpropagation Networks. Technical Report CMU-CS-88-162, Carnegie Mellon University.

[3] Fullmer, B. and Miikkulainen, R. (1992). Using Marker-Based Genetic Encoding of Neural Networks to Evolve Finite-State Behaviour. In *Proceedings of the First European Conference on Artificial Life (ECAL'91)*, 255-262.

[4] Goldberg, D.E. (1989). *Genetic Algorithms in Search, Optimization and Machine Learning*. Addison-

Wesley Publishing Company.

[5] Harp, S.A., Samad, T. and Guha, A. (1989). Towards the Genetic Synthesis of ANN. In *Proceedings of the Third International Conference on Genetic Algorithms (ICGA'89)*, 360-369.

[6] Holland, J. (1975). *Adaptation in Natural and Artificial Systems*. The University of Michigan Press, Ann Arbor, MI.

[7] Jacobs, R.A. (1988). Increased Rates of Convergence Through Learning Rate Adaptation. *Neural Networks*, 1(4):295-307.

[8] Koza, J.R. (1992). *Genetic Programming: On the Programming of Computers by Means of Natural Selection*. MIT Press.

[9] Lock, D.F. (1998). Using Genetic Algorithms to Build, train and Optimize Neural Networks. MSc Thesis, Department of Computer Science, University of Bristol.

[10] Merz, C.J. and Murphy, P.M. (1996). *UCI Repository of Machine Learning Databases*. Department of Information and Computer Science, University of California, Irvine.

[11] Miller, G.F., Todd, P.M. and Hedge, S.U. (1991). Designing Neural Networks. *Neural Networks*, 4:53-60.}.

[12] Plaut, D.C. and Hinton, G.E. (1987). Learning Sets of Filters Using Backpropagation. *Computer Speech and Language*, 2:35-61.

[13] Rumelhart, D.E., Hinton, G.E. and Williams, R.J. (1986). Learning Internal Representations by Error Propagation. In Rumelhart, D.E. and McClelland, J.L. (Eds.), *Parallel Distributed Processing*, Vol. 1, MIT Press.

Exploring the Relationship between Neural Network Topology and Optimal Training Set by Means of Genetic Algorithms.

J. Dávila

Graduate School and University Center, City University of New York
33 West 42 Street New York NY 10036
jdavila@broadway.gc.cuny.edu

Abstract

In a previous paper I have presented the results of optimizing Neural Network (NN) topology for the task of Natural Language Processing (NLP). In that research, all NN were trained with a fixed 20% of the total language. In this paper I present results of optimizing a set of configuration values that have proven to affect NN performance. For example, Elman has reported improved performance when the NN were trained with simple sentences first, and complex sentences later. On the other hand, Lawrence, Giles, and Fong have reported better results when the training data was presented in a single set. Lawrence, Giles, and Fong have also studied the effect of different learning algorithms on natural language tasks. Because of the ability of GA to search a problem space for minima without using knowledge about the problem itself, they are well suited for problems that might contain more than one possible solution. Finding different minima becomes important for real-life applications, since variables such as number of hidden nodes, number of hidden layers, number of connections, and size of training set all can affect training and response time for NN.

1 Introduction

In a previous paper I have presented the results of optimizing Neural Network (NN) topology for the task of Natural Language Processing (NLP), [1]. Some topologies clearly demonstrate a better ability to learn the task at hand. In that research, all NN were trained with a fixed 20% of the total language. If instead of always using the same 20% of the language for training we use a random 20%, the networks begin to exhibit different behavior; some topologies do not perform as well, while others that did not seem to learn with the fixed training set do manage to perform better. Other researchers have explored the relationship between training data and NN performance on NLP. Elman reported better performance when training with simple sentences before using more complex ones, [2]. Laurence, Giles, and Fong, on the other hand, report that sectioning the training data in this manner decreased performance in their studies, [4]. The question then

becomes how to choose a subset of the training data that will allow the NN to better learn the intended task, and how to present this given data.

2 The Task for the NN

The task the networks attack is very similar to the one described by Jain in [3]. A sentence is presented to the NN one word at a time by activating the corresponding nodes at the input layer. The NN incrementally generates a description of the input by correctly identifying the parts of the sentence and how they relate to each other. For example, in the sentence <the boy runs> the NN should respond by (1) identifying each word in the sentence, e.g. <runs> is a movement verb; (2) identifying phrases, e.g.<the boy> is a noun phrase; and (3) identifying the relationships among phrases, e.g. <the boy> is the agent of <runs>. This task will require the NN to store time-dependant information, since it needs to react to the word <john> differently when seen after <mary saw> than when seeing it as the first word of the sentence. The training set is presented for 600 epochs. Testing of the trained NN is done by presenting all sentences of the language (419 in total). All experiments were repeated 50 times with different random number seeds.

3 Previous Results

My previous experiments in topology optimization have produced evidence indicating that some NN topologies are better able to solve NLP problems. In particular, NN with four or more layers between the input and output layers outperformed those with fewer hidden layers. Another characteristic of NN that managed to learn the task at hand was that they had only one path between the input and output layers (Ex: "Fig. 1(a)"). This seems to point to a marked advantage for pipeline-like topologies. In terms of the connectivity between the hidden layers, topologies that were successful in solving the task presented to them had feedback loops ending at the first hidden layer (Ex.: "Fig. 2(a)"). These feedback loops could be divided into two different types. One of them is a feedback loop formed by layers that do not form part of the main path between input

308

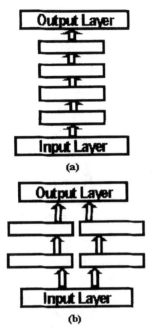

Fig. 1. (a) Pipeline-like topology; (b) multiple-paths topology

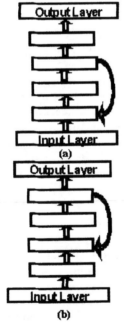

Fig. 2. (a) feedback loop that terminates at the first hidden layer; (b) feedback loop that terminates at a hidden layer other than the first.

and output layers (Ex.: "Fig. 3(a)"). I call this type of topology NMPL, for Non-Main Path Loop. In the other type the feedback loop was formed by the same hidden layers

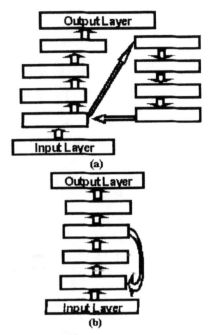

Fig. 3. (a) Feedback that uses hidden layers not in the main input-output path. (b) feedback that uses hidden layers in the main input-output path

that are part of the main path between input and output (Ex.: "Fig. 3(b)"). I call this type of topology MPL, for Main Path Loop. MPL networks had longer feedback loops, and also needed to be trained with lower values for learning parameter than NMPL. In addition, feedback loops in MPL networks were always longer than those in NMPL networks.

Two other topologies were found to be effective in learning the NLP task at hand. Both of them have only one hidden layer between input and output. One has 30 or more nodes in it (1HMN, for 1 hidden layer, many nodes). The other has very few nodes in its hidden layer (1HFN, for 1 hidden layer, few nodes). When the value for backstep was fixed at 1, both 1HMN and 1HFN networks failed to learn from the training data (backstep is a parameter used when training recurrent NN that indicates how far back into the activation history of the network to look when computing error gradients).

All of these results were obtained with a fixed training set, formed of 20% of the complete language. If, instead of using this same training set, training was done with a random 20% of the language, the networks began to exhibit radically different performance. In particular, 1HFN and 1HMN networks had difficulty learning the intended task. In addition, MPL and NMPL networks, although they always managed to learn the task, had different performance depending on the training set. Because of these differences, caused by the training data, the main concern of this paper is the optimization of the training set for those topologies that had previously demonstrated the ability to learn the task at hand.

4 Genetic Definition of the Training Set

In order to test what effect the training data has on the NN performance, I have used Genetic Algorithms to determine the size, composition, and mode of presentation of the training set for each of the NN topologies described in the preceding section. The language used has a total of 419 sentences. There are 64 sentences in group 1, 184 sentences in group 2, and 171 sentences in group 3. Sentences in group 1 have one verb phrase and one noun phrase. Sentences in group 2 have two noun phrases and one verb phrase. Sentences in group 3 have three noun phrases and either one or two
verb phrases.

The basic size of the training set will consist of 20% of the sentences. These 84 sentences will be chosen from the three groups in the following way. Three genes of the network's chromosome will be used to represent a relative worth for each of the sentence groups. The total number of sentences to be used from each group will be 84 X (relative worth of this group)/(sum of the three relative worth). Therefore, we obtain the formula

$$sen.\text{-}group(a) = 84x \frac{weight\text{-}sen.\text{-}group(a)}{\sum_{i=1}^{3} weight\text{-}sen.\text{-}group(i)} \quad (1)$$

The system can also increase the number of sentences from any of the three sentence groups by requesting so via three additional genes. Since having more training data should naturally lead to better performance, networks that use additional sentences have their fitness value decreased by a percentage which will be equal to the percentage of additional sentences requested. That is, if a network asks for 17% more sentences, the observed performance value will be decreased by 17%. In this way I will be able to measure how much of a positive effect additional data might have on the network's performance.

There is a gene in the network's chromosome that determines how many groups the training set is divided into. A value between 0 and .25 means all sentences are presented to the network as part of a single training set (training mode 1). A value between .26 and .5 means the network is first presented with sentences from groups 1 and 2. Then, once that training is done, the network is trained with sentences from group 3(training mode 2). A value

between .51 and .75 means the network is first presented with sentences from group 1. Then, once that training is done, the network is trained with sentences from groups 2 and 3 (mode 3). A value between .76 and 1 means the network is trained first with sentences from group 1, then with sentences from group 2, and finally with sentences from group 3 (mode 4).

5 Genetic Definition of NN Parameters

In addition to optimizing the training set, the GA is responsible for optimizing learning parameters such as the learning algorithm to be used, learning parameter (LP), momentum (M), and weight decay (WD). Each of these variables obtains its value from a particular gene in the network's genotype. These genes have values between 0 and 1, which get mapped into the options available for the corresponding variable. In the case of the learning algorithm, a value between 0 and .33 indicates Backpropagation Through Time. A value between .34 and .66 indicates Batch Backpropagation Through Time. A value between .67 and 1.0 indicates Quickpropagation through time. Both QPTT and BBPTT update weights after all patterns are presented, while BPTT updates them after each pattern is presented. QPTT assumes that the error curve is quadratic, trying to jump directly to the bottom of the parabola. For the other genes, its floating-point value is mapped as follows: for LP: 0 - 2; Momentum: 1.25 - 1.75; weight decay: .00005 - .0005; backstep: 1 - 10.

6 Results

For each of the topologies mentioned above, the GA was allowed to run for 300 generations. Each run was repeated 48 times; the values reported here are the average of those 48 runs. During these generations, the GA was optimizing all parameters except the topology to be used, which remained fixed. For each of the topologies, after letting the GA run for 300 generations, the best 48 elements of the population were able to process the sentences of the language quite effectively. The average error per pattern for these networks was of 5 incorrect values in the output layer. Since the output layer consists of 106 nodes, this gives an average error of 0.047 per node for the worst of these networks. Following is an analysis of the configurations chosen by the GA, with descriptions of further experiments I have performed in order to corroborate these results.

6.1 Size of Training Set

All networks forming part of the genetic populations after 300 generations were being trained with training sets bigger than 20% of the complete language. This indicates that, even with the "increase-in-sentences" penalty defined above, networks benefit from being trained with more sentences. The smallest amount of sentences used by any genotype was 140, or 33% of the complete language. Of all topologies, NMPL had the highest percentage of genotypes defining small training sets, followed by MPL topologies. Of a total of 26 genotypes across the 4 topologies that used less than 188 sentences (45% of the complete language), 9 of them where using NMPL topologies, followed by 8 MPL networks. 1HFN networks proved to be the worst at being trained with few sentences, having only 3 genotypes using less than 45% of the complete language.

6.2 Learning Algorithm

Of the three learning functions available, the one most commonly used was QPTT. Of the 192 networks that formed part of the populations after 300 generations, 129 (or 67%) were being trained with QPTT. The only topology that did not show a marked preference for QPTT was 1HMN. In fact, only 21 networks (or less than 44%) were being trained with it. This not only indicates 1HMN networks preference for learning algorithms other than QPTT, but it also points to a higher preference among other topologies for QPTT itself; recomputing percentages without including the data for 1HMN networks, 108 of 144 networks (or 75%) were using QPTT as their training function.

The preference for QPTT as learning function was higher when looking at networks trained with less than 45% of the complete language. For these networks, 18 out of 26 (or 69%) used QPTT. As was the case when considering training sets of all sizes, 1HMN networks did not show this preference; only 3 out of 6 networks trained with less than 45% of the language used QPTT. This brings the percentage of the other three topology types using QPTT when trained with the smaller training set to 75%.

As an additional experiment, in order to test this preference for QPTT as learning function, the genetic experiments were repeated, but this time allowing the genotype to select between either BPTT or BBPTT as its learning function. After iterating for 600 generations (twice as many generations as before) the average number of output nodes with an incorrect value was of 7.33, for an increase of 46% when compared with the results obtained when QPTT was an option for the genotype. This clearly points to the difficulty of finding minima in the error space when QPTT is not an available option. When this last experiment is repeated allowing the GA to choose one of the four topologies described above, 36 out of 48 networks (or 75%) are either 1HFN or 1HMN. This signals to a

dependency of MPL and NMPL networks for being trained with QPTT. For genotypes that use less than 45% of the complete language for training, the ratio is only 60%.

6.3 Mode of Training

All four training modes were being used by all topologies, and in almost identical proportions. There was some difference, though, when looking at topology type in conjunction with the type of sentence being used for training. NMPL, 1HMN and 1HFN networks that used few sentences in its training set had sentences presented in one group only (mode 1) in 55% of the cases, when a completely random distribution should have had this happening only 25% of the time. Genotypes that defined bigger training sets chose to use training modes that divided the training set in subgroups based on sentence complexity (training modes 2, 3, or 4).

The situation is different, though, when we look at networks trained with few sentences from complexity group 1. When the complete training set had less than 30 sentences from complexity group 1, the networks were successfully trained only when being presented with sentences from group 1 by themselves before training with more complicated sentences (modes 3 and 4).

7 On the Effect of Genetic Parameters

In order to check the effect of mutation on the search for effective NN configurations, experiments were repeated with different mutation rates. Researchers have found that there is an inverse relationship between population size and mutation rates. That is, the higher the population size, the lower the ideal mutation rate is, [5].

Each mutation rate value tested was used a in a total of 48 experiments. Runs performed with mutation rates of 0% and 10% converged to the same types of configurations, but runs using a value of 10% took close to 15% more generations to reach those configurations. The fact that runs using a mutation rate of 0% reached the same configurations as runs using a higher mutation value indicates that a population size of 48 elements contains enough genetic diversity to avoid early convergence.

Tanese, [6], has described another way in which population size can affect the performance of a genetic algorithm. If a population contains elements representing more than one minimum, crossover operations between these different elements can lead to offspring with worst fitness either parent. Although such an element would be discarded as part of the normal GA operations, such an occurrence would slow the evolution process, and would force all but one minimum to be discarded. As can be seen with the experiments reported in this paper, this has not been the case with this experiment using a population size of 48 elements, since it allows for the independent development of several minima simultaneously.

8 Conclusions

The procedure outlined in this paper has shown to be valuable in distinguishing which parameters are important in order to efficiently train neural networks for a simple natural language processing task. Quick Propagation Through Time seems to be optimal for topologies with feedback loops, but not for topologies without them.

It is obvious from looking at the resulting data that the NN gain an advantage in training when trained with more sentences. This is still true when the networks are penalized when trained with bigger training sets. At the same time, the GA was able to find good solutions to the problem when the NN were being trained with less than 45% of the complete language. In these cases, there was a preference for networks with feedback loops.

The GA was also able to find solutions regardless of how the constitution of the data, or how it was presented. There was a marked preference, though, for showing sentences from difficulty group 1 first when the training set had few of these sentences. When the total size of the training set was small, the networks achieved better performance when the sentences where presented as a single group. With bigger training sets, the genotypes chose to divide sentences into at least two different groups.

Because of the ability of GA to search a problem space for minima, it is well suited for problems that might contain more than one possible solution. Finding different minima becomes important for real-life applications, since variables such as number of hidden nodes, number of hidden layers, number of connections, size of training set, and backstep value all can affect training and response time for NN.

9 Future Research

The topologies used in this research were chosen because they had been successful when using a fixed 20% of the language in their training data. As mentioned previously, when given the opportunity to do so, all genotypes choose to define training sets bigger than that. I am currently investigating topologies evolved by the GA when given the opportunity to optimize topologies, training data, and learning parameters simultaneously. Initial results indicate that the GA converges to topologies with some commonalities with the ones reported previously, but they also contain important differences..

References

[1] Dávila, J. Co-evolution of neural network parameters for the task of natural language processing. Available at http://csp.gc.cuny.edu:8080/Mosaic_pages/nn_param_evol.ps . 1998.

[2] Elman, J. L. Distributed Representations, Simple Recurrent Networks, and Grammatical Structure. *Machine Learning,* volume 3, pp.195-224. 1991.

[3] Jain, A. Parsing Complex Sentences with Structured Connectionist Networks. *Neural Computation, 3,* pp.110-120. 1991.

[4] Lawrence, S., Giles, C., and Fong, S. Natural Language Grammatical Inference with Recurrent Neural Networks. Accepted for Publication, *IEEE Transactions on Knowledge and Data Engineering.* 1998.

[5] Schaffer, J., Caruana, R., Eshelman, L., and Das, R. A Study if Control Parameters Affecting Online Performance of Genetic Algorithms for Function Optimization. *Proceedings of the third International Conference on Genetic Algorithms.* San Mateo, California. Morgan Kaufmann. 1989.

[6] Tanese, R. Distributed Genetic Algorithms. *Proceedings of the third International Conference on Genetic Algorithms.* San Mateo, California. Morgan Kaufmann. 1989.

Development of Fuzzy Learning Vector Quantization Neural Network for Artificial Odor Discrimination System

B. Kusumoputro
Faculty of Computer Science, University of Indonesia
Depok Campus, PO BOX 3442, Jakarta 10002, INDONESIA

Abstract

The author had developed an artificial odor discrimination system for mimicking a function of human odor experts. The system used a back-propagation neural network and shows high recognition capability, however, the system work efficiently if it is used to discriminate a limited number of odors. The back-propagation learning algorithm will force the unlearned odor to the one of the already learned class-category. To improve the system's capability, a fuzzy learning vector quantization (FLVQ) neural network is developed, in which LVQ neural network will be used together with fuzzy theory. In the experiments on four different ethanol concentrations and three different kinds of fragrance odor from Martha Tilaar Cosmetics, it is found that the FLVQ shows high recognition capability, comparable with the back propagation neural network, however, the FLVQ can cluster the unlearned sample to different class of odor.

1 Introduction

Artificial odor discrimination system is now being developed as systems for the automated detection and classification of odor vapors and gases. This system is generally composed of a chemical sensing system and a pattern recognition system. The artificial odor discrimination system is proposed to mimic the human sensory test that is often performed to obtain the sensory quantities of odors. As the inspector's state of health inevitably affected the performance of the recognition results, some discrepancies among panelists of expert can not be avoided. It is therefore essential to develop an objective evaluation method for practical uses. We are developing the artificial odor discrimination system for the automated identification of different concentration of alcohol base odors, and various aromas in cosmetics industries [1]. The developed system shows high recognition probability to discriminate various odors, however, the back-propagation neural system is working effectively if there is a limited number of odors that should be recognized. This system also have disadvantages, such as the unknown-category of odor may accidentally be classified as the known-category of odor.

In order to improve the performance of the proposed system, we used Kohonen's learning vector quantization (LVQ) neural model as the pattern recognition system. In this improved odor recognition system, LVQ neural network will be used together with fuzzy theory [2][3] in Fuzzy LVQ neural network system to have the ability to distinguish the unknown odor from the known ones. The system is then used to recognize odors from ethanol gas with four different concentrations and three aromas from Martha Tilaars Cosmetics products. Using the same data sets, we have compared the recognition capabilities of FLVQ and the multilayer perceptron with back-propagation algorithm. The result shows that the FLVQ has a comparable ability in the experiments for determining the already known-category of odors. However, as the backpropagation algorithm can not recognized the unknown-category of odor, the FLVQ algorithm clustered the unknown-odor, to a different new class of odor.

2 Fuzzy Learning Vector Quantization

The learning vector quantization (LVQ) proposed by Kohonen is a powerful method for realizing a neural network, since the neuron in LVQ learning is non-linear, localized updated and the network does not take much time to converge [4][5]. The output neuron in LVQ has a reference vector, which is allocated to the one category of the recognition data. All of the neurons in the output layer calculate the Euclidean distance between its reference vector and an input vector, and the reference vector which is the nearest to the input vector is defined as the winner. The network is then locally updated, with only the nearest reference vector is modified. The learning rule is obtained by finding such that the expected value of the square discretization error between two vectors, or the objective function, is minimized [6]. However, the LVQ updating strategy is very localized that ignores the global relationships between the winner reference vector and the rest of the reference vector. To include the other non-winner vector in the updating procedure, Bezdek *et.al* used a fuzzy c-means algorithm in order to minimized the objective function of LVQ and called it as Fuzzy-LVQ [7], however, the neuron behavior is the same as the conventional LVQ.

Another Fuzzy-LVQ is developed by Sakuraba et.al [8]. In this fuzzy type of the network, neuron activation is expressed in terms of fuzzy numbers and by doing this; the spread of the measurement data taken from the odor recognition system can be defined as fuzziness. All of the data taken are then be normalized to form a triangular fuzzy numbers with the maximum membership value equal to 1. The triangular fuzzy number is expressed in term of its center of position c, the left and right fuzziness *l* and *r* and the height *v*. Figure1 show the fuzziness of citrus fragrance (Martha Tilaar Cosmetics) taken from frequency counter of sensor 1. If the frequency data taken from the experiments are varies significantly, than the *l* and *r* will be larger means that the membership function is wider.

Mean (f)	0.673
Std	0.003
Max (f_r)	0.678
Min (f_l)	0.670

Fig.1. Fuzziness of citrus fragrance taken from sensor 1

The architecture of fuzzy LVQ network consists of three-layered structure, with each neuron in the input layer corresponds to the number of sensors, and each neuron in the output layer corresponds to the number of classes of the odor in question. Hidden layer consists of several groups of neurons that performed calculation on finding the closest between the reference vector and the input vector. Every group of neurons in the hidden layer is connected to only one neuron in the output layer. An output neuron in FLVQ has a reference vector that is allocated to the category of the predetermined odor. Neuron calculates the similarity between reference vector and the input vector. The

reference vector, which is the nearest to the input vector, is defined as the winner, and the network output the category of the winner. In FLVQ, the relationship between reference vector and the input vector is expressed in fuzzy similarity, instead of Euclidean distance like in LVQ.

As the learning in FLVQ is supervised, the desired category of the input vector is given to the network for learning, and the network locally updates only the winner reference vector. The winner reference vector will approach the input vector, while the others will be forced to move away from that of input vector. Fuzziness of the reference vector is developed during learning and depends on the statistical distribution of the input data.

Suppose *x* is an n-dimensional fuzzy training vector that can be expressed by:

$$x_{kj}(t) = (x_{k1}(t), x_{k2}(t), .., x_{kj}(t), .., x_{kn}(t)) \quad (1)$$

with $k = 1, 2, 3,.. N$ denotes the total number of training vector for all odor classes, and $j = 1, 2, 3,.. n$ total number of sensors used in the system. The membership function of this training vector can be written as:

$$h_{xkj}(t) = (h_{xk1}(t), h_{xk2}(t), ..,h_{xkj}(t), ..,h_{xkn}(t)) \quad (2)$$

The fuzzy reference vector for known category *i*, m_i, for example, is

$$m_{ij}(t) = (m_{i1}(t), m_{i2}(t), .., m_{ij}(t), ..., m_{in}(t)) \quad (3)$$

with $i = 1, 2, 3, .., m$ denotes the class category of odors. The membership function of this reference vector:

$$h_{mij}(t) = (h_{mi1}(t), h_{mi2}(t), ...,h_{mij}(t), ...,h_{min}(t) \quad (4)$$

For the description of the relationship between input vectors and the reference vectors, we utilized the fuzzy similarity (μ) of the two fuzzy numbers through:

$$\mu_{ij}(t) = max_{ij}\{h_{xj}(\mu_j(t)) \wedge h_{mij}(\mu_j(t))\} \quad (5)$$

where the operation $\wedge$ means: $c \wedge d = min \{c \wedge d\}$. For all the components between the two vectors, the similarity between them is defined as the minimum among all the axial similarities by:

$$\mu_i(t) = min(\mu_{ij}(t)) \quad (6)$$

When fuzzy similarity value $\mu_i(t)$ is 1, the reference vector and the input vector are exactly resemble; and if $\mu_i(t)$ is 0, fuzzy reference and the input has no resemblance at all.

314

3   Learning Methodology

The learning method in FLVQ is accomplished by presenting a sequence of training vector with an associated reference vector for the same category. Fuzzy processing element is then calculates the similarities between them, and updating the reference vectors repeatedly according to the difference between the training vector and the desired output vector. By using the FLVQ algorithm, we encounter three cases, that are: when the network outputs a right category answer, a wrong category answer, and when the reference and the input vector have no intersection of their fuzziness.

The learning rule of the FLVQ has two steps, first is by shifting the center of position (f) of the reference vector, and secondly, by modifies the left and right fuzziness (f_l and f_r). According to this, for the first case, when the network outputs the right category, the closest reference vector is updated by moving the reference vector m_{ij} approaches x through:

$$m_{ij}(t+1) = m_{ij}(t) + \alpha(t)\{(1- \mu_{ij}(t)) *(x_j (t)-m_{ij}(t))\} \quad (7)$$

where α denotes the learning speed. The second step is done by modify the left and right fuzziness through constantly increase:

$$m(t+1) = (l - \beta(c-l); \quad c \quad ; r + \beta(r-c)) \quad (8)$$

where β is a constant variable with $1< \beta <2$.

For the second case, when networks outputs a wrong category answer, the closest reference vector is updated by moving the reference vector m_{ij} to go away from x through:

$$m_{ij}(t+1) = m_{ij}(t) - \alpha(t)\{(1- \mu_{ij}(t)) *(x_j (t)-m_{ij}(t))\} \quad (9)$$

and modifying the left and right fuzziness by constantly decrease:

$$m(t+1))=(l+(1-\gamma)(c-l); c ; r-(1-\gamma)(r-c)) \quad (10)$$

where γ is a constant variable with $0< \gamma <1$.

For the third case, the fuzziness of the reference vector is increased in order to have a possibility of being crossing the input vector. Learning in the third case is just increasing the entire reference vector through:

$$m_{ij}(t+1) = \delta * m_{ij}(t) \quad (11)$$

where δ is a constant variable with value of 1.1.

4   Measurement System

The odor recognition system consists of the sensory system, a frequency counter for measuring the characteristic frequency of the sensors and a computer performing the neural network. The sensory system used 4 quartz-resonator sensors; each one is covered by a sensitive thin membrane on the two surfaces. They were AT-cut with fundamental resonance frequencies of about 10 MHz. The sensors are mounted on a 1.2-L glass chamber, and the air on the chamber can be exchanged from dry one to aroma-contained air through computer-controlled valve. When odorant molecules are absorbed onto the membrane, the resonance frequency of the crystals will decrease significantly, and returned to the normal resonance frequency after deadsorbtion process. This phenomenon is called mass-loading effect [9][10]. The shift of the frequency (ΔF) is proportional to the total mass of adsorbed odorant molecules, which is given by:

$$\Delta F = -2,3 \times 10^{6} \times F^{2} \times \frac{\Delta M}{A} \quad (12)$$

where F denotes the characteristics frequency (MHz), ΔM the total mass of the absorbed molecules (g) and A electrode area (cm^2). As the responses of the sensors with different membrane to an input odor are different slightly, the output pattern from the sensor array is also specific and can be used to identify the odor.

5   Experimental Design

Experiments are designed, firstly, to know the ability of the system to distinguish between each of the odors and to separate the unknown category, which is not included in the training, from the known categories. Secondly, the recognition capabilities of the FLVQ neural system will be compared with the BP based multi layer perceptron [11]. The data used for learning and recognition are obtained from 10 experiments of each samples, with the first seven sets are used for learning, and the last three for recognition. The sensor data are first converted to the fuzzy quantities after normalization procedure. The first group of experiments are performed to discriminate odors by determining the correct percentage of four gradient concentrations, i.e. 0% (C1), 15% (C2), 25% (C3) and 35% (C4) of ethanol (*Ethaex*). The second groups of experiments are done to discriminate three fragrances of Martha Tilaar Cosmetics products, namely, citrus, rose and canangga (*Marthaex*).

For Ethaex and Marthaex, training vector set *(70%)* for each of aroma are utilized in the learning process with learning rate *α: 0.05*, increasing factor *β: 1.1* and decreasing factor *γ: 0.1*. Those parameters are ones optimized after several trials, for obtaining *100%* of recognition ability. For comparison, the same training set is also used in the back-propagation MLP network. The multi layer perceptron is composed of four neurons in the input layer, nine neurons in the hidden layer and several neurons in the output layer that correspond to the number of odors to be recognized. For the total of the experiments of the aromas, BP-MLP shows perfect recognition ability, and converges to a minimum error accuracy of *1.2%* and *1%*, for Ethaex and Marthaex, respectively. The two networks are then tested by a test set, which contains *30%* of the remaining data sets, and the results are shown in Table 1. It is found that the recognition capability of the FLVQ is as high as the BP-MLP. The minimum error accuracy of BP-MLP for the test sets of both Ethaex and Marthaex increase to a higher value of *30.2%* and *6%*, respectively. The higher value of minimum error accuracy for the Ethaex test set, suggest that the separability of the sensor system to discriminate various different concentration from the same type of odor much more difficult compare with different type of odors.

Table 1. Comparison of the recognition probability of BP-MLP and FuzzyLVQ

	BP- MLP		Fuzzy -LVQ	
	Training	Testing	Training	Testing
Ethaex	100%	83.3%	100%	83.3%
Mathaex	100%	100%	100%	!00%

5.1. Recognition of the Unknown Category

For determining the unknown category in the Ethaex, 100% of input vectors of C2 (which is excluded in the training stage) and 30% of the remaining input vectors from another each concentration (C1, C3, and C4) are utilized as the test set. The test set is designed to simulate the condition when the data of the known-category of odors are concealed within the data of the unknown-category of odor. The recognition results of the FLVQ are shown in Table 2. The percentage numbers in the matrix of the Table, indicate the percentage of the input data that are identified as the corresponding category. Unknown category in this Table shows the percentage of the input data that is not included in the predefined category. It is clearly seen, that FLVQ can determine correctly *(100%)* the C1, C3

and C4, while for C2, the unknown category, the recognition probability is *90%*, with other *10%* is recognized as C1. The average recognition is about *97.5%*. When the C1 (or C3 or C4) is excluded in the training stage, the average recognition probability are depicted in Table 3, makes the total average of recognition probability for Ethaex is *79.38%*.

Table 2. Recognition probability of the unknown category of C2 in Ethaex and Marthaex, respectively.

Output	Input	Data		
	C1	C2	C3	C4
C1(0%)	100%	10%	-	-
C2(15%)	-	-	-	-
C3(25%)	-	-	100 %	-
C4(35%)	-	-	-	100 %
Unknown	-	90 %	-	-

Note: The C2 *(15%)* category is not included in the training stage.

Output	Input	Data	
	Rose	Citrus	Canangga
Rose	100 %	-	-
Citrus	-	-	-
Canangga	-	-	100 %
Unknown	-	100 %	-

Note: The citrus category is not included in the training stage.

Table 3. Average recognition probability of the FLVQ

Unknown	C1	C2	C3	C4	Average
Ethaex	71.70%	97.50%	65.00%	83.33%	79.38%

Unknown	Rose	Citrus	Canangga	Average
MathaEx	83.30%	100%	75.60%	86.30%

Similar experiments are also conducted for the Martha Tilaar Cosmetic Products, and the results are also depicted in Table 2. When the citrus fragrance is adopted as that of the unknown sample, the recognition rate is high, i.e. 100%, and the recognition probability for the other two samples are also perfect, makes the average recognition probability is *100%*. When the rose or canangga is excluded in the training stage, the recognition probability is *83,30%* and *76,50%*, respectively. As can be seen in Table 3, the total average recognition probability of the FLVQ for Marthaex is *86.30%*.

6 Conclusion

We have utilized fuzzy learning vector quantization method to improve the capability of the electronically developed odor recognition system. The learning mechanism is done by moving toward or moving away the fuzziness-reference vector to the fuzziness-input vector, and by increasing or decreasing the fuzziness of the reference vector. It is confirmed in the experiments that the system can distinguish the unknown odors from the known ones. It is also found that the FLVQ had the same recognition capability as the BP-MLP, however, FLVQ can separate the unlearned sample to a different class. Further improvement is necessary to raise the identification capability; one of them is by increasing the number of the sensors, and by using other sensitive membrane that is most suitable to different type of odor, especially for different grade of concentrations of the same type of odor.

Acknowledgments

The author would like to express their gratitude to Prof. Dr. T. Moriizumi and Assoc. Prof. T. Nakamoto, from Tokyo Institute of Technology, Tokyo, Japan, for their support and useful advice. This work is supported by National Research Council of Indonesia through RUT VI under contract number 45/SP/RUT/1998.

References

[1] B. Kusumoputro and M. Rivai, 'Discrimination of fragrance odor by arrayed quartz resonator and a neural network', ICCIMA-98, H. Selvaraj and B. Verma (Eds), Singapore: Word Scientific, pp. 264-270, 1998.

[2] L. A. Zadeh, 'Similarity relations and fuzzy ordering', Information Sciences, 3, pp.177-200, 1971.

[3] L. T. Koczy, 'Fuzzy if ..then rule models and their transformation into one another', IEEE Trans. Syst.Man, Cybern., vol. 26, 5, pp.621-637, 1996.

[4] T. Kohonen, G.Barna and R. Chrisley, 'Statistical pattern recognition with neural networks: Benchmarking studies', IEEE Proc. of ICNN, pp.61-68, 1987.

[5] T. Kohonen, 'Improved versions of learning vector quantization', IEEE Proc. of IJCNN, I, pp. 545-550, 1990.

[6] N. R. Pal, J. C. Bezdek and E. Tsao, 'Generalized clustering networks and Kohonen's self-organizing sheme', IEEE Trans. Neural Networks, vol. 4, 4, pp. 549-558, 1993.

[7] J. C. Bezdek and N. R. Pal, 'Two soft relatives of learning vector quantization', Neural Networks, vol. 8, no. 5, pp. 729-743, 1995.

[8] Y. Sakuraba, T. Nakamoto and T. Moriizumi, 'New method of learning vector quantization', Systems and Computers in Japan, vol. 22, 13, pp.93-102, 1991.

[9] G. Sauerbrey, 'Vermendung von schwingquaren zur wagung dunner schichten und zur wagung', Z.Phys., 155, pp. 206-209, 1959.

[10] W.H.King, 'Piezoelectric sorption detector', Anal. Chem., pp. 206-222, 1964.

[11] D.E. Rumelhart, J.L. Mc.Cleland, and The PDP Research Group, Parallel Distributed Processing, MIT Press, 1996

Differential Inclusions Mapping Simplexes to Simplexes

David William Pearson and Gérard Dray

Nonlinear and Uncertain Systems Group,
LGI2P, EMA - Site EERIE, Parc Scientifique Georges Besse,
30035 Nîmes Cedex 1, France.
E-mail - pearson@site-eerie.ema.fr, dray@site-eerie.ema.fr

Abstract

When trying to model and analyse uncertainty in a dynamical system, differential inclusions are a natural tool. In this article we look at the situation where uncertain vector fields and uncertain initial points are both modelled by simplexes. The simplex representation lends itself to the analysis of the uncertain dynamical system, in particular the numerical simulation of the system.

1 Introduction

When analysing real applications concerning dynamical systems, one is frequently confronted by uncertainty. One way to overcome the problems associated with uncertainty is to use differential inclusions rather than differential equations and eventually to extend these to fuzzy differential inclusions [1, 2].

In this paper, we restrict ourselves to the problem where a differential inclusion is described by a simplex. The simplified structure of a simplex facilitates the modelling process and the numerical analysis. We also assume that the initial state of the system is uncertain and can be modelled by a simplex. Our objective is to find conditions on the type of inclusions such that the simplex describing the initial state is mapped to another simplex of the same dimension for a particular choice of vector field belonging to the inclusion.

2 Differential Inclusions and Simplexes

We are concerned with differential inclusions of the form

$$\dot{x} \in V(x) \tag{1}$$

where $x \in X$ is the system state, X is an n-dimensional manifold and $V : X \rightarrow X$ is a multivalued mapping.

A simplex is an example of a convex set [5], given that a set of scalar quantities $\tau^k, k = 0, \ldots, p$ satisfy the following constraints

1. $\tau^k \geq 0, k = 0, \ldots, p$

2. $\sum_{k=0}^{p} \tau^k = 1$

then the set corresponds to a p-dimensional simplex. If $\sum_{k=0}^{p} \tau^k = 1$ then obviously $\tau^0 = 1 - \sum_{k=1}^{p} \tau^k$ and so we can replace the above by a p-dimensional vector

$$\tau = \begin{bmatrix} \tau^1 \\ \vdots \\ \tau^p \end{bmatrix}$$

where $\tau^k \geq 0$ and $\sum_{k=1}^{p} \tau^k \leq 1$, we can then describe a mapping from the p-dimensional simplex into a p-dimensional simplex in $\mathbb{R}^n$ by

$$x = \tau^0 v_0 + \cdots + \tau^p v_p$$

where the v_k are $p + 1$ vectors such that the set of vectors $\{v_1 - v_0, \ldots, v_p - v_0\}$ are linearly independent. From the previous paragraph we see that this is equivalent to

$$x = v_0 + \tau^k(v_k - v_0) \tag{2}$$

where here and throughout the rest of the paper we make use of the summation convention on repeated indices such that $\tau^k(v_k - v_0) = \sum_{k=1}^{p} \tau^k(v_k - v_0)$. The set of points in Euclidian space mapped to from the p-dimensional simplex represented by the vector τ is a p-dimensional simplex, or simply p-simplex which we denote by s^p. Due to the fact that (2) is an injection into X we also abuse notation a little by referring to the abstract simplex denoted by the vector τ as s^p as well.

Now, if the v_k in (2) are vector fields, ie mappings $v_k : X \to X$, then, for some fixed but arbitrary value of τ, we have a simplicial representation of the differential inclusion (1) via the following expression

$$\dot{x} = v_0(x) + \tau^k(v_k(x) - v_0(x)) \qquad (3)$$

and thus for varying values of τ satisfying the above listed constraints

$$\dot{x} \in s^p$$

We make use of the exponential notation for a solution of (3) passing through the point x_0 [6]

$$x(t) = \exp(t(v_0 + \tau^k(v_k - v_0)))x_0 \qquad (4)$$

It can be proved that for a fixed but arbitrary value of t which is sufficiently small (4) can be decomposed into a sequence of concatenated exponentials [6]

$$x(t) = \exp(t\tau^p(v_p - v_0)) \cdots \\ \exp(t\tau^p(v_1 - v_0)) \exp(tv_0)x_0 \qquad (5)$$

In [4] we showed that if the following criterion is imposed on the vector fields v_k

- in a neighbourhood of the point x_0 the rank of the system of vector fields $\{v_1 - v_0, \ldots, v_p - v_0\}$ plus the derived system $\{[v_i, v_j], [v_i, [v_j, v_k]], etc\}$ is equal to p

then (5) is an injective immersion from the p-simplex represented by the vector τ into a p-simplex in $\mathbb{R}^n$ and via a chart into X, for simplicity we supress reference to X and work locally in $\mathbb{R}^n$ throuhgout the paper. For references to Lie algebras, derived systems and Jacobi brackets see [6].

We now extend this idea to the point where the initial state of the system is modelled by a q-dimensional simplex, where $p + q = n$. In this preliminary analysis the $p + q = n$ condition is necessary, in the future we intend to investigate other possibilities. Let

$$x_0 = u_0 + \sigma^i(u_i - u_0) \ , \ i = 1, \ldots, q$$

where the σ^i satisfy the same conditions as the τ^k above, but for a q-simplex, thus $x_0 \in s^q$. In this way we can define a mapping $\phi : s^p \times s^q \to X$ such that

$$\phi(\tau, \sigma) = \exp(t\tau^p(v_p - v_0)) \cdots \\ \exp(t\tau^1(v_1 - v_0)) \exp(tv_0)(u_0 + \sigma^i(u_i - u_0))$$

For a fixed but arbitrary value of $\tau \in s^p$ we define

$$\phi_\tau(\sigma) := \phi(\tau, \sigma) \qquad (6)$$

we require $\phi_\tau : s^q \to s^q$, $\forall \tau \in s^p$, where this has to be interpreted as ϕ_τ maps a space homeomorphic to a q-simplex into a space homeomorphic to a q-simplex.

3 Possible Solution

In order to obtain a first solution to our problem we assume that the vector fields v_k are analytic. If the vector fields also satisfy the criterion listed above then the dimension of the tangent space to X at a point x, noted X_x is equal to p. Thus, this is one of the conditions required for the solution proposed.

Secondly, we require the tangent space to the initial simplex $\sigma \in s^q$ to lie in a subspace complementary to the initial simplex. In other words we require that the vectors

$$\{(v_1 - v_0)(u_0 + \sigma^i(u_i - u_0)), \ldots, \\ (v_p - v_0)(u_0 + \sigma^i(u_i - u_0)), \qquad (7) \\ u_1 - u_0, \ldots, u_q - u_0\}$$

form a linearly independent set of dimension n, where $(v_k - v_0)(u_0 + \sigma^i(u_i - u_0)) := v_k(u_0 + \sigma^i(u_i - u_0)) - v_0(u_0 + \sigma^i(u_i - u_0))$ etc, for $\sigma \in s^q$.

Now, move the initial simplex σ in a direction specified by a choice of τ via (6). In order to compare the tangent space at the new point with the image of the simplex we need to pull the points back to the initial points via the inverse differential [3]. Define

$$ad_{v_i}^0 v_j = v_j \\ ad_{v_i}^1 v_j = [v_i, v_j] \\ ad_{v_i}^k v_j = [v_i, ad_{v_i}^{k-1} v_j]$$

then we have the Campbell-Baker-Hausdorff formula [3]

$$\exp(-tv_0)_\star \cdots \exp(-t\tau^p(v_p - v_0))_\star \\ (v_0 + \tau^k(v_k - v_0))(\exp(t\tau^p(v_p - v_0)) \cdots \exp(tv_0))x_0 \\ = \frac{t^{k_0}}{k_0!} ad_{v_0}^{k_0} \left(\frac{(t\tau^1)^{k_1}}{k_1!} ad_{v_1 - v_0}^{k_1} (\cdots \\ \left(\frac{(t\tau^p)^{k_p}}{k_p!} ad_{v_p - v_0}^{k_p} \left(v_0 + \tau^k(v_k - v_0) \right) \right) \right) \cdots \right) (x_0) \qquad (8)$$

where $x_0 \in s^q$, $\exp(-tv_0)_\star$ denotes the differential of the mapping and remembering the summation convention. It goes without saying that (8) is only valid locally for sufficiently small values of t and for a sufficiently small initial simplex.

We see from (8) that if (7) is to be satisfied at the image points of the initial simplex then we require

$$\mathrm{Im}(\exp(-tv_0)_\star \cdots \exp(-t\tau^p(v_p - v_0))_\star$$
$$(v_0 + \tau^k(v_k - v_0))(\exp(t\tau^p(v_p - v_0)) \cdots \exp(tv_0))x_0)$$
$$\subset \mathrm{Im}\{(v_1 - v_0)(u_0 + \sigma^i(u_i - u_0)), \ldots,$$
$$(v_p - v_0)(u_0 + \sigma^i(u_i - u_0)), u_1 - u_0, \ldots, u_q - u_0\}$$

where Im denotes image. One way for this condition to be satisfied is that

$$[v_i, v_j] = c_{ij}^k (v_k - v_0) \,, \text{ for } 0 \le i, j \le p \qquad (9)$$

where the c_{ij}^k are scalars.

Thus conditions (7) and (9) are the ones we are looking for in order to guarantee that the mapping (6) maps simplexes onto simplexes, at least locally. Condition (9) means that the vector fields form a Lie subalgebra, which is a strong condition and future research will hopefully lead us to weaken this condition.

4 Example

Consider the following system

$$v_0(x) = \begin{bmatrix} x^1 \\ 0 \\ 0 \end{bmatrix} \,, \; v_1(x) = \begin{bmatrix} 0 \\ x^2 \\ (x^2)^2 \end{bmatrix} \text{ and}$$

it is easily verified that

$$[v_0, v_1] = 0$$

and so the Lie subalgebra condition (9) is trivially satisfied. We choose

$$u_0 = \begin{bmatrix} 1 \\ 1 \\ 1 \end{bmatrix} \,, \; u_1 = \begin{bmatrix} 2 \\ 1 \\ 1 \end{bmatrix} \text{ and } u_2 = \begin{bmatrix} 1 \\ 2 \\ 1 \end{bmatrix}$$

and it can be verified that the rank condition (7) is also satisfied.

For simulation purposes we set $t = 0.5$ and $\tau = 0.1$ and then chose at random 300 points belonging to a 2-simplex, the equation (3) was then solved numerically for each of the 300 initial points. The results can be seen in figure 4, where the lower simplex represents the initial points and the upper simplex their images. We can see that, in this example at least as far as the experimental results show, the initial simplex is indeed mappped to a simplex.

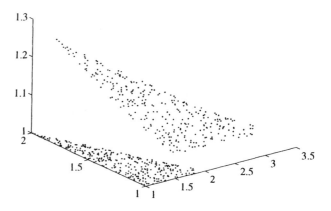

Fig. 1. Lower simplex is mapped to upper simplex

5 Conclusion

In this paper we have presented a practical method of simulating image sets of differential inclusions. The conditions for a solution presented are very strong and we are continuing our research in order to try and weaken these conditions and, if possible, look for different conditions and solutions to the problem.

References

[1] J P Aubin, Fuzzy Differential Inclusions, *Problems of Control and Information Theory*, Vol 19, No 1, pp 55-67, (1990).

[2] V A Baidosov, Fuzzy Differential Inclusions, *Journal of Applied Mathematics and Mechanics*, Vol 54, No 1, pp 8-13, (1990).

[3] A J Krener, Local Approximation of Control Systems, *Journal of Differential Equations*, Vol 19, pp 125-133, (1975).

[4] D W Pearson, Simplex Type Differential Inclusions, Submitted to *Applied Mathematics Letters*.

[5] E Spanier, *Algebraic Topology*, Springer-Verlag, (1982).

[6] H J Sussmann and V Jurdjevic, Controllability of Nonlinear Systems, *Journal of Differential Equations*, Vol 12, pp 95-116, (1972).

Crossbar Adaptive Array: The first connectionist network that solved the delayed reinforcement learning problem

S. Bozinovski

Electrical Engineering Faculty, University of Skopje, Karpos II, 91000 Skopje, Macedonia
Email: bozinovs@rea.etf.ukim.edu.mk

Abstract

The paper discusses important issues for reinforcement learning agents, the issue of delayed reinforcement learning (DRL). It points out that an early agent, the Crossbar Adaptive Array (CAA) architecture, not widely known in connectionist and reinforcement learning community, was the first to solve the DRL problem among connectionist agents. The work contributes toward understanding the initial neuron-like computational efforts to solve the DRL problem, giving a comparison between CAA and the well-known Actor/Critic (AC) architecture. It also points out relevant contemporary issues of autonomous agents, the issue of genetic/behavioral environment and the issue of emotion based learning architectures.

1   Introduction

It can be said that the work of Watkins [11] entitled "Learning from Delayed Rewards" had an effect to Reinforcement Learning (RL) community analogous to the effect of the work of the Rumelhart, McClelland and the PDP Group [10] to the connectionist community. Both works induced a great number of researchers to turn their attention toward RL and artificial neural networks (ANN) problems and issues. Although known as a learning paradigm (e.g. [1], [3]), before the Watkins' work, within the ANN community the delayed reinforcement learning (DRL) paradigm was mainly pursued by the Adaptive Networks (ANW) Group (e.g. [8], [6],[7]).

Watkins' work induced a reevaluation of the previous ANW work using new perspective (Dynamic Programming), new concept (Q-values) and new notation (more intuitive $W(a,s)$ rather than traditional, W_{ij} notation). Along with that reevaluation, it becomes evident that an almost forgotten connectionist architecture, the Crossbar Adaptive Array (CAA) architecture, has to be placed in the space of research efforts for solving the DRL problem. Recently [9], there has been pointers toward the CAA learning as a forerunner of the Watkins' Q-learning. However, it is our belief that the appearance of the CAA architecture has far reaching consequences for the reinforcement learning effort that should be mentioned explicitly. One of them is that it was the first ANN agent that solved the DRL problem, before the well-known Actor/Critic (AC) architecture. This paper gives a comparison between the CAA and AC architectures, two different philosophies for solving the DRL problem. Such a comparison has not been done in literature, and we believe it gives very useful knowledge for a reader interested in neural based reinforcement learning agents.

2   CAA and AC architecture: A Comparison

The DRL problem has many instances, and actually there is a class of DRL problems. In all the problems, an agent performs an action and there is no immediate reinforcement for that action. A reinforcement will appear later, and it is not given to which previous action it should be assigned. Learning path in a maze and learning to control to balance a pole are such problems from the DRL class. Both of them were considered within ANW group in 1981, resulting in two different agent architectures to solve them: the CAA and the AC architecture. The architectures were developed independently from each other and reported in [6,7] and [9] respectively. Here we will give short description of both the CAA and the AC architecture.

2.1   Crossbar Adaptive Array architecture

Figure 1 shows the CAA architecture. It is assumed that the architecture drives an agent which is connected to two environments: 1) the *behavioral environment* in which the agent behaves and learns, and 2) *genetic environment* from which it receives genome vector containing information about initial memory values and other initial parameters; it is also able to export a genome toward the genetic environment [12]. It is assumed that the environment can represent itself as a graph having *m* distinct states, environment situations. The situations are received by a crossbar connectionist memory, which is used to compute two types of actions: 1) the physical actions toward the environment (this part of the CAA resembles some type of motor system), and 2) emotions about the received situation; the emotions are understood as internal state evaluation, internal feeling produced by a received situation (this part resembles some kind of hormonal system).

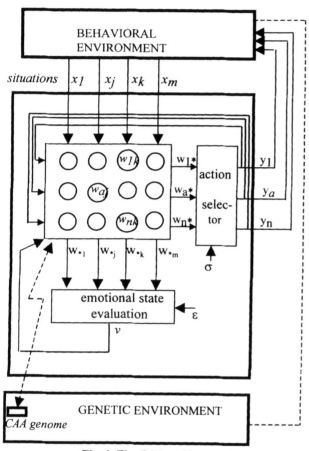

Fig. 1. The CAA architecture

The CAA learning algorithm is based on the crossbar computation over the crossbar memory elements w_{aj}, denoted as *SAE* elements *(situation-action-emotion)*. In general terms, CAA has the following behavioral and learning routine: Given current situation k, the CAA system will compute the current emotion $v_k=Vfunc\{w*_k\}$ and will update the crossbar element $w_{aj}=Wfunc(v_k)$, which produced action a in the situation j which in turn lead to the situation k; it will also compute the action b to be taken from the state k, using some action taking function $b=Afunc\{w*_k\}$. Here $w*_k$ means the k-th column vector of the matrix W. In other

words, the CAA learning routine in each time step has four activity steps :

1) state j: perform action a biasing on $w*_j$
   environment gives the state k
2) state k: compute emotion v_k using $w*_k$.
   then compute overall emotion v
3) state j: increment w_{aj} using v
4) change state: $j=k$; goto 1

Being in situation x_j, CAA computes its action as

$$y_a = sgnmax\{\Sigma w_{aj} + \sigma_a\} \qquad (1)$$
$$a=1,..,n$$

where σ_a is a random number from a uniform distribution between -0.5 and ,+0.5. It represents the nature of the CAA searching strategy. The function *sgnmax{.}* gives 1 for nonnegative and 0 for negative argument.

After receiving the next situation, x_k, the emotion in that situation is computed as

$$v_k = sign\{\Sigma w_{ak} + \varepsilon_k\} \qquad (2)$$
$$k=1,..,m$$

where *sign{.}* gives 1 for positive, 0 for zero, and -1 for negative arguments. The emotional sensitivity ε_k can be a random number or a number received by the initial genome, and it is used in environments where a system should learn to pass an unpleasant situation in order to reach a goal situation. The overall emotion is computed as

$$v = sum\{v_k\} \qquad (3)$$
$$k=1,..,m$$

which actually reduces to

$$v=v_k \qquad (4)$$

meaning that the overall emotion in situation x_k depends only of the emotion obtained from that situation k with no influence of previous situations. It is possible to define other computation functions for v_k and v [12].

The learning function is

$$w_{aj} = w_{aj} + v_k \qquad (5)$$

which, written as difference equation is

$$w_{aj}(t) = w_{aj}(t-1) + x_j(t-1)y_a(t-1)v(t) \qquad (6)$$

Equations (5) and (6) are two forms of the *CAA learning rule*.

322

2.2 The Actor/Critic architecture

The Actor/Critic architecture is given in Figure 2.

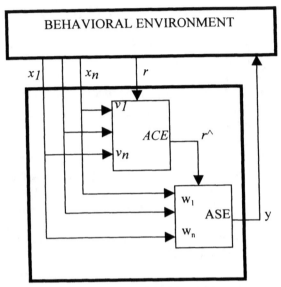

Fig. 2. The Actor/Critic architecture

The Actor/Critic architecture [9] has two main functional units, named Associative Search Element (ASE) and Adaptive Critic Element (ACE). Functioning of the AC architecture can be described by following equations

Action, computed by ASE, is given by

$$y = f\left(\Sigma w_i x_i + noise_i\right) \qquad (7)$$

where $f(.)$ is some neural activation function, in this case a threshold function giving 1 for nonnegative and -1 for negative arguments. Here $noise_i$ is a *gaussian* noise for ensuring random movement. The learning rule of ASE is given by

$$w_i(t+1) = w_i(t) + \alpha r^\wedge(t) e_i(t) \qquad (8)$$

where α is positive constant defining learning step size, $r^\wedge$ is the internal reinforcement, and e_i is the eligibility trace of the i-th input pathway, computed as

$$e_i(t+1) = \delta e_i(t) + (1-\delta) y(t) x_i(t) \qquad (9)$$

where δ, $0 \leq \delta < 1$, is the forgetting parameter of this learning equation.

The internal reinforcement $r^\wedge$ is computed by ACE as

$$r^\wedge(t) = r(t) + \gamma p(t) - p(t-1) \qquad (10)$$

where r is the external reinforcement given by the environment, p is a prediction of the external reinforcement, which has discount factor γ (here $\gamma=.95$). The reinforcement prediction is computed as $p = \Sigma v_i x_i$, and the ACE memory elements v_i are computed as

$$v_i(t+1) = v_i(t) + \beta\left(r(t) + \gamma p(t) - p(t-1) \right) \tau_i(t) \qquad (11)$$

where τ_i is a trace of the input x_i, with dynamics described as

$$\tau_i(t+1) = \lambda \tau_i(t) + (1-\lambda) x_i(t) \qquad (12)$$

where λ, $0 \leq \lambda < 1$, is forgetting factor of this learning equation.

2.3 A comparison

Figure 3 gives a comparison between the architectures. First to note in Figure 3 is that AC architecture has two identical memory structures, one for computing internal reinforcement, and one for computing actions. On the other side, CAA has only one memory structure of the same size as one of the AC memory structures. In a crossbar fashion, CAA computes both actions and performs state evaluation. CAA receives from the environment only the situation vector, whereas AC, in addition, receives an external reinforcement input, in each step of its performance routine. Further, comparing the functional description one can see that CAA uses only one incremental relation, its first order learning rule (6). On the other hand, AC uses four incremental relations, two for learning rules, (8) and (11), and two for traces, (9) and (12) . Also AC has a second order learning rule, (11).

Besides the apparent technical differences between AC and CAA, we believe that the most important difference between CAA and AC is the philosophy of computing the learning rule of the memory structure for evaluation of the current state. CAA uses the concept of emotional *state evaluation v*, desirability of being in some situation, whereas AC uses concept of *internal reinforcement $r^\wedge$* and *reinforcement prediction p* of the *external reinforcement r*.

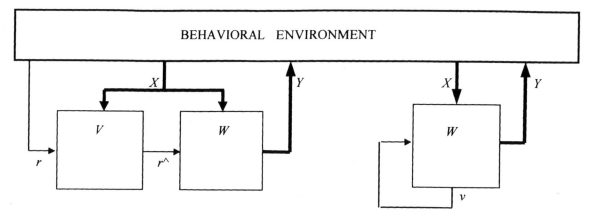

Fig. 3.a. Actor/Critic architecture   Fig. 3.b. Crossbar Adaptive Array architecture

Fig. 3. A comparison between AC and CAA architecture

CAA does not use the concept external reinforcement in each time step. In each time step it receives only the encountered situation. CAA uses the concept of genetic vector, an artificial genome which comes from the genetic environment only once, at the beginning of the learning experiment. That vector defines initial value $W(0)$ of W, which in turn defines emotional preference toward the states of the environment. $W(0)$ wstablishes connection between behavioral and genetic environment. In $W(0)$, an internal state is defined which will produce an emotional value if addressed by an environment situation.

That emotional value is backpropagated using *secondary reinforcement principle*: the computed emotional value $v(X)$ of the current state X is backpropagated to the state from which X is reached, and that

state is becoming a new reinforcing state. The importance of the concept of *state value* in reinforcement learning becomes evident after the work [11] which pointed a relation to Dynamic Programming [2].

3 Solving some DRL problems

Here we will consider the maze learning problem and the pole balancing learning problem.

3.1 The maze learning problem

The maze learning problem was the first one considered in the design of AC and CAA architectures. It is interesting that different types of mazes challenged the design of the AC and CAA agents. Figure 4 shows the challenging problems for AC and CAA.

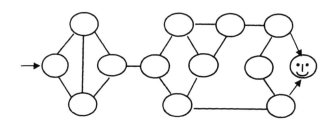

Fig. 4.a. A typical maze challenging the AC architecture

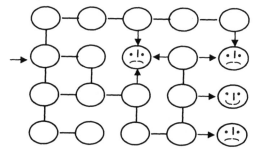

Fig. 4..b. A typical maze challenging the CAA architecture

Fig. 4. Two types of mazes considered during the design of the AC and CAA architectures

Figure 4.a shows an AC challenge and Figure 4.b a CAA challenge. A typical maze that the AC architecture should learn had one goal state and all other neutral states. In contrast, the CAA architecture from the very beginning had the challenge of states with different state values in terms of desirability. The CAA agent had to learn to go toward a goal state while avoiding the unpleasant states.

The solution of the maze learning problem using CAA architecture was achieved during 1981 and reported in [6,7], while solution using the AC architecture was achieved later. The two examples shown in Figure 4 were solved by CAA in 1981.

3.2   The pole balancing learning problem

The pole balancing learning is another illustrative problem from the DRL class. An agent inserting force F on the cart learns to balance the pole mounted on a cart, while a (negative) reinforcement is only given when the pole is fallen down. Here, applying an action in a situation does not completely specify the next state: the problem includes random environment responses, it is a *non-deterministic environment*.

For this problem AC approach was to use the Michie-Chambers representation [4] with 162 environment states. The CAA approach used simpler representation, with only 10 states, taking two important

assumptions: 1) there is no need to consider optimization of the cart displacement, (within $-x_{lim}$ *and* $+x_{lim}$), *in order to demonstrate learning*; it suffices to consider only the pole displacement optimization (between $-\theta_{lim}$ and $+\theta_{lim}$), and 2) it is more convenient to assume a ternary value for the action force, $F \in \{-F,0,+F\}$ rather then Michie-Chambers binary value $F \in \{-F,+F\}$. With those assumptions the pole balancing problem can be represented as a partially defined emotional Petri graph [12] where only the actions of the agent are represented by the Petri bars. It is assumed that only one output from a Petri bar is allowed, but it is not determined which one. With this design, a simple delayed reinforcement learning controller $F = g(\theta, \omega)$, where θ is angle and ω is angular velocity, was designed which learned a simple control heuristics:

if $sign\theta sign\omega < 0$ then $F = sign\theta$, otherwise $F = 0$

We believe that the choice of ternary action space $\{-F,0,+F\}$ is important improvement over the classical binary $\{-F,+F\}$ action space.

CAA solved this problem in 1981 and a report was given in [5,8]. The AC solution was achieved after 1981 and solution was described in [9].

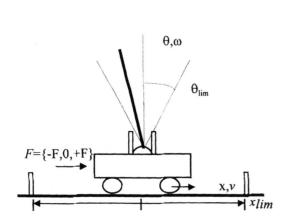
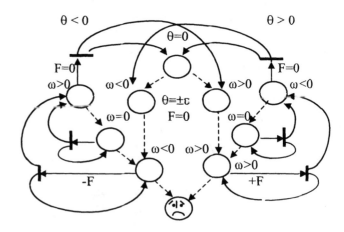

Fig. 5. The pole-balancing problem represented as an emotional Petri net

4 Discussion

The CAA architecture is a compact designed, recurrent, crossbar computing, connectionist structure. It computes both actions and emotions (internal state evaluations). It is rather old connectionist architecture, relative to the new connectionist movement started in 1986 with the book [10]. As a connectionist architecture, to the best of our knowledge, CAA was first one to solve the DRL problem. It certainly solved it before well known AC architecture did.

Viewing it from a time distance, it is an architecture which has interesting and far reaching features. Two important features of the early CAA makes it relevant for contemporary consideration: 1) the species (genetic) vectors and genetic environment, in addition to the behavioral environment, which makes interesting relation to genetic algorithms and artificial life research, and 2) the concept of *artificial emotion as internal state evaluation*, used in the system learning rule, which is relevant to contemporary issues of emotion based architectures and affective computing.

5 Conclusion

The PDP Group published the work done within the group in early eighties [10]. The ANW Group never did that. That has advantage, since after the work of Watkins [11], the early ANW group work can be evaluated more accurately. It has also disadvantage, since some potentially significant piece of work was left forgotten. This paper, for the benefit of researchers in ANN and RL, points attention to a significant 1981 event, relevant for the common effort of solving the DRL class of problems using a connectionist network It also points to some issues (e.g. genetic environment and artificial emotion) which are of contemporary interest to the RL and connectionist community.

References

[1] Keller, F., Schoenfeld, W.: Principles of Psychology. Appleton-Century-Croffts, 1950.

[2] Bellman R.: Dynamic Programming. Princeton University Press 1957

[3] Minsky, M.: Steps toward artificial intelligence, Proceedings of the IRE, pp. 8-30, 1961

[4] Michie, D., Chambers, R.: BOXES: An experiment in adaptive control. In E. Dale, and D. Michie, eds. Machine Intelligence 2: pp. 137-152. Oliver and Boyd, 1968

[5] Bozinovski S.: Inverted pendulum learning control. ANW Memo, December 10, COINS Department, University of Massachusetts, Amherst, 1981

[6] Bozinovski S. A self-learning system using secondary reinforcement. Published Abstracts of the Sixth European Meeting on Cybernetics and Systems, Vienna, April 1982a

[7] Bozinovski, S. A self-learning system using secondary reinforcement. In R. Trappl, ed. Cybernetics and Systems Research, pp. 397-402, North Holland. 1982b

[8] Bozinovski, S.; Anderson C.: Associative memory as a controller of an unstable system: Simulation of a learning control. In Proceedings of the IEEE Mediterranean Electrotechnical Conference, C5.11, Athens, Greece, 1983

[9] Barto, A.; Sutton, R.; Anderson, C.: Neuronlike elements that can solve difficult learning control problems. IEEE Trans. Systems, Man, and Cybernetics 13: pp. 834-846, 1983.

[10] Rumelhart D., McClelland J., and the PDP group: Parallel Distributed Processing, MIT Press, 1986

[11] Watkins, C.: Learning from Delayed Rewards. Ph. D. Thesis, Kings College, Cambridge, England, 1989

[12] Bozinovski, S.: Consequence Driven Systems, Gocmar Press 1995

[13] Barto A.: Reinforcement learning, In O. Omidvar and D. Elliot (Eds.) Neural Systems for Control, pp. 7-29, Academic Press 1997

An Evolutionary Algorithm for a Non-standard Scheduling Problem

Bogdan Filipič

Department of Intelligent Systems, Jožef Stefan Institute, Jamova 39, SI-1000 Ljubljana, Slovenia, and
Faculty of Mechanical Engineering, University of Ljubljana, Aškerčeva 6, SI-1000 Ljubljana, Slovenia
E-mail: bogdan.filipic@ijs.si

Abstract

The paper describes an evolutionary computation approach to solving a real-world scheduling problem for a group of production lines. The problem is non-standard in that it requires scheduling of process interruptions rather than processes themselves, and the schedule cost is determined by energy consumption on the machines. A scheduling system was developed to cope with instances of the problem on a daily basis. It consists of a heuristic procedure to generate initial schedules and an evolutionary algorithm to iteratively improve the schedules. Experimental evaluation of the system on numerous problem instances shows that, through proper scheduling, energy costs on the production lines can be substantially reduced.

1 Introduction

The task of scheduling is to allocate activities to resources over time in such a way that given objectives are optimized, while temporal constraints and resource limitations are satisfied. Problems of this type appear in manufacturing, timetabling, vehicle routing, design of computer operating systems and other domains. For its great practical importance, scheduling has permanently attracted research interests. Following the attempts in the fields of Operations Research and Artificial Intelligence with limited success in practice, Evolutionary Computation has recently offered means of generating near-optimal schedules for complex problems at reasonable computational costs [1]. A number of successful applications of evolutionary algorithms have been reported in scheduling [2, 4]. However, there are still challenging issues to be considered in the development of evolutionary scheduling systems. Above all, real-world problems should be dealt with and realistic criteria for schedule optimization taken into account [3].

Our work in evolutionary scheduling is oriented towards complex problem situations appearing in industrial environments. Previous application oriented studies include scheduling of operations in a production unit of a textile factory, where the objective was to ensure optimal energy consumption [5], and scheduling activities in ship repair in order to provide equable work load for workers of various trades [6]. In this paper we describe production scheduling on a group of production lines of an automobile factory. The addressed problem is non-typical in two respects. First, it requires process interruptions to be scheduled rather than processes themselves. Second, the optimality criterion is not based on a traditional schedule performance measure, such as overall processing time, but related to energy consumption. The objective is to schedule interruptions of the running processes in such a manner that energy consumption over the peak demand periods is minimized. In addition, the schedules are subject to time and resource constraints that have to be strictly satisfied.

Design of a scheduling system for this problem and its algoritmic details were presented in [7], while here the focus is on systematic evaluation of the approach. The paper reviews the problem and the scheduling system, and gives the results obtained on a set of test problem instances.

2 Problem Description

The considered production unit includes six lines of hydraulic presses for production of car body parts from sheet. The lines consist of 37 machines and are partly robotized. A line in operation is regarded as a single work process. Energy consumption is treated at the level of processes rather than individual machines. Power demands of the processes vary from 20kW to 370kW. The focus of the energy consumption management at the plant is on power peak demand leveling.

The pressing unit operates according to a daily production plan that specifies which of the lines are in operation and what is their work time. Operation of the lines results in power demand that is a sum of power demands of the running processes. There are also other energy consumers at the plant whose power demand represents the so called background power demand. The total power demand of the plant consists of the demand of the pressing unit and the background demand. It is constrained by a time-varying maximum load limit, also called the target load, prescribed for the plant by the energy supplier.

Exceeding the target load is paid at a higher rate, hence the company strives to reduce it as much as possible. Since the background energy consumption cannot be affected significantly, the contribution of the pressing unit to the target load excess should be minimized. A way to achieve this goal is to interrupt some of the processes running during the peak demand periods. Process interruptions are intended as regular breaks for the workers, or can be spent to change machine tools and perform maintenance on the lines. An example daily production plan and typical power demand profiles are shown in Figure 1.

To accomplish the production plan, the interruptions cannot be scheduled arbitrarily. Time and resource constraints have been introduced to balance between the conflicting requirements for high productivity and reduction of the target load excess. The constraints prescribe:

– maximum duration of process interruptions,

– minimum period of time between two interruptions of a process,

– maximum number of processes that can be interrupted simultaneously.

Taking into account these constraints, process interruptions have to be scheduled so as to minimize the target load excess contributed by the production lines.

3 Schedule Cost Function

To define the schedule cost function, let us introduce the following notation. Let $P_{\max}(t)$ be the target load for the plant and $P_b(t)$ the background power demand. Let $P_i(t)$, $i = 1, \ldots, 6$, denote the power demand of an individual process in the pressing unit. Then the overall power demand of the unit is

$$P_{\mathrm{proc}}(t) = \sum_{i=1}^{6} P_i(t) \tag{1}$$

and the total power demand of the plant

$$P(t) = P_{\mathrm{proc}}(t) + P_b(t). \tag{2}$$

We can now express the contribution of the production processes in the pressing unit to the the target load excess at time t, as

$$P_{\mathrm{exc}}(t) = \begin{cases} P_{\mathrm{proc}}(t); & P_b(t) \geq P_{\max}(t) \\ P(t) - P_{\max}(t); & P_b(t) < P_{\max}(t) \ \& \\ & P(t) > P_{\max}(t) \\ 0; & \text{otherwise} \end{cases} \tag{3}$$

and the energy consumption resulting from the target load excess as

$$W_{\mathrm{exc}} = \int_{t_{\mathrm{start}}}^{t_{\mathrm{end}}} P_{\mathrm{exc}}(t) \, dt. \tag{4}$$

Here t_{start} and t_{end} are starting and finishing times of the process execution. The objective is to minimize W_{exc} over all possible schedules of process interruptions.

4 Hybrid Algorithm for Schedule Optimization

We have developed a hybrid optimization algorithm to schedule process interruptions on the considered production lines. The algorithm generates a daily interruption schedule and calculates the achieved reduction of the target load excess. It accepts the following information as input:

– estimates of power demand profiles for processes to be executed,

– an estimate of the background demand profile,

– the target load profile, and

– schedule construction constraints.

Schedule optimization is performed in two phases. First, a greedy heuristic procedure is activated to generate an initial schedule. The schedule is further improved using the evolutionary algorithm.

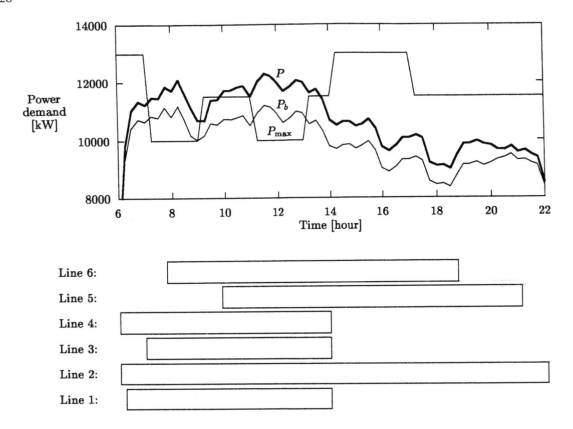

Figure 1. Processes under execution on the production lines and the related power demand profiles: background demand P_b, total power demand P, and the target load P_{max}

The heuristic procedure for generating the initial solution starts with an empty interruption schedule and progressively inserts interruptions into it. At each step exhaustive search is performed over all running processes to find the interruption that most reduces the target load excess and satisfies the constraints. The interruption is then inserted and the power demand recalculated. This is repeated until no additional interruption can be added due to the constraints.

The evolutionary algorithm used to improve the schedules is a $(1, \lambda)$-type evolution strategy [9]. It assumes the schedule obtained by the heuristic procedure as the parent of the initial population. It then generates λ descendants from the parent by introducing small changes of interruption starting times. The changes are implemented in such a way that the constraints remain satisfied. From the generated descendants, the one representing the schedule that most reduces the target load excess becomes the parent for the next generation. The improvement algorithm proceeds until the given number of generations is reached.

5 Experimental Evaluation and Results

The hybrid scheduling algorithm was experimentally evaluated on a set of 30 problem instances. Real data recorded in the factory within a month of operation was used as input to optimize daily schedules for the production lines. The constraints for schedule construction were set as follows. Duration of process interruptions was 30 minutes. Each process had to run continuously for at least four hours between two interruptions, while at most three process interruptions were permitted to take place simultaneously.

Time step used during search for assigning starting times to interruptions was 5 minutes. The improvement phase of the scheduling algorithm was run for 100 generations and 10 descendants were generated from a parent in each generation. Table 1 shows an example of an interruption schedule generated by the algorithm.

Table 1. An example of the optimized production schedule for the production lines

Line number	Number of interruptions	Interruption times	
1	2	8:00–8:30	12:15–12:45
2	2	7:25–7:55	11:55–12:25
3	2	7:00–7:30	12:20–12:50
4	2	7:15–7:45	11:45–12:15
5	1		11:00–11:30
6	1		11:30–12:00

The evaluation confirmed that schedule optimization can substantially contribute to the decrease of energy costs at the plant. Energy consumption resulting from the target load excess on the production lines was reduced by at least 25% on workdays, but in most cases by about 30%. Table 2 shows the achieved reduction averaged over 10 runs of the optimization algorithm for each problem instance.

6 Conclusion

A simple hybrid optimization algorithm consisting of a heuristic to generate initial solutions and an evolutionary procedure to improve the initial solutions has proved powerful in scheduling process interruptions on a group of real-world production lines. Similar experience regarding the methodology was gained in [8] where a simple hybrid of genetic algorithm and local heuristic search are combined into an efficient procedure for block assembly scheduling in shipbuilding.

The approach presented in this paper was implemented as a process scheduling module within a system for energy consumption management at the plant. The role of the scheduling module is to assist the process supervisor in preparing daily schedules for the pressing unit. Through its application it is possible to analyse the impact of various production plans and constraint settings on energy consumption. When used regularly, it is expected to make an important contribution to the decrease of production costs at the plant.

Table 2. Average reduction of the target load excess on the production lines obtained by the hybrid scheduling system

Day	Date	Target load excess [kWh]	Reduction	
			[kWh]	[%]
Sat	1.6.	0.0	0.0	0.0
Sun	2.6.	0.0	0.0	0.0
Mon	3.6.	2616.5	1000.4	38.2
Tue	4.6.	2569.6	970.5	37.8
Wed	5.6.	3218.3	1012.6	31.5
Thu	6.6.	2892.2	926.2	32.0
Fri	7.6.	3055.1	931.6	30.5
Sat	8.6.	655.0	413.0	63.0
Sun	9.6.	0.0	0.0	0.0
Mon	10.6.	2461.2	810.1	32.9
Tue	11.6.	2117.7	636.6	30.1
Wed	12.6.	2910.3	836.8	28.7
Thu	13.6.	2752.8	803.3	29.2
Fri	14.6.	2523.5	869.8	34.5
Sat	15.6.	0.0	0.0	0.0
Sun	16.6.	0.0	0.0	0.0
Mon	17.6.	2262.0	847.4	37.4
Tue	18.6.	3662.8	1068.5	29.2
Wed	19.6.	3353.2	1012.3	30.2
Thu	20.6.	2277.1	813.4	35.7
Fri	21.6.	2814.6	924.1	32.8
Sat	22.6.	238.1	238.1	100.0
Sun	23.6.	0.0	0.0	0.0
Mon	24.6.	2527.2	862.3	34.1
Tue	25.6.	0.0	0.0	0.0
Wed	26.6.	4108.9	1063.4	25.9
Thu	27.6.	3074.1	995.0	32.4
Fri	28.6.	2425.3	671.4	27.7
Sat	29.6.	46.7	46.7	100.0
Sun	30.6.	0.0	0.0	0.0

References

[1] Bruns, R.: Scheduling. In Bäck, Th., Fogel, D. B., Michalewicz, Z. (Eds.): *Handbook of Evolutionary Computing*, Chapter F1.5. Bristol: Institute of Physics Publishing, and New York: Oxford University Press, 1997.

[2] Biethahn, J., Nissen, V. (Eds.): *Evolutionary Algorithms in Management Applications*. Berlin: Springer-Verlag, 1995.

[3] Corne, D. Ross, P.: Practical issues and recent advances in job- and open-shop scheduling. In

Dasgupta, D., Michalewicz Z. (Eds.): *Evolutionary Algorithms in Engineering Applications*, pp. 531–546. Berlin: Springer-Verlag, 1997.

[4] Dasgupta, D., Michalewicz Z. (Eds.): *Evolutionary Algorithms in Engineering Applications.* Berlin: Springer-Verlag, 1997.

[5] Filipič, B.: Enhancing genetic search to schedule a production unit. In Neumann, B. (Ed.): *Proceedings of the 10th European Conference on Artificial Intelligence ECAI '92*, pp. 603–607, Vienna, Austria, 1992. Chichester: John Wiley. Also published in Dorn, J., Froeschl, K. A. (Eds.): *Scheduling of Production Processes*, pp. 61–69. Chichester: Ellis Horwood, 1993.

[6] Filipič, B.: A genetic algorithm applied to resource management in production systems. In Biethahn, J., Nissen, V. (Eds.): *Evolutionary Algorithms in Management Applications*, pp. 101–111. Berlin: Springer-Verlag, 1995.

[7] Filipič, B.: A hybrid optimization algorithm for energy consumption management at a motor plant. *Proceedings of the 5th European Congress on Intelligent Techniques and Soft Computing EUFIT '97*, Vol. 1, pp. 717–721, Aachen, Germany, 1997.

[8] Ryu, K. R., Hwang, J., Choi, H. R., Cho, K. K.: A genetic algorithm hybrid for hierarhical reactive scheduling. In Bäck, Th. (Ed.): *Proceedings of the Seventh International Conference on Genetic Algorithms ICGA '97*, pp. 497–504, East Lansing, MI, 1997. San Francisco: Morgan Kaufmann.

[9] Schwefel, H.-P.: *Evolution and Optimum Seeking*, New York: John Wiley, 1995.

An Evolutionary Approach to Concept Learning with Structured Data

Claire J. Kennedy and Christophe Giraud-Carrier
Department of Computer Science,
University of Bristol,
Bristol BS8 1UB, U.K.
{kennedy,cgc}@cs.bris.ac.uk
fax: +44-117-9545208

Abstract

This paper details the implementation of a strongly-typed evolutionary programming system (STEPS) and its application to concept learning from highly-structured examples. STEPS evolves concept descriptions in the form of program trees. Predictive accuracy is used as the fitness function to be optimised through genetic operations. Empirical results with representative applications demonstrate promise.

1 Introduction

The aim of concept learning is to induce a general description of a concept from a set of specific examples. The examples and the concept description are expressed in some representation language (e.g., attribute-value language, Horn clauses) and the learning task can be viewed as a search, through the space of all possible concept descriptions, for a description that both characterises the examples provided and generalises to new ones [9].

As concept learning problems of increasing complexity are being tackled, increasingly expressive representation languages are becoming necessary. The original concept learning systems use the attribute-value language, where examples and concepts are represented as conjunctions of attribute-value pairs. The simplicity of this (propositional) representation allows efficient learners to be implemented, but also inhibits their ability to induce descriptions involving complex relations. Research in the area of inductive logic programming [12] extends the propositional framework to the first-order by designing systems around the Prolog language. In this context, examples and concepts take the form of Horn clauses. Most recently, higher-order representations, based on the Escher language, have been proposed [3]. In this most expressive context, examples are closed terms and concepts are arbitrary Escher programs. Whilst the greater expressiveness of the representation language extends the applicability of concept learn-

ing, it also results in the explosion of the search space. In addition, higher-order concept learning still lacks a counterpart to the clean refinement methods of the first order and propositional settings; subsequently the search for a solution to the problem can become intractable.

Evolutionary techniques have been successfully applied to concept learning in both a propositional [6] and first-order setting [14]. The idea of and basic assumptions for the application of evolutionary higher-order concept learning are presented in [4]. This paper details the implementation of a strongly-typed evolutionary programming system (STEPS) and its application to the problem of concept learning within Escher. In STEPS, examples are closed terms and concept descriptions take the form of program trees. STEPS starts from a randomly generated, initial population of program trees and iteratively manipulates it by genetic operators until an optimal solution is found.

The paper is organised as follows. Section 2 briefly introduces Escher and the closed term representation used by STEPS. Section 3 details the implementation of STEPS. Section 4 presents the results of experiments with some representative concept learning problems. Finally, section 5 concludes the paper.

2 STEPS Representation

The program trees evolved by STEPS use constructs from the Escher language [8]. Escher is a new strongly typed declarative programming language that integrates the best features of both functional and logic programming languages. Its syntax is based on Haskell, and it features higher-order constructs such as set processing to provide a facility for learning in a higher order context.

This paper focuses on the use of STEPS to evolve concept descriptions from examples. Here, the concept descriptions (or program trees) take the form of

```
IF Cond THEN Ci ELSE S,
```

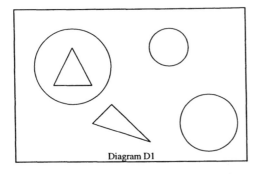 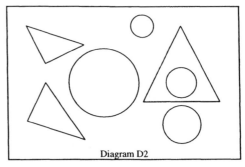

Diagram D1   Diagram D2

```
Data Shape = Circle | Triangle | Inside(Shape, Shape);
type Diagram = {(Shape, Int)};
D1 = {(Circle,2), (Triangle,1), (Inside(Triangle, Circle),1)};
D2 = {(Circle,3), (Triangle,2), (Inside(Circle, Triangle),1)};
```

Fig. 1. Example diagrams and their corresponding Escher representations

where Cond is a Boolean expression, Ci is a class label and S is either a class label or another if-then-else expression. The examples are represented as closed terms, which give a compact and self-contained description of each example.

Figure 1 illustrates the closed term representation used here on two diagrams taken from Problem 47 of [2]. Each diagram contains a number of shapes, where each shape can be a circle, a triangle or a shape inside another shape. A diagram in the Escher closed term representation is a set of pairs, each consisting of a shape and a number indicating the number of times the shape appears in the diagram. Thus, D1 contains 2 circles, 1 triangle and 1 triangle inside a circle.

In order to induce descriptions, it is necessary to extract parts of the individual closed terms so as to make inferences about them. This is accomplished by selector functions, which "pull" individual components out of terms. Each structure (e.g., list, set) that is used in a term has it own set of associated selector functions. The selector functions for tuples, lists and sets:

v = Tuple Type
```
        proji(v)
```
v = List Type
```
        exists \v2 -> v2 'elem' v
        length(v)
```
v = Set Type
```
        exists \v2 -> v2 'in' v
        card(v)
```
For example, the number of occurrences of some shape in D1 above is obtained with

```
exists \x -> x 'in' D1 && proj2(x),
```

which appears as Figure 2(a) in tree form.

Once the components of the data structures have been extracted, conditions can be made on them or they can be compared to values or other data types. For example, the following expression tests whether the number of circles in D1 is equal to 2.

```
exists \x -> x 'in' D1 &&
            (proj1(x) == Circle &&
            proj2(x) == 2),
```

The equivalent tree form appears as Figure 2(b). An algorithm has been designed to automatically generate the appropriate selector function associated with a set of types [1].

3   Evolutionary Approach

Since Escher is a strongly typed language, an evolutionary paradigm that incorporates type information is necessary so that only type-correct programs are generated during learning. Traditional program tree based evolutionary paradigms, such as Genetic Programming (GP), assume the closure of all functions in the body of the program trees [7]. This means that every function in the function set must be able to take any value or data type that can be returned by any other function in the function set. While this characteristic simplifies the genetic operators, it limits the applicability of the learning technique and can lead to artificially formed solutions. In order to overcome this problem, a type system was introduced to standard GP to give Strongly Typed Genetic Programming (STGP) [10]. STGP helps to constrain the search space by allowing only type correct programs to be considered. STEPS extends the STGP approach to allow the vast space of highly ex-

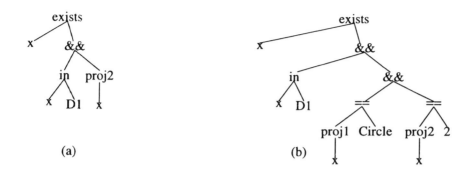

Fig. 2. (a) A sample selector in tree form, (b) A sample condition in tree form

pressive Escher concept descriptions to be explored efficiently.

3.1 Population Creation

The if-then-else form of the descriptions to be evolved by STEPS provides a template for all individuals in the population. In tree form, the template is as follows.

if_then_else
Cond Ci S

Trees in the initial population are formed by randomly selecting subtrees from the problem alphabet. The total alphabet for a problem consists of the appropriate selector function subtrees, any additional functions provided by the user, and the domain-derived constants. The function set provided by the user typically includes the connective functions && and || (the boolean functions conjunction and disjunction) so that a number of comparisons can be made on the components of the data types.

However, subtrees selected to fill in a blank slot in a partially created program tree must satisfy certain constraints so that only valid Escher programs are produced. These constraints are type and variable consistency. In order to maintain type consistency, each node in a subtree in the alphabet is annotated with a type signature indicating its argument and return types. A subtree selected to fill in a blank slot must be of the appropriate return type. The program tree in Figure 3 provides an example of type consistency violation.

The type signatures in Figure 3 are in curried form and dotted-lines indicate where a subtree has been added. The addition of the

```
Circle :: Shape
```

subtree violates type consistency, as it is of type Shape and the function

```
== :: Int -> Int -> Bool
```

requires a subtree returning type Int as its second argument.

In order to maintain variable consistency, the local variables in a subtree selected to fill in a blank slot in the partially created program tree must be within the scope of a quantifier. In addition, all quantified variables in a program tree must be used in the conditions of their descendant subtrees to avoid redundancy. The program tree in Figure 4 provides an example of variable consistency violation.

The addition of the subtree rooted at

```
== :: Shape -> Shape -> Bool
```

in Figure 4 violates variable consistency as the variable

```
v4 :: Shape
```

is not within the scope of a quantifier. In addition variable consistency is violated by not using the quantified variable

```
v2 :: (Shape, Int).
```

3.2 Modified Crossover

The requirement for type and variable consistent program trees needs to be maintained during the evolution of the programs so that only syntactically correct programs are evolved. In addition to this, it is necessary to preserve the structure of the selector function subtrees. This results in a situation where crossover can only be applied to certain nodes within a program tree. These crossover points correspond to the roots of the subtrees in the function set. Once a crossover point has been randomly selected from the first parent, a crossover point that will maintain type and variable consistency can be randomly selected in the second parent. If no such crossover point is available then an alternative crossover point is selected from the first parent and the process is repeated.

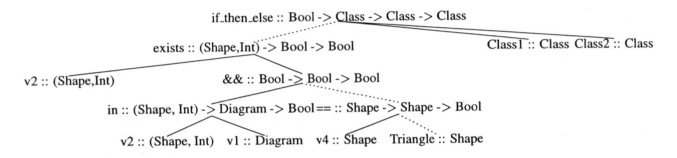

Fig. 3. A program tree exhibiting type consistency violation

Fig. 4. A program tree exhibiting variable consistency violation

3.3 Mutation

During successive iterations of the evolutionary process, the amount of genetic variation in a population decreases. In an extreme case, this can lead to the loss of genetic material that is essential to the search for an optimal solution and a method for reintroducing such lost material is required. STEPS ensures the preservation of genetic diversity through six distinct forms of mutation. These mutation operators are the terminal and functional mutation operators of conventional GP and four specialisations of functional mutation. The various functional mutations can only be applied at the crossover points in a program tree and must preserve type and variable consistency. The specialised functional mutations include AddConjunction, DropConjunction, AddDisjunction and DropDisjunction. AddConjunction and AddDisjuction involve inserting an && or || respectively at the node to be mutated. Its first argument is the subtree originally rooted at that node and its second argument is randomly grown. For example, if we apply the AddConjunction operator to the == node in the tree of Figure 5(a), then we could obtain the tree of Figure 5(b).

The DropConjunction and DropDisjunction operators involve randomly selecting an && or || crosspoint respectively, replacing it with the subtree that makes up its first argument.

3.4 Learning Strategy

STEPS creates an initial population of a specified size ensuring that each tree preserves the necessary constraints and is unique. In order to perform population updates steady state replacement is used. Parent program trees are selected by the tournament selection technique and are recombined using both crossover and mutation. Fitness is evaluated as the predictive accuracy of a program tree over the set of examples.

The choice of genetic operator is determined by the depth of the program tree. If the depth of the tree is greater than a specified maximum depth, then the tree is considered to be too big so a mutation operator that drops a conjunction or a disjunction is used. If the depth of the selected tree is less than a specified minimum depth then a conjunction or disjunction is added. If the depth of the program tree lies within the specified depth constraints then any genetic operator can be applied to it. If the offspring of a program tree already exists within the population then the tree is mutated.

4 Experiments

4.1 Michalski's train

Problem Description: The objective of Michalski's train problem is to generate a concept description that distinguishes trains that are travelling East from trains travelling West [11]. A train in the Escher closed term representa-

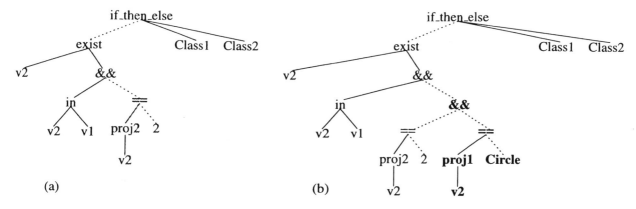

Fig. 5. Sample AddConjunction Mutation

tion is a list of cars with each car represented by a tuple of characteristics including shape, length, wheel, roof and load.

```
type Car =
  (Shape,Length,Wheels,Roof,Load)
type Train = [Car]

direction :: Train -> direction
direction([
  (Rectangular,Short,2,Open,(Circle,1)),
  (Hexagonal,Short,2,Flat,(Triangle,1)),
  (Rectangular,Short,2,Flat,(Circle,2))
  ]) = East;
```

Learning Parameters: Examples: 10, Population size: 300, Minimum depth of trees: 4, Maximum depth of trees: 7.

Results: The experiment was carried out for 10 runs with an optimal solution found in the initial population for 2 out of the 10 runs. For the remaining runs, the optimal solution was found in an average of 1187 evaluations. The following is the optimal solution found in one of the runs.

```
direction(v1) =
    if exists \v2 ->
       v2 'elem' v1 &&
       proj2(v2) == Short &&
       proj4(v2) /= Open
    then East else West;
```

i.e. A train is travelling East if it contains a short closed car.

4.2 Mutagenicity

Problem Description: The aim of the Mutagenicity problem is to generate a concept description that distinguishes mutagenic chemical compounds from non-mutagenic chemical compounds [13, 5]. A chemical com-

pound in the Escher closed term representation is a highly-structured term consisting of the atom and bonds that make up the structure of the compounds and some chemical attributes thought to be relevant to the problem by domain experts, as illustrated below.

```
type Atom =
  (Label,Element,AtomType,Charge);
type Bond = ({Label},BondType);
type Molecule =
  (Ind1,IndA,Lumo,{Atom},{Bond});

mutagenic :: Molecule -> Bool;
mutagenic (True,False,-1.487,
  {1,C,22,-0.188) ... (28,O,40,-0.389)},
  {{1,2}, 7) ... ({26,28},2)})
  = True;
```

Learning Parameters: Examples: 188, Population size: 300, Minimum depth of trees: 5, Maximum depth of trees: 12.

Results: A 10-fold cross validation was carried out to give an average accuracy of 87.3% with a standard deviation of 5.8%. This is comparable to the results of other learning systems on the same data set (e.g., see [13]). The following is a solution (87.8% accurate with respect to all examples) found in 3337 evaluations.

```
mutagenic(v1) =
    if (proj1(v1) == True ||
    proj3(v1) < -2.368) &&
       (not
            (exists \v2 ->
                v2 'in' proj4(v1) &&
                proj3(v2) == 93))
    then True else False;
```

5 Conclusion

This paper details the implementation of a strongly-typed evolutionary programming system (STEPS) and its application to the problem of learning concept descriptions from highly-structured examples. The type system allows efficient use of genetic operators to evolve complex program trees. Preliminary experiments with STEPS on two representative, higher-order concept learning tasks demonstrate promise.

Acknowledgements

This work is funded by EPSRC grant GR/L21884. The authors would like to thank the members of the Machine Learning Research Group at Bristol University for many interesting discussions relating to this work. Special thanks to Tony Bowers for his implementation of the Escher interpreter.

References

[1] C. Giraud-Carrier A.F. Bowers and J.W. Lloyd. Higher-order logic for knowledge representation in inductive learning. In preparation, 1999.

[2] M. Bongard. *Pattern Recognition*. Spartan Books, 1970.

[3] P. Flach, C. Giraud-Carrier, and J.W. Lloyd. Strongly typed inductive concept learning. In *Proceedings of the International Conference on Inductive Logic Programming (ILP'98)*, pages 185–194, 1998.

[4] C.J. Kennedy. Evolutionary higher-order concept learning. In John R. Koza, editor, *Late Breaking Papers at the Genetic Programming 1998 Conference*, University of Wisconsin, Madison, Wisconsin, USA, 22-25 July 1998. Stanford University Bookstore.

[5] R. King, S. Muggleton, S. Srinivasan, and M. Sternberg. Structure-activity relationships derived by machine learning: The use of atoms and their bond connectivities to predict mutagenicity in inductive logic programming. *Proceedings of the National Academy of Sciences*, 93:438–442, 1996.

[6] J.R. Koza. Concept formation and decision tree induction using the genetic programming paradigm. In H.-P. Schwefel and R. Männer, editors, *Parallel Problem Solving from Nature*, pages 124–128, 1990.

[7] J.R. Koza. *Genetic Programming: On the Programming of Computers by Means of Natural Selection*. The MIT Press, Cambridge, Massachusetts, 1992.

[8] J.W. Lloyd. Declarative programming in escher. Technical Report CSTR-95-013, Department of Computer Science, University of Bristol, 1995.

[9] T.M. Mitchell. Generalization as search. *Artificial Intelligence*, 18:203–206, 1982.

[10] D.J. Montana. Strongly typed genetic programming. *Evolutionary Computation*, 3(2):199–230, 1995.

[11] S. Muggleton and C.D. Page. Beyond first-order learning: Inductive learning with higher order logic. Technical Report PRG-TR-13-94, Oxford University Computing Laboratory, 1994.

[12] S. Muggleton and L. De Raedt. Inductive logic programming: Theory and methods. *Journal of Logic Programming*, 19/20:629–679, 1994.

[13] A. Srinivasan, S. Muggleton, R. King, and M. Sternberg. Mutagenesis: ILP experiments in a nondeterminate biological domain. In S. Wrobel, editor, *Proceedings of Fourth Inductive Logic Programming Workshop*. Gesellschaft für Mathematik und Datenverarbeitung MBH, 1994. GMD-Studien Nr 237.

[14] M. L. Wong and K. S. Leung. Genetic logic programming and applications. *IEEE Expert*, October 1995.

Genetic Redundancy: Desirable or Problematic for Evolutionary Adaptation?

R. Shipman

Future Technologies Group, Complex Systems Laboratory,
BT Laboratories, Ipswich, IP5 3RE, England.
Email: rob.shipman@bt.com

Abstract

Evolution is commonly viewed as a process of hill climbing on a fitness landscape. A major problem with such a view is the presence of local optima; sub-optimal regions of the landscape from which no further progress is possible. There is an increasing amount of evidence [3,4,7], however, that the presence of large degrees of redundancy in the genome may alleviate this problem through the creation of *neutral networks*; sets of genotypes at the same level of fitness that are connected by single point mutations. These networks allow drift at the same fitness level and hence may increase the reliability of the evolutionary process by allowing the exploration of larger portions of genotype space. The presence, or otherwise, of genetic redundancy could thus be an important concern in the design of artificial evolutionary systems. This paper explores the effects of genetic redundancy in the context of an evolutionary robotics experiment. Neural network control systems are evolved for a simple navigation task and the speed and reliability of the evolutionary process ascertained for differing levels of redundancy. Evolutionary progress is found to halt far more readily as the degree of redundancy is reduced indicating a greater probability of entrapment at local optima.

1 Introduction

The dominant view of adaptive evolution is one of an uphill walk by a population on a fitness landscape. The peaks of this landscape represent phenotypes of high fitness value and are separated by valleys of relatively low fitness. Selection pressures are seen to pull the population up to the peaks and genetic operators to distribute the population allowing exploration of the space. A major problem with this view of evolution is the potential for local optima. A population will climb any hill in its vicinity; however, if the peak turns out to be sub-optimal further adaptation is very difficult. Careful balancing between selection and mutation pressures may allow some continued exploration of genotype space but in general adaptive evolution will cease. This scenario does not seem to be compatible with the open-ended creativity of the biological world. One reason for this is likely to be due to the effects of

co-evolution where other species effectively change the nature of the landscape on which a population resides. However even without co-evolution recent work in, for example, molecular evolution suggests that the view of evolutionary progression as a hill climbing process may be incomplete [1,2].

A great deal of attention has been given to the simple genotype-phenotype mapping defined by the folding of an RNA molecule from its primary nucleotide sequence into a secondary structure [3,4]. This is a biologically realistic mapping and gives some insight into the kind of mappings that may be common in natural systems. It thus also gives insight into the kind of artificial mappings that we may want to construct. Its properties are summarised below:

1. *Large scale redundancy*. There is a many-to-one genotype-phenotype mapping with many sequences folding into the same structure.

2. *Common structures*. A relatively small number of structures occur significantly more frequently than the remainder. There are relatively few common structures and many more rare structures.

3. *Neutral networks*. The sequences folding into a common structure form connected paths that percolate throughout sequence space. That is, it is possible to traverse the set of sequences through a series of one or two point mutations without changing the structure.

4. *Shape Space Covering*. Examples of sequences folding into all the common structures can be found within a small neighbourhood of any random sequence. Thus, from any arbitrary position in sequence space the neutral networks of the common structures are within relatively close proximity.

The rare structures are effectively inaccessible and thus movement on, and transitions between, the neutral networks associated with the common structures become the dominant evolutionary dynamic. For such a scenario to be desirable the following two conditions should hold; the networks should percolate throughout large regions of genotype space and there should be high accessibility between networks. The former protects against a population becoming isolated in a sub-optimal region of genotype space and the latter encourages beneficial transitions reducing the amount of time spent randomly drifting along a network. An aim of this work is to begin to investigate whether these desirable properties of redundancy can be encouraged and exploited in artificial systems.

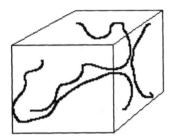

Fig. 1. Percolating neutral networks in genetic hyperspace. Large areas of the space can be explored through fitness-neutral mutations. The networks are in close proximity in a number of places allowing transitions to occur.

A prerequisite for the generation of these properties is a genotype-fitness mapping containing large degrees of *potentially useful* redundancy. This presents something of a dilemma for artificial evolution practitioners. If the dominant view of hill climbing on fitness landscapes were subscribed to, adding redundant genetic material would surely be foolhardy. The size of the search space would be increased making it increasingly unlikely that the global optimum would be found. However, if the speed and reliability of the evolutionary process can be improved then the addition of redundancy would be wise indeed. Can these properties be encouraged in our artificial systems? Can we increase the reliability of evolution through the removal of local optima, or are we simply decreasing the likelihood of finding the global optimum? It is the aim of this work to explore such questions in the context of a real world evolutionary robotics experiment with varying degrees of redundancy.

The structure of this report is as follows. Section 2 reviews related work in artificial evolution, section 3 details the experimental methods including the evolutionary robotics task that was used in this work and the ways in which redundancy was encouraged in the robot control system. Section 4 presents the results of the experiments, section 5 discusses their implications and section 5 concludes.

2   Related Work

In the most closely related work to that presented in this paper, Hoshino and Tsuchida [6] employed a graph-based chromosome that represented the neural network controller for a mobile robot. Regulatory genes were used that effectively switched on and off portions of the chromosome. Thus, "switched off" neurones where under no selection pressure and could accumulate hopefully beneficial modifications before being switched on again. There was thus a potential for the formation of neutral networks. The biological plausibility of this model is questionable. A gene would have to play no role throughout the entire development cycle of an organism in order to be free from selection pressure. Many genes are switched off in some cells but it is unlikely that any are switched off in all. If this were the case, it is probable that a non-functional pseudogene would be formed and quickly "blend" into its non-functional surroundings; a process termed compositional assimilation. Nevertheless, the work revealed interesting properties resulting from the manifestation of neutral genes. The robot was found to approach competence at its simple navigation task in sporadic, large fitness jumps. Punctuated equilibrium dynamics, the hallmark of redundant mappings, were evident. The evolution of chromosomes with and without neutral genes was compared, the former was found to approach target fitness more quickly initially; the latter however was found to be much more reliable. This is perhaps not a surprising conclusion, when genes are switched off they experience no guiding influence at all, effectively resulting in a random search. The process of encouraging neutral networks through these kinds of mechanisms is thus inherently unreliable. This is an important concern in general for neutral network research and will be discussed further in section 5.

An important contribution from the artificial evolution community resulted from the hardware evolution work carried out at the University of Sussex. Most of the discussion of neutral networks centred on an experiment to evolve a Field Programmable Gate Array (FPGA) for a tone-recognition task [7]. The FPGA consisted of 100 logic blocks and was specified by a binary genotype of length 1800. The successful evolved circuit used only 30 to 40% of the functionality available to it. The genome thus contained large amounts of redundancy. It is important to note that the type of this redundancy is crucial. Although a large proportion of the circuit was redundant in the final circuit it could all, given changes elsewhere on the genotype, have affected its functionality. The entire chip was potentially useful. Analysis was carried out showing that some non-functional parts of the final chip had been used during evolution, thus suggesting that these areas may indeed have had an important role to play in allowing progression to the final circuit. This led the authors to advocate the use of many more components than are required in evolvable hardware experiments. In order to investigate the fruitfulness of such an approach, it would be useful to re-run the evolutionary process with the non-functional parts of the circuit completely removed. This may be problematic when evolving integrated circuits. It is more straightforward, however, in other domains and indeed this form of experimentation played an important role in this work. The exact mechanisms employed are described in section 3.

Recently several contributions have been made in the context of abstract fitness landscapes [8,9]. Both of these contributions extend a model of evolution known as an NK model in which the two parameters, N and K, can be used to modify the structure of the fitness landscape and in particular its ruggedness. The extended models add an extra parameter that controls the degree of neutrality. Even in very generic landscapes, such as these, the properties described above are evident. This work suggests that the desirable properties found in the genotype-phenotype maps of biopolymers are not exclusive to that domain but may be extensible to many others. The potential benefits certainly justify the research effort in ascertaining the extent to which this is true.

3   Experimental Methods

3.1   The Evolutionary Robotics Task

In order to experiment with realistic noisy fitness landscapes without the cost of actual real world evolution, a minimal simulation was employed for this work [10]. The simulated task, shown in Figure 2, was chosen to be non-trivial but simple enough to lend itself well to analysis. As the robot control systems were evolved facets of their environment were varied within a range so as to encourage robust solutions. In addition, noise was added to the robot's sensors and motors for the same purpose. Signals from the robots infra red and ambient light sensors were fed into the control system which consisted of 8 binary threshold sensory neurones and 2 motor neurones with a saturated ramp output function. All the neurones had 3 associated links that could connect to any other neurone in the network with genetically modifiable positive or negative weighting. The networks were thus arbitrarily recurrent. This task was originally explored in [10] and the reader is referred to this work for full

details of the minimal simulation and robot control system.

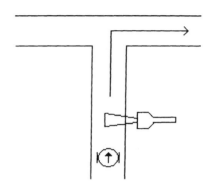

Fig. 2. The T-junction task. The robot must turn left or right at the junction in correspondence with the position of the light in the approaching corridor.

These neural networks are relatively small and it may be argued that they are not likely to encourage redundancy. However, experimentation revealed that only a small amount of this functionality was required to successfully perform the T-junction task. The genomes of many evolved controllers were 60 to 70% redundant. Redundancy was encouraged in another way; the connecting neurone of a link was specified by a 4-bit number and could therefore reference a total of $2^4 = 16$ neurones. Thus, for a 10-neurone network, 6 link values referenced non-existent neurones and resulted in redundant links. All areas of the network always remained potentially useful; a mutation could easily cause a link to reference a different neurone and return previously redundant areas of the network to functionality. In addition the probability of such an occurrence remained high, typically only a change in a single link index was required and thus non-functional areas of the network were only an absolute maximum of 4 mutations away from becoming functional.

3.2   Evolutionary Algorithm

Bilateral symmetry was employed in the genetic encoding of the robot control system and thus it consisted of a total of 15 links and 5 neurone thresholds. Each link was specified by a total of 8 bits, 4 to specify the neurone to connect to and 4 the weight. The neurone thresholds were also specified by a 4-bit value and thus the resulting genome was 140 bits long. A steady-state distributed genetic algorithm [11] was employed to search the genotype space. A population of 100 was distributed on a virtual 10x10 grid. An offspring event consisted of randomly selecting an individual in this grid and constructing a 3x3 breeding pool around it. Linear rank based selection was used to select a probabilistically strong parent from the breeding pool and with a probability of 0.7 a second parent. A new individual was constructed subject to a per bit mutation rate of 0.02 and single point crossover in the case of two parents. The offspring replaced a probabilistically unfit member from the same breeding pool. A generation consisted of 100 offspring events.

An individual's fitness was a function of the distance travelled and the direction of turn at the junction. The distances travelled down the approaching corridor, d1, and the departing corridor, d2, were summed and a fitness bonus of 100 was assigned if the robot turned in the correct direction. Thus, for a correct turn $Fitness = d1 + d2 + 100$ and for an incorrect turn $Fitness = d1 + d2$. To account for the noise and environmental variability fitness assessments consisted of ten trials. The side of the light was initialised in such a way as to force five trials to be performed on either side. The final fitness value was an average of these ten trials.

3.3 Network Pruning

In order to test the degree of redundancy in a controller it was necessary to isolate the functional areas through pruning neurones and links that were playing no role in the network performance. The noise and environmental variability that was an inherent part of the minimal simulation complicated this process. It was necessary to isolate insignificant fitness changes due to noise from those resulting from the removal of vital controller components. In order to achieve such a task the fact that the noise and variability of the simulation could be controlled was exploited. It was possible to add *exactly* the same noise to two simulations by ensuring the random number generator was seeded with the same value before the trial began. In this way a networks performance over a range of trials could be compared to that of a pruned equivalent with the knowledge that any change in fitness was due to the removal of network components and not noise.

A network was pruned by removing each of its neurones and links in turn and ascertaining whether there was any resulting change in fitness. Only single components were removed at any one time, a change in fitness indicated that the component was part of at least one subset of components that had an effect on the controller's performance. The singular nature of this process meant that certain combinations of redundant components could not be revealed, for example if two neurones could not be removed in isolation without

effecting fitness but could be removed together. Situations such as these were adjudged to be sufficiently unlikely for the controllers used in this work to be neglected from the analysis.

Once the functional area of a controller had been ascertained attempts were made to evolve a successful controller using only that functionality. The performance of the evolutionary process was then compared for controllers with the full complement of components and controllers with varying degrees of removed components. In addition to removal of redundant network components the allowance for a link to reference a non-existent neurone was also removed. Thus, every link in the network was forced to attach to a neurone. This was achieved by randomly assigning each redundant link index to an existing neurone at the start of each run. Thus for a ten-neurone network, link indexes of 0 to 9 would reference existing neurones, link indexes from 10 to 15 would no longer result in redundant links but would reference a randomly chosen neurone.

3.4 Controller Categorisation

In order to visualise the population's progress during an evolutionary run, in particular with regard to movement on and transitions between neutral networks, it was necessary to have some mechanism of ascertaining whether two controllers were on the same neutral network. In order to do this a measure of the qualitative behaviour of the robots was required. Small changes in fitness were typically indistinguishable to selection in the context of the minimal simulation and thus it was necessary to change the point of view from the precise fitness value of a controller to the *underlying discovery*. Current environmental conditions made changes to the fitness value but could not change the underlying capabilities of the controller. The controller was unable to maintain any fortuitously high fitness value unless the underlying discovery allowed for robust, successful behaviour. Thus, the fitness values themselves clouded the real issue of controller capability. In order to assess such capability noise-free, median variability fitness assessments were performed. All environmental parameters were set to be at the centre of their allowable ranges and the sensor noise was not added. Clearly, controllers successful under these conditions may not have been successful under normal, noisy conditions but they must *at least* be capable of performing this median task. Success here was necessary but not sufficient for robust, successful controllers. Analysis revealed that there were six stages involved in progression from initial, random controllers to successful controllers. It was not necessary to move through all these stages on any one evolutionary run; a population could bypass a number of them. However, controllers always fell into one of the following categories:

0. Unable to reach the junction.
1. Able to reach the junction but unable to make the turn.
2. Able to make 50% correct turns but not navigate down the departing corridor.
3. Able to make 50% correct turns and navigate down the departing corridor.
4. Able to make 100% correct turns but not navigate down the departing corridor.
5. Able to make 100% correct turns and navigate down the departing corridor.

It is important to keep in mind that this categorisation does not tell the whole story, it says nothing about whether the controllers could robustly perform the same task in the presence of noise. In fact a controller of a particular category may *never be able* to perform the task robustly given any amount of time in which to evolve; it may not have the required functionality. Nevertheless, categorising the population in this way proved to be very informative.

3.5 Random Neutral Walks

The categories presented in the previous subsection equate to neutral networks in the genotype-fitness mapping of the controllers. Mutations making small, robust changes within a category may eventually be propagated through selection pressures. However, they will de disguised by the noise and this process may take a long time. Ohta pointed out [12] that if this length of time were significantly longer than the time-scale of mutation occurrence the effect would not be established. Thus, mutations that do not change the category are likely to be *effectively* neutral and it is mutations that do result in a change of category that are likely to have the most significant impact on the evolutionary process. Punctuated equilibrium was by far the most common evolutionary dynamic witnessed in this work. The periods of stasis corresponded to a population lying on the neutral network of a particular category of controller and the rapid gains to transitions between these categories.

In order to ascertain the degree to which these neutral networks percolated genotype space and the accessibility of one network from another a technique, inspired by Huynen [1], was employed that allowed a random walk to be performed on the neutral networks associated with each of the categories. To be of use in evolutionary adaptation random drift on the neutral networks must allow greater access to *different* controllers of a higher or the same category. It is important that the controllers had different functional components (as revealed by pruning) as only these controllers were potentially better adapted. Such discoveries were termed *innovations*.

The walks began with a controller of a known category. All one-point mutants of this controller were generated and their categories ascertained. The number of mutants in each category was stored including the number of neutral neighbours i.e. mutants of the same category. One of the neutral neighbours was then chosen at random and the procedure repeated. The walk was terminated after one hundred steps.

4 Results

4.1 Controller Pruning

A series of 50 runs was performed in which the initial 10–neurone controller was evolved for the T-junction task. Successfully evolved controllers were then pruned to reveal the functional areas of the network. The pruned networks were categorised according to the amount of genome that was removed. For each category a further 50 experiments were performed in which an example pruned controller from that category was chosen at random and the evolutionary process restarted using the genome representing this controller. A run was classed as a failure after 2000 generations, many more than were normally required to produce a successful controller. The average number of generations for successful runs is shown in Figure 3 together with the number of failures.

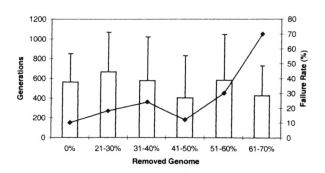

Fig. 3. Results for 50 runs of example controllers from each level of redundancy. The per bit mutation rate was held constant for all sizes of genome. The columns represent the average generations for successful runs and the error bars plus one standard deviation. The line represents the percentage failure rate.

As the degree of redundancy is reduced the graph suggests a corresponding increase in the failure rate. However, controllers of 41-50% the size of the original fair better than all others apart from the original controllers. This anomaly may be due to statistical fluctuations or to a mutation rate that was particularly appropriate for controllers of this size.

As components were removed from the controller the genome size was also reduced, however, the per-bit mutation rate was held constant. This had the effect of altering the per-genome mutation rate. To ensure that the results were not a manifestation of modifications to the effective mutation rate per genome a series of 50 runs were performed for the 61-70% category at a range of mutation rates from 0.2 to 3.0 expected mutations per genotype. These results showed a greatly increased number of failures independently of the mutation rate.

The standard deviation reveals a very high variability in the average number of generations required to produce a successful controller. This suggests that no reliance can be placed on this measure to compare the ease of evolution for differing degrees of redundancy. The implications of this result will be discussed in section 5.

4.2 Random Neutral Walks

Random neutral walks were performed for 10 independent examples of all controller categories apart from category 0. The latter was omitted from the analysis, as it played no significant role in the evolutionary process; examples of category 1 controllers were almost always found in the initial random population. The results in figure 4 show the number of innovations, i.e. controllers of equal or higher category and differing functionality, for walks on neutral networks associated with each category. Each walk consisted of a total of 100 steps.

The figure shows a rapidly decreasing innovation rate with decreasing redundancy. For non-pruned controllers around 0.8 innovations per step are made on average, this number reduces to around 0.03 for the smallest controllers. The reduction of the innovation rate is particularly severe for highest category controllers where there is often *no* innovation for the smallest controllers. Thus, if the controller forming the starting point of the run were not robust it would be very difficult to continue adaptation. There may be some possibility of several concurrent mutations allowing discovery of new controllers but the probability of this is low. Under these circumstances evolutionary adaptation is inevitably unreliable. In contrast movement on neutral networks for higher

degrees of redundancy allowed the discovery of many different, potentially better adapted, controllers.

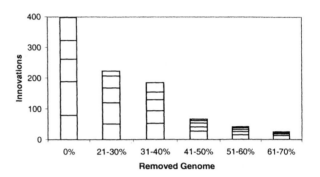

Fig. 4. Average number of innovations found on 10 neutral walks of length 100. The accumulation of innovations for all controller categories is shown. The sections within the columns show the contribution of category 1 at the bottom to category 5 at the top.

5 Discussion

The results presented in the previous section strongly suggest significant benefits arising from the presence of redundancy. From an engineering perspective, however, an important concern in evolutionary systems is the speed at which a solution can be found. The results did not reveal a reduction in the number of generations required to discover successful controllers with increasing redundancy. However, the average number of generations was calculated using only successful runs. If the failures were incorporated into the measure, by for example using the number of generations at the cut-off point, a trend similar to that observed for the failure rate would be evident in the average generations measure. The relatively good performance of successful runs without redundancy is not unexpected; the presence of neutral networks results in drift at the same category level. With reduced redundancy this drift would also be reduced and hence the population could move to higher categories of controller more quickly; assuming that it was fortunate enough to be in a part of the space that allowed such progression. The fact that, with an increased degree of drift, the results for higher degrees of redundancy are comparable could thus be considered encouraging; there was no discernible speed penalty for increased reliability.

In order to reduce the time spent searching a neutral network the accessibility of new and potentially more capable controller's needs to be increased; an outcome that would be encouraged through ensuring percolating neutral networks and shape space covering. This is actually easy to do, as an example consider expanding the genome of the minimal controller[1] to result in a new genome consisting of 90% junk DNA, highly repetitive tandem sequences perhaps, and the original genome that is interpreted in exactly the same way to produce the controller. The junk DNA can be changed at will without affecting the operation of the controller and hence the resulting neutral networks will percolate widely throughout at least 90% of the space. In addition, it is now *guaranteed* that from any arbitrary position in the space all the neutral networks associated with each category of controller are within a neighbourhood that is no greater than 10% the radius of the search space. Thus, percolating neutral networks and shape space covering have been achieved. However, in this case their presence does not aid evolutionary adaptation in the slightest. The problem is unaltered, as the constraints on the parts of the genome corresponding to the initial genome remain exactly the same. This process can not remove any local optima.

The addition of redundancy, however, can change the nature of the problem if it allows a facet of the controllers' functionality to be carried out in a number of different ways. Essentially, it is useful if it allows for the relaxation of constraints on the genome. As a trivial example, consider a situation in which two neurones must be connected together and only one possible link exists capable of playing such a role. During earlier stages of an evolutionary run that link may contribute to the controllers functionality in another way and it may subsequently be impossible for it to be modified into the required state without reducing the controllers fitness. This situation would correspond to a local optimum. However, if the controller contained several links capable of making the required connection it is more likely that one of them will not be crucial to the current controller and could be modified in order to make the necessary connection. Thus the constraints on the initial link have been reduced; it is no longer required to be in exactly the same configuration. Analysis revealed that over 40% of the minimal controllers' genome had to be in exactly the same state to result in a successful controller. This percentage included every link index and thus all the successful minimal controllers employed the same basic network configuration. In contrast, not a single bit of the non-pruned genome was required to be in exactly the same state; many more basic configurations were capable of performing the T-junction task. Thus, the presence of redundancy greatly decreased the constraints on the genome and hence increased the reliability of the evolutionary process.

It would seem sensible, therefore, to advocate a focus on constraint minimisation when designing genetic systems. Such a focus would inevitably lead to redundancy and, potentially, the beneficial effects of percolating neutral networks. However, this process needs to be performed in a particular manner. It is trivially easy to minimise constraints in such a way as to ensure that none of the genome is required to be in exactly the same configuration. In [13] a theoretical encoding scheme was described in which the entire genome was duplicated and a single "switch" bit introduced to determine which half of the genome was currently interpreted. Thus, either half could produce the required behaviour; no individual bit was required to be in exactly the same state. Clearly, however this is unlikely to aid evolutionary adaptation; the switched out genome is restricted to random search and there is an infinitesimal probability of it proving beneficial.

This is equivalent to the use of neutral genes in the work reviewed in section 2 with the alteration that a smaller subset of the genome was switched off at any one time. Indeed, such a mechanism was also employed in this work by allowing link indexes to reference non-existent neurones and hence be switched off. However, a series of runs was performed in which such a possibility was disallowed for non-pruned controllers and allowed for minimal controllers. The allowance for a link to be switched off was not found to be beneficial in either case. In fact, the number of failures showed a slight increase. Of course, links are still effectively switched off due to the implicit redundancy in the network. However, these results lead to the conclusion that thinking about minimising constraints through switching off portions of the genome and relying on random search is not likely to be beneficial, even for small areas of the genome.

The point of view advocated here, therefore, for encouraging the beneficial effects of redundancy is to minimise constraints on the genome through constructing systems of many components that are capable of combining and interacting in a number of different ways to form the functional phenotype. Such a system would increase the number of possible configurations capable of performing the required task and hence encourage percolating neutral networks. An outcome that would be further encouraged if all the components retained a relatively high probability of forming part of the functional phenotype.

---

[1] The minimal controllers consisted of only 6 neurones and 8 links. The resulting genome size was 31.4% that of the original.

6   Conclusions

This paper has presented evidence that redundancy can increase the reliability of the evolutionary process. The ability of evolution to discover successful control systems was shown to decrease as the degree of redundancy was decreased. The removal of redundant controller components also reduced the possibility of a mutation being fitness-neutral which restricted neutral drift. Random neutral walks were employed to reveal that this severely reduced the probability of discovering new and potentially more capable controllers.

Few would consider the control system used in this work to be highly redundant; it contained only 10 neurones each with 3 links. However, it was shown to contain a relatively high degree of redundancy, which had real benefits for evolutionary adaptation. The potential exists for these properties to be exploited and enhanced in other artificial evolutionary systems through the consideration of redundancy.

Acknowledgements

The author is grateful to the Future Technologies Group at BT Labs[2] for supporting the preparation of this paper. The work benefits from a number of discussions at the University of Sussex in particular with Inman Harvey[3]. Thanks also go to Nick Jakobi[4] for assistance in constructing the minimal simulation.

References

[1]   Huynen, M.: Exploring phenotype Space through neutral evolution. Santa Fe Institute Working Paper, 95-10-100, 1995.

[2]   Huynen, M., Stadler, P., Fontana, W.: Smoothness within ruggedness: the role of neutrality in adaptation. Proc. of the National Academy of Science, vol. 93, pp. 397-401, 1995.

[3]   Gruner, W., Giegerich, R., Strothmann, D., Reidys, C., Weber, J., Hofacker, I., Stadler, P., Schuster, P.: Analysis of RNA Sequence Structure Maps by Exhaustive Enumeration. Part One: Neutral Networks. University of Vienna, Theoretical Biochemistry Group Working Paper, 95-10-16, 1995.

[4]   Gruner, W., Giegerich, R., Strothmann, D., Reidys, C., Weber, J., Hofacker, I., Stadler, P., Schuster, P.: Analysis of RNA Sequence Structure Maps by Exhaustive Enumeration. Part Two: Structures of Neutral Networks and Shape Space Covering. University of Vienna, Theoretical Biochemistry Group Working Paper, 95-10-16, 1995.

[5]   van Nimwegen, E., Crutchfield, J.: Optimizing Epochal Evolutionary Search: Population-Size Independent Theory. Santa Fe Institute Working Paper, 98-06-046, 1998.

[6]   Hoshino, T., Tsuchida, M.: Manifestation of Neutral Genes in Evolving Robot Navigation. Proc. of Artifical Life V, pp. 408-415, 1996.

[7]   Harvey, I., Thompson, A.: Through the Labyrinth Evolution Finds a Way: A Silicon Ridge. Proc. of the First International Conference on Evolvable Systems: From Biology to Hardware (ICES96). Berlin: Springer Verlag, 1996.

[8]   Barnett, L.: Tangled Webs: Evolutionary Dynamics on Fitness Landscapes with Neutrality. University of Sussex, MSc Dissertation, 1997.

[9]   Newman, M., Engelhardt, R.: Effects of Neutral Selection on the Evolution of Molecular Species. Santa Fe Institute Working Paper, 98-01-001, 1998.

[10] Jakobi, N.: Evolutionary Robotics and the Radical Envelope of Noise Hypothesis. University of Sussex, Cognitive Science Research Paper 457, 1997.

[11] Collins, R., Jefferson, D.: Selection in Massively Parallel Genetic Algorithms. Proc. of the Fourth International Conference of Genetic Algorithms (ICGA-91), pp. 249-256, Morgan Kauffman, 1991.

[12] Ohta, T.: Population Size and rate of evolution. Journal of Molecular Evolution, vol. 1, 1972.

[13] Jakobi, N.: Encoding Scheme Issues for Open-Ended Artificial Evolution. Proc. of Parallel Processing in Nature, pp. 52-61, Springer-Verlag, 1996.

---

[2] http://www.labs.bt.com/projects/ftg/index.htm

[3] http://www.cogs.susx.ac.uk/users/inmanh/

[4] http://www.cogs.susx.ac.uk/users/nickja/

Dynamic Systems Modelling with Evolving Cellular Automata

A. Dobnikar[1], A. Likar[2], S. Vavpotič[1]

[1] Faculty of Computer and Information Science, University of Ljubljana, Tržaška 25, 1001 Ljubljana, Slovenia
[2] Faculty of Mathematics and Physics, University of Ljubljana, Jadranska 19, 1001 Ljubljana, Slovenia
Email: {Andrej.Dobnikar,Simon.Vavpotic}@fri.uni-lj.si, Andrej.Likar@fmf@fri.uni-lj.si

Abstract

Dynamic systems modelling with evolving cellular automata is described in this paper. The basic idea is to use stochastic cellular automata together with local evolving algorithms in order to model dynamic systems representing certain physical phenomena and on the other hand to search for the optimal local parameters that enable the close fit of the model processing with the actual sequence of the phenomena. With the help of a case study on the problem of the spread of forest fires, we show the value of the approach.

1 Introduction

The modelling of unknown systems dynamics is one of the most exciting challenges of science today. Discovering the model of a certain natural phenomenon enables one to gain deeper insight into the observed appearance and, on the other hand, gives an opportunity to forecast its activity. Although well-known phenomena already have some reasonable physical interpretations, this usually does not allow one to use them for prediction purposes. The reason is that natural phenomena are time- and space-dependent, where time dependencies are normally given with corresponding differential equations while the influences of space are captured within space-dependent parameters. They are usually not known explicitly. What we expect to be known is the sequence of the states of the phenomena under observation (also called the measuring data). Typical examples of the fields that fit into this kind of reasoning are meteorology, water or air pollution, spread of fires, etc.

In this paper we want to demonstrate a new approach to the problem of process modelling. We suppose that the physical interpretation of the natural phenomena under investigation is known. We also presume that the sequence of the states describing the dynamics of the phenomenon is at our disposal. The idea is to use cellular automata (CA), together with a Local Evolving Algorithm (LEA) based on the local genetic operators. With CA we want to realise the modelling of a dynamic system representing certain natural phenomena and with LEA the search for the proper values of the space-dependent parameters. In this way we want to achieve the optimal alignment and synchronisation between the processing of the model which is based on physical equations and the sequence of the measured data (states of a phenomenon). As a case study we use the forest-fire prediction problem, where a diffusion equation as a physical interpretation of the spread of fire is used and where we expect to have the sequence of the pictures describing the actual process.

Several new ideas are introduced in this paper. For example, we show how a diffusion differential equation is transformed into cellular automata processing, how space parameters are incorporated into the CA structure, and how it is possible to optimise the values of the local parameters.

The content of this paper is as follows: Chapter 2 describes the basic idea of the paper and gives some necessary background. Chapter 3 shows the physical interpretation of the natural phenomena modelling that results in CA implementation. As a case study the Forest Fire Spread (FFS) problem is used. In Chapter 4 the Local Evolutionary Algorithm (LEA) is briefly outlined. Chapter 5 describes the experimental work. The results that relate to the FFS problem are shown at the end of it. Finally, we conclude with some comments and ideas about future work.

2 Problem Definition

"Cellular automata are discrete dynamic systems whose behaviour is completely specified in terms of a local relation, much as is the case for a large class of continuous dynamic systems defined by partial differential equations. In this sense, cellular automata are the computer scientist's counterpart to the physicist's concept of 'field '," [1].

We want to follow this basic idea. Our goal is therefore to implement a physical model of a dynamic system (for example, a certain natural phenomenon) with the help of CA. We chose to use an evolution-based procedure (LEA) in order to optimise the space dependent parameters. We show that it is possible to realise a certain physical law (such as diffusion in our case study) with CA and that such interpretation enables the search for optimal local parameters with an evolution-based procedure. Such an approach differs significantly from the one described in [2]. There, the idea is to use the so-called Cellular Programming Algorithm (CPA) for searching for the most appropriate non-homogenous and deterministic CA structure. In our approach, CA is supposed to be stochastic and homogenous, which is due to the nature of the phenomena we want to model. The probability functions (which generally depend on local and global parameters) of the cells influence the transitions of the states depending on the neighbouring states. With the help of LEA, originating in CPA but with some important distinctions, the local parameters are changed indirectly through the changes in the transition probabilities. The goal is to achieve the best possible match between the CA processing and the measured data of the phenomenon under investigation. We give some details that relate to our approach below.

2.1 Stochastic CA (SCA)

SCA is defined with the two-dimensional array of finite automata cells, $SCA = (SCA(r); 0<=r<MaxX, MaxY)$. Each cell is described with $SCA(r)=(S,p,L,G,f,F)$ where S is the set of all possible states of the cell indicated by 2D vector r, p is the probability vector of the state transitions for each neighbouring (5-cell neighbourhood) combination, L is the set of local and G the set of global parameters that belong to cell r, f is a current cell's fitting value, and F is the set of flags of the cell, which are problem-dependent. In the most commonly used 5-cell neighbourhood, there are 32 combinations, and in that case p has 32 components. Each component corresponds to a neighbourhood configuration and acts as a gene. The set of all probability components that belongs to one particular cell represents the rule genome. There are as many genomes as cells in the CA structure. The complete number of genomes corresponds to the so-called blueprint or chromosome of the SCA.

2.2 Local Evolving Algorithm (LEA)

As mentioned above, the LEA originates in the CPA, which means that it follows the same basic "big loop", [2]. Its local features are essentially the same as within the CPA. There are however some important differences. The mutation operator in the LEA replaces the probability value (gene) that belongs to the current neighbourhood combination with a randomly selected probability (value within a closed interval [0..1]) instead of inverting a binary value attached to a randomly chosen combination within a CA rule (genome). The mutation rate within LEA is also an important parameter that should be picked up so that the convergence is appropriate and that the results are accurate enough. The crossover operator changes genomes in the same way as the CPA does. The only difference is that it moves probabilities instead of binary values. It first selects a dividing point for rules

from the two better-fitted neighbouring cells and then makes the exchange in a similar way as in the CPA.

2.3 Fitting Function

The searching of the optimal local parameters with an evolving procedure (LEA) is controlled with the help of a so-called fitting function. The optimal local parameters, when used within the physical model and/or its cellular implementation, should fit best to the sequence of the measured data of the observed phenomenon. The fitting function belongs to every cell within SCA and is a cumulative one, which means that for each comparison of the cell's state with the corresponding pixel taken from the observed picture within the measured sequence, a weighed contribution is calculated. The order of the pictures in the sequence, as well as the state of the cell, influence its contribution.

3 SCA Model of Diffusion (Forest Fire Spread problem)

3.1 Original SCA Diffusion Model

To test the approach described above, we tackled as a case study the important yet still unresolved problem of wild land fire spread. The starting-point is the basic equation of energy law:

$$\frac{\partial h}{\partial t} = -div\ \bar{j} + \sigma \qquad (1)$$

where h is specific enthalpy of the medium, j energy flux density and σ the specific power of heat generation. Assuming that the flux density is proportionate to the sum of the gradient of the temperature and the advective term from the bulk motion of hot air due to wind with velocity ω [4],

$$\bar{j} = -\alpha(\bar{r})\ grad\ T + \rho\ c_p T \bar{\omega} \qquad (2)$$

where ρ is air density, $\alpha(r)$ conductivity field, c_p specific heat of air at constant pressure and that the specific enthalpy h and heat generation are functions of temperature as well as co-ordinates, one arrives at the simplest diffusion model for the temperature T as a function of time and co-ordinates r, as follows:

$$\frac{\partial T}{\partial t} = \frac{1}{\rho c_p}\frac{\partial h}{\partial t} = \frac{1}{\rho c_p}div\left(\alpha(\bar{r})\ grad\ T\right)$$

$$-div\left(\bar{\omega}T\right) + \frac{\sigma}{\rho c_p} \qquad (3)$$

Several essential features can be learned from the last differential equation. First, we observe the locality principle, which means that the temperature profile is determined from the vicinity of the observed cell. In the simplest case we take into account the first neighbours of the cell at *r*, which we denote as *N(r)*. We later increase this vicinity to the next ones simply in order to enlarge the propagation velocity span of the CA. Next, we observe that equation (3) possesses so-called "diffusive wave solutions" in which local disturbances spread in time and space with a finite speed.

$$\frac{\partial T}{\partial t} = \beta \, div \, (\alpha(\vec{r}) \, grad \, T) - div \, (\vec{\omega} \, T) + Q \qquad (4)$$

The equation (4) was derived from equation (3) and serves as a guide for the construction of a CA structure and the rules appropriate for modelling the forest fire. It is important to realise that the parameter fields α, ß and Q are not known and, furthermore, that the equation is only a rough representation of the actual circumstances, which may require non-linear dynamic equations. The precise temperature field is never required in fire spread forecasting. What is important is the propagation velocity of the fire front in different areas, which depends on fuel distribution, wind direction and the geographical details of the terrain [3]. The CA should therefore mimic front propagation with different speeds at different areas, and take into account the influence of the wind. In our approach we also tried to avoid the well-known co-ordinate ghosting [3], where the form of the fire front closely follows the form of the mesh representing the cells of the fuel bed.

The following CA incorporates most of the basic facts built into the equation (3), as well as arguments based on common sense. The fuel bed is discretised and could be envisaged as trees in the forest. Instead of observing temperature, we observe a tree as our cellular automaton (cell), which has three states represented by variable *s* values 0, 1 and 2. The value 0 represents a tree that has not yet been reached by the forest fire (low temperature, fuel available). The value 1 represents a burning tree (temperature is rising, fuel is being burned, possible consequential ignition of neighbouring trees). The last state which is represented by the value 2 marks a burned tree (no fuel left, temperature is falling, consequential ignition of neighbouring trees impossible).

The ignition is regulated by probabilistic criteria based on the state of the vicinity *N(r)*. The fire from a burning tree ignites a tree in its vicinity with a probability *v(r)*. The probabilities *v(r)* form a field and incorporate the fuel resources, as well as the diffusivity

and influence of the wind. The ignition is supposed to be certain if more than three neighbouring trees are on fire.

The velocity of a fire front is regulated by the proper choice of the probability field *v(r)*. In real-time applications the field is regulated by several global parameters in order to synchronise with the observed patterns of the fire fronts. With low probabilities *v(r)* the fire front can extinguish spontaneously.

The effects of the global wind are introduced through a local re-calibration of the field *v(r)*. The probability for the ignition of trees in *N(r)* is increased in the direction along the wind and decreased in the opposite direction. Smooth interpolation is assumed for other directions. The velocity of the wind is another global parameter to be adjusted during the synchronisation period. Due to the probabilistic formulation of the CA, the approach avoids the problems of co-ordinate ghosting for probabilities, which are not close to unity.

3.2 Inverted *SCA* Diffusion Model

The above diffusion model that spreads fire from a burning tree to neighbouring trees is far from suitable for cellular implementation. We can obtain a cellular diffusion algorithm by inverting the original diffusion model. Instead of spreading fire (original diffusion model), the cell's probability of being ignited from its vicinity is calculated. After ignition, no further investigation of its vicinity is needed. This is due to the fact that we are mostly interested in the dynamics of the fire front.

Although the effect of both algorithms is virtually the same, there are two important differences. The probability of ignition *v(r)* for each cell in the inverted model must be higher, because the spreading of the fire from one (distant tree) cell gives the tree less chance to ignite than to be ignited from the whole vicinity. Second, we have already mentioned that the *SCA* cell has 3 states. Observing the states in the original model, state 1 (burning) is an active state in the sense that it influences the ignition of the neighbouring cells with some probability, while in the inverted model, state 0 (unburned) is active due to the possibility of its being burned. State 2 in both models means a burned cell.

Vicinity is defined with 8 cells that surround the observed cell with radius 1, together with the cell itself, plus the limited number of cells that surround the observed cell with radius 2, as is seen from Figure 1. The neighbourhood consists of distant *(d)*, close *(c)* neighbouring cells and the observed cell *(x)*. Since distant cells usually ignite an observed cell less frequently, we have to use additional ignition condition for the close vicinity. The observed tree ignites regardless of the probability $v(n)$ if at least three of the neighbouring trees are already burning.

d		d		d
	c	c	c	
d	c	x	c	d
	c	c	c	
d		d		d

Fig. 1. Vicinity defined with 8 cells

To put it in more compact form, the inverted diffusion algorithm of the $SCA(r)$ is as follows:

```
1:if s(r)>=1 then s(r):=2
{a tree burns in 1 simulation step}
2:else begin
3:count:=0;
4:for ∀n ∈ N(r,radius=1) do begin
5: if s(n)=1 then begin
6: if random(0..1)<v(n) then s(r):=1;
7: count=count+1;
8:end;
9:if count>3 then s(r):=1;
10:end;
11:for ∀n ∈ N(r,radius=2) do begin
12: if s(n)=1 then
if random(0..1)<v(n) then s(r):=1;
13:end;
14:end;
```

4 Local Evolving Algorithm (LEA) for FFS Problem

The main feature of the LEA for the FFS problem is in combining the inverted diffusion model with a stochastic version of a Cellular Programming Algorithm (SCPA). This approach enables one to adapt the theoretical diffusion model to the actual sequence of events.

This is achieved by replacing the default SCA ignition probabilities $v(r)$ with adapted ignition probabilities $w(r)$, which are defined as follows:

$$q(\bar{r}) = v(\bar{r}) + 2 \cdot (p(\bar{r})_z - 1/2)$$

$$w(\bar{r}) = \begin{cases} 0 & q(\bar{r}) < 0 \\ q(\bar{r}) & 0 \le q(\bar{r}) \le 1 \\ 1 & q(\bar{r}) > 1 \end{cases} \qquad (5)$$

$w(r)$ is obtained from $q(r)$ by constraining $q(r)$ to the interval [0..1]. $q(r)$ depends on a constant probability of ignition $v(r)$ and the dynamic, neighbourhood-dependent probability of ignition $p(r)_z$, where index z denotes neighbourhood dependence.

Every cell in SCA has the same rule, which means that SCA has an homogenous structure. The rule consists of 32 probabilities. It is obvious that even in a 5-neighbor neighbourhood of 3-state SCA there are 3^5 different combinations. In order to reduce the number of combinations to 32, we use a transformation, which reduces the number of cell states to two, denoting the old states 0 and 2 by 0. This gives us a total of 2^5 or 32 combinations.

SCPA is an extension of the binary standard Cellular Programming Algorithm (CPA), [2] commonly used for obtaining optimal cellular automata rules. SCPA deals with probabilities from the interval [0..1] instead of using binary values 0 and 1. The two genetic operators (mutation and crossover) are adapted in the same way, although it is worth noting that the mutation has to undergo additional changes, which is due to the scaling of the changes of probability values instead of just mutating 0 with 1 or 1 with 0. One possible solution is given with the equation:

$$p(\bar{r})_z = p(\bar{r})_z \cdot (1 - scale) + (random(0..1) \\ - (1 - scale)/2) \cdot scale \qquad (6)$$

Since the parameter *scale* is constrained to the interval [0..1], we can observe two extreme cases. When *scale* is 0 there is no change to the rule. On the other hand, when *scale* is 1 the new rule combination value is randomly chosen regardless of the previous value.

As we said, LEA is a combination of the inverted diffusion model and SCPA. It works in two modes. In the first (learning mode), evolution takes place together with inverted diffusion, and in the second (testing mode) only inverted diffusion is active. The learning mode starts with the copying of the initial picture from the measured sequence into SCA. SCA starts to change states according to the rules of the cells and with initial local parameters. Each time the SCA cell's state is changed, the new state is compared with the corresponding pixel of the next picture from the measured data set. The weighed difference is added to the cumulative fitting function of the cell. At the end of the sequence the evolution is activated, which changes the local parameters of the SCA cells due to the mutation and crossover operators within the "big loop". Then the whole procedure is repeated. It is worth noting that the fitting function of the SCA cell is crucial. It enables a convergence of the local parameters towards the optimal ones - those enabling maximal coverage of the measured sequence with the

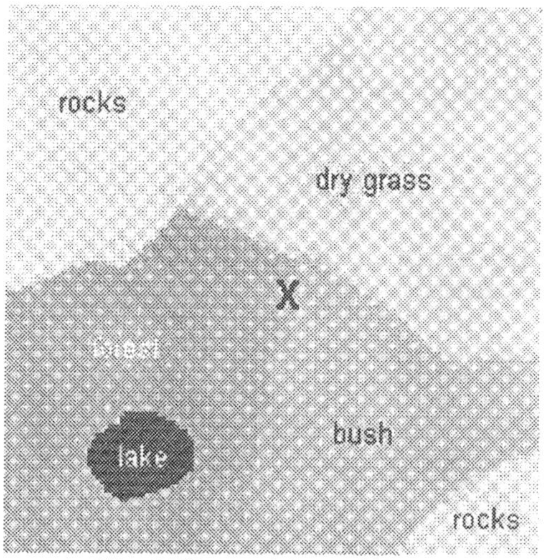

349

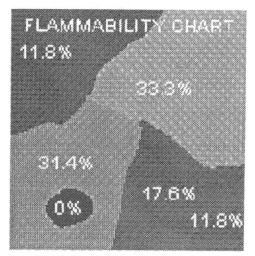

Fig. 2. The actual terrain map

Fig. 5. Fire spread by SCA and geographical
flammability chart

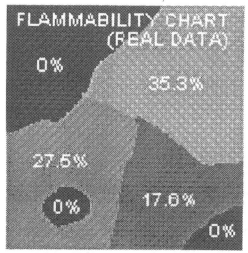

Fig. 3. Typical flammability chart

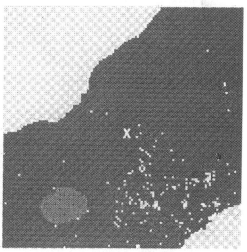

Fig. 6. Result of SCA with help of LEA

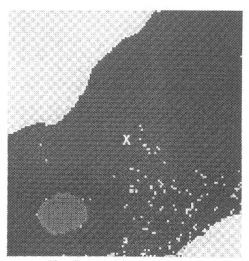

Fig. 4. Realistic flammability chart

Fig. 7. Actual fire devastation

350

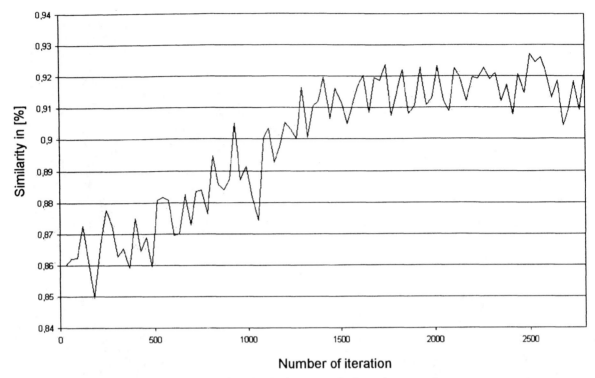

Fig. 8. Increase of similarity between the last picture and state

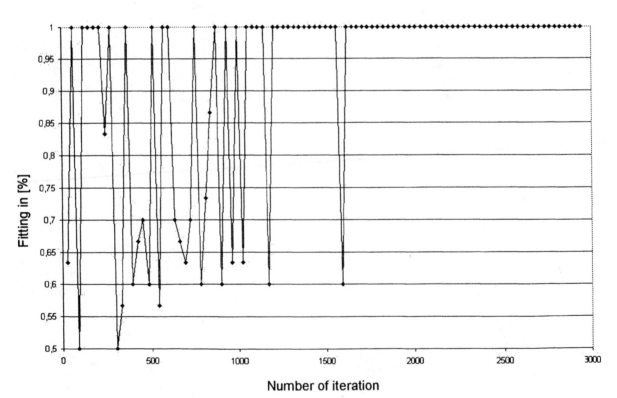

Fig. 9. The fitting value of a typical SCA cell

calculated. In the testing mode, *SCA* only changes its states by following an inverted diffusion algorithm without any changes of the local parameters of the cells.

Before the experimental results are given, one important detail has to be explained – that is, the results of the learning mode have to be spread into the area where one expects fire propagation. This is done so for different regions which have similar geographical and/or meteorological data (or, in short, local parameters); the average of the learned parameters are correspondingly extrapolated.

The last step is prediction. It loads the last picture from the measured sequence into the SCA, which is followed by the test mode of the LEA. The number of simulation steps depends on how far we want to forecast. The time period between the two states corresponds to the time between the two pictures from the measured sequence. Nevertheless, the test mode can also be used to establish future fire spread or its possible extinguishing.

5 Experimental Results

We test the above idea on the forest fire problem, where we presume that the sequence of pictures showing the time propagation is at our disposal. We also suppose that the individual regions of the area of the fire are equipped with typical local data such as humidity, type of vegetation and therefore the degree of flammability or probability of ignition, average wind, etc. The actual sequence is obtained in an artificial way, with the help of original diffusion and a number of realistic local parameters. Our goal is to use *SCA* with typical (geographical) parameters and LEA in order to find the parameters that make a close fit between the measured sequence and the sequence obtained with *SCA* and an inverse diffusion algorithm. We show also that the convergence of the local parameters strongly indicates the feature of learning, which is due to the LEA, based on local genetic operators.

Figure 2 shows the actual terrain map, where there are five areas with different flammability coloured with natural colours. Place of the initial fire ignition is marked with a black cross.

Figure 3 shows a typical (geographical) flammability chart where all the influential data is combined into the percentage (probability) of ignition.

Figure 4 gives the realistic (actual) data which is not known to us.

In Figure 5 the result of the SCA using the inverted diffusion algorithm with typical local parameters is shown. The site of the initial ignition of the fire is marked with a white cross.

Figure 6 gives the result of the SCA with the learned data obtained with the help of the LEA.

Figure 7 indicates the fire spread obtained with the inverted diffusion based on the actual data. It is evident that the last two figures are almost identical and that the processing using inverted diffusion on the typical parameters does not guarantee good prediction.

Figure 8 shows the fire spread during the learning sequence (black area), when the local parameters were taught. An increase in the similarity between the last picture of the learning sequence and the last state of the evolving *SCA* in percentages during the learning sequence can be observed in Figure 9.

Figure 10 finally outlines the convergence of the fitting value of a typical cell within *SCA*. It confirms the feature of learning during evolutionary processing. Although the curve fluctuates, it can clearly be seen that it quickly converges to the final state of 100% fitness. Fluctuations are also becoming less common and smaller before they completely disappear.

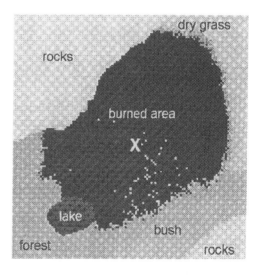

Fig. 10. Fire spread during the learning sequence

6 Conclusion

The problem of dynamic system modelling with Stochastic Cellular Automata and a Local Evolving Algorithm is shown in this paper. With the help of a Forest Fire Spread case study, we demonstrate that the approach can be useful, particularly in the sphere of interest in which it is difficult to get the exact local parameters or where they are too complex to be calculated. On the other hand, it enables us to bring together theoretical and experimental work related to natural phenomena, which are almost always very complex dynamic systems.

7 References

[1] T. Toffoli, N. Margolus, Cellular Automata Machines, MIT Press, 1988

[2] M. Sipper, Evolution of Parallel Cellular Machines, Springer, 1997

[3] D. D. Richards, Int. J. Wildland Fire 5(2): 63-72, 1995

[4] R. O. Weber, Int. J. Wildland Fire 1(4): 245-248, 1991

[5] E. Sanchez, M. Tomassini, Towards Evolvable Hardware, Springer, 1996

SpringerComputerScience

George D. Smith, Nigel C. Steele,
Rudolf Albrecht (eds.)

Artificial Neural Nets and Genetic Algorithms

Proceedings of the International Conference in Norwich, U.K., 1997

1998. XVI, 634 pages. 384 figures.
Softcover DM 228,–, öS 1596,–
(recommended retail price)
ISBN 3-211-83087-1

Contents:

Robotics and Sensors • ANN (Artificial Neural Networks) Architectures • Power Systems • Evolware • Vision • Speech/Hearing • Signal/Image Processing and Recognition • Medical Applications • GA (Genetic Algorithms) Theory and Operators • GA Models/Representation • GA Applications • Parallel GAs • Combinatorial Optimisation • Scheduling/Timetabling • Telecommunications – General and Frequency Assignment Problem • Applications – General Heuristics • Evolutionary ANNs • Reinforcement Learning • Genetic Programming • ANN Applications • Sequences/Time Series • ANN Theory, Training and Models • Classification • Intelligent Data Analysis/Evolution Strategies • Coevolution and Control • Process Control/Modelling • Learning Classifier Systems and Prisoner's Dilemma

SpringerWienNewYork

Sachsenplatz 4–6, P.O.Box 89, A-1201 Wien, Fax +43-1-330 24 26, e-mail: books@springer.at, Internet: http://www.springer.at
New York, NY 10010, 175 Fifth Avenue • D-14197 Berlin, Heidelberger Platz 3 • Tokyo 113, 3–13, Hongo 3-chome, Bunkyo-ku

SpringerComputerScience

Rudolf Albrecht (ed.)

Systems: Theory and Practice

1998. VII, 314 pages. 143 figures.
Softcover DM 118,–, öS 826,–
(recommended retail price)
ISBN 3-211-83206-8
Advances in Computing Science

The notion of system is common to a great number of scientific fields. This book provides insight into present theoretical approaches to systems and demonstrates relationships between concepts and methods developed in the diverse fields where system theory is applied.

Contents:
- C. Rattray: Abstract modelling complex systems
- G. Hill: Formal specification
- R. F. Albrecht: On mathematical systems theory
- F. J. Barros, B. P. Zeigler: An introduction to discrete-event modelling formalisms
- F. Pichler: Design of microsystems: systems-theoretical aspects
- Y. Takahara, X. Chen: A formal representation of DSS generator
- H. D. Wettstein: Structure and functions of operating systems
- W. Kreutzer: Object-oriented system development – concepts and tools
- C. Märtin: Model-based software engineering for interactive systems
- M. Dal Cin: Modelling fault-tolerant system behavior
- D. W. Pearson, G. Dray: Applications of artificial neural networks
- K. Tchoń: The method of equivalence in robotics
- E. Canuto, F. Donati, M. Vallauri: Manufacturing Algebra: a new mathematical tool for discrete-event modelling of manufacturing systems

SpringerWienNewYork

Sachsenplatz 4–6, P.O.Box 89, A-1201 Wien, Fax +43-1-330 24 26, e-mail: books@springer.at, Internet: http://www.springer.at
New York, NY 10010, 175 Fifth Avenue • D-14197 Berlin, Heidelberger Platz 3 • Tokyo 113, 3–13, Hongo 3-chome, Bunkyo-ku

Springer Texts and Monographs in Symbolic Computation

Bob F. Caviness,
Jeremy R. Johnson (eds.)

Quantifier Elimination and Cylindrical Algebraic Decomposition

1998. XIX, 431 pages. 20 figures.

Softcover DM 118,–, öS 826,–

ISBN 3-211-82794-3

Alfonso Miola,
Marco Temperini (eds.)

Advances in the Design of Symbolic Computation Systems

1997. X, 259 pages. 39 figures.

Softcover DM 98,–, öS 682,–

ISBN 3-211-82844-3

Norbert Kajler (ed.)

Computer-Human Interaction in Symbolic Computation

1998. XI, 212 pages. 68 figures.

Softcover DM 89,–, öS 625,–

ISBN 3-211-82843-5

George Collins' discovery of Cylindrical Algebraic Decomposition (CAD) as a method for Quantifier Elimination (QE) for the elementary theory of real closed fields brought a major breakthrough in automating mathematics with recent important applications in high-tech areas (e.g. robot motion), also stimulating fundamental research in computer algebra over the past three decades.

This volume is a state-of-the-art collection of important papers on CAD and QE and on the related area of algorithmic aspects of real geometry. It contains papers from a symposium held in Linz in 1993, reprints of seminal papers from the area including Tarski's landmark paper as well as a survey outlining the developments in CAD-based QE that have taken place in the last twenty years.

New methodological aspects related to design and implementation of symbolic computation systems are considered in this volume aiming at integrating such aspects into a homogeneous software environment for scientific computation. The proposed methodology is based on a combination of different techniques: algebraic specification through modular approach and completion algorithms, approximated and exact algebraic computing methods, object-oriented programming paradigm, automated theorem proving through methods à la Hilbert and methods of natural deduction. In particular the proposed treatment of mathematical objects, via techniques for method abstraction, structures classification, and exact representation, the programming methodology which supports the design and implementation issues, and reasoning capabilities supported by the whole framework are described.

There are many problems which current user interfaces either do not handle well or do not address at all. The contributions to this volume concentrate on three main areas: interactive books, computer-aided instruction, and visualization. They range from a description of a framework for authoring and browsing mathematical books and of a tool for the direct manipulation of equations and graphs to the presentation of new techniques, such as the use of chains of recurrences for expediting the visualization of mathematical functions. Students, researchers, and developers involved in the design and implementation of scientific software will be able to draw upon the presented research material here to create ever-more powerful and user-friendly applications.

All prices are recommended retail prices

 SpringerWienNewYork

Sachsenplatz 4–6, P.O.Box 89, A-1201 Wien, Fax +43-1-330 24 26, e-mail: books@springer.at, **Internet: http://www.springer.at**
New York, NY 10010, 175 Fifth Avenue • D-14197 Berlin, Heidelberger Platz 3 • Tokyo 113, 3–13, Hongo 3-chome, Bunkyo-ku